Advances in
HETEROCYCLIC CHEMISTRY

VOLUME **97**

Advances in
HETEROCYCLIC CHEMISTRY

VOLUME **97**

Editor

ALAN R. KATRITZKY, FRS
Kenan Professor of Chemistry
Department of Chemistry
University of Florida
Gainesville, Florida

Amsterdam • Boston • Heidelberg • London
New York • Oxford • Paris • San Diego
San Francisco • Singapore • Sydney • Tokyo
Academic Press is an imprint of Elsevier

Academic Press is an imprint of Elsevier
Linacre House, Jordan Hill, Oxford OX2 8DP, UK
84 Theobald's Road, London WC1X 8RR, UK
Radarweg 29, PO Box 211, 1000 AE Amsterdam, The Netherlands
30 Corporate Drive, Suite 400, Burlington, MA 01803, USA
525 B Street, Suite 1900, San Diego, CA 92101-4495, USA

First edition 2009

ISBN: 978-0-12-374733-4
ISSN: 0065-2725

For information on all Academic Press publications
visit our web site at books.elsevier.com

Printed and bound in USA

09 10 11 12 13   10 9 8 7 6 5 4 3 2 1

CONTENTS

Numbers in parentheses indicate the pages on which the author's contribution begins.

Ismail Abdelshafy Abdelhamid (1)
Department of Chemistry, Faculty of Science, Cairo University, Giza, A. R. Egypt

Nouria A. Al-Awadi (1)
Department of Chemistry, Faculty of Science, Kuwait University, P.O. Box 5969, Safat 1360, Kuwait

Mohamed Hilmy Elnagdi (1)
Department of Chemistry, Faculty of Science, Kuwait University, P.O. Box 5969, Safat 1360, Kuwait

Gunther Fischer (131)
Geibelstraße 15, D-04129 Leipzig, Germany

Alexander D. Garnovskii (291)
Institute of Physical and Organic Chemistry of the Southern Federal University, Stachka Avenue, 194/2, Rostov on Don 344090, Russia

Dmitry A. Garnovskii (291)
Institute of Physical and Organic Chemistry of the Southern Federal University, Stachka Avenue, 194/2, Rostov on Don 344090, Russia

Subodh Kumar (219)
Department of Chemistry, Guru Nanak Dev University, Amritsar-143005, India

Vijay Luxami (219)
Department of Chemistry, Guru Nanak Dev University, Amritsar-143005, India

Vladimir I. Minkin (291)
Institute of Physical and Organic Chemistry of the Southern Federal
University, Stachka Avenue, 194/2, Rostov on Don 344090, Russia

Alexander P. Sadimenko (45, 291)
Department of Chemistry, University of Fort Hare, Alice 5700, Republic
of South Africa

Evgeniya V. Sennikova (291)
Institute of Physical and Organic Chemistry of the Southern Federal
University, Stachka Avenue, 194/2, Rostov on Don 344090, Russia

Harjit Singh (219)
Department of Chemistry, Guru Nanak Dev University, Amritsar-
143005, India

Igor S. Vasilchenko (291)
Institute of Physical and Organic Chemistry of the Southern Federal
University, Stachka Avenue, 194/2, Rostov on Don 344090, Russia

Volume 97 of Advances in Heterocyclic Chemistry commences with a chapter describing recent developments in syntheses of pyridazines and condensed pyridazines authored by Professor M.H. Elnagdi and Professor N.A. Al-Awadi of the University of Kuwait, and Professor I.A. Abdelhamid of the University of Cairo, Egypt. The chemistry of these important classes of compounds has been summarized previously, most recently by Professor B. Stanovnik and his colleagues in the year 2000. The many further advances of the last nine years are now covered thoroughly by three leaders in this field.

Professor A.P. Sadimenko of the University of Fort Hare, Republic of South Africa, has continued his series on organometallic complexes by summarizing the organometallic complexes formed by metals of the cobalt (cobalt, rhodium, iridium) and nickel (nickel, palladium, platinum) groups with polypyridines. This is the first summary of the numerous organic complexes that these metals form with polypyridines.

The third chapter in this volume by Dr. G. Fischer (Leipzig, Germany) provides the first compilation of our knowledge of azulenes fused to heterocycles.

Volume 97 continues with a survey authored by Dr. S. Kumar, Dr. V. Luxami, and Professor H. Singh of conjugates of calixarenes with heterocycles and especially the role of such compounds in the design of new chemical entities.

In the final chapter of volume 97, Professor V. Minkin and four colleagues at the University of Rostov, Russia, together with Professor A.P. Sadimenko of Fort Hare University, South Africa cover metal complexes of heteroaryl azomethines, compounds that have become of considerable importance recently in a wide range of industrial applications. This survey emphasizes specific functions in the structure of azomethine ligands and heterocyclic frameworks.

Alan R. Katritzky
Gainesville, Florida

CHAPTER 1

Recent Developments in Pyridazine and Condensed Pyridazine Synthesis

Mohamed Hilmy Elnagdi[*], **Nouria A. Al-Awadi**[*] and **Ismail Abdelshafy Abdelhamid**[**]

Contents		

1. INTRODUCTION

Pyridazine and its derivatives, although known for almost a century, received little attention until the recent discovery of medicinally useful

[*] Department of Chemistry, Faculty of Science, Kuwait University, P.O. Box 5969, Safat 1360, Kuwait
[**] Department of Chemistry, Faculty of Science, Cairo University, Giza, A. R. Egypt

Advances in Heterocyclic Chemistry, Volume 97
ISSN 0065-2725, DOI 10.1016/S0065-2725(08)00201-8

natural products (1899MI398, 1981MI471, 1981MI133, 1986MI2199, 2006TL149). Today, the pyridazine nucleus and its 3-oxo derivatives have been recognized as versatile pharmacophores. This key subunit is constituted in many biologically active substances with a broad range of biological and pharmaceutical activities including inhibitors of PED III, IV (1996JME297), α_1- and α_2- adrenoceptors (2001JMC2118), antibacterial and antifungal activities (1992CHJ475), 5-lipoxygenase inhibitors (1992BML1357) and inhibitors of interleukin, β-production (2001BML2373). As a result in recent years, pyridazine ring synthesis has attracted much attention (2001JMC2188, 2001MOL959, 2002JHC869, 2002MI2833, 2002MI481, 2002MI989, 2002T2227, 2003TL7799, 2004BMCL321, 2004TL3459, 2004TL8781, 2005T4785, 2006BMCL1080, 2006EJM101, 2006T8966, 2006TL1853). Developments in pyridazine chemistry have been from time to time surveyed by Stanovnik and co-workers (1990MI385). Since then many advances have taken place, including developing totally new synthetic approaches from new precursors or utilizing green synthetic methodologies. Several of these approaches were first noted in our laboratories. We now survey recent reported synthetic approaches to pyridazines and highlight new methodologies.

2. SYNTHESIS OF PYRIDAZINES *VIA* TWO BOND FORMATION

2.1 Synthesis *via* 3+3 atom combination

Combining 3+3 atoms to constitute a pyridazine ring is a rather new approach. Two methodologies have been reported. In the first, cyclopropenes and diazoalkenes are used as the 3 atom component, while the other utilizes a reaction between aldehyde hydrazones and α,β-unsaturated nitrile as the 3+3 atom components.

2.1.1 Combining cyclopropenes and diazoalkenes

Initially, Al Dulayymi reported that reaction of 1,2-dibromocyclopropene **2**, generated by treatment of the tetrabromocyclopropane **1** with methyllithium at $-78\,^{\circ}$C, affects diene cycloaddition by initial 3+2 dipolar addition of the reagents at the strained double bond with diazomethane at $0\,^{\circ}$C, to yield the pyrazoline **3** which is stable at 0–5 $^{\circ}$C as neat liquid. However, **3** completely rearranged into **4** in chloroform. Treatment of salt **4** with base at room temperature for 5 min yielded pyridazine **5**. On the other hand, the reaction of **2** with ethyl diazoacetate involves an initial dipolar addition to yield the unstable pyrazoline **6**, which undergoes a rapid rearrangement to pyridazines **8** through intermediate **7** (1998T12897; Scheme 1).

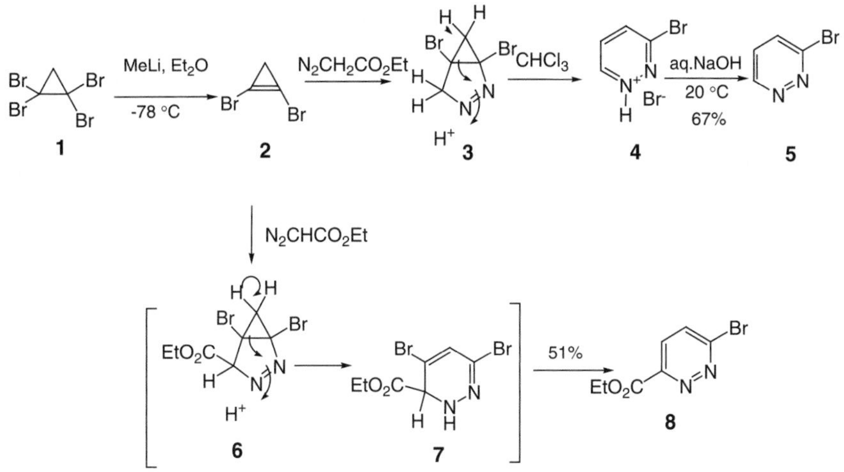

Scheme 1

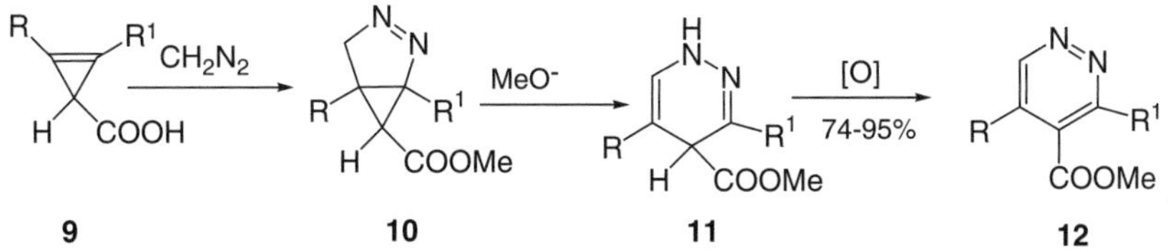

Scheme 2

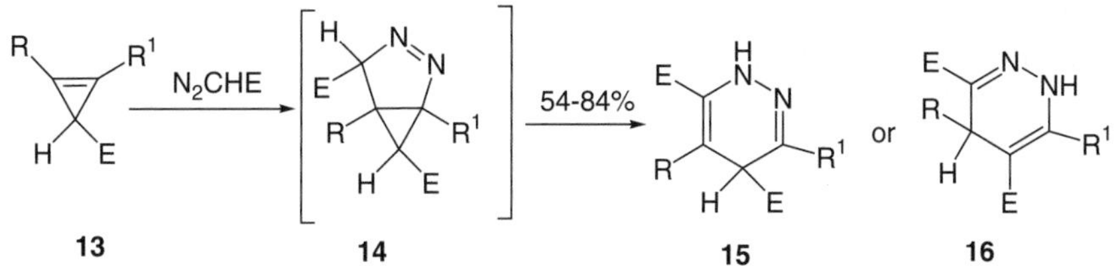

Scheme 3

Cycloaddition of diazomethane to the double bond of cyclopropene **9** yields substituted 2,3-diazabicyclo[3.1.0]hex-2-enes **10**, which isomerizes to 1,4-dihydropyridazine derivatives **11**. The latter is then oxidized into the substituted pyridazines **12** (2004RJO1027; Scheme 2).

Similarly, diazomethane adds to **13** to yield **14**, which then isomerizes to either **15** or its tautomeric form **16**. This synthesis is of theoretical rather than practical value as reactions required almost three months to go to completion (2004RJO1033; Scheme 3).

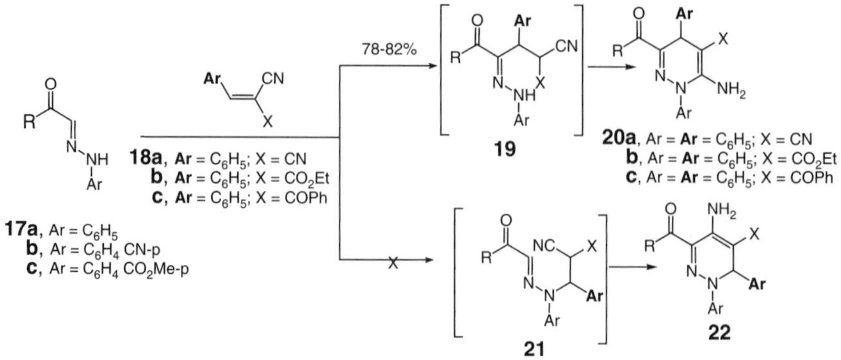

Scheme 4

2.1.2 Combining aldehyde hydrazones and α,β-unsaturated nitrile

A synthesis of 1,4-dihydropyridazines has been reported by reacting pyruvaldehyde-1-arylhydrazones **17** with α,β-unsaturated nitriles **18**. Thus, **17a** reacts readily with α-substituted cinnamonitriles **18** through Michael addition at the CH site to yield dihydroamino pyridazines **20**. NMR cannot however exclude a possible initial addition of the hydrazone NH site to the activated double bond in **18** yielding **21** that then would cyclize to isomeric 6*H*-pyridazine derivatives **22**. The structure of the products was shown to be 4*H*-pyridazines **20** by an X-ray crystal structure of compound **20c** (2007JHC105, 2007JHC877; Scheme 4).

Hydrazone **17** was reacted with **18** yielding acyclic **19d,e** which then cyclized into **23** or **24** through intermediate of **20d,e** depending on the substituent (2006ARK147). And **17a** with ethoxyethylidenemalononitrile **25** resulted in the formation of a product that was formulated as cinnoline **27**. Formation of **27** is assumed to take place by condensing **17** with **25** to yield intermediate **26** which then condenses with malononitrile that results from hydrolysis of unreacted **25** under the reaction conditions (2007MIPh.D thesis; Scheme 5).

2.2 Synthesis *via* 4+2 atom combination

2.2.1 Combining hydrazines with 1,4-bifunctional reagents

This is the most generally used approach for the synthesis of pyridazines, and has been adopted for the synthesis of phthalazines and cinnolines. In addition, several other condensed pyridazines were synthesized in this way.

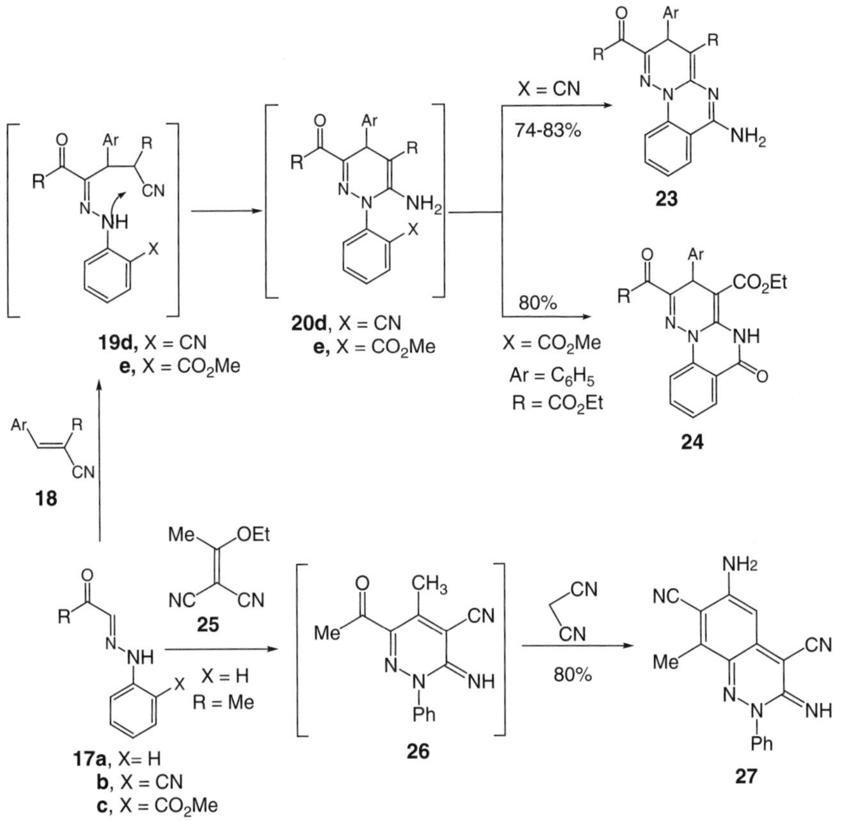

Scheme 5

Classically, hydrazines cyclocondense with 1,4-difunctional reagents 28 to yield pyridazines 29 (2006TL149).

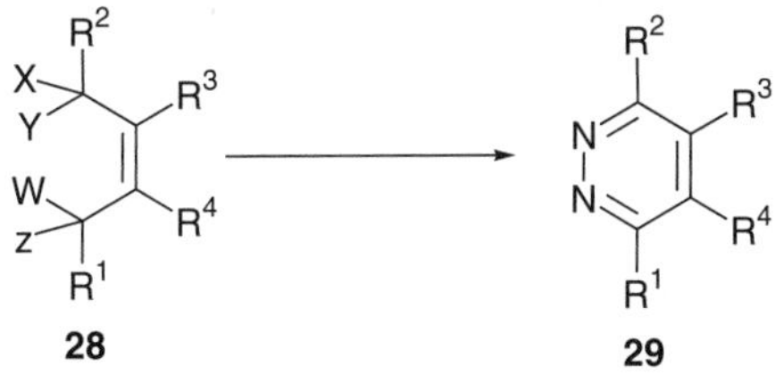

Condensed pyridazines can be produced if R3 and R4 are parts of a ring system. 1,4-Diketones afford pyridazines, whereas 1,4-oxoenoic acid derivatives afford pyridazinones. 4-Oxo acid derivatives afford hexahydropyridazediones.

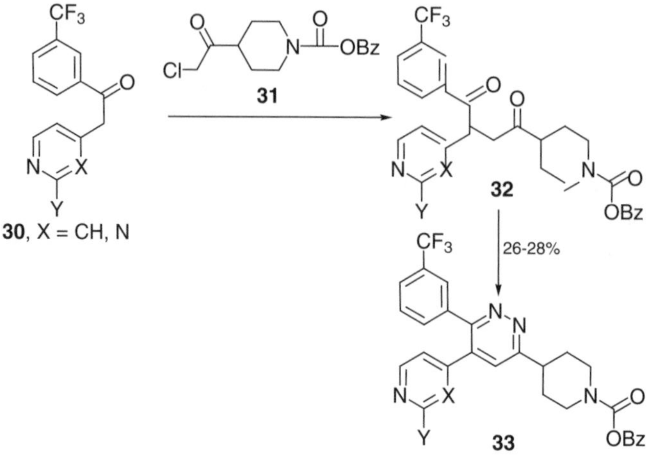

Scheme 6

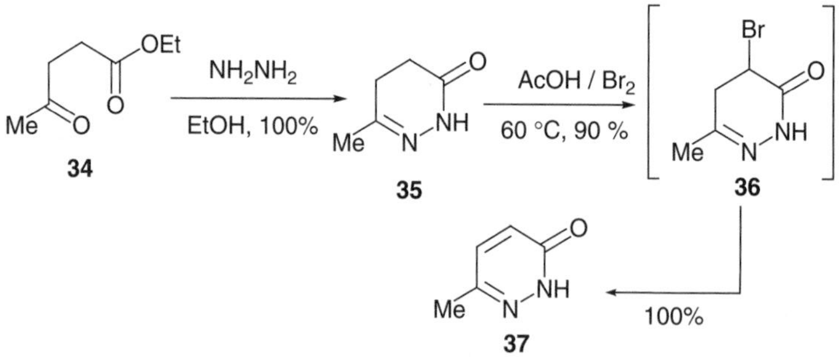

Scheme 7

Reaction of **32** (generated from **30** with **31** in DMSO in the presence of sodium hydride for 8 h) with hydrazine hydrate afforded **33** in 26–28% yields (2002BML689, 2005BML2409; Scheme 6).

Also, a 1,4-keto-ester yields dihydropyridazinone that can be dehydrogenated to the fully aromatic pyridazinone, first by C-bromination then dehydrobromination, as in the conversion of **34** into **37** *via* **35** and **36** (1947JCS239; Scheme 7).

Addition of antipyrene **39** to **38** affords the carboxylic acid derivative **40** that reacted with hydrazine and with phenylhydrazine in refluxing butanol to yield **42**. The acyclic phenylhydrazone **41** was isolated from the reaction of **40** with hydrazine at room temperature (2003MOL322; Scheme 8).

The methylester of 2-oxo-3-pentenoic acid **43** with hydrazine derivatives was used to synthesize new pyridazinone derivatives through intermediate hydrazides **44** or **47** with subsequent cyclization.

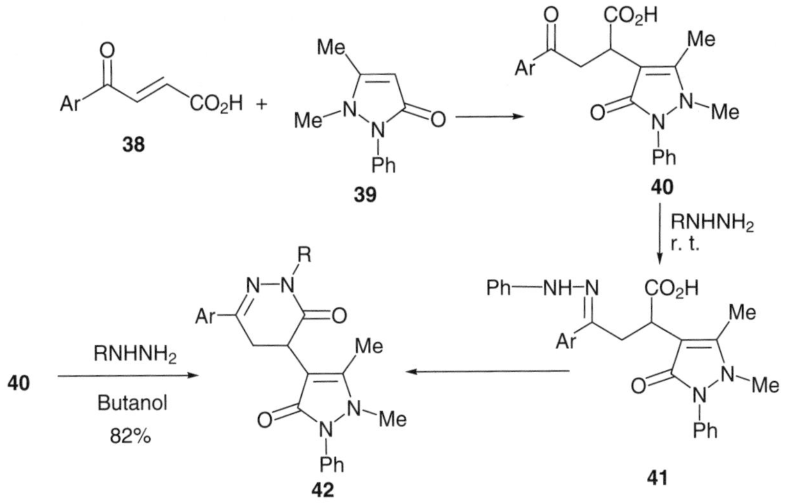

Scheme 8

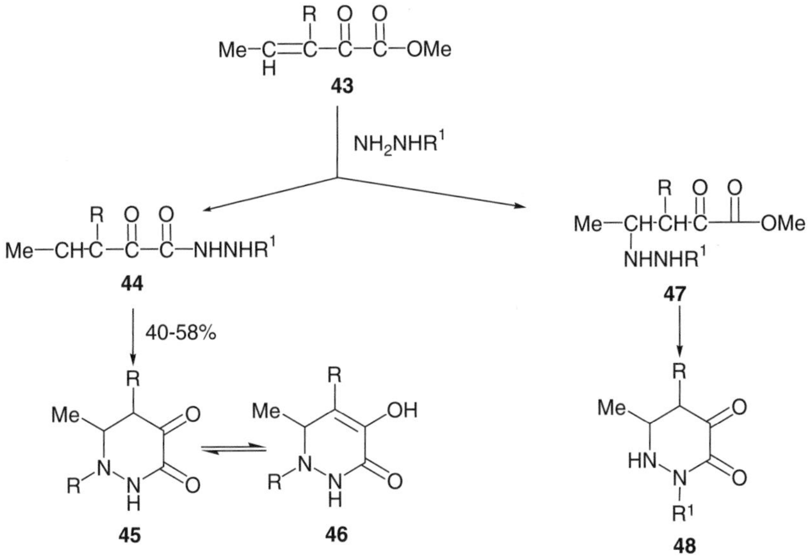

Scheme 9

4-Hydroxy-3-oxotetrahydropyridazines **46** are mainly formed using equimolar amounts of starting reagents, while the corresponding 3,4-dioxohexahydropyridazines **48** are formed when using a two-fold excess of hydrazine derivatives (2004CHE1047; Scheme 9).

Almost all azolo[x,y-*d*]pyridazines have been prepared from the corresponding azole with two adjacent acyl, aroyl or carboxy derivatives. Thus, pyrrolo[3,4-*d*]pyridazines **50** are produced from hydrazine hydrate with **49** in ethanol at room temperature (2004BML2031).

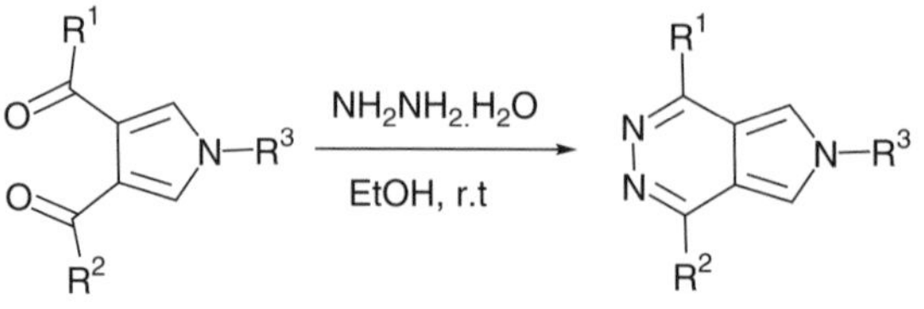

49   **50**

Pyrazolo[3,4-*d*]pyridazines are readily obtained from pyrazoles having 4,5-dicarbonyl or carboxy substituents (1997OPPI285, 2003T5869). This reaction has been used to elucidate the structure of products from **51** with **52** or **53**. Thus, the produced **54** and **55** afforded pyrazolo[3,4-*d*]pyridazines **56** and isoxazolo[3,4-*d*]pyridazines **57** on reaction with hydrazine (2004JHC647; Scheme 10).

In further example, **58** with hydrazines affords **59**. Reaction of **58** with 2-pyridylhydrazine **60** afforded only a carboxylic hydrazone **61** that is believed to be stabilized by hydrogen bonding in its tautomeric form **62** (2005EJM401; Scheme 11).

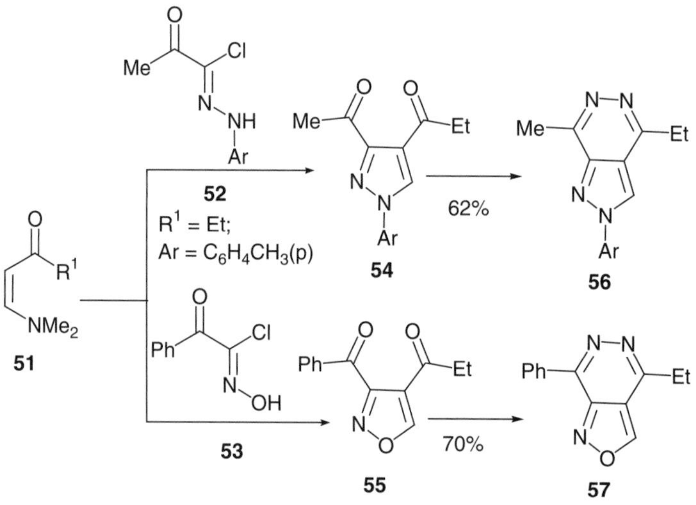

Scheme 10

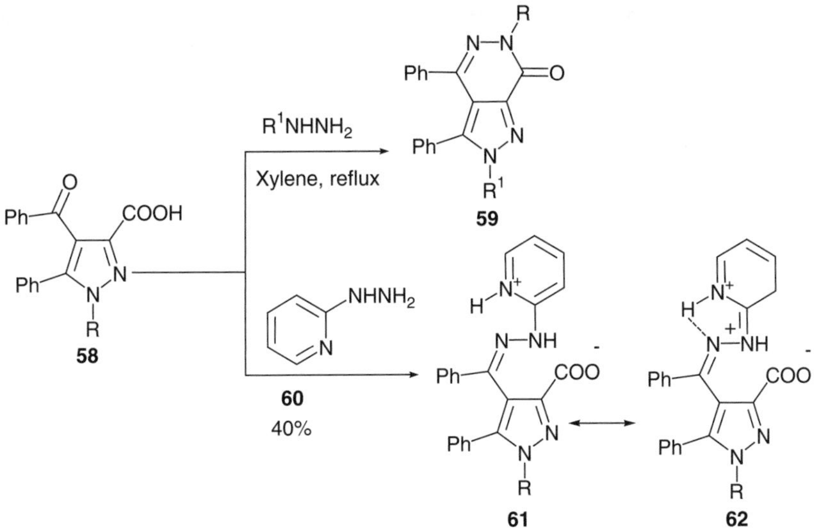

Scheme 11

Similarly, **63** yields **64** when treated with hydrazine hydrate (2005EJM401, 2002JHC869).

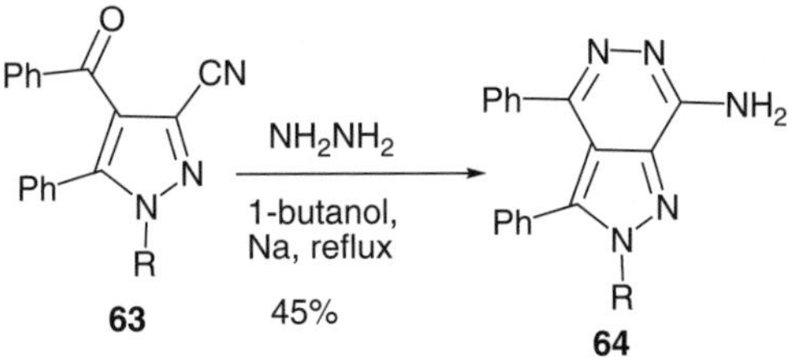

Isoxazoloylpyrazole ester **65** reacts with hydrazine hydrate to yield the pyrazolo[3,4-*d*]pyridazines **66** (1997MI617).

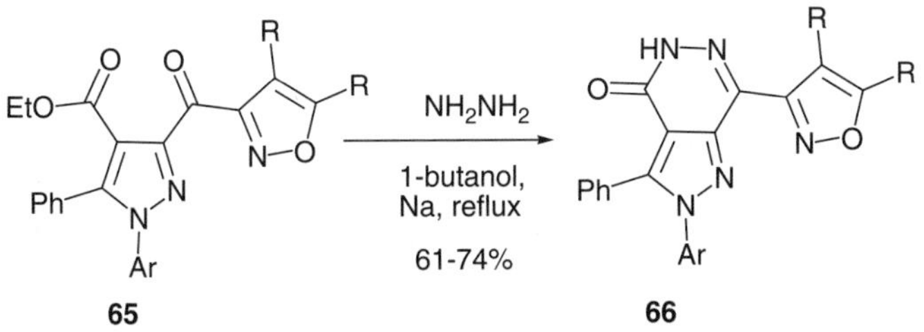

An interesting application of these synthetic methods is the conversion of **67** into **68** upon reflux with hydrazine hydrate (2004ZNB1132).

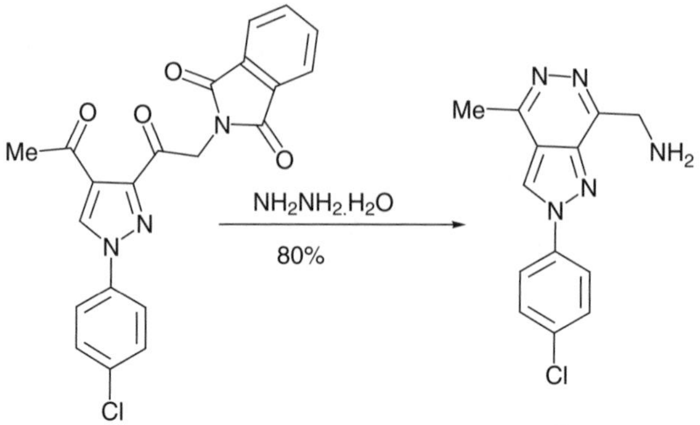

Thiazolo[4,5-*d*]pyridazines are similarly obtained; for example reaction of **69a** with hydrazine hydrate yields thiazolo[4,5-*d*]pyridazines **70**. Refluxing the hydrazide **69b** in aqueous acetic acid yields the dione **71** that was used as precursor to a variety of thiazolo[4,5-*d*]pyridazine derivatives (2000AJC905, 1994CPB2219; Scheme 12).

Similarly, 1,2,3-triazolo[4,5-*d*]pyridazines **73**, 1,2,5-oxadiazolo[3,4-*d*] pyridazines **75** and 1,2,5-thiadiazolo[3,4-*d*]pyridazines **80** are obtained from the corresponding diesters **72**, **74**, and **76**, respectively. The dihydrazide is isolated in some cases (2002JHC889, 1989CS45; Scheme 13).

Diacyl and diaroyl derivatives afforded similar azolopyridazines. Although quite unexpectedly, treatment of **74** with N_2H_4 in ether yields **75** (1988LA1017).

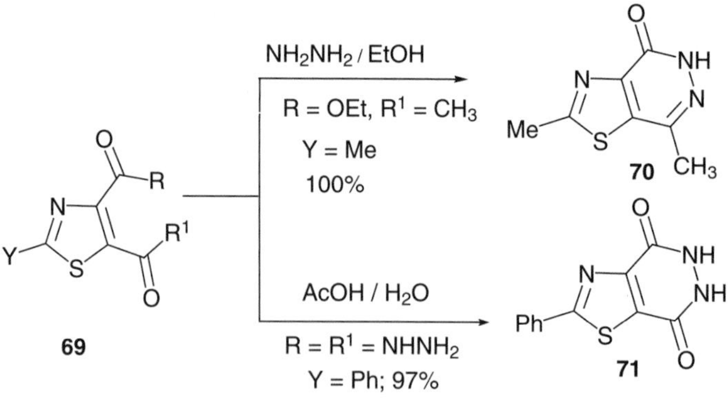

Scheme 12

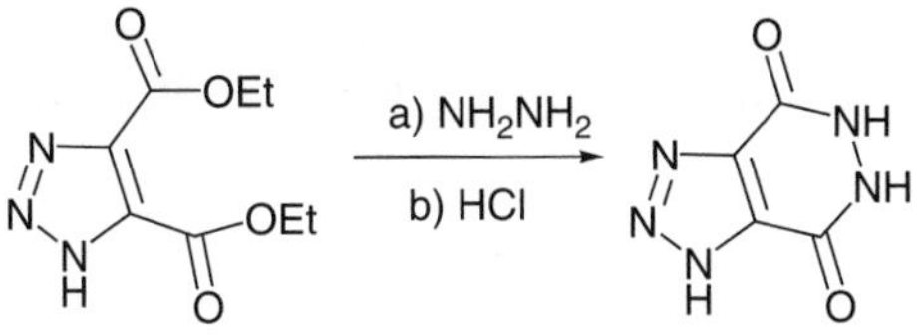

72 73

Scheme 13

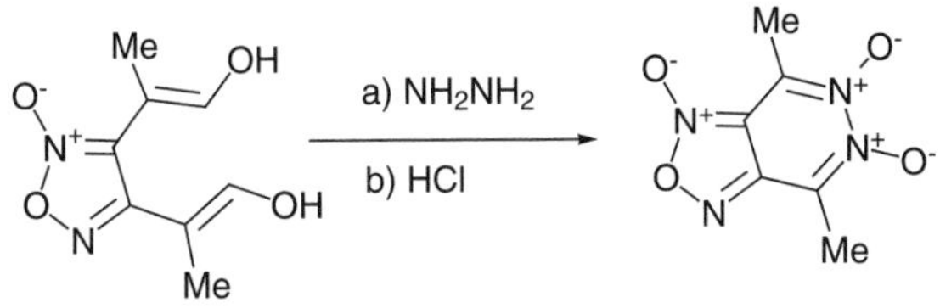

74 75

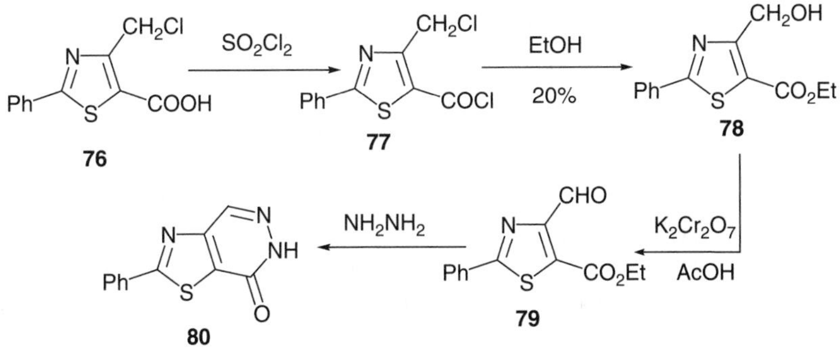

Scheme 14

Chloromethylthiazole **76** reacts with sulfuryl chloride to yield **77**, which then reacts with ethanol to yield **78**. Alcohol **78** is oxidized by $K_2Cr_2O_7$ to **79**, which then treated with hydrazine hydrate to yield **80** (1983AP1013; Scheme 14).

Synthones for unsaturated 1,4-dicarbonyl compounds are readily obtained from furan derivatives. The reaction of 2,3-furandione **81** with hydrazide **82** afforded the pyridazinones **83** (2006MI691).

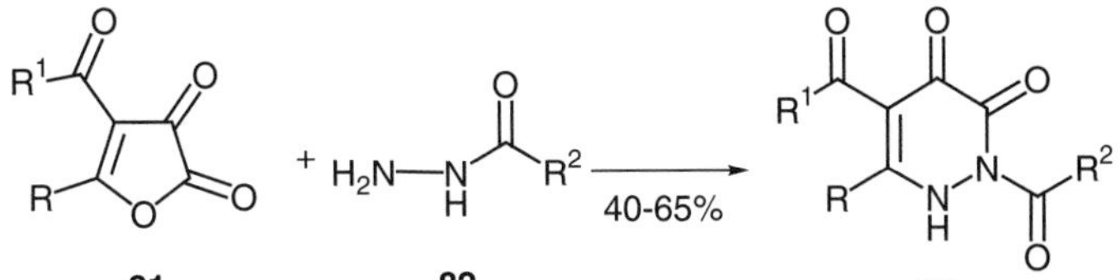

81 82 83

Similarly, treatment of antipyrene derivative **84** with hydrazine hydrate affords the corresponding pyridazines **85** (2003CHE541).

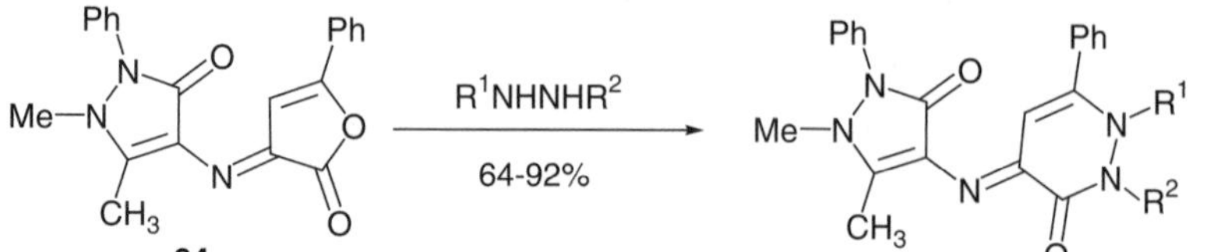

The 4,5-dichloropyridazine-3,6-dione **87** was obtained from the available dichloromaleic acid anhydride **86** (2006T9002).

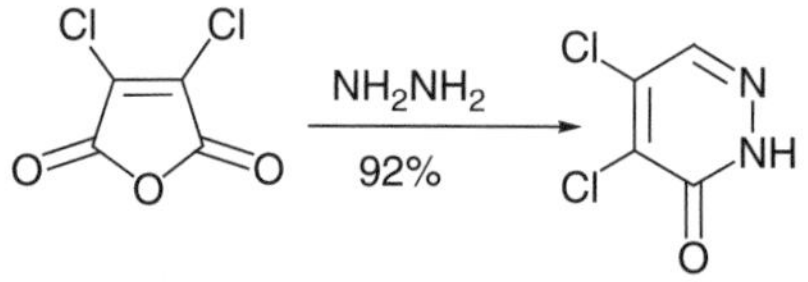

Treatment of furanones **88** with hydrazine hydrate in ethanol led to the formation of acid hydrazide **89**. The latter hydrazides **89** were utilized as the key starting materials for the synthesis of pyridazinones **90** and **92** (2007EMJonline; Scheme 15).

Tartaric acid was oxidized with sodium periodate to generate glyoxalic acid *in situ*, and the reaction mixture was added to a suspension

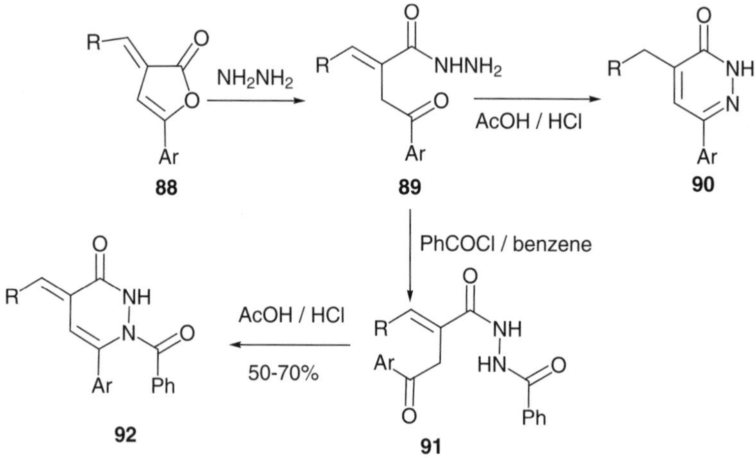

Scheme 15

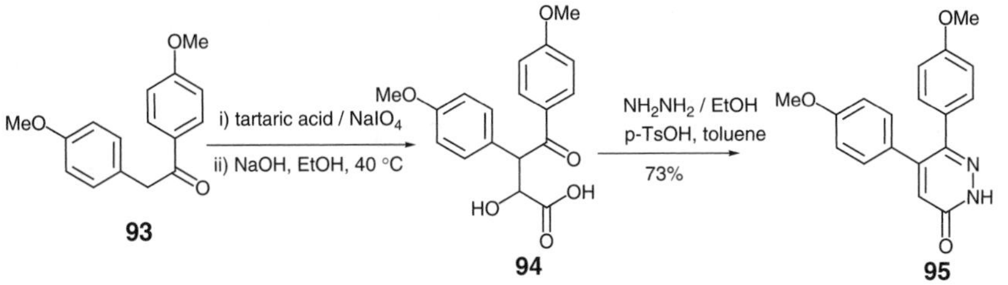

Scheme 16

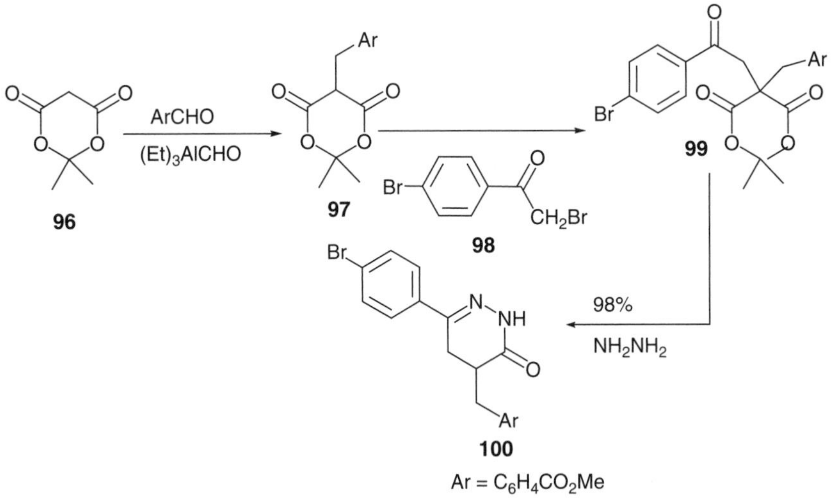

Scheme 17

of desoxyanisoin **93** under basic conditions to provide aldol adduct **94**. Compound **94** was cyclized with hydrazine, followed by dehydrogenation with *p*-toluene sulfonic acid to yield pyridazinone **95** (2001BML2373; Scheme 16).

Another synthon to 4-oxocarboxylic acids has been recently described and utilized for the synthesis of pyridazinones of potential antihypertensive activity. Thus, arylmethylene Meldrum acid derivatives **97** (prepared by reacting Meldrum acid **96** with aldehydes and subsequent reduction of the formed arylidene derivative using triethylammonium formate, TEAF) could be alkylated with 4-bromophenacyl bromide **98** to yield **99** that then reacted with hydrazine hydrate to yield **100** (2004SC783; Scheme 17).

3-Amino-2*H*-pyran-2-ones and their fused derivatives **101** react with hydrazine hydrate to yield the corresponding pyridazine derivatives **106**.

It is suggested that **102–105** are intermediates. This reaction has been subsequently extended to enable the synthesis of differently condensed pyridazines (2006T9718; Scheme 18).

Fusion of phthalic acid anhydride **107** with α-amino acids **108** in the presence of anhydrous sodium acetate yielded **109**, which reacted readily with **107** to yield **110**. The latter with hydrazine hydrate yielded **111** in 60–70% yields (2006ARK133; Scheme 19).

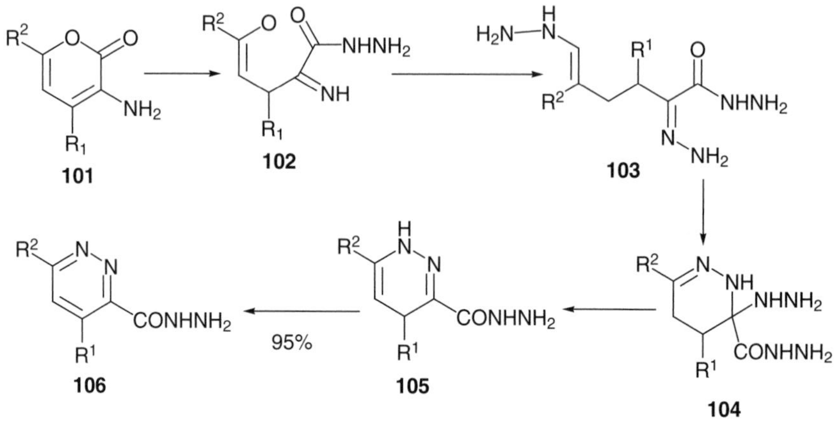

Scheme 18

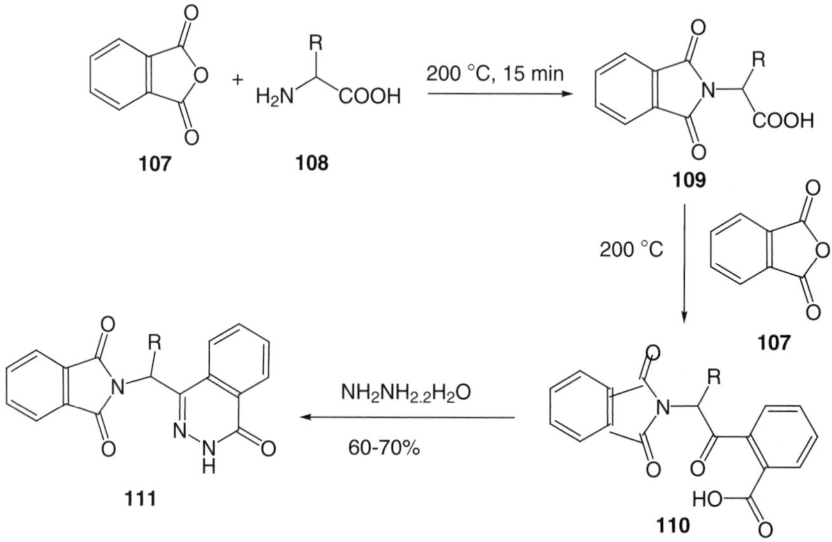

Scheme 19

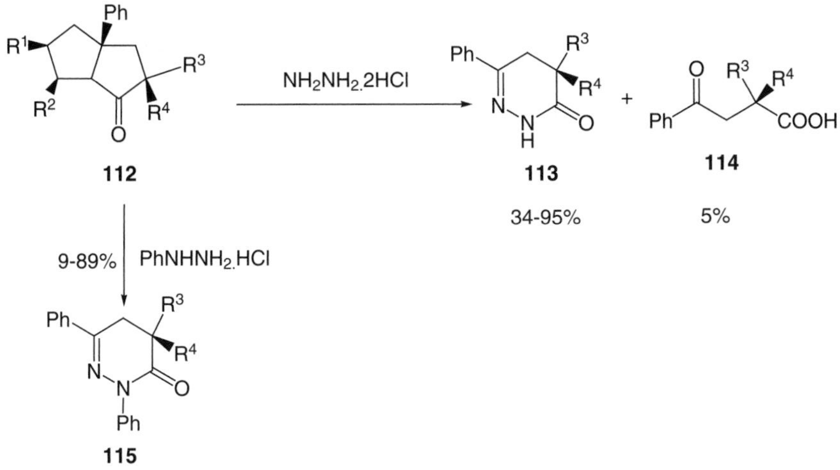

Scheme 20

Bicyclic lactams are uniquely suited as precursors for the synthesis of chiral substituted 4,5-dihydro-2*H*-pyridazinones. Thus, bicyclic lactams **112** with hydrazine dichloride in a dioxan/water mixture produced the desired 6-phenyl-4,5-dihydropyridazin-3-ones **113**. However, reacting the bicyclic lactam with phenylhydrazine hydrochloride yields 2,6-diphenyl-4,5-dihydropyridazin-3-one **115** as a sole product (2004TL7799; Scheme 20).

Aldol condensation of arylmethylketones **116** with α-ketoacids followed by reaction with hydrazine hydrate leads to the formation of the corresponding pyridazine **117** (2006TL6125).

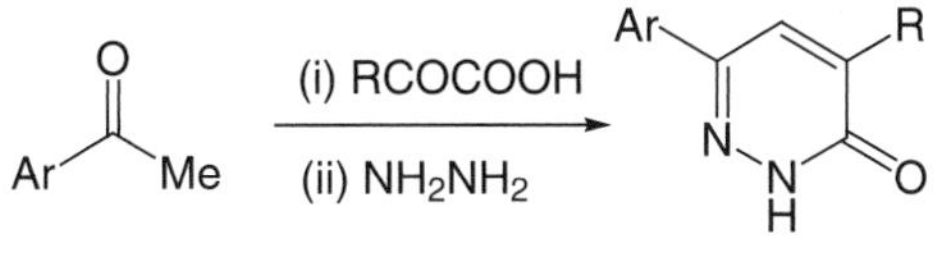

Treatment of the dithioketal **118** with hydrazine hydrate and phenylglyoxal in basic medium afforded the corresponding 4-nitro-6-phenyl-3-methylthiopyridazine **119**, while the double amination of **118** yielded aminal **120** that was again reacted with phenylglyoxal to yield the corresponding 3-amino-4-nitropyridazine **121** (2006COR277; Scheme 21).

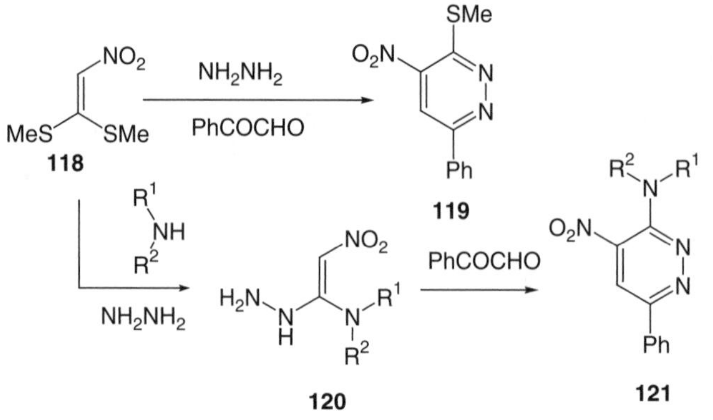

Scheme 21

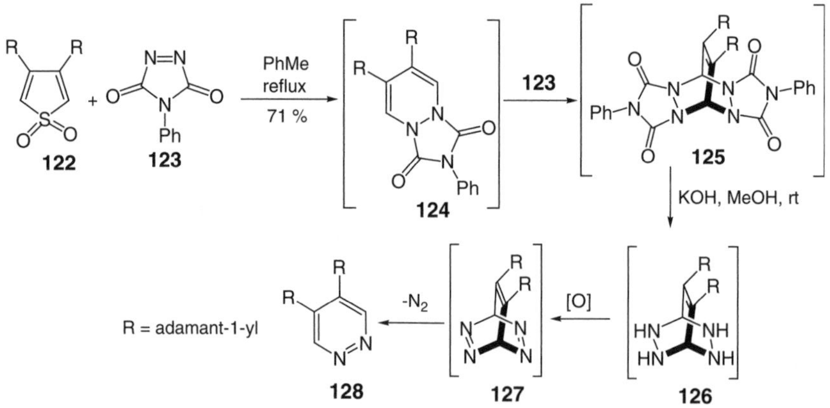

R = adamant-1-yl

Scheme 22

2.2.2 Diels–Alder cycloaddition as a way of combining 4+2 atoms to form pyridazines and condensed pyridazines

Diels–Alder cycloadditions and Diels–Alder inverse electron demand cycloadditions have been extensively utilized for the preparation of pyridazines and condensed pyridazines.

2.2.2.1 Normal electron demand Diels–Alder cycloaddition. Thiophene *S,S*-dioxides **122** with bulky 3,4-disubstitution undergo Diels–Alder additions with 4-phenyl-1,2,4-triazoline-3,5-dione (PTAD) **123** to yield the *bis*-adduct **125** by further addition of PTAD **123** to the initial adduct **122**. Treatment of the adduct **125** with methanolic KOH, followed by air oxidation, yields the 4,5-di(1-adamantyl)-pyridazine **128** (1998JOC4912; Scheme 22).

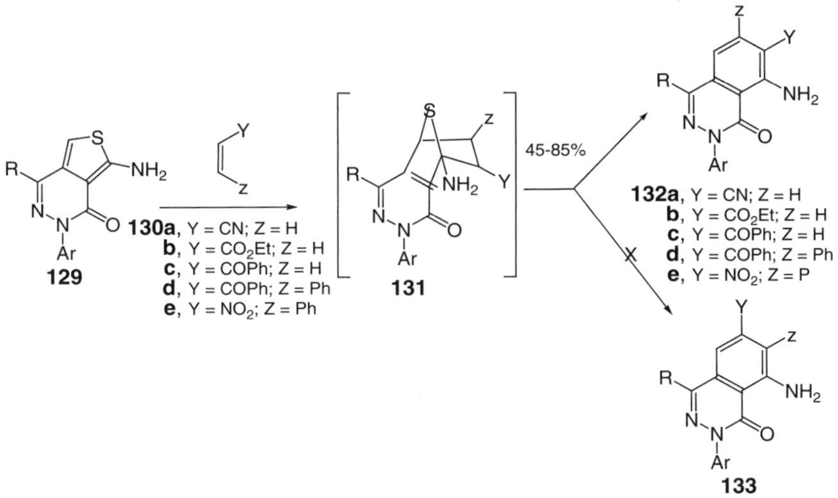

Scheme 23

In 1989, Elnagdi et al. (1989LA1255) reported a general synthesis of benzofused pyridazines. Treating **129** with acrylonitrile **130a** and ethyl acrylate **130b** yielded products that may be formulated as **132a,b** by the intermediacy of nonisolable cycloadducts **131**. The alternate possible formation of isomeric **133** was excluded based on spectral evidence. Similarly, **130b–e** reacted in the same manner (1997HAC29, 1998CJC1038, 2004HAC502; Scheme 23).

Similar to the previously mentioned reactions, pyridazines **129a–c** reacted with di-t-butylacetylene dicarboxylate **134**, tetracyanoethylene **136** and Mannich compound **138** to yield **135**, **137** and **140**, respectively (1997HAC29; Scheme 24).

Enaminones **51** afforded C-1 alkylation products **141** with **129** (2005JRM23, 1999JRM151; Scheme 25).

2.2.2.2 Inverse electron demand cycloaddition of 1,2,4,5-tetrazine with alkenes and alkynes.   Inverse electron demand Diels–Alder addition has also been employed for the synthesis of pyridazines and condensed pyridazines. The reaction of olefinic and acetylenic compounds with 3,6-disubstituted 1,2,4,5-tetrazines **142** to yield substituted pyridazines **144** by the intermediacy of **143** was first reported by Carboni and Lindsey (1959JA4342). Analogous reaction of **142** with a variety of aldehydes and ketones **145** in base at room temperature proceeded smoothly to yield the corresponding pyridazines **144**. Compounds **146–148** are proposed nonisolable intermediates (1979JOC629; Scheme 26).

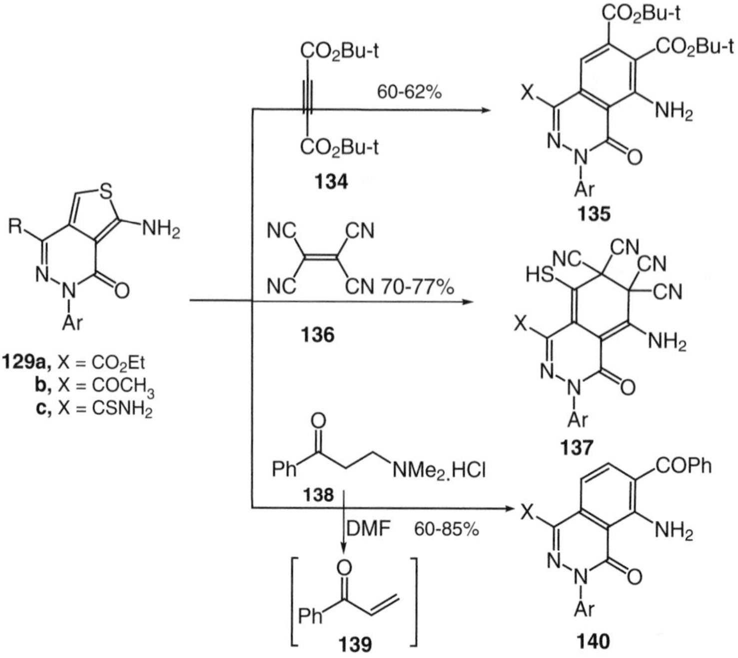

Scheme 24

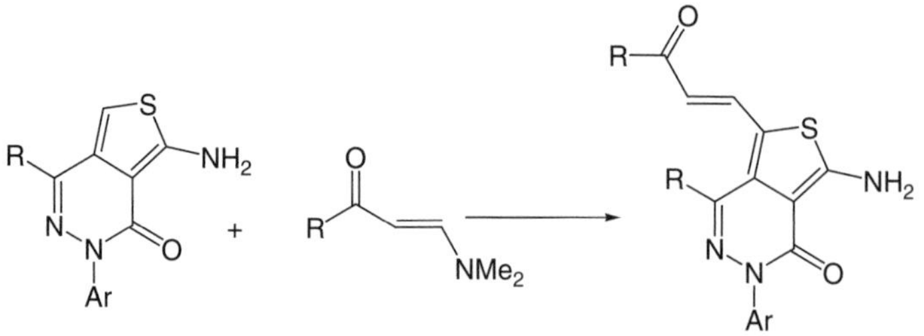

Scheme 25

Triazolyl-dienamine **151**, formed by the reaction of morpholine with the triazolopyridinium salt **149**, reacts readily with tetrazine **142** to form triazolylvinylpyridazine **152** (2004T3421; Scheme 27).

The inverse electron demand Diels–Alder [4+2]-cycloaddition of imidazoles to electron-poor dienes to yield imidazo[4,5-d]pyridazines, reported in Comprehensive in heterocyclic chemistry II (CHEC-2), has been further developed. The reaction of **153** with tetrazines **142** was fruitless. However, **153** reacted with an excess of **142** to yield aromatic **155** along with 1,4-dihydrotetrazine **156**. Most likely, **155** arose from

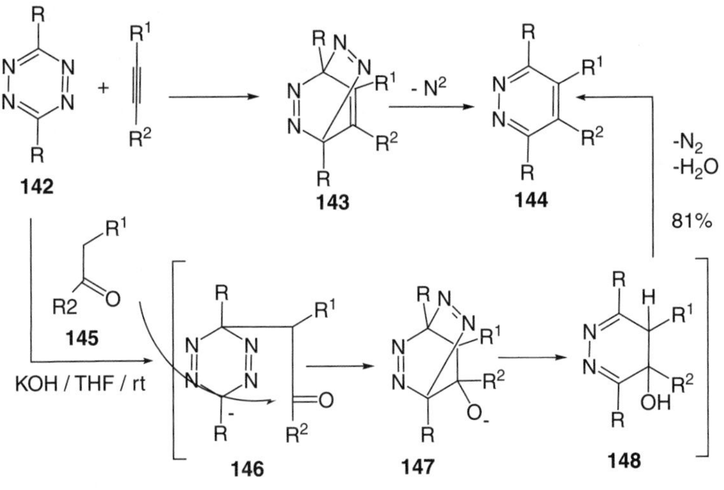

Scheme 26

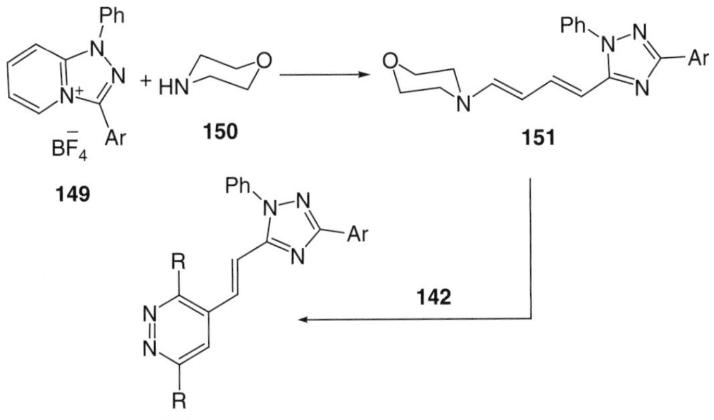

Scheme 27

dehydrogenation of the first-formed **154** by an extra equivalent of **142** (2001T5497). Adopting this chemistry enabled the synthesis of **157** from **153b** and **142**. Thus, reacting **153b** and **142** yields **154b** that was then hydrolyzed and decarboxylated to yield **157**. Structure **157** was believed to be the cytotoxic marine natural product zarzisine, but, differences in the ^{1}H and ^{13}C NMR spectra of **157** and those reported for zarzisine led to structure **158** being proposed for the latter (2001T5497; Scheme 28).

Tetrahydrobenzodifuran **161** was prepared from resorcinol **159**. Selective oxidation of the dihydrofuran ring of **161** was accomplished using one equivalent of 2,3-dichloro-5,6-dicyanobenzoquinone (DDQ) in

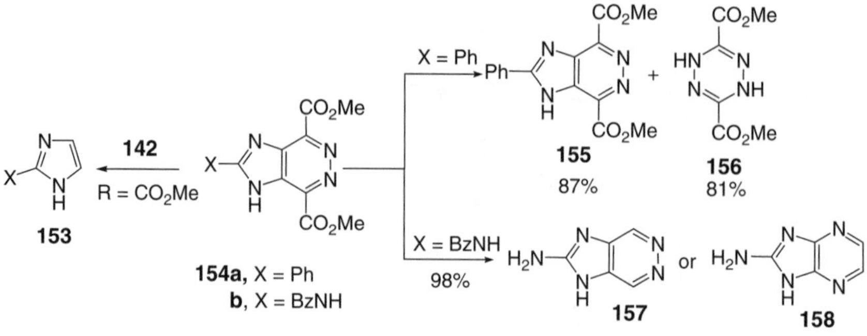

Scheme 28

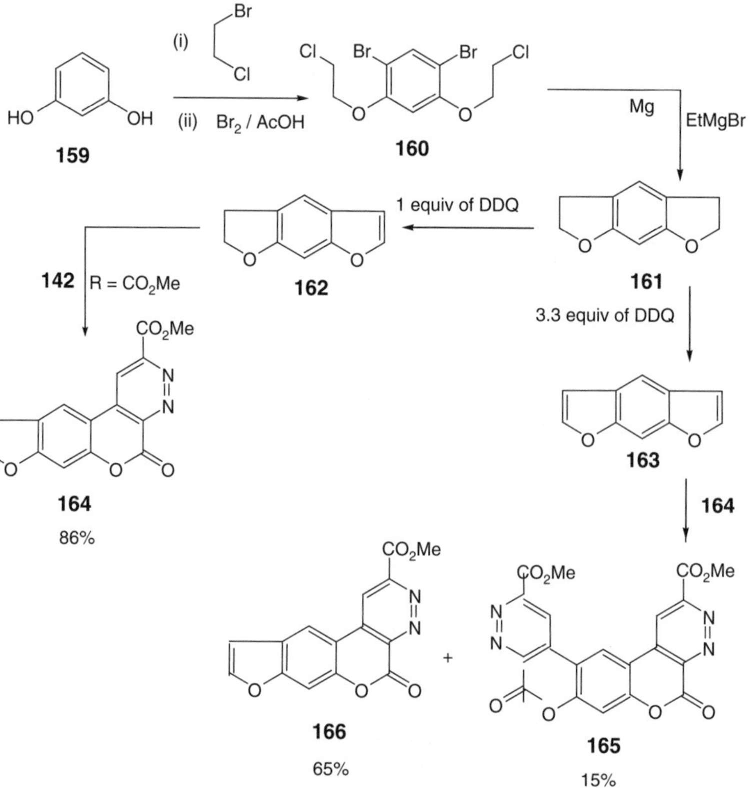

Scheme 29

dioxan at room temperature to afford **162**. However, oxidation of both dihydrofuran rings of **161** was performed using an excess of DDQ in refluxing dioxan and afforded **163**. Diels–Alder reaction between tetrazine **142** and **162** and **164** led to the expected **164** and **165**, respectively (2005T4805; Scheme 29).

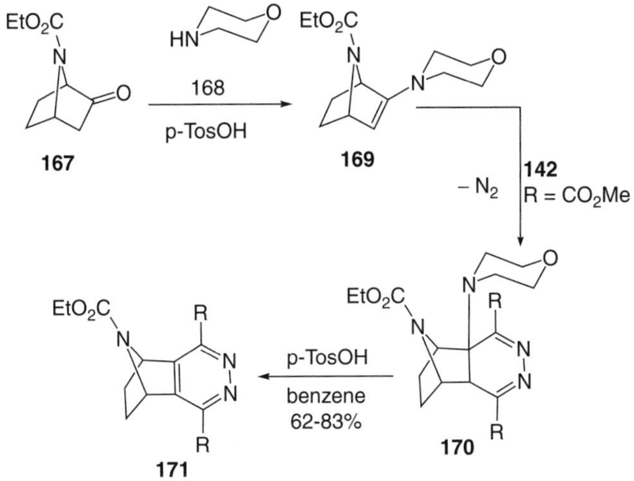

Scheme 30

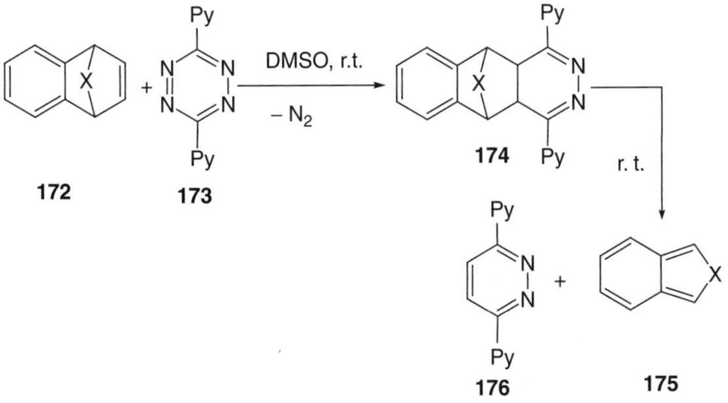

Scheme 31

Ketone **167** reacted with morpholine and p-TosOH to yield enamine **169**, which undergoes Diels–Alder reaction with tetrazine **142** to yield the isolable intermediate **170**. Aromatization occurs easily on further treatment of **170** with p-TosOH in boiling benzene leading to N-protected tricyclic pyridazines **171** (2002BMC1; Scheme 30).

The retro Diels–Alder reaction (cycloreversion) under mild conditions can be used for the preparation of pyridazines. Bridged 1,4-epimino-naphthalenes **172** react readily with 3,6-di(2-pyridyl)-s-tetrazine **173** in an exothermic reaction involving nitrogen elimination to yield pyridazines **176** (2003COR1423; Scheme 31).

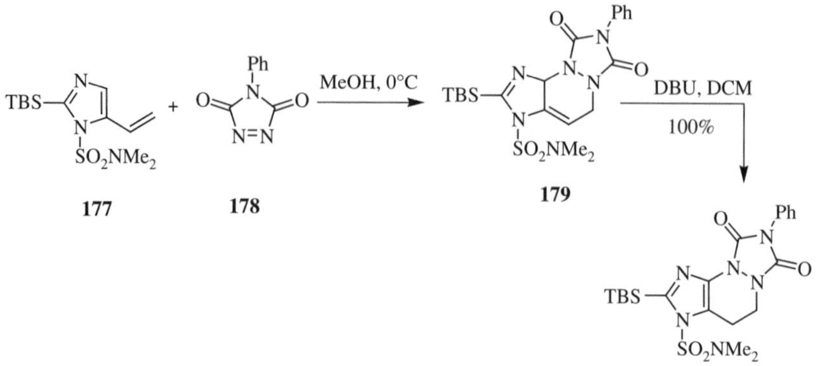

Scheme 32

The addition of 1-phenyl-3,5-triazolidendione **178** to vinylimidazole **177** afforded **179** which can be converted using a multistep sequence into **180** (1998T4561; Scheme 32).

2.3 Synthesis *via* combining alkylpyridazinylcarbonitrile with α,β-unsaturated nitriles

This route was invented by Elnagdi et al. to convert pyridazines into substituted pyridazines and cinnolines (2006JHC1575, 2005JHC307, 1995T12745, 1992JPR723, 1990JRM1124, 1990ZNB389, 1989T3597, 1989ZNB683, 1988LA1005). Alkylpyridazinylcarbonitriles **181** with cinnamonitriles **18** affords phthalazines **184**. Here alkylpyridazinyl carbonitriles were added to α-substituted cinnamonitriles. The adducts cyclized and lost hydrogen cyanide to yield benzofused derivatives, which is an extension of our earlier synthesis of aromatic amines from crotononitriles. On several occasions it was possible to isolate the nonaromatic dicyano compound **183**, although aromaticity of the final product is the deriving force for hydrogen cyanide elimination. The structure of tetrahydrophthalazine **184**, prepared from **181** with **18**, was established based on ^{1}H NMR spectra that revealed the presence of an amino group and the absence of CH multiplets in the range of $\delta = 2$–4 ppm. Formation of **184** from **181** is assumed to occur by initial formation of Michael adducts **182** (2006JHC1575, 2005JHC307, 1995T12745, 1992JPR723, 1989T3597; Scheme 33).

Similar results have been recently reported by Al-Omran et al. (2006JRM237) and Abdelrazek (2006AP305) who failed to isolate intermediates **188** as previously isolated by Elnagdi et al. (1999JCM96; Scheme 34).

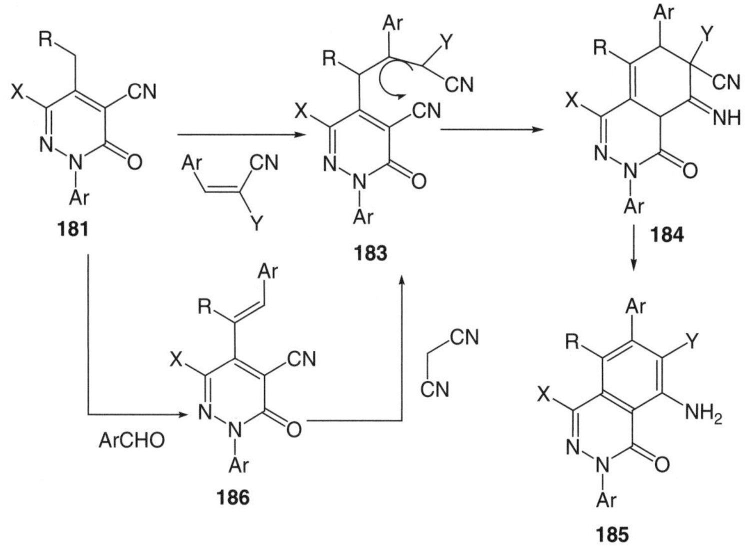

Scheme 33

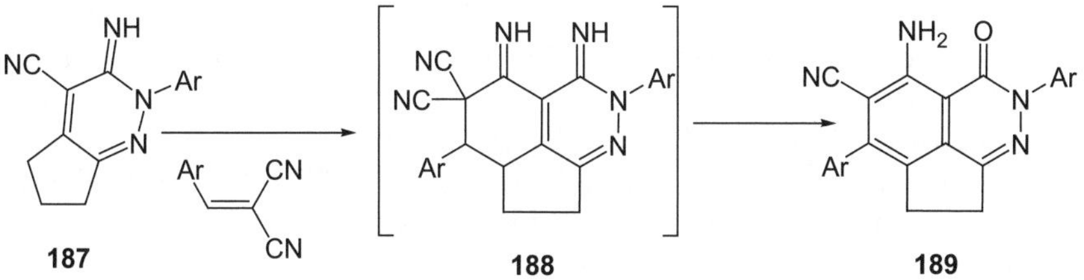

Scheme 34

2.4 Synthesis *via* coupling aromatic diazonium salts with carbon nucleophilic 4 atom fragments

The first reported synthesis of a pyridazine derivative by this route was conducted in our laboratories 25 years ago when **190**, produced by coupling diethyl 3-amino-2-cyano-2-pentene-1,5-dicarboxylate **189** with aromatic diazonium salts, was readily cyclized into pyridazine-6-one derivative **191** on treatment with ethanolic potassium hydroxide. However, **190** cyclized readily into the pyridazin-6-imine **192** on treatment with an acetic/hydrochloric acid mixture. And **190** also afforded 4-acetylaminopyridazin-6-one derivative **193** on treatment with acetic anhydride (1982S490; Scheme 35).

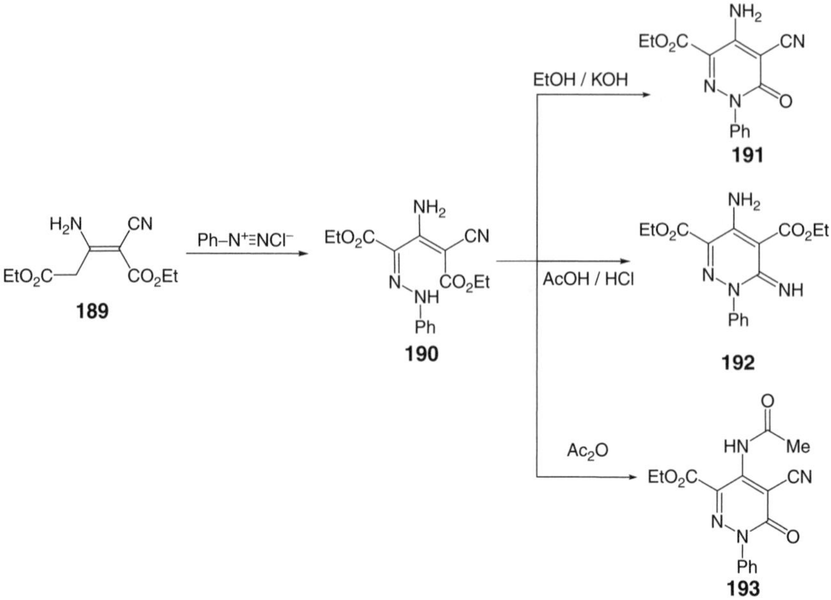

Scheme 35

Subsequently, Elnagdi et al. (1987H(26)899, 1994CCC1753) and Mittelbach et al. (1987LA889) coupled **194a,b** with aromatic diazonium salts. Elnagdi et al. isolated the arylhydrazones **195** as indicated by the presence of appropriate three and two carbonitrile signals in the ^{13}C NMR of **195a,b**, respectively. Mittelbatch et al. (1987LA889) isolated **198**, whose structure has been supported by an X-ray crystal determination (Scheme 36).

The structure of the coupling product of **199** with an aryldiazonium chloride was found to depend on the reaction conditions. Thus, compound **199** coupled with aryldiazonium salts in acetic acid/sodium acetate to yield the arylhydrazones **200** that then afforded pyridazine-6-imines **201** on heating above their melting points. In contrast, compound **199** coupled with aryldiazonium salts in ethanolic sodium acetate to afford the arylhydrazones **202** that then cyclized into pyridazine-6-one derivatives **204** in refluxing acetic acid. This same product could be obtained by coupling **205** with an aryldiazonium salt followed by refluxing the formed arylhydrazone **206** in acetic acid (1986H(24)1219). Coupling diene **207** with phenyldiazonium chloride yielded the pyridazinal **208** (1997S91; Scheme 37).

Other examples include cyclization of the products of coupling **209a,b** with aromatic diazonium salts into pyridazinimines **211a,b** (1985H(23)1999, 1999MI571). Quite unexpectedly and most probably incorrect, Abdelrazek and Fadda (1986ZN(B)499) reported that the product

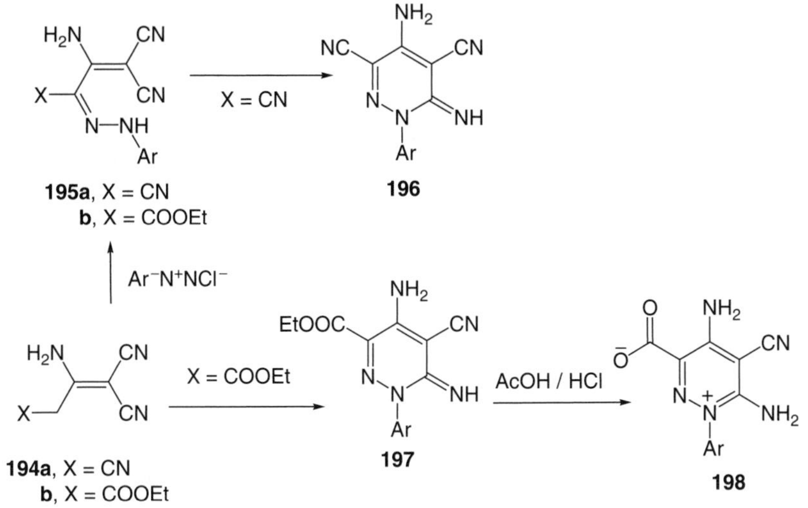

Scheme 36

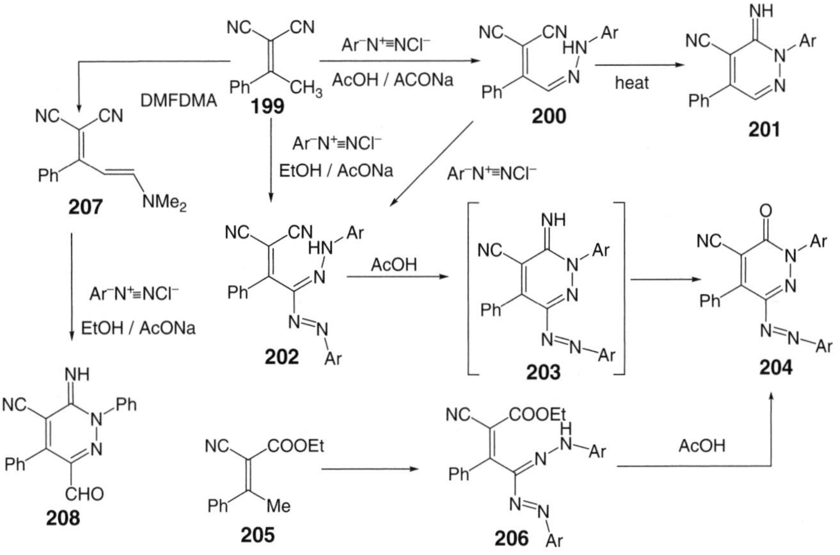

Scheme 37

of coupling **209c** with an aromatic diazonium salt was **212**, that has an azo structure and proved stable for cyclization under different acidic and basic conditions. The proposed structure of **212** should be rechecked because it would be a rare case in which arylhydrazones prefer an azo form and do

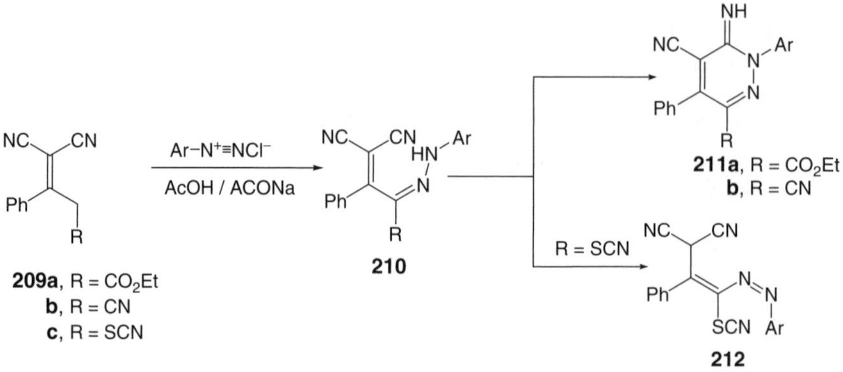

Scheme 38

not cyclize into thiadiazoles as established by Shawali et al. (1994T5091) or pyridazinimines as discussed in the chapter (Scheme 38).

Similarly, coupling **213** with aromatic diazonium salts yields **214** (2000JCM20).

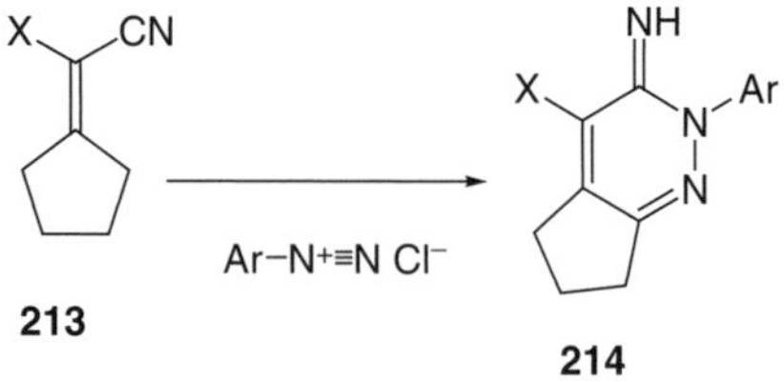

Diazonium salts of methyl anthranilate **215a** or anthranilonitrile **215b** with substituted nitriles yields hydrazones that simultaneously cyclize into pyridazinimines and then to pyridazinoquinazolines. The first example of this synthetic methodology apparently was reported by Abdelhamid et al. (1988AQ22) when **218** was formed by reacting **194** with **215a** by way of intermediates **216** and **217** (Scheme 39).

2.5 Synthesis via condensing arylhydrazones with active methylenes

Ethyl 1-aryl-4-methyl-6-oxo-5-cyanopyridazine-3-carboxylate **221** was synthesized from ethyl 2-arylhydrazono-3-oxobutanoate **220** and ethyl cyanoacetate in acetic acid in the presence of ammonium acetate (1986H(24)1219, 1989T3597, 1988LA1005). Gewald and Hain (1984S62) could prepare these same pyridazinones **221** by condensing ethyl acetoacetate and malononitrile and subsequently coupling **224**. However,

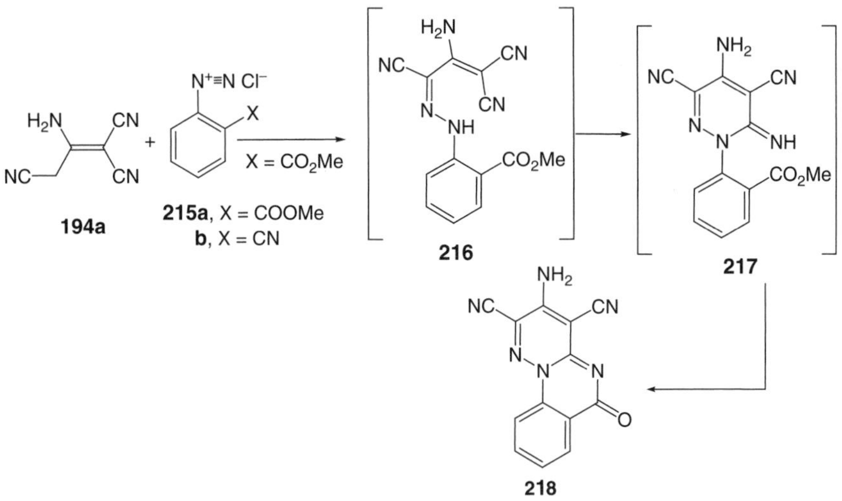

Scheme 39

this is tedious and produces the targeted pyridazines in much lower yields. Moreover, **224** is difficult to keep for a long time. Esters **221** in refluxing acetic acid–hydrochloric acid mixture were hydrolyzed to **222** that were then decarboxylated to pyridazinone **223** when heated above their melting points (1990JRM1124). Pyridazinone **223** can also be simply obtained by heating arylhydrazone **17** with ethyl cyanoacetate in the presence of ammonium acetate at 160 °C for 2 h (2006ARK147). Recently, this synthesis has been conducted in a much shorter time on heating in microwave oven (2006JHC1575; Scheme 40).

Condensation of formazans **226a** with ethyl cyanoacetate in a microwave oven afforded **227**. However, quite unexpectedly **226b** yields pyridazine **228** by nitrogen extrusion. Its structure was established by an X-ray crystal structure determination (2008H1151; Scheme 41).

Hydrazone **220** condensed with malononitrile to yield products that may be formulated as **229** or **229A**. ^{1}H NMR revealed the existence of an NH$_2$ group and this led to assigning zwitter-ionic structure **229A** as the main product (1990CCC734, 1994SL27, 1995T12745). Hydrazone **220** also condensed with diethyl malonate, benzoylacetonitrile and cyclohexanone to yield **231**, **236** and **238**, respectively (1990CCC2977; Scheme 42).

Condensing **239** with malononitrile afforded the polycyclic pyridazine derivative **243** where malononitrile dimerized into **194a** prior to its condensation with **239** (1995T12745; Scheme 43).

Hydrazone **244a** and ethyl cyanoacetate yielded **245a** but required long reflux with azeotropic elimination of water. When water was not eliminated amide **245b** was the sole isolable product. Condensing **242**

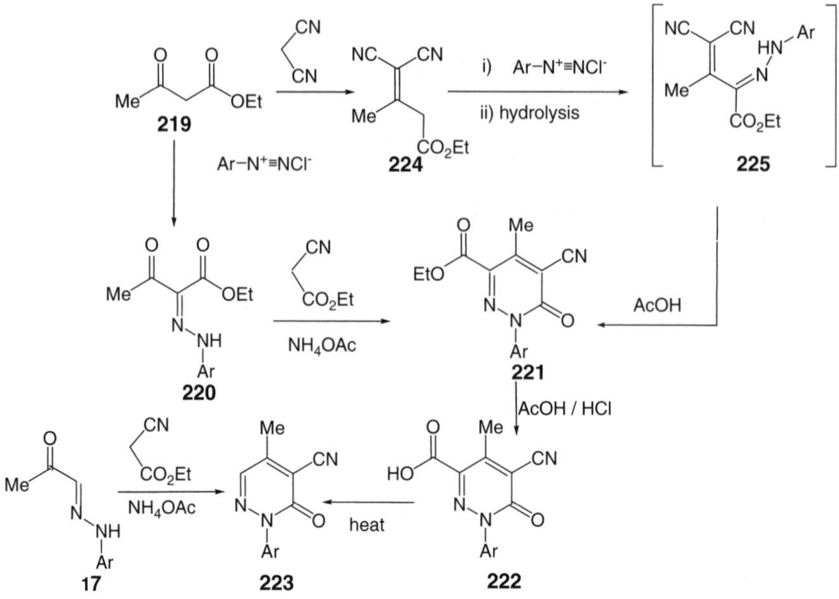

Scheme 40

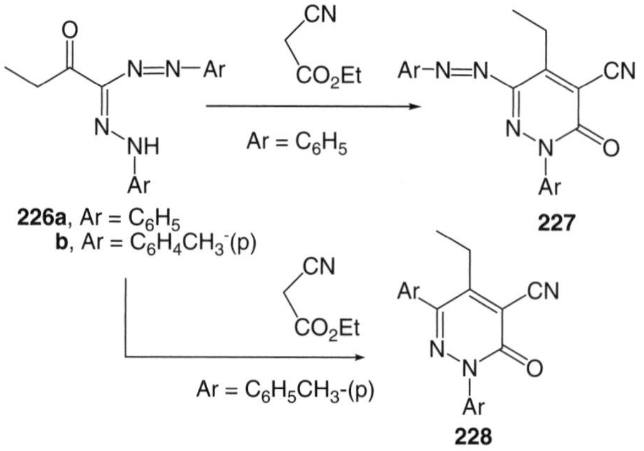

Scheme 41

with dimethylformamide dimethylacetal (DMFDMA) yields pyridazi-nones **246** whose IR spectra did not show band for a ring CO, indicating that it is best postulated as **247** (1995T12745; Scheme 44).

Arylhydrazomesoxalononitrile **248** with ethyl cyanoacetate, ethyl cyanoacetamide and ethyl acetoacetate afforded the corresponding pyri-dazines **250**, whereas **248** reacted with malononitrile in ethanol/Et$_3$N to

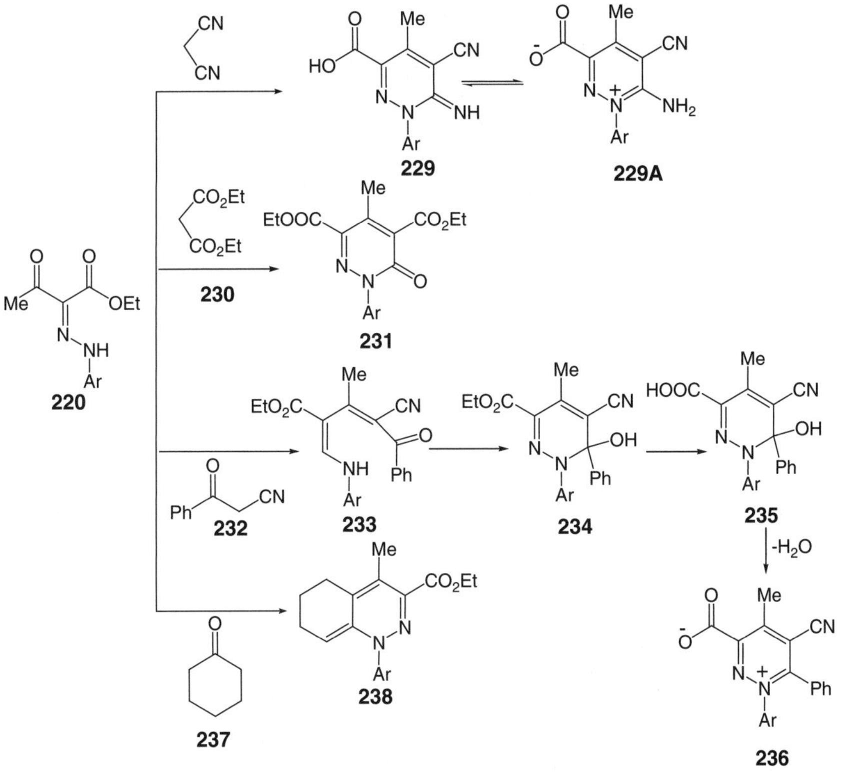

Scheme 42

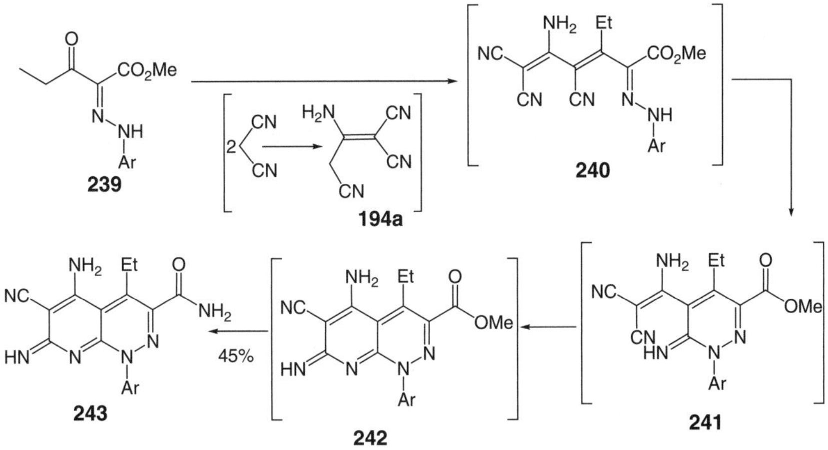

Scheme 43

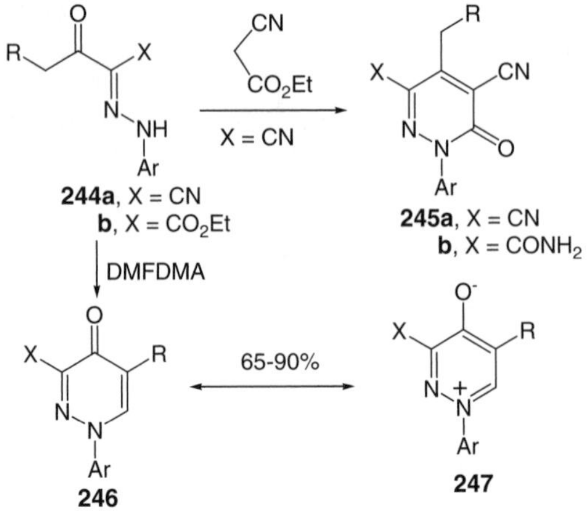

Scheme 44

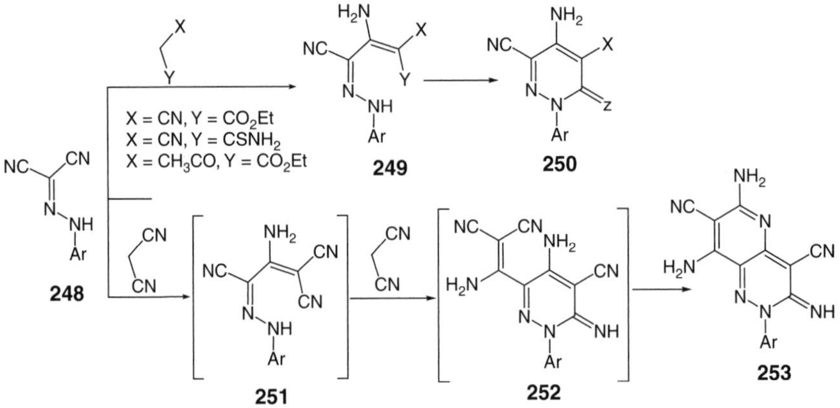

Scheme 45

yield the corresponding pyridopyridazines **253** by way of intermediates **251** and **252** (1999JRM129; Scheme 45).

Hydrazone **254** reacted with malononitrile in ethanolic triethylamine to yield thiazolylpyridazin-6-imine derivative **255** (1994EJC509).

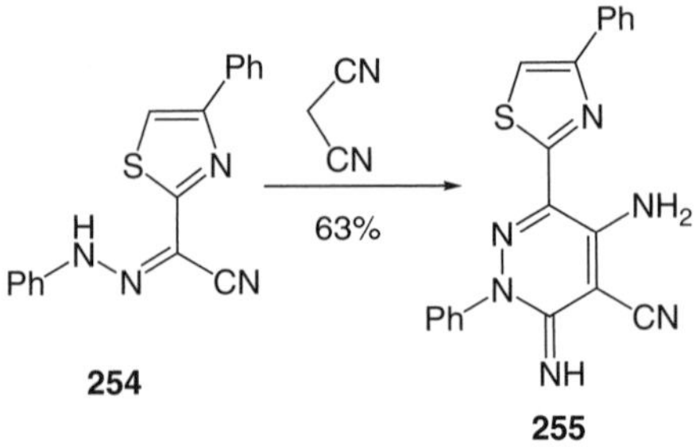

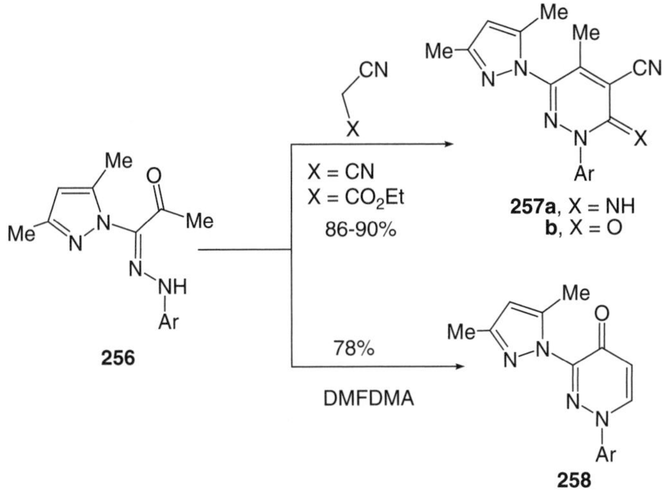

Scheme 46

Arylhydrazones **256** condensed readily with malononitrile and ethyl cyanoacetate to yield the pyridazine derivative **257** and **256** also condensed with DMFDMA to yield pyrazolylpyridazines **258** (2001JHC685; Scheme 46).

Similarly, arylhydrazones **259** were converted into the corresponding pyridazines **260** on treatment with ethyl cyanoacetate in the presence of AcOH and ammonium acetate (1999JRM2848, 2000S1166).

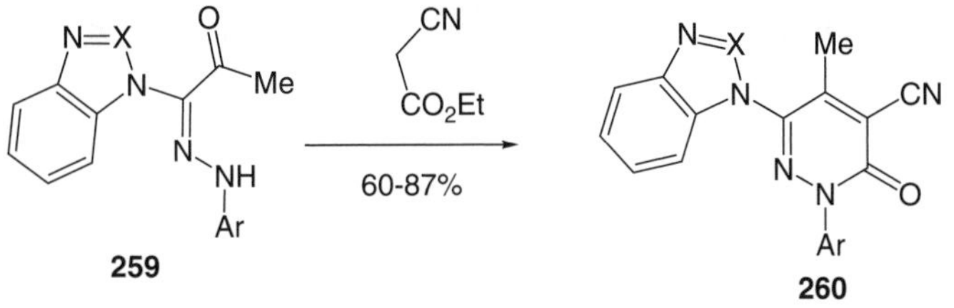

Ketone **261** condenses with ethyl format to yield intermediate **262** that couples with aromatic diazonium salts forming **263**. The latter with malononitrile yielded **264** that is proposed to exist in equilibrium with **265** (2002BMC3197; Scheme 47).

Methylquinoxaline derivative **266** reacted with bromine to yield bromomethylquinoxaline **267** and then reacted with NaCN to yield cyanomethylquinoxaline **268**. The latter readily coupled with aromatic diazonium salts to yield hydrazone derivative **269** that then reacted with active methylene reagents to yield the corresponding quinoxalinylpyridazine **270**. Quinoxaline **266** also reacted with DMFDMA to yield the enamine **271** that then coupled with a diazonium salt to yield aldehydehydrazone **272**. Subsequently, **272** with active methylene compounds yielded the corresponding pyridazines **273** (2006MI901; Scheme 48).

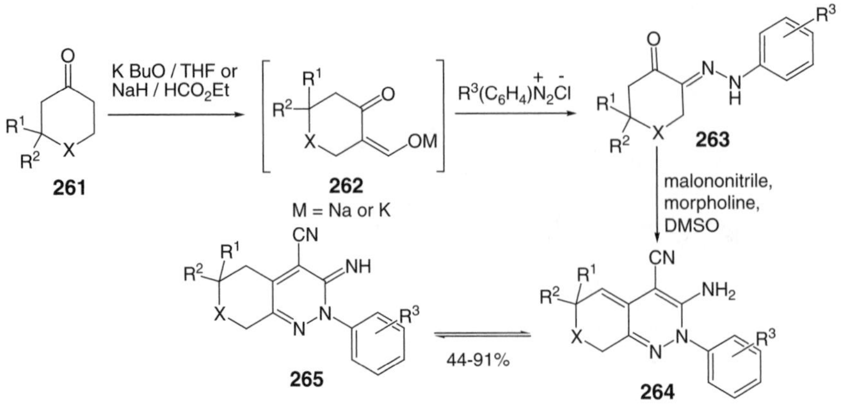

Scheme 47

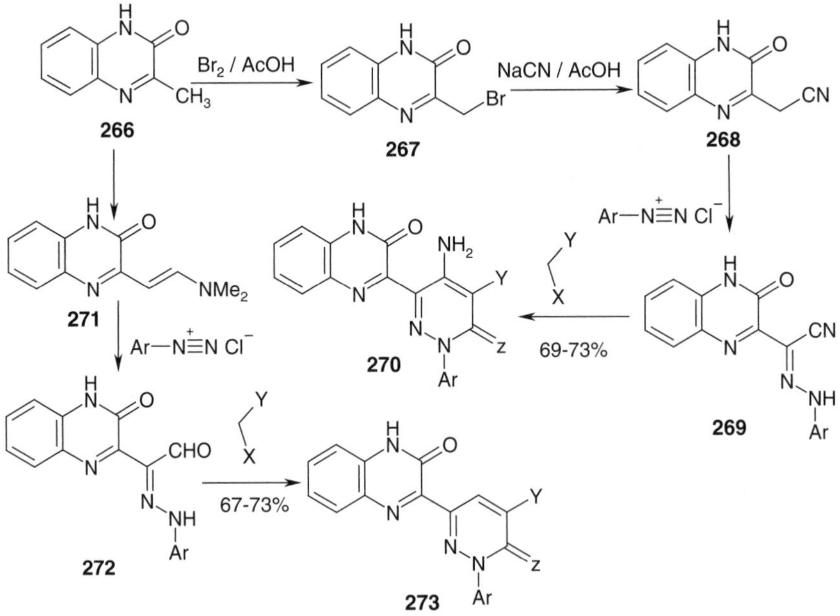

Scheme 48

The conversion of arylhydrazonals into pyridazines and condensed pyridazines has been extensively investigated by Elnagdi et al. (1997S91, 1999OPPI551). Thus, **275** with active methylene reagents yields **276**, while reaction with hippuric acid or glycine in presence of acetic anhydride affords **279**. In perhaps the first example of a Baylis–Hillman reaction with arylhydrazones **275** reacted with carbethoxymethyl triphenylphosphonium chloride to yield pyridazine **280** in 67% (Scheme 49).

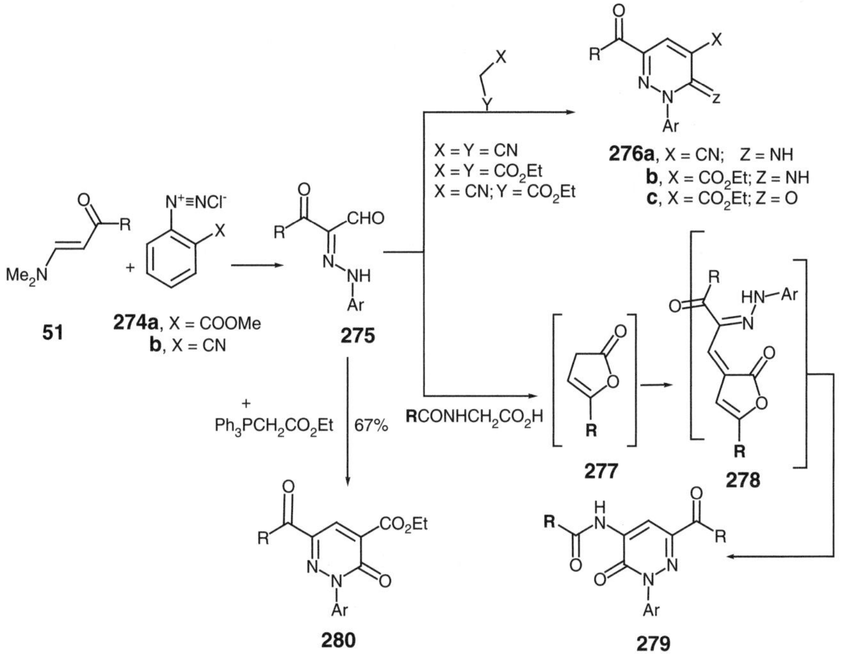

Scheme 49

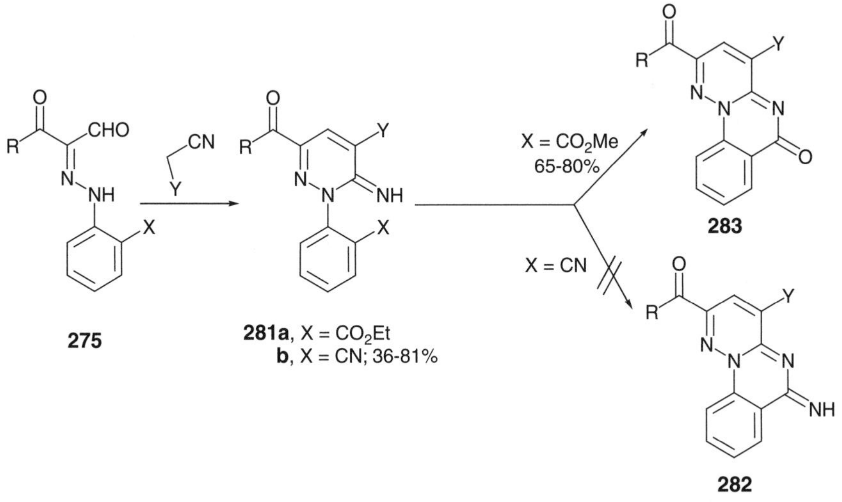

Scheme 50

Hydrazonals **275a,b** condensed with active methylene nitriles to yield intermediates **281a,b** that cyclized readily into **282** or **283** depending on the substituent (2000JCR510) and **281a** cyclized readily into **283**, while **281b** failed to cyclize into **282** (Scheme 50).

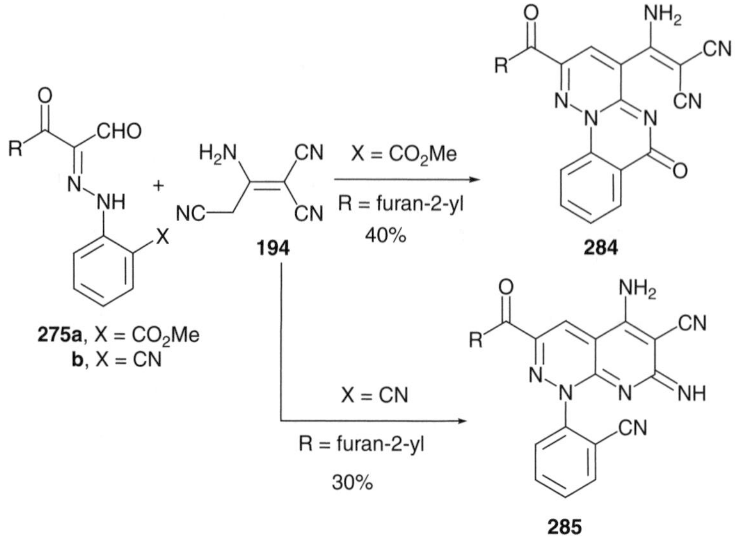

Scheme 51

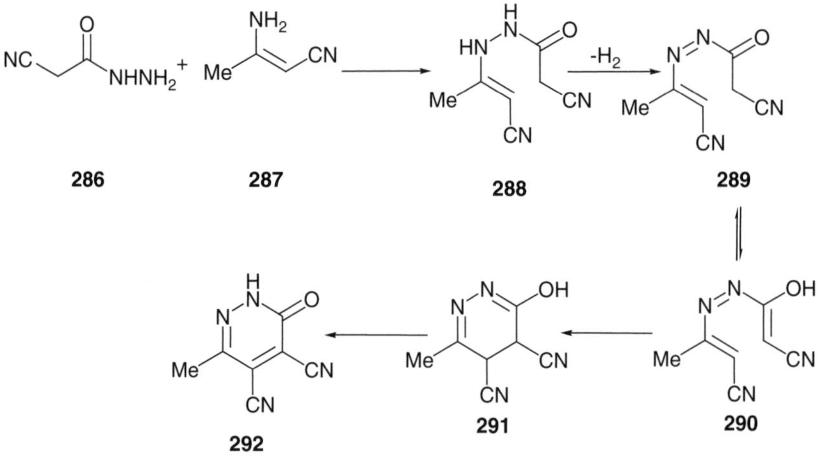

Scheme 52

Hydrazonal **275a** readily condensed with 3-amino-2-cyano-pent-2-enedinitrile **194** to yield the pyridazinoquinazoline **284**, while **275b** with **194** yielded pyridopyridazine **285** (2004ZNB721, 2003MOL910; Scheme 51).

A new route to pyridazinones enabled the synthesis of **292** on reacting **286** with **287** to yield **289** that afforded **292** when heated above

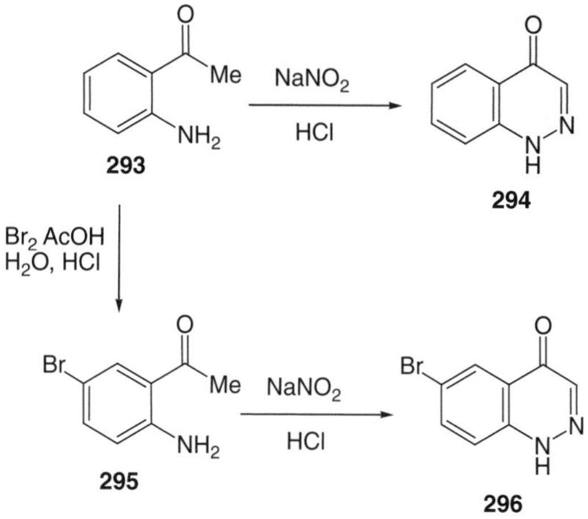

Scheme 53

its melting point. The postulated mechanism is indicated in Scheme 52 (1994SL27, 1995BSF920).

3. SYNTHESIS OF CONDENSED PYRIDAZINES

3.1 Synthesis *via* one bond formation (5+1 combination)

This route has been extensively used for more than 50 years to prepare cinnolines from 2-aminoacetophenone derivatives. For example, diazotizing **293** afforded **294** (1945JSC512, 1949JCS2393, 1981AJC2619). The recent literature reported several applications that employ substituted acetophenones (2006BMC6832). For example, bromo derivative **295** was prepared by initial bromination of **293** and subsequent diazotization (Scheme 53).

Alternatively, diazotization of acetylenes **297** affords cinnoline. This old synthesis (1945JSCS512, 1949JCS2393, 1994SC1733) finds recent applications (2004T7983, 2003TL5453, 1995LA775).

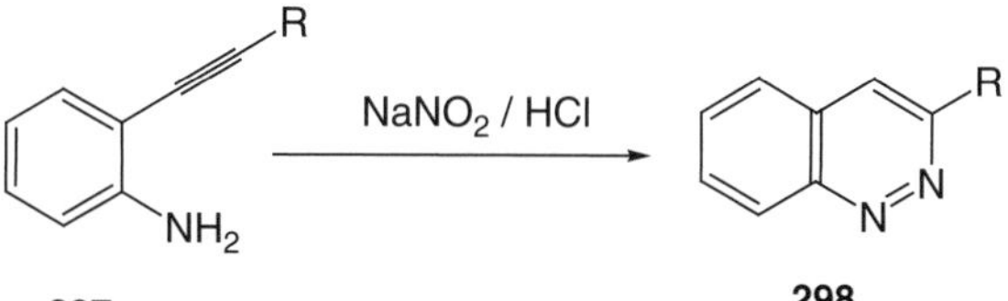

Pyrido[3,4-c]pyridazines **301** was synthesized by condensing **299** with DMFDMA, where compound **300** was assumed to be an intermediate (2007ARK213).

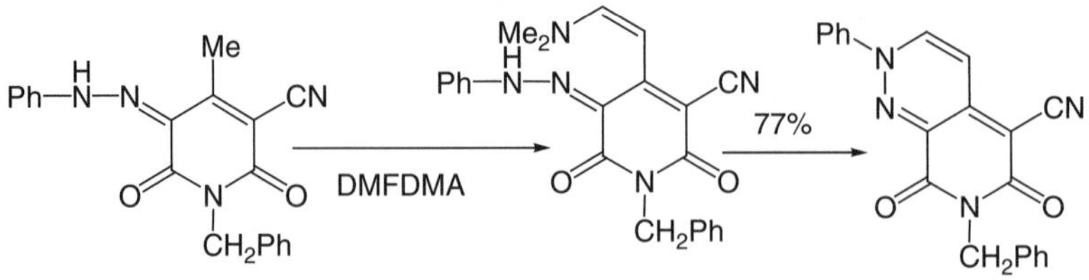

299 → DMFDMA → **300** → 77% → **301**

3.2 Synthesis utilizing functionally substituted arylhydrazones precursors *via* one bond formation (γ- to heteroatom)

More than 50 years ago cinnolines **303** were synthesized by cyclizing **302** in the presence of Lewis acids (1961JCS2828). Some 30 years latter Gewald et al. reported the cyclization of **304** in the presence of nitrobenzene and anhydrous aluminum chloride (Friedel–Craft's conditions) into cinnolines **305** (1984LA1390, 2004M595; Scheme 54).

Attempted coupling of enaminoesters **306** with benzenediazonium fluroborate **307** directly afforded cinnolines **309**, perhaps through intermediate **308** (1981T3513, 1975T1325). Coupling **310** with aromatic diazonium salts affords only **310** and formazans **311** (2007ZN529; Scheme 55).

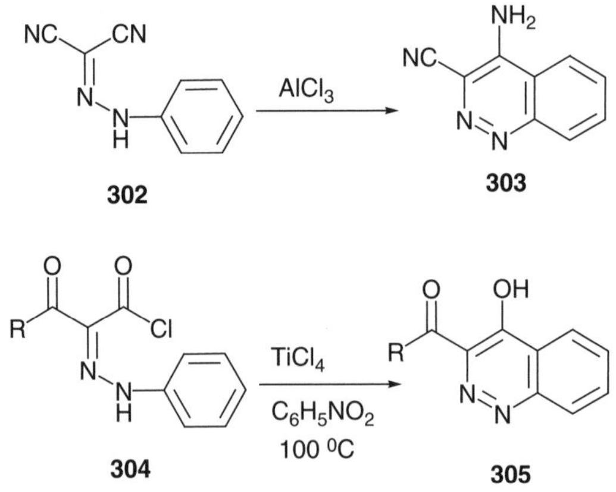

302 → AlCl₃ → **303**

304 → TiCl₄ / C₆H₅NO₂ / 100 °C → **305**

Scheme 54

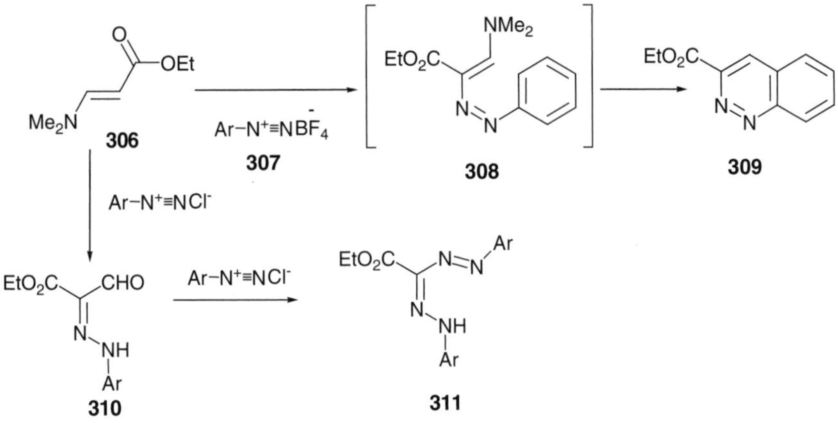

Scheme 55

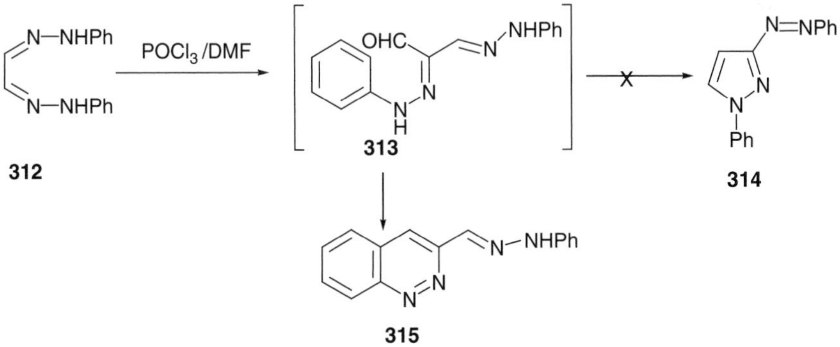

Scheme 56

Glyoxal *bis*(phenylhydrazone) **312** with $POCl_3/DMF$ yielded a product that may be formulated as cinnoline **315** or phenylazopyrazole **314**, by common intermediate **313**. Structure **315** was confirmed based on ^1H NMR that revealed an exchangeable signal in D_2O at δ 3.33 ppm for an NH (2006JCR729; Scheme 56).

Heating aryl hydrazonal **275** in sulfuric acid afforded only 3-aroyl-cinnolines **318** through the postulated mechanism (2001T1609, 2005JHC1185; Scheme 57).

Attempts to affect this cyclization thermally by flash vacuum pyrolysis of **275** resulted in the formation of a mixture of amines, acyl amines and aroyl nitriles (2001T10171).

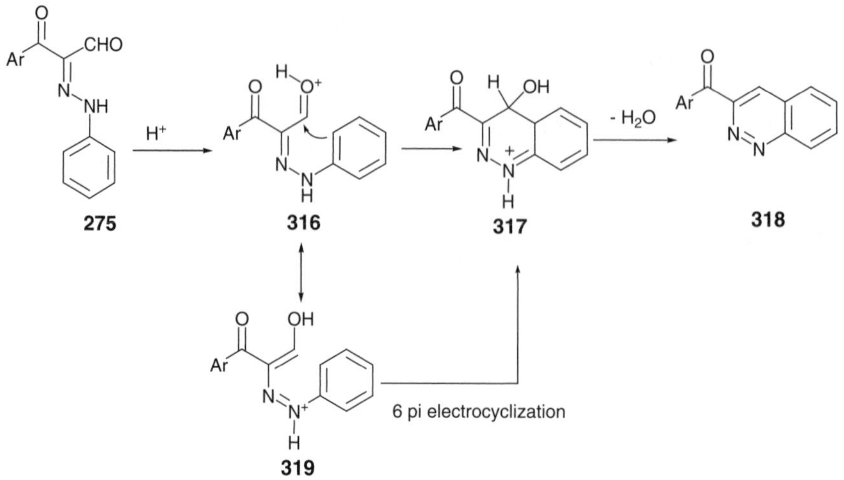

Scheme 57

REFERENCES

1899MI398	S. Gabriel and J. Colman, *Chem. Ber.*, **32**, 398 (1899).
1945JSC512	K. Schofield and J. C. E. Simpson, *J. Chem. Soc.*, 512–520 (1945).
1947JCS239	W. G. Overend and L. F. Wiggins, *J. Chem. Soc.*, 239 (1947).
1949JCS2393	K. Schofield and T. Swain, *J. Chem. Soc.*, 2393–2399 (1949).
1959JA4342	R. A. Carboni and R. V. Lindsey, *J. Am. Chem. Soc.*, **81**, 4342 (1959).
1961JCS2828	H. J. Barber, K. Washbourn, W. R. Wragg, and E. Lunt, *J. Chem. Soc.*, 2828 (1961).
1975T1325	M. S. Manhas, J. W. Brown, U. K. Pandit, and P. Houdewind, *Tetrahedron*, **31**, 1325 (1975).
1979JOC629	M. J. Haddadin, S. J. Firsan, and B. S. Nader, *J. Org. Chem.*, **44**, 629 (1979).
1981AJC2619	P. A. Marshall, B. A. Mooney, R. H. Prager, and A. D. Ward, *Aust. J. Chem.*, **34**(12), 2619–2627 (1981).
1981MI133	H. F. Oates and L. M. Stoker, *Clin. Exp. Pharmacol. Physiol.*, **8**, 133–139 (1981).
1981MI471	G. M. Maxwell, D. Ness, and V. Rencis, *Eur. J. Pharmacol.*, **69**, 471–476 (1981).
1981T3513	C. B. Kanne and U. K. Pandit, *Tetrahedron*, **37**, 3513 (1981).
1982S490	S. M. Fahmy, N. M. Abed, R. M. Mohareb, and M. H. Elnagdi, *Synthesis*, 490–493 (1982).
1983AP1013	I. Simiti and M. Coman, *Arch. Pharm.*, **316**(12), 1013–1017 (1983).
1984LA1390	K. Gewald, O. Calderon, H. Schaefer, and U. Hain, *Liebigs Ann. Chem.*, **7**, 1390 (1984).
1984S62	K. Gewald and U. Hain, *Synthesis*, 62 (1984).
1985H(23)1999	G. E. H. Elgemeie, H. A. Elfahham, and M. H. Elnagdi, *Heterocycles*, **23**, 1999 (1985).
1986H(24)1219	N. S. Ibraheim, F. M. Abdel-Galil, R. M. Abdel-Motaleb, and M. H. Elnagdi, *Heterocycles*, **24**, 1219 (1986).
1986MI2199	P. Worms, C. Gueudet, and K. Biziere, *Life Sci.*, **39**, 2199 (1986).

1986ZN(B)499	F. M. Abdelrazek and A. A. Fadda, *Z. Naturforsch. B.*, **41b**, 499 (1986).
1987H(26)899	M. H. Mohamed, N. S. Ibraheim, and M. H. Elnagdi, *Heterocycles*, **26**, 899 (1987).
1987LA889	M. Mittelbach, U. Wanjer, and C. Karrky, *Liebigs Ann. Chem.*, 889 (1987).
1988LA1017	R. Fruttero, B. Ferrarotti, A. Gasco, G. Calestani, and C. Rizzoli, *Liebigs Ann. Chem.*, **11**, 1017–1023 (1988).
1988LA1005	M. H. Elnagdi, N. S. Ibrahim, K. U. Sadek, and M. H. Mohamed, *Liebigs Ann. Chem.*, 1005 (1988).
1988AQ22	A. O. Abdelhamid, N. M. Abed, and A. M. Farag, *An. Quim.*, **84C**, 22 (1988).
1989CS45	L. Andersen, F. E. Nielsen, and E. B. Pedersen, *Chem. Scr.*, **29**(1), 45–49 (1989).
1989LA1255	M. H. Elnagdi, A. M. Negm, and A. W. Erian, *Liebigs Ann. Chem.*, 1255 (1989).
1989T3597	M. H. Elnagdi, F. M. Abdelrazek, N. S. Ibraheim, and A. W. Erian, *Tetrahedron*, **45**, 3597 (1989).
1989ZNB683	M. H. Elnagdi, F. Maksoud, A. Aal, E. A. Hafez, and Y. M. Yassin, *Z. Naturforsch. B.*, **44**(6), 683–689 (1989).
1990CCC734	M. H. Elnagdi, K. U. Sadek, N. M. Taha, and Y. M. Yassin, *Collect. Czech. Chem. Commun.*, **55**, 734 (1990).
1990CCC2977	S. M. El-Kousy, E. El-Sakka, A. M. El-Torgoman, H. Rashdy, and M. H. Elnagdi, *Collect. Czech. Chem. Commun.*, **55**, 2977 (1990).
1990JRM1124	M. H. Elnagdi, A. W. Erian, K. U. Sadek, and M. A. Selim, *J. Chem. Res. (M)*, 1124–1142 (1990).
1990MI385	M. TiSler and B. Stanovnik, *Adv. Heterocycl. Chem.*, **49**, 385–474 (1990).
1990ZNB389	M. H. Elnagdi, F. A. Abdul-Aal, N. M. Taha, and Y. M. Yassin, *Z. Naturforsch. B.*, **45**(3), 389–392 (1990).
1992BML1357	D. W. Brooks, A. Basha, F. A. J. Kerdesky, J. A. Holms, J. D. Ratajcyk, P. Bhatia, J. L. Moore, J. G. Martin, S. P. Schmidt, D. H. Albert, R. D. Dyer, P. Young, and G. W. Carter, *Bioorg. Med. Chem. Lett.*, **2**, 1357–1360 (1992).
1992CHJ475	G. H. Sayed, A. Radwan, S. M. Mohamed, S. A. Shiba, and M. Khalil, *Chin. J. Chem*, **10**, 475–480 (1992).
1992JPR723	A. H. H. Elghandour, A. H. M. Hussein, and M. H. Elnagdi, *J. Parakt. Chemi.*, 723 (1992).
1994CCC1753	A. M. A. Helmy, M. A. Morsi, and M. H. Elnagdi, *Collect. Czech. Chem. Commun.*, **59**, 1753 (1994).
1994CPB2219	E. Oishi, K.-I. Iwamoto, T. Okada, S. Suzuki, K.-I. Tanji, A. Miyashita, and T. Higashino, *Chem. Pharm. Bull.*, **42**(11), 2219–2224 (1994).
1994EJC509	A. M. Negm, F. M. Abdelrazek, M. H. Elnagdi, and L. H. Shaaban, *Egypt. J. Chem.*, **37**, 509–518 (1994).
1994SC1733	S. F. Vasilevsky, E. V. Tretyakov, and H. D. Verkruijsse, *Synth. Commun.*, **24**(12), 1733–1736 (1994).
1994SL27	M. H. Elnagdi, A. M. Negm, and K. U. Sadek, *Synlett*, 27 (1994).
1994T5091	A. M. Farag, A. S. Shawali, M. S. Algharib, and K. M. Dawood, *Tetrahedron*, **50**(17), 5091 (1994).
1995BSF920	M. H. Elnagdi and A. W. Erian, *Bull. Chim. Soc. Fr.*, **132**, 920 (1995).

1995LA775   S. F. Vasilevsky and E. V. Tretyakov, *Liebigs Ann. Chem.*, (5), 775–779 (1995).

1995T12745   H. Al-Awadi, F. Al-Omran, M. H. Elnagdi, L. Infantes, C. Foces-foces, N. Jagerovic, and J. Elguero, *Tetrahedron*, **51**, 12745 (1995).

1996JME297   Y. Nomoto, H. Takai, T. Ohno, K. Nagashima, K. Yao, K. Yamada, K. Kubo, M. Ichimura, A. Mihara, and H. Kase, *J. Med. Chem.*, **39**, 297–303 (1996).

1997HAC29   A. M. Hussein, A. A. Atalla, I. S. A. Hafez, and M. H. Elnagdi, *Heteroatom Chem.*, **8**, 29 (1997).

1997MI617   H. F. Zohdi, T. A. Osman, and A. O. Abdelhamid, *J. Chin. Chem. Soc.*, **44**, 617 (1997).

1997OPPI285   F. Al-Omran, N. Al-Awadi, A. A. El-Khair, and M. H. Elnagdi, *Org. Prep. Proced. Int.*, 285 (1997).

1997S91   F. Al-Omran, M. M. Abdel Khalik, A. A. El-Khair, and M. H. Elnagdi, *Synthesis*, 91–94 (1997).

1998CJC1038   A. W. Erian, E. A. A. Hafez, E. S. Darwish, and M. H. Elnagdi, *Can. J. Chem.*, **76**, 1038 (1998).

1998JOC4912   J. Nakayama, R. Hasemi, K. Yoshimura, Y. Sugihara, S. Yamaoka, and N. Nakamura, *J. Org. Chem.*, **63**, 4912 (1998).

1998T4561   P. Y. F. Deghati, M. J. Wanner, and G. Koomen, *Tetrahedron*, **39**, 4561 (1998).

1998T12897   A. R. Al Dulayymi and M. S. Baird, *Tetrahedron*, **54**, 12897 (1998).

1999JCM96   A. A. Elassar, *J. Chem. Res. (S)*, **5**, 96–97 (1999).

1999JRM129   A. Z. A. Hassanien, I. S. A. Hafiz, and M. H. Elnagdi, *J. Chem. Res. (M)*, 129 (1999).

1999JRM151   A. Al-Etaibi, N. Al-Awadi, F. Al-Omran, M. M. Abdel Khalik, and M. H. Elnagdi, *J. Chem. Res. (M)*, 151 (1999).

1999JRM2848   A. Al-Naggar, M. M. Abdel-Khalik, and M. H. Elnagdi, *J. Chem. Res. (M)*, 2848 (1999).

1999MI571   S. Al-Mousawi, K. S. George, and M. H. Elnagdi, *Die Pharm.*, **54**, 571 (1999).

1999OPPI551   H. Behbehani, M. M. Abdel-Khalik, and M. H. Elnagdi, *Org. Prep. Proced. Int.*, **31**, 551–557 (1999).

2000AJC905   A. McCluskey, M. Finn, M. Bowman, R. Smith, and P. A. Keller, *Aust. J. Chem.*, **53**(11 and 12), 905–908 (2000).

2000JCM20   F. Al-Omran, N. Al-Awadi, A. Elassar, and A. A. El-Khair, *J. Chem. Res. (S)*, 20–21 (2000).

2000JCR510   K. M. Al-Zaydi, E. A. Hafiz, and M. H. Elnagdi, *J. Chem. Res. (M)*, 510–527 (2000).

2000S1166   M. M. Abdel-Khalik, M. H. Elnagdi, and S. M. Agamy, *Synthesis*, **8**, 1166 (2000).

2001BML2373   T. Matsuda, T. Aoki, T. Koshi, M. Ohkuchi, and H. Shigyo, *Bioorg. Med. Chem. Lett.*, **11**, 2373 (2001).

2001JHC685   M. H. Mohamed, M. M. Abdel-Khalik, and M. H. Elnagdi, *J. Heterocycl. Chem.*, **38**, 685 (2001).

2001JMC2188   R. Barbaro, L. Betti, M. Botta, F. Corelli, G. Giannaccini, L. Maccari, F. Manetti, G. Strappaghetti, and S. Corsano, *J. Med. Chem.*, **44**, 2118–2132 (2001).

2001MOL959   M. Treu, U. Jordis, and V. J. Lee, *Molecules*, **6**, 959 (2001).

2001T1609   N. A. Al-Awadi, M. H. Elnagdi, Y. A. Ibrahim, K. Kaul, and A. Kumar, *Tetrahedron*, **57**, 1609 (2001).

2001T5497	Z. Wan, G. H. C. Woo, and J. K. Snyder, *Tetrahedron*, **57**, 5497 (2001).
2001T10171	Y. A. Ibrahim, K. Kamini, and N. A. Al-Awadi, *Tetrahedron*, **57**, 10171 (2001).
2002BMC1	D. Gundisch, T. Kampchen, S. Schwartz, G. Seitz, J. Siegl, and T. Wegge, *Bioorg. Med. Chem.*, **10**, 1 (2002).
2002BMC3197	C. Liljebris, J. Martinsson, L. Tedenborg, M. Williams, E. Barker, J. E. S. Duffy, A. Nygren, and S. James, *Bioorg. Med. Chem.*, **10**, 3197 (2002).
2002BML689	C. J. McIntyre, G. S. Ponticello, N. J. Liverton, S. J. O'Keefe, E. A. O'Neill, M. Pang, C. D. Schwartz, and D. A. Claremon, *Bioorg. Med. Chem. Lett.*, **12**, 689 (2002).
2002JHC869	A. Sener, R. Kasimogullari, M. K. Sener, I. Bildirici, and Y. Akcamur, *J. Heterocycl. Chem.*, **39**(5), 869–875 (2002).
2002JHC889	G. Biagi, F. Ciambrone, I. Giorgi, O. Livi, V. Scartoni, and P. L. Barili, *J. Heterocycl. Chem.*, **39**(5), 889–893 (2002).
2002MI2833	G. T. Manh, R. Hazard, A. Tallec, J. P. Pradere, D. Dubreuil, M. Thiam, and L. Toupet, *Electrochim. Acta*, **47**, 2833 (2002).
2002MI481	S. M. Sayed, M. A. Khalil, M. A. Ahmed, and M. A. Raslan, *Synth. Commun.*, **32**(3), 481 (2002).
2002MI989	A. A. Shalaby, M. M. El-Shahawai, N. A. Shams, and S. Batterjee, *Synth. Commun.*, **32**(7), 989 (2002).
2002T2227	M. Gnanadeepam, S. Selvaraj, S. Perumal, and S. Renuga, *Tetrahedron*, **58**, 2227 (2002).
2003COR1423	G. Stajer, F. Csende, and F. Fulop, *Curr. Org. Chem.*, **7**, 1423–1432 (2003).
2003MOL910	K. M. Al-Zaydi and M. H. Elnagdi, *Molecules*, **5**, 910 (2003).
2003MOL322	G. H. Sayed, A. A. Hamed, G. A. Meligi, W. E. Boraie, and M. Shafik, *Molecules*, **8**, 322 (2003).
2003CHE541	A. E. Rubtsov and V. V. Zalesov, *Chem. Heterocycl. Compd. (Engl. Transl.)*, **39**(4), 541 (2003).
2003T5869	J. M. Chezal, E. Moreau, O. Chavignon, C. Lartigue, Y. Blache, and J. C. Teulade, *Tetrahedron*, **59**, 5869 (2003).
2003TL5453	L. G. Fedenok and N. A. Zolnikova, *Tetrahedron Lett.*, **45**(29), 3459 (2004).
2003TL7799	Y. J. Lim, M. Angela, and P. T. Buonora, *Tetrahedron Lett.*, **44**, 7799 (2003).
2004BMCL321	A. Coelho, E. Sotelo, N. Friaz, M. Yanez, R. Laguna, E. Cano, and E. Ravina, *Bioorg. Med. Chem. Lett.*, **14**, 321 (2004).
2004BML2031	T. Hu, B. A. Stearns, B. T. Campbell, J. M. Arruda, C. Chen, J. Aiyar, R. E. Bezverkov, A. Santini, H. Schaffhauser, W. Liu, S. Venkatraman, and B. Munoz, *Bioorg. Med. Chem. Lett.*, **14**(9), 2031–2034 (2004).
2004CHE1047	S. A. Hovakimyan, A. V. Babakhanyan, V. S. Voskanyan, V. E. Karapetian, G. A. Panosyan, and S. T. Kocharian, *Chem. Heterocycl. Compd. (Engl. Transl.)*, **40**(8), 1047 (2004).
2004HAC502	M. M. Abdelkhalik, A. M. Negm, A. I. Elkhouly, and M. H. Elnagdi, *Heteroatom Chem.*, **15**, 502 (2004).
2004JHC647	M. A. Al-Shiekh, A. M. Salah El-Din, E. A. Hafez, and M. H. Elnagdi, *J. Heterocycl. Chem.*, **41**, 647–654 (2004).
2004M595	A. M. Amer, M. M. El-Mobayed, and S. Asker, *Monatsh. Chem.*, **135**(5), 595 (2004).

2004RJO1027	V. V. Razin, M. E. Yakovlev, K. V. Shataev, and S. I. Selivanov, *Russ. J. Org. Chem.*, **40**(7), 1027 (2004).
2004RJO1033	M. E. Yakovlev and V. V. Razin, *Russ. J. Org. Chem.*, **40**(7), 1033 (2004).
2004SC783	E. Meyer, A. C. Joussef, J. H. Gallardo, and de. B. P. de Souza, *Synth. Commun.*, **34**(5), 783–793 (2004).
2004T3421	A. Kotschy, J. Farago, A. Csampai, and D. M. Smith, *Tetrahedron*, **60**, 3421 (2004).
2004T7983	N. Le Fur, L. Mojovic, A. Turck, N. Ple, G. Queguiner, V. Reboul, S. Perrio, and P. Metzner, *Tetrahedron*, **60**(36), 7983–7994 (2004).
2004TL3459	A. Coelho, E. Sotelo, H. Novoa, O. M. Peeters, N. Blaton, and E. Ravina, *Tetrahedron Lett.*, **45**, 3459 (2004).
2004TL7799	Y. J. Lim, M. Angela, and P. T. Buonora, *Tetrahedron Lett.*, **44**, 7799–7801 (2003).
2004TL8781	J.-J. Kim, Y.-D. Park, S.-D. Cho, H.-K. Kim, H. A. Chung, S.-G. Lee, J. R. Falck, and Y.-J. Yoon, *Tetrahedron Lett.*, **45**, 8781 (2004).
2004ZNB721	K. M. Al-Zaydi and M. H. Elnagdi, *Z. Naturforsch. B.*, **59b**, 721 (2004).
2004ZNB1132	H. M. E. Hassaneen, H. M. Hassaneen, and M. H. Elnagdi, *Z. Naturforsch. B.*, **59b**, 1132–1136 (2004).
2005BML2409	N. Tamayo, L. Liao, M. Goldberg, D. Powers, Y.-Y. Tudor, V. Yu, L. M. Wong, B. Henkle, S. Middleton, R. Syed, T. Harvey, G. Jang, R. Hungate, and C. Dominguez, *Bioorg. Med. Chem. Lett.*, **15**, 2409 (2005).
2005EJM401	E. Akbas and I. Derber, *Eur. J. Med. Chem.*, **40**, 401 (2005).
2005JHC307	F. Al-Omran and A. El-Khair, *J. Heterocycl. Chem.*, **42**(2), 307 (2005).
2005JHC1185	S. A. S. Ghozlan, I. A. Abdelhamid, H. M. Gaber, and M. H. Elnagdi, *J. Heterocycl. Chem.*, **42**(6), 1185–1189 (2005).
2005JRM23	B. Al-Saleh, M. M. Abdelkhalaik, M. A. El-Apasery, and M. H. Elnagdi, *J. Chem. Res. (M)*, 23 (2005).
2005T4785	A. Coelho, H. Novoa, O. M. Peeters, N. Blaton, M. Avarado, and E. Sotelo, *Tetrahedron*, **61**, 4785 (2005).
2005T4805	J. C. Gonzalez-Gomez, L. Santana, and E. Uriarte, *Tetrahedron*, **61**, 4805 (2005).
2006AP305	F. M. Abdelrazek, F. A. Michael, and A. E. Mohamed, *Arch. Pharm.*, **339**(6), 305 (2006).
2006ARK133	V. K. Salvi, D. Bhambi, J. L. Jat, and G. L. Talesara, *Arkivok*, **xiv**, 133–140 (2006).
2006ARK147	S. A. S. Ghozlan, I. A. Abdelhamid, and M. H. Elnagdi, *Arkivoc*, **xiii**, 147 (2006).
2006BMC6832	K. W. Woods, J. P. Fischer, A. Claiborne, T. Li, S. A. Thomas, G.-D. Zhu, R. B. Diebold, X. Liu, Y. Shi, V. Klinghofer, E. K. Han, R. Guan, S. R. Magnone, E. F. Johnson, J. J. Bouska, A. M. Olson, R. de Jong, T. Oltersdorf, Y. Luo, S. H. Rosenberg, V. L. Giranda, and Q. Li, *Bioorg. Med. Chem.*, **14**(20), 6832–6846 (2006).
2006BMCL1080	A. Crespo, C. Meyers, A. Coelho, M. Yanez, N. Fraiz, E. Sotelo, B. U. W. Maes, R. Laguna, E. Cano, G. L. F. Lemiere, and E. Ravina, *Bioorg. Med. Chem. Lett.*, **16**, 1080 (2006).

2006COR277	J. J. Bourguignon, S. Oumouch, and M. Schmitt, *Curr. Org. Chem.*, **10**(3), 277–295 (2006).
2006EJM101	M. Sonmez, I. Berber, and E. Akbas, *Eur. J. Med. Chem.*, **41**, 101 (2006).
2006JCR729	H. M. E. Hassaneen, T. A. Abdalla, H. M. Hassaneen, and M. H. Elnagdi, *J. Chem. Res. (M)*, 729 (2006).
2006JHC1575	B. Al-Saleh, N. M. Hilmy, M. A. El-Apasery, and M. H. Elnagdi, *J. Heterocycl. Chem.*, **43**, 1575 (2006).
2006JRM237	F. Al-Omran, N. Al-Awadi, A. A. Elassar, and A. A. El-Khair, *J. Chem. Res.*, 237 (2000).
2006MI691	D. Unal, E. Saripinar, and Y. Akcamur, *Tur. J. Chem.*, **30**, 691–701 (2006).
2006MI901	A. A. Elassar, *J. Chin. Chem. Soc.*, **53**, 901–907 (2006).
2006T8966	P. J. Crowley, S. E. Russell, and L. G. Reynolds, *Tetrahedron*, **62**, 8966 (2006).
2006T9002	C. Ma, S.-J. Liu, L. Xin, J. R. Falck, and D.-S. Shin, *Tetrahedron*, **62**, 9002–9009 (2006).
2006T9718	F. Pozgan, S. Polanc, and M. Kocevar, *Tetrahedron*, **62**, 9718 (2006).
2006TL149	Y.-M. Pu, Y.-Y. Ku, T. Grieme, R. Henry, and A. V. Bhatia, *Tetrahedron Lett.*, **47**, 149–153 (2006).
2006TL1853	T. G. Deryabina, N. P. Belskaia, M. I. Kodess, S. T. Dehaen, and V. A. Bakulev, *Tetrahedron Lett.*, **47**, 1853 (2006).
2006TL6125	J. X. De Araujo-Junior, M. Schmitt, P. Benderitter, and J.-J. Bourguignon, *Tetrahedron Lett.*, **47**, 6125 (2006).
2007ARK213	S. M. Al-Mousawia, I. A. Abdelhamid, and S. M. Moustafa, *Arkivoc*, (i), 213–221 (2007).
2007EMJonline	A. I. Hashem, A. S. A. Youssef, K. A. Kandeel, and W. S. I. Abou-Elmagd, *Eur. J. Med. Chem.*, (2007), online.
2007JHC105	S. A. S. Ghozlan, I. A. Abdelhamid, H. M. Hassaneen, and M. H Elnagdi, *J. Heterocycl. Chem.*, **44**, 105–108 (2007).
2007JHC877	S. I. Aziz, H. F. Anwar, M. A. El-Apasery, and M. H. Elnagdi, *J. Heterocycl. Chem.*, **44**(4), 877–881 (2007).
2007MIPh.D thesis	I. A. Abdelhamid, Ph.D thesis, Cairo University (2007).
2007ZN529	S. Makhseed, H. M. Hassaneen, and M. H. Elnagdi, Accepted for publication in Z. *Naturforsch.*, **62**(4), 529–539.
2008H1151	S. M. Al-Mousawi, M. A. EL-Apasery, and M. H. Elnagdi, *Heterocycles*, **75**(5), 1151–1161 (2008).

CHAPTER 2

Organometallic Complexes of Polypyridine Ligands IV

Alexander P. Sadimenko

1. INTRODUCTION

This chapter is a continuation of a series on organometallic complexes of polypyridine ligands. The previous three chapters (07AHC(93)179, 07AHC(94)109, 08AHC(95)219) concerned the earlier organotransition

Department of Chemistry, University of Fort Hare, Alice 5700, Republic of South Africa

Advances in Heterocyclic Chemistry, Volume 97
ISSN 0065-2725, DOI 10.1016/S0065-2725(08)00202-X

metal complexes and highlighted the role of the polypyridine ligands as spectator ligands. This role allowed modeling compounds with interesting reactivity patterns, creating useful catalytic systems, modern materials, especially with interesting photochemical properties. Herein the organometallic complexes of the cobalt and nickel groups are discussed, and the new aspect that is discussed here is a variety of coordination modes of polypyridines themselves. They include the bridging function of the 2,2'-bipyridine ligand, roll-over N, C coordination mode of polypyridines, multiple cyclometalation, and creation of the C, C-coordinated polypyridine in the process of synthesis of organometallic compounds. This adds a certain interest to the problem and inspired extensive research especially during the past few years.

2. COMPLEXES WITH COBALT GROUP METALS

2.1 M(I) (M = Co, Rh, Ir) complexes

Most rhodium(I) complexes with a 2,2'-bipyridine ligand represent cationic species [Rh(bipy)L$_2$]X (L = CO (65IC161, 70CC54, 72JOM(35)389, 74JOM(65)119, 74JOM(67)443, 75JCS(D)133), L$_2$ = cod (57JCS4735, 71JCS(A)2334, 72JOM(35)389, 74JOM(65)119, 76JOM(105)365, 78JOM(157) 345, 99EJI27), COT (72JOM(35)389), nbd (71JCS(A)2334, 74JOM(65)119, 75JOM(91)379, 76JOM(116)C35, 78JOM(148)81, 79TMC55, 85TMC288, 89IC2097), 1,5-hexadiene (77JOM(140)63)), [Rh(bipy)(CO)(PR$_3$)]X (72JOM(35)389), [Rh(bipy)(CO)$_3$]X (74JOM(65)119), [Rh(bipy)(L)(L$_2'$)]X (L = CO, L' = PR$_3$ (72JOM(35)389, 81TMC45, 86ZNK413), L = PR$_3$, L$_2'$ = COT) (72JOM(35)389). In the carbonyl complexes of the type [Rh(CO)$_2$(LL)][RhX$_2$(CO)$_2$] (LL = bipy, phen; X = Cl, Br), there is a substantial cation–anion association (79JCS(D)1569). Complexes [(η^4-LL)M(bipy)] (LL = cod, nbd; M = Rh, Ir) are characterized by substantial M→L π-back bonding effect (68JPC1853, 82IC1023, 82IC1027). The same refers to the iridium(I) complexes (75INCL359, 76JCS(D)762). Complexes [(η^4-cod)Ir(LL)]Cl (LL = bipy, phen) follow from the reaction of the polypyridine ligands with [(η^4-cod)Ir(Cl)(2-picoline)$_2$] (80IC7). Reaction of the dimer [(η^4-nbd)Rh(Cl)]$_2$ with silver perchlorate in acetone and further with 6,7-dihydropyrido[2,3-b:3',2'-j]-1,10-phenanthroline, 7,8-dihydro-6H-cyclohepta[2,1-b:3,4-b]-di-1,8-naphthyridine, or 2,2'-bi (3-methyl)-1,8-naphthyriodine (LL) gives the cationic complexes of composition [Rh(nbd)(LL)](ClO$_4$) (96POL1823). Rhodium(I) complexes of 2,2'-bipyridine are often implicated as the catalysts of hydrogenation and hydrogen-transfer reactions of alkenes and ketones (92CRV1051, 97SCI2100) and other catalytic reactions (81NJC543, 85NJC225, 88IC4582, 91RKCL185, 94AG(E)497, 99ZK463, 02EJO1685, 03JA11430, 05ICC94).

4,4'-Disubstituted 2,2'-bipyridines (R = OMe, Me, H, Cl, NO_2) react with $[(\eta^4\text{-cod})Rh(THF)_2](BF_4)$ to yield complexes $[(\eta^4\text{-cod})Rh(4,4'\text{-}R_2bipy)]$ catalytically active in hydroformylation of terpenes (06JOM2037). 1,1'-*Bis*(2-pyridyl)ferrocene with $[(\eta^4\text{-cod})Rh(Cl)]_2$ and silver perchlorate gives the cationic complex **1** (06JOM4573). Similarly complex **2** based on 1,1'-*bis*(4-pyridyl)ferrocene was prepared.

1   **2**

1,10-Phenanthroline readily reacts with $[(\eta^4\text{-diolefin})_2Rh_2(C_2O_4)]$ to afford the ionic complexes $[(\eta^4\text{-diolefin})Rh(phen)][(\eta^4\text{-diolefin})Rh(C_2O_4)]$ (diolefin = cod, tfb) (85POL325). 2,2'-Bipyridine with $[(\eta^4\text{-tfb})_2Rh_2(C_2O_4)]$ gives $[(\eta^4\text{-tfb})Rh(bipy)][(\eta^4\text{-tfb}) Rh(C_2O_4)]$. $[Rh(\eta^4\text{-diolefin})(LL)][Rh(C_2O_4)(\eta^4\text{-diolefin})]$ (LL = phen, bipy) reacts with carbon monoxide to afford $[Rh(CO)_2(LL)][Rh(C_2O_4)(CO)_2]$ (80JOM(197)87, 85POL325). 1,10-Phenanthroline reacts with $[(\eta^4\text{-cod})_2Rh_2(\mu\text{-}C_4O_4)]$ to yield $[(\eta^4\text{-cod})Rh(phen)][(\eta^4\text{-cod})Rh(C_4O_4)]$ where $C_4O_4^{2-}$ is dihydroxycyclobutenedionate (97ICA(255)351). The product can be carbonylated to provide $[Rh(CO)_2(phen)][Rh(C_4O_4)(CO)_2]$. 3,3'-Disubstituted 2,2'-bipyridine depicted in **3** reacts with $[(\eta^4\text{-cod})Rh(THF)_2]BF_4$ to yield cationic **3** (00OM622). This complex was immobilized on the surface of SiO_2 or Pd/SiO_2 to produce efficient heterogeneous catalysts for hydrogenation of arenes. Complexes $[(\eta^4\text{-diene})M(Cl)]_2$ (M = Rh, Ir; diene = cod, nbd, 1,5-hexadiene) interact with polymeric films of pyrrole-substituted 2,2'-bipyridine or 4,4'-dimethyl-2,2'-bipyridine (LL) to yield the polymeric films of $[(\eta^4\text{-diene})M(LL)]^+$ (96JCS(D)2503, 97JOM(532)31, 00ICC620).

3

The cationic allyl complexes of rhodium(I) can be prepared by the reaction of 2,2'-bipyridine with $[(\eta^3\text{-}C_3H_5)_4Rh_2Cl_2]$ or $[(\eta^3\text{-}C_4H_7)_4Rh_2Cl_2]$ to yield $[(\eta^3\text{-}C_3H_5)_2Rh(bipy)]^+$ or $[(\eta^3\text{-}C_4H_7)_2Rh(bipy)]^+$, respectively (68JCS(A)583). 2,2'-Bipyridine, 4,4'-dimethyl-2,2'-bipyridine, 1,10-phenanthroline, 2,9-dimethyl-1,10-phenanthroline, or 4,7-diphenyl-1,10-phenanthroline (LL) react with $[(\eta^3\text{-allyl})RhCl_2]_n$ (allyl $= C_3H_5$, C_4H_7) in methanol or ethanol to yield $[(\eta^3\text{-allyl})RhCl_2(allyl)(LL)]$ serving as catalysts for hydrogenation of alkenes (96TMC305). In methanol the process leading to $[(\eta^3\text{-allyl})Rh(LL)(MeOH)_2]^{2+}$ is possible.

The general synthesis of the complexes of polypyridine ligands (LL) of composition $[Rh(CO)Cl(LL)]$ is based on the dimer $[Rh(CO)_2Cl]_2$ (92ZOB821). The product first contains two carbonyl groups and then gradually decarbonylates to yield the monocarbonyl complex. The latter may dissociate into an ion pair of composition $[Rh(CO)_2(LL)]Cl$ (75JCS(D)133, 81ICA(51)241, 84POL799, 91ICA(184)73). 4,4'-Dimethyl-2,2'-bipyridine reacts with the dimer $[Rh(CO)_2Cl]_2$ in tetrahydrofuran (THF) to yield $[Rh(CO)Cl(4,4'\text{-}Me_2bipy)]$ (01JOM(626)118). The product reacts with silver nitrate in THF to produce the nitrate complex $[Rh(CO)(NO_3)(4,4'\text{-}Me_2bipy)]$, which in turn on reaction with molecular iodine in THF gives $[Rh(CO)(I)_2(NO_3)(4,4'\text{-}Me_2bipy)]$. The latter with methyl or iso-propyl magnesium bromide in THF yields $[Rh(CO)R_2I(4,4'\text{-}Me_2bpy)]$ (R $=$ Me, i-Pr). 2,2'-Bipyridine and $[Rh(CO)_2I]_2$ give $[Rh(bipy)(CO)I]$ (03OM1047). The product oxidatively adds methyl iodide to yield the rhodium(III) complex $[Rh(CO)(bipy)I_2Me]$.

Examples of neutral complexes are $[(\eta^4\text{-nbd})Rh(bipy)Cl]$ (89IC2097), $[Rh(bipy)(CO)X]$ (X $=$ Cl, Br, I) (74JOM(65)119, 75JCS(D)133, 85JOM(282)123), $[(\eta^4\text{-cod})Ir(LL)X]$ (LL $=$ phen; 3,4,7,8-Me$_4$phen X $=$ Cl, I) (07JCS(D)133). 2,2'-Bipyridine reacts with $[(\eta^2\text{-}C_2H_4)_2RhCl_2]$ to yield $[(\eta^2\text{-}C_2H_4)Rh(bipy)Cl]$ (93JOM(463)215). The product reacts with sodium tetraphenylborate in an unusual way to afford $[(\eta^6\text{-PhBPh}_3)Rh(bipy)]$. With formaldehyde in THF, the ethylene complex gives $[Rh(bipy)(CO)Cl]$. Reaction of $[(\eta^2\text{-}C_2H_4)Rh(4,4'\text{-}Me_2bipy)Cl]$ with dimethylsulfoxide (DMSO) generates $[Rh(4,4'\text{-}Me_2bipy)(DMSO)Cl]$ with the S-coordination of the DMSO ligand, and with t-butylisocyanide the product is $[Rh(4,4'\text{-}Me_2bipy)(CNBu\text{-}t)Cl]$ (02EJI1827). $[(\eta^4\text{-1,5-hexadiene})Rh(4,4'\text{-}Me_2bipy)](BF_4)$ with t-butylisocyanide gives $[Rh(4,4'\text{-}Me_2bipy)(CNBu\text{-}t)_2](BF_4)$. The t-butylisocyanide products are capable of the oxidative addition of organyl halides to afford the rhodium(III) complexes. Thus, $[Rh(4,4'\text{-}Me_2bipy)(CNBu\text{-}t)Cl]$ oxidatively adds benzyl chloride to give $[Rh(4,4'\text{-}Me_2bipy)(CNBu\text{-}t)(PhCH_2)Cl_2]$, and $[Rh(4,4'\text{-}Me_2bipy)(CNBu\text{-}t)_2](BF_4)$ in an identical reaction produces $[Rh(4,4'\text{-}Me_2bipy)(CNBu\text{-}t)_2(PhCH_2)Cl](BF_4)$.

Complex $\{[(\eta^5\text{-Cp*})Co(Br)]_2\}$ enters disproportionation reaction with 2,2'-bipyridine to give $[(\eta^5\text{-Cp*})Co(Br)(bipy)]Br$ and $[(\eta^5\text{-Cp*})Co(bipy)]$

(86OM980). Potassium reduction of $[(\eta^5\text{-Cp*})Co(Cl)(bipy)]Cl$ gives $[(\eta^5\text{-Cp*})Co(bipy)]$ (96JOM(524)195). The third way of synthesis of this product is the interaction of 2,2′-bipyridine with $[(\eta^5\text{-Cp*})Co$ $(\eta^2\text{-C}_2\text{H}_3\text{SiMe}_3)_2]$ in acetone (00OM1247). Structural data of this apparently cobalt(I) product are indicative of the contribution of the mesomeric form of the cobalt(III) species as depicted below, **4**. Complex **5** with 2,2′-bipyridine affords the cobalt(I) product **6** (96CB319, 00CRV1527).

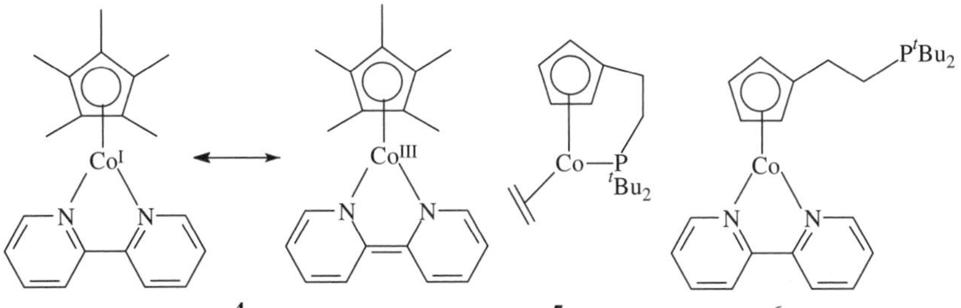

4   **5**   **6**

2.2. M(III) (M = Co, Rh, Ir) complexes

6-(2-Diphenylphosphinoethyl)-2,2′-bipyridine reacts with $\{[(\eta^2\text{-C}_2\text{H}_4)_2$ $RhCl]_2\}$ in dichloromethane to yield neutral **7**, which further interacts with silver triflate in ethanol to form **8** (97JCS(D)3777). Complex $[(\eta^4\text{-cod})Ir(4,4′-t\text{-Bu}_2\text{bipy})](OTf)$ reacts with *bis*(pinacolato)diboron to yield the iridium(III) *bis*(boryl) complex $\{(\eta^4\text{-cod})Ir(4,4′-t\text{-Bu}_2\text{bipy})$ $[B(O_2C_2Me_4)]_2\}(OTf)$ (02JA390). A product of different composition $[(\eta^2\text{-COE})Ir(4,4′-t\text{-Bu}_2\text{bipy})\{B(O_2X_2Me_4)\}_3]$ follows from $[(\eta^2\text{-COE})_2IrCl]_2$, 4,4′-di-$t$-butyl-2,2′-bipyridine, and *bis*(pinacolato)diboron. Iridium(III) complexes based on 4,4′-di-t-butylpyridine and $[(\eta^4\text{-cod})Ir(OMe)]_2$ are active catalysts for aromatic C–H borylation (02AG(E)3056). 2,2′-Bipyridine with $[IrI_4(CO)_2]^-$ in the presence of the methoxide anion gives the neutral iridium(III) complex of composition $[IrI_2(COOMe)$ $(bipy)(CO)]$ (69IC298).

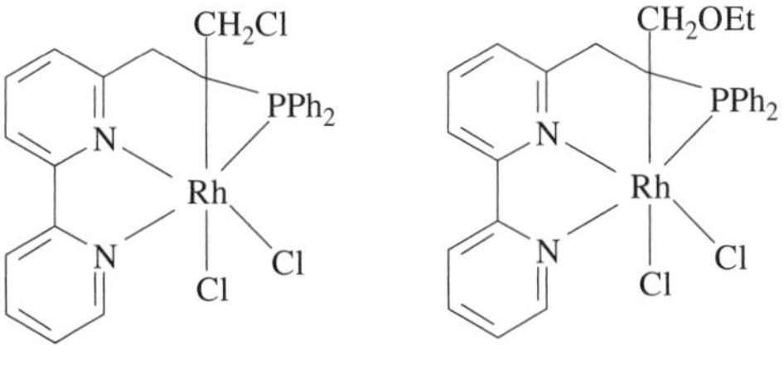

7   **8**

$[(\eta^5\text{-Cp})M(H)(\text{bipy})]^+$ (M = Rh, Ir) intermediates are described in a number of hydride transfer processes (87AG(E)568, 88CC16, 88CC1150, 89CB1869, 89CC1259, 89JOM(363)197, 90AG(E)388, 91AG(E)844, 91OM1568, 92AG(E)1529, 92JOM(439)79, 93JEAC213, 94IC4453, 01CCC207, 01EJI69, 01EJI613, 01IC4150, 01JPC(B)4801, 01OM1668, 03IC5185). They, for example, may result from $[(\eta^5\text{-Cp*})\text{Ir}(\text{Cl})(\text{LL})]^+$ (LL = derivatives of bipy and phen) with the hydride agent $NaBH_3(CN)$ (92AX(B)515, 93JA118, 96JEAC189). The reduced species with the formal oxidation state of a metal of +1 are known $[(\eta^5\text{-Cp})M^{1+}(\text{bipy})]$, but they tend to be in mesomerism with $[(\eta^5\text{-Cp})M^{2+}(\text{bipy}^-)]$, as indicated by electrochemical methods (91JOM(419)233). The metals in these species are characterized by an M→L π-electron back donation in the ground state (93ZAAC1998). Complexes $[(\eta^5\text{-Cp*})M(\text{LL})\text{Cl}]\text{Cl}$ (M = Rh, Ir; LL = bipy, phen, 4,7-$(HO)_2$phen) catalyze hydrogenation of carbon dioxide and bicarbonates (03JMC(A)95, 04OM1480). 4,4'-Dihydroxy-2,2'-bipyridine and 4,7-dihydroxy-1,10-phenanthroline (LL) react with $[(\eta^5\text{-Cp*})MCl_2]_2$ (M = Rh, Ir) to yield $[(\eta^5\text{-Cp*})M(\text{LL})\text{Cl}]\text{Cl}$ (07OM702). The phenanthroline iridium complex can be methylated using methyl iodide in the presence of potassium carbonate in acetone to give $[(\eta^5\text{-Cp*})\text{Ir}(4,7\text{-}(MeO)_2\text{phen})I]I$. Iridium(III) complexes $[(\eta^5\text{-Cp*})\text{Ir}(\text{LLL})(OH_2)]^{2+}$ (LL = bipy, 4,4'-$(OMe)_2$bpy) are catalysts for the hydrogenation of carbon dioxide to yield formic acid (06JCS(D)4657).

4'-Phenyl-2,2':6',2''-terpyridine reacts with $[\{\text{Cp*}M(\mu\text{-Cl})\text{Cl}\}_2]$ (M = Rh, Ir) in methanol with sodium tetrafluoroborate to yield mononuclear cationic **9** (M = Rh, Ir) with the bidentate coordination mode and fluxional behavior of the ligand (98POL299). 1,4-*Bis*(2,2':6',2''-terpyridyn-4'-yl)benzene with the same rhodium(III) and iridium(III) precursors forms dinuclear **10** (M = Rh, Ir) where the polypyridine ligand is still bidentately coordinated and fluxional.

9

10

$[(\eta^5\text{-Cp*})M(bipy)Cl](BF_4)$ (M = Rh, Ir) reacts with $[(n\text{-Bu})_3MeN]_2$ $(SnB_{11}H_{11})$ to yield the zwitterionic products $[(\eta^5\text{-Cp*})M(bipy)(SnB_{11}H_{11})]$ (01ZAAC1146). $[(\eta^5\text{-Cp*})Ir(H_2O)_3](OTf)_2$ reacts with 2,2'-bipyridine to yield $[(\eta^5\text{-Cp*})Ir(H_2O)(bipy)](OTf)_2$ (01OM4903), which serve as promoters of transfer hydrogenation, reductive amination, and dehalogenation processes. 1,8-Naphthyridine reacts with $[(\eta^5\text{-Cp*})IrCl_2]_2$ to yield **11** with monodentate coordination of the heterocyclic ligand (06ICA2431). The product reacts with silver hexafluorophosphate to give chelating **12**. Reaction of $[(\eta^5\text{-Cp*})IrCl_2]_2$, 1,8-naphthyridine, and silver hexafluorophosphate leads to dicationic chelating **13** characterized by solution dynamics. Complex **12** on reaction with sodium tetrahydroborate gives dinuclear **14** with the bridging mode of coordination of the 1,8-naphthyridine ligand.

| **11** | **12** | **13** | **14** |

2.3 Cyclometalation

4'-Functionalized 6'-phenyl-2,2'-bipyridine ligands react with dimer $[(2\text{-Phpy})_2IrCl]_2$ and further with excess ammonium hexafluorophosphate to yield the cationic *ortho*-metalated iridium(III) **15** (R = $OC_{12}H_{25}$,

$OC_{18}H_{37}$, $OCOC_6H_4OCOC_6H_4OC_6H_{13}$, $OCOC_6H_4OCOC_6H_4OC_{12}H_{25}$) (00EJI1039). Such properties of the products as shape anisotropy and amphiphilic character make them attractive for the creation of new materials. Complexes containing transition metal oxalate anions [Ir (2-Phpy)$_2$(bipy)][$M^{II}Cr^{III}(C_2O_4)_3$] · 0.5H$_2$O (M^{II} = Ni, Mn, Co, Fe, Zn) and [Ir(2-Phpy)$_2$(bipy)][$M^{II}Fe^{III}(C_2O_4)_3$] · 0.5H$_2$O (M^{II} = Fe, Mn) are molecular magnets (06IC5653). 2,2'-Bipyridine (84JA6647) or 4,4'-di-t-butyl-2,2'-bipyridine (LL) (04JA2763) reacts with [(2-Phpy)$_2$Ir(μ-Cl)$_2$Ir(2-Phpy)$_2$] and ammonium hexafluorophosphate to yield [Ir(2-Phpy)$_2$(LL)](PF$_6$) with electroluminescent properties. Other cases where 2,2'-bipyridine serves as an ancillary ligand in the cyclometalated iridium(III) species are described (87JPC1047, 93IC3081, 93IC3088). When functionalized 2,2'-bipyridine ligands are used in such complexes, polymetallic species with interesting luminescent properties follow (01IC1093, 03CEJ475, 03IC686, 03JCS(D)2080, 04OM5856, 05CC230, 05EJI110, 05IC8723, 05JMTC2820, 05ZOB705). Interaction of [(2-Phpy)$_2$Rh(μ-Cl)$_2$Rh(2-Phpy)$_2$] with 1,10-phenanthroline-1,10-dione (LL) and further with potassium hexafluorophosphate gives the mononuclear cationic cyclometalated rhodium(III) complex [Rh(2-Phpy)$_2$(LL)](PF$_6$) (02IC6521, 07JOM3810). The coordination of the LL ligand is by the pyridine nitrogen atoms. Functionalized dipyridophenazines with a bis(dichloro)-bridged iridium(III) precursor give heteroleptic cyclometalated **16** (R = H, O(CH$_2$)$_5$COOMe, COOEt, C≡C(CH$_2$)$_3$COOEt) (07IC10187). Related complexes are [Ir(2-Phpy)$_2$ (5-R-phen)](PF$_6$) (R = H, Me, NMe$_2$, NO$_2$) and [Ir(2-Phpy)$_2$(4,7-R$_2$phen)] (PF$_6$) (R = Me, Ph) (07CC4116), [IrL$_2$(LL)](PF$_6$) (HL = 2-(4-(N-((2-biotinamido)ethyl)aminomethyl)phenyl)pyridine, 2-(4-(N-((6-biotinamido)hexyl) aminomethyl)phenyl)pyridine; LL = 3,4,7,8-Me$_4$phen, 4,7-Ph$_2$phen) (07IC700), [Ir(1-phenylisoquinoline)$_2$(LL)](PF$_6$) (LL = bipy, phen, 2-pyridyl-quinoline, 2,2'-biquinoline, 1,1'-biisoquinoline) (06IC6152). The crown ether-linked iridium(III) **17** exhibits a notable luminescence enhancement in the presence of aqueous silver cations (07IC9139). The focus of the study of the photophysical properties of the complexes [Ir(2-Phpy)$_2$ (5-X-1,10-phen)](PF$_6$) (X = NMe$_2$, NO$_2$), [Ir(2-phenylquinoline)$_2$(5-X-1,10-phen)](PF$_6$) (X = H, Me, NMe$_2$, NO$_2$), [Ir(2-Phpy)$_2$(4,7-Me$_2$phen)] (PF$_6$), [Ir(2-Phpy)$_2$(5,6-Me$_2$phen)](PF$_6$), [Ir(2-Phpy)$_2$(2,9-Me$_2$phen)](PF$_6$), [Ir(2-phenylquinoline)$_2$(4,7-Ph$_2$phen)](PF$_6$), [Ir(2-Phpy)$_2$(4-Mephen)](PF$_6$), [Ir(2-Phpy)$_2$(5-Rphen)](PF$_6$) (R = H, Me), and [Ir(2-Phpy)$_2$(4,7-Ph$_2$phen)] (PF$_6$) was tuning the color of the emission and other photochemical properties by various substituents at the phenanthroline ring (07IC8533). Their photophysical and electrochemical properties of the complexes of type **18** change drastically under protonation and addition of various anions (07OM5922). A cyclometalated iridium(III) **19** contains DNA oligonucleotides (07JA14733).

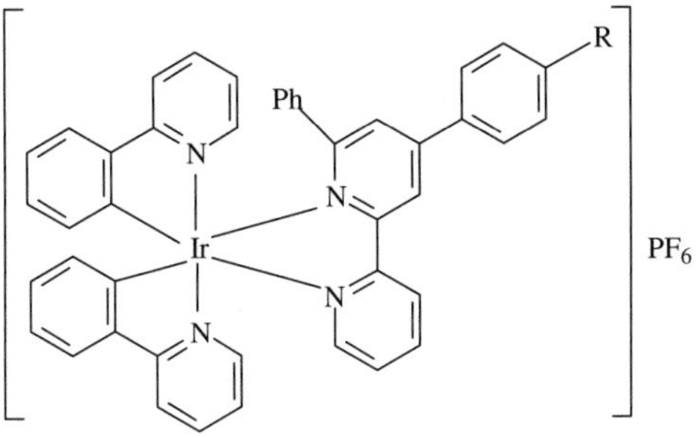

15

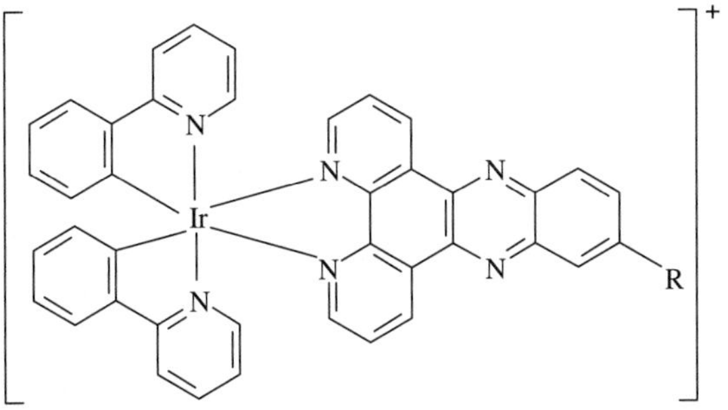

16

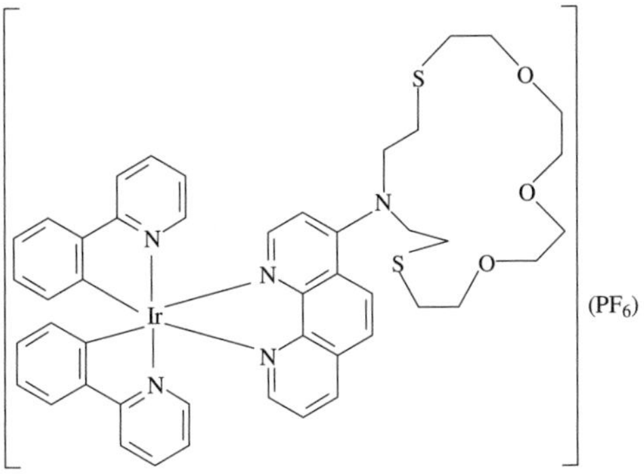

17

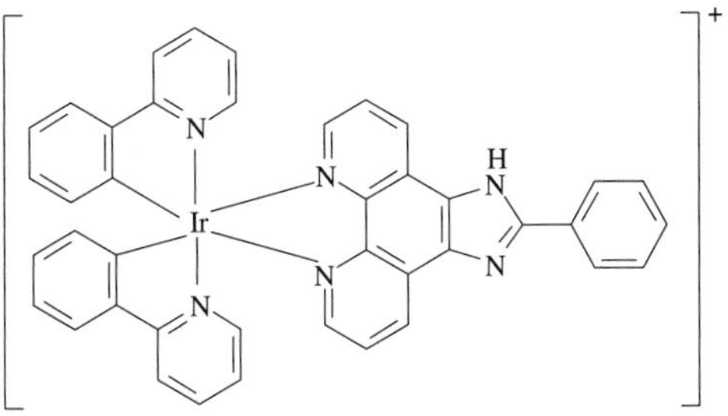

18

19

4-[4'-(4-Phenyloxy)-6'-phenyl-2,2'-bipyridyl]butane reacts with (2-Phpy)$_2$ Ir(μ-Cl)Ir(2-Phpy)$_2$ and further with ammonium hexafluorophosphate to yield **20** (06ICA1666). 4'-(4-carboxyphenyl)-2,2':6',2''-terpyridine similarly gives **21**, where the terpyridine ligand is η^2(N)-coordinated. Application of 2,2':6',2''-terpyridine and its derivatives as the η^2(N)-coordinated co-ligand in the cyclometalated iridium(III) complexes is known (99IC2250, 04MRC1491). 2-(3,5-*Bis*(trifluoromethyl)phenyl)-4-trifluoromethylpyridine undergoes cyclometalation by iridium(III) chloride in the presence of silver triflate to yield *tris*-cyclometalated **22** where the polypyridine ligand is generated in the process of cyclometalation (06JCS(D)2468).

20   **21**

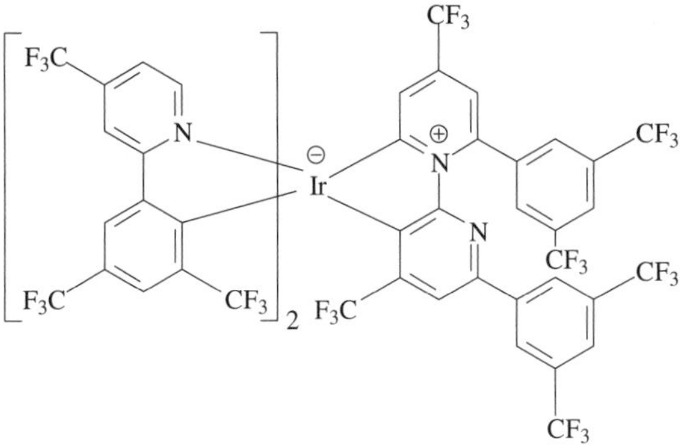

22

1,3-Di(2-pyridyl)-4,6-dimethylbenzene (LLH) reacts with $IrCl_3 \cdot 3H_2O$ to generate $[Ir(\eta^3(N,C,N)\text{-}LL)Cl(\mu\text{-}Cl)]_2$ (06IC8685). With DMSO the mononuclear complex $[Ir(LL)(DMSO)Cl_2]$ is formed. The dimer reacts with 2,6-diphenylpyridine to yield $[Ir(\eta^3(N,C,N)\text{-}LL)(\eta^3(C,N,C)\text{-}2\text{-}Phpy)Cl]$. Treatment of the dimer with 2,2':6',2''-terpyridine leads to the dicationic $[Ir(\eta^3(N,C,N)\text{-}LL)(\eta^3(N,N,N)\text{-}terpy)]^{2+}$. The dinuclear complex $[Ir(LL)Cl_2]_2$ (LLH = 1,3-bipyridyl-4,6-dimethylbenzene) and the corresponding bis-2,2',6',2''-terpyridine ligand give dinuclear iridium(III) complexes consisting of a terpyridine bridging ligand and cyclometalated LL, **23** (n = 0–2) (06IC10990).

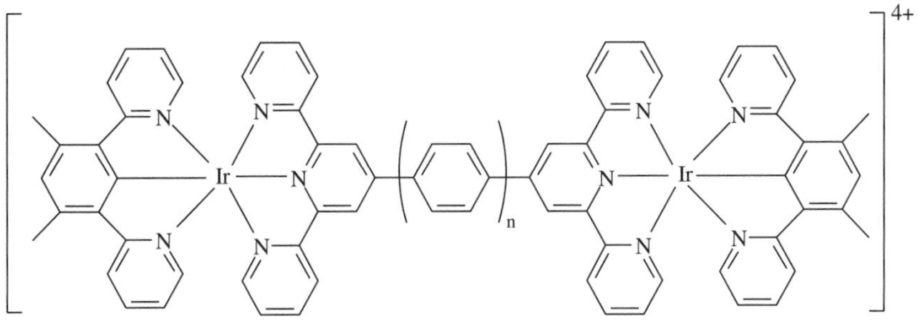

23

A unique coordination type of 2,2'-bipyridine ligand is observed in the tris(2,2'-bipyridine) iridium(III) hydrated complexes $[Ir(Hbipy\text{-}C^3,N')(bipy)_2]^{3+}$ and $[Ir(Hbipy\text{-}C^3,N')(bipy)_2(OH)]^{2+}$, when one of the polypyridine ligands is C,N-coordinated (83IC3429, 83IC4060, 84IC3425). 6-Phenyl-2,2'-bipyridine reacts with iridium(III) chloride in 2-methoxyethanol and further with pyridine to yield unusual cyclometalation dinuclear **24**, where one of the pyridine rings of the bipyridine ligand is C–H activated (07OM2137). The product with 4,4'-di-t-butyl-2,2'-bipyridine gives mononuclear **25** (R = X = Cl), where the coordination mode of 6-phenyl-2,2'-bipyridine ligand is retained. Dimethyl zinc in THF causes substitution of one of the chloride ligands by methyl and formation of **25** (R = Me, X = Cl). Silver triflate in methylene chloride leads to **25** (R = Me, X = OTf). In benzene the product is converted to **25** (R = Ph, X = OTf) followed by the elimination of methane.

24   **25**

Another ligand of interest in rhodium(III) and iridium(III) chemistry is *bis*(8-quinolyl)silane (LLH). Thus, [(η^2-COE)Ir(η^2-LL)(H)Cl] was prepared (01CC1200). This ligand also reacts with [Rh(PPh$_3$)$_3$Cl] in dichloromethane by oxidative addition to yield **26** (06OM1607). On treatment of the latter with LiB(C$_6$F$_5$)$_4$, cationic **27** follows, which can enter the H/Cl exchange reaction in dichloromethane. Complex **26** reacts with benzylmagnesium chloride to yield neutral **28**, which can be converted into **29** with trimethylsilyl triflate. *Bis*(8-quinolyl)silane (LLH) with [RhCl$_3$(AN)$_3$] in the presence of triethylamine yields [RhCl$_2$(LL)]. Further reaction with benzylmagnesium chloride in benzene allows the preparation of the neutral **30**. The starting ligand (LLH) with [(η^4-cod) Rh(η^3-CH$_2$Ph)] gives [(η^4-cod)Rh(LL)], which can oxidatively add triphenylsilane to yield [(η^4-cod)Rh(LL)(SiPh$_3$)(H)], which on further interaction with 1,5-cyclooctadiene gives the insertion product **31**. Complex **32** was prepared from the corresponding chloride complex with silver triflate (07OM5557). On addition of triphenylsilane in benzene to **32** Si-H bond activation, octane elimination, and α-phenyl migration occur to form **33**. Complex **26** reacts with various silanes in acetonitrile to give cationic iridium(III) complexes [(LL)Ir(SiR$_3$)(AN)$_2$](OTf) (R = OSiMe$_3$, Et, Ph) or [(LL)Ir{SiH(R)Ph}(AN)$_2$] (OTf) (R = H, Ph).

26   **27**

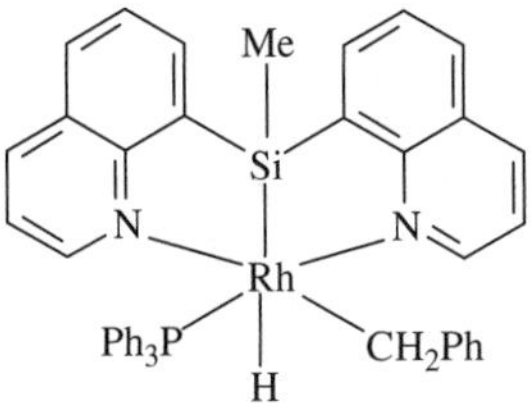

28

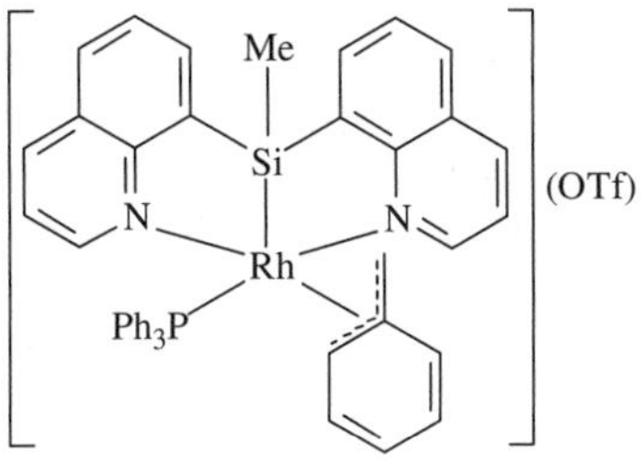

29 (OTf)

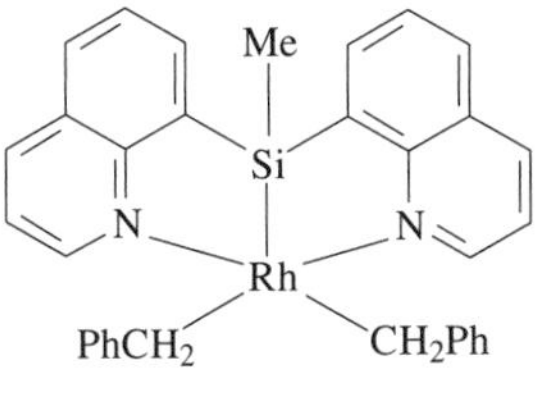

30

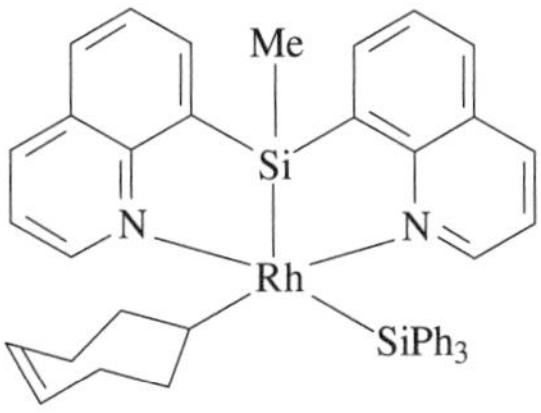

31

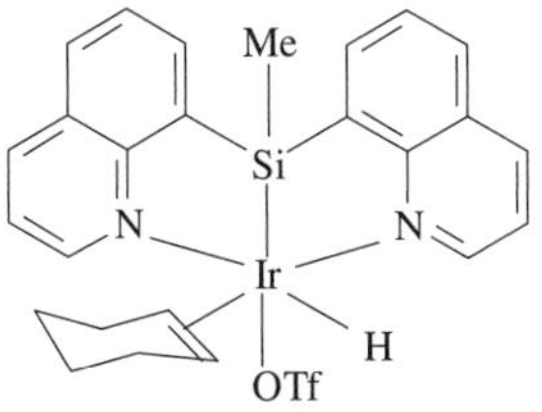

32

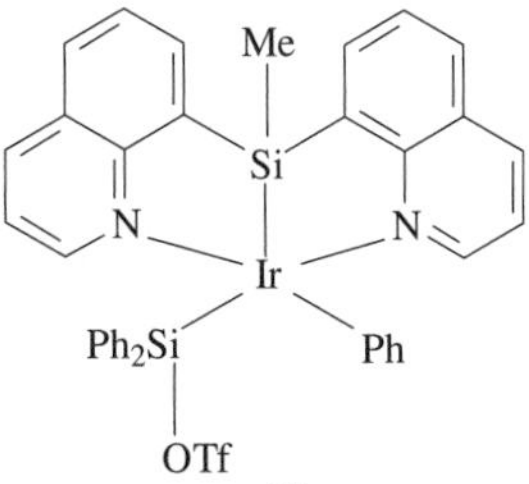

33

2.4 Polynuclear complexes and clusters

2,2'-Bipyridine, 4,4'-dimethyl-2,2'-bipyridine, 1,10-phenanthroline, 4,7-di-methyl-1,10-phenanthroline, 5,6-dimethyl-1,10-phenanthroline, or 3,4,7,8-tetramethyl-1,10-phenanthroline (LL) react with $(NEt_4)[Ir_4(CO)_{11}X]$ (X = Br, I) in dichloromethane in the presence of silver salts to yield $[Ir_4(CO)_{11}(\eta^1\text{-LL})]$, which on heating can be transformed to $[Ir_4(CO)_{10}$ $(\eta^2\text{-LL})]$ (96ICA(244)11). Examples of dinuclear complexes where 2,2'-bipyridine is a spectator ligand include the species containing a bridging triazenide ligand $[Rh_2(CO)_2(bipy)(\mu\text{-}N^1,N^3\text{-}p\text{-TolNNNTol-}p)_2]$, $[Rh_2(CO)_2(bipy)(\mu\text{-}N^1,N^3\text{-}p\text{-TolNNNTol-}p)_2](BF_4)$, $[Rh_2I(CO)(bipy)(\mu\text{-}N^1,$ $N^3\text{-}p\text{-TolNNNTol-}p)_2](PF_6)_2$ (87CC246, 89JCS(D)2049), $[Rh_2(\mu\text{-CO})(bipy)$ $(Ph_2PCH_2CH_2PPh_2)(\mu\text{-}N^1,N^3\text{-}p\text{-TolNNNTol-}p)_2]^{2+}$ (92CC143, 92JCS(D) 2907).

Bis(2,2'-bipyridine)xylene reacts with $[(\eta^4\text{-hexadiene})RhCl]_2$ and silver hexafluoroborate, $[(\eta^4\text{-nbd})_2Rh](BF_4)$ or $[(\eta^4\text{-nbd})_2Rh(AN)_2](BF_4)$ to yield dinuclear **34** (L_2 = hexadiene, nbd) (04IC7180). Tris(2,2'-bipyr-idine)bis(xylene) similarly leads to the trinuclear products **35**

(L_2 = hexadiene, nbd). *Bis*(2,2′-bipyridine)binaphtholate reacts with [(η^2-COE)$_2$Rh]$_2$ followed by hexadiene and then silver tetrafluoroborate or with [(η^4-nbd)$_2$Rh](BF$_4$) to give cationic dinuclear **36** (L_2 = hexadiene, nbd). With [(η^4-cod)IrCl]$_2$ this ligand produces neutral **37**. Into this category falls the complex similar to **34** with L_2 = nbd and polypyridine ligand *bis*(2,2′-bipyridine)calyx[4]arene (03IC3160).

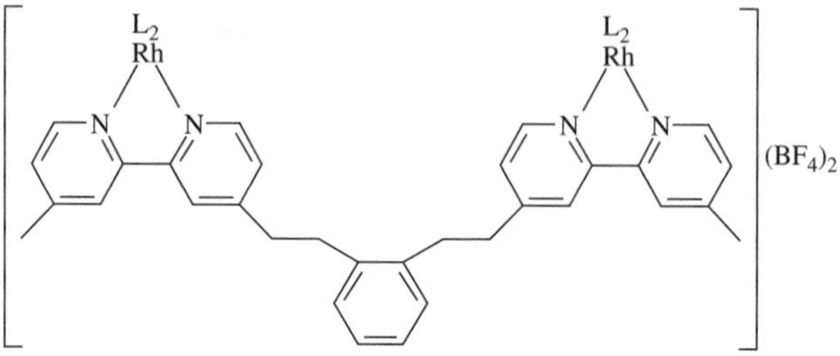

34

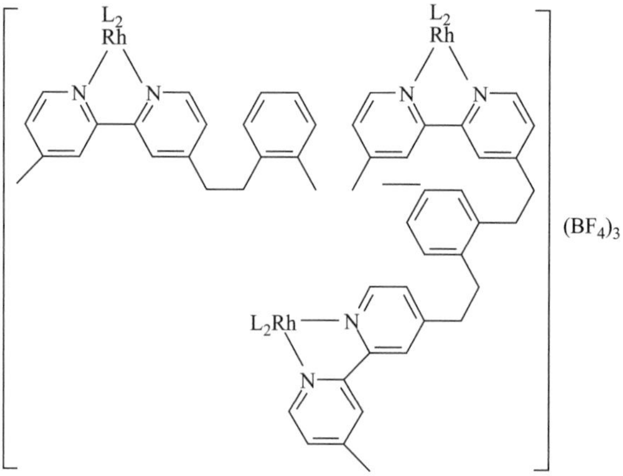

35

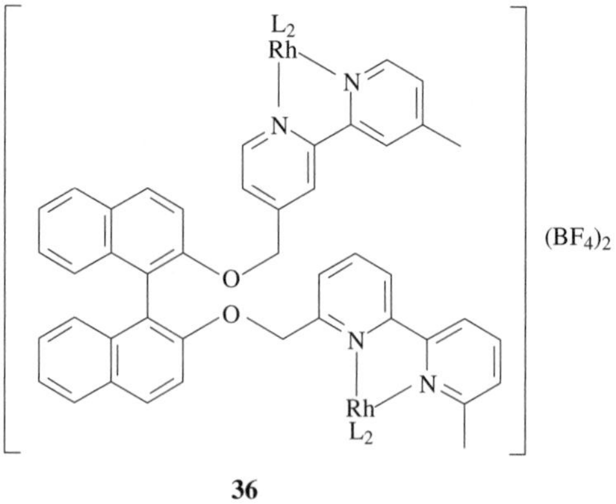

36

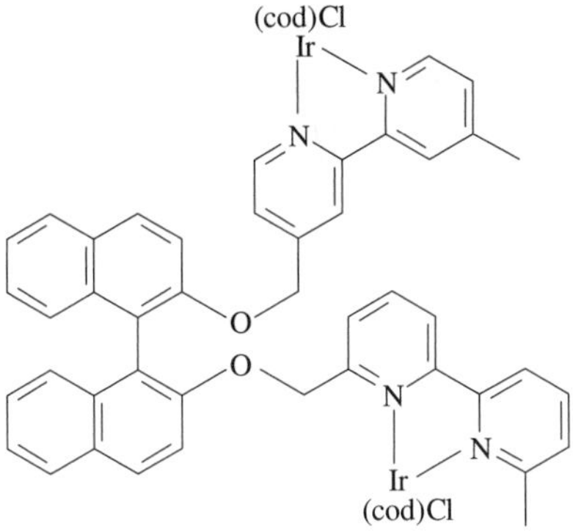

37

4,4'-Bipyridine with $[(\eta^5\text{-Cp}')ClM(\mu\text{-Cl})_2MCl(\eta^5\text{-Cp}')]$ (M = Rh, Cp' = 1,2-t-Bu$_2$Cp; M = Ir, Cp' = Cp*) yields the dinuclear complexes $[(\eta^5\text{-Cp}')Cl_2M(\mu\text{-4,4}'\text{-bipy})MCl_2(\eta^5\text{-Cp}')]$ (06OM74). The products with silver tetrafluoroborate yield the tetranuclear complexes $[\{(\eta^5\text{-Cp}')M\}_4(\mu\text{-Cl})_4(\mu\text{-4,4}'\text{-bipy})_2](BF_4)_4$. Their structure was described (02CEJ372, 05CCR1085). Further reaction of the rhodium complex with sodium tere-phthalate causes substitution of the chloride bridges with the formation of **38**. [Ir$_2$(2-Phppy)$_4$Cl$_2$] with 3,8-dipyridyl-4,7-phenanthroline (LL) in the presence of silver perchlorate affords the cyclometalated iridium dinuclear complex [(2-Phpy)$_2$Ir(μ-LL)Ir(2-Phpy)$_2$](ClO$_4$)$_2$ (07IC6911).

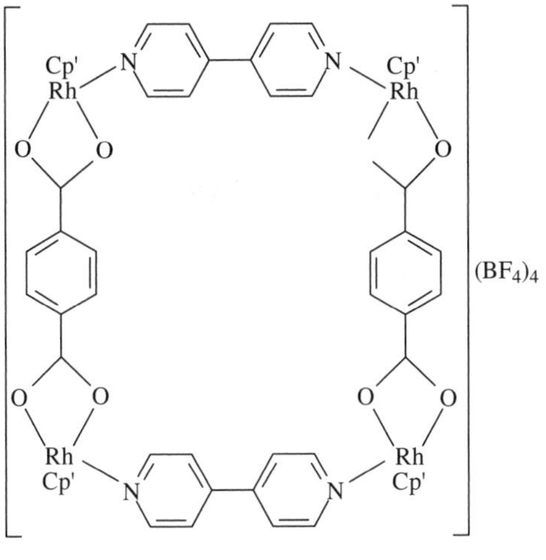

38

2,6-Bis(4-pyridyl)-1,4,5,8-tetrathiafulvalene (LL) with $[(\eta^5\text{-Cp*})IrCl_2]_2$ in methylene chloride yields the dinuclear complex with the bridging LL ligand $[(\eta^5\text{-Cp*})Cl_2Ir(\mu\text{-LL})Ir(\eta^5\text{-Cp*})Cl_2]$ (07JOM4545). Chloride ligands in the product can be replaced by 1,2-dicarba-*closo* – dodecaborane(12)-1,2-dichalogenolates by reaction with the dilithium salt of this dianion, and the products have the composition $[(\eta^5\text{-Cp*})\{E_2C_2(B_{10}H_{10})\}Ir$ $(\mu\text{-LL})Ir(\eta^5\text{-Cp*})\{E_2C_2(B_{10}H_{10})\}]$ (E = S, Se). $[((\eta^5\text{-Cp*})Co\{S_2C_2(B_{10}H_{10})\})_2$ $(\mu\text{-bipy})]$ is the related complex (06JCS(D)5225). Various polypyridines with $[(\eta^5\text{-Cp*})M\{S_2C_2(B_{10}H_{10})\}]$ (M = Rh, Ir) give clusters $[(\eta^5\text{-Cp*})MS_2C_2$ $(B_{10}H_{10})]_n(LL)$ (M = Rh, Ir, LL = N,N'-*bis*(4-pyridinylmethylene)biphenyl-4,4'-diamine 2,5-di(4-pyridyl)-1,3,4-oxadiazole, 1,2-di(4-pyridyl)ethylene, 4,4'-bipyridine) (06EJI3274). Complex $[W(CO)_3(LL)]$ (LL = *trans*-1,2-*bis* (4-pyridyl)ethylene) reacts with the dimer $[Rh(CO)_2Cl]_2$ to yield hetero-dinuclear **39** (02OM5830). The homodinuclear complex of this ligand of composition $[cis\text{-Cl}(OC)_2Rh(\mu\text{-LL})Rh(CO)_2Cl\text{-}cis]$ is also known (02ICA(330)128).

$$(OC)_5W—N \quad\quad N—Rh(CO)_2Cl$$

39

2,2'-*Bis*(diphenylphosphino)-4,4'-bipyridine (LL) reacts with excess $[Rh_6(CO)_{15}(AN)]$ in methylene chloride to yield the cluster complex $[\{Rh_6(CO)_{14}\}_2(\mu\text{-LL})]$ (05OM3516). 2,4,6-Tri(pyridine-4-yl)-1,3,5-triazine (LLL) with $[(\eta^5\text{-Cp*})_2M_2(\mu\text{-C}_2O_4)Cl_2]$ (M = Rh, Ir) in the presence of silver triflate yields the cationic hexarhodium and hexairidium clusters $[(\eta^5\text{-Cp*})_6M_6(\mu_3\text{-LL})_2(\mu\text{-C}_2O_4)_3](OTf)_6$ (M = Rh, Ir) (07JCS(D)4457).

3. COMPLEXES WITH NICKEL GROUP METALS

3.1 Metal carbonyl complexes

Under electrochemical conditions, the nickel(0) complexes $[Ni(LL)_2]$ (LL = bipy, 4,4'-Me$_2$bipy, phen) react with carbon dioxide to yield $[Ni(LL)(CO)_2]$ (79CL1513, 95JEAC195). Nickel(II) complexes of 2,2'-bipyridine catalyze the electroreductive coupling of organic halides with carbon monoxide to produce ketones (95CC2331, 97TL17089). The reaction may include insertion of carbon monoxide into the oxidative addition adduct (68JOM(15)209, 76CL1217, 76JA4115, 81BCJ2161). In the electrocatalytic scheme, the active species are $[Ni(CO)_2(bipy)]$ (98JOM (560)103), $[Ni^0(bipy)]$ (96NJC659), or $[Ni^0(CO)(bipy)]$ (98JOM(571)37). The nickel(0) species $[Ni(CO)_2(bipy)]$ can also be prepared by electro-reduction of $[Ni(bipy)]^{2+}$ in the presence of carbon dioxide

(89JOM(367)347, 89NJC53). In the complexes [Pt(CO)X$_2$(2,9-Me$_2$phen)], there is monodentate coordination of the polypyridine ligand (92CC333).

3.2 Metal–alkyl complexes

3.2.1 Synthesis and reactivity

2,9-Dimethyl-1,10-phenanthroline with *trans*-[Pt(Cl)(Me)(DMSO)$_2$] gives [Pt(Cl)(Me)(2,9-Me$_2$phen)] and further with AgPF$_6$, AgBF$_4$, AgOTf, or AgClO$_4$, a series of complexes [Pt(Me)(DMSO)(2,9-Me$_2$phen)]X (X = PF$_6$, BF$_4$, OTf, ClO$_4$) followed (01IC3293). The hexafluorophosphate complex with NaBPh$_4$ or Na[B(3,5-(CF$_3$)$_2$C$_6$H$_3$)$_4$] extended the series. An exchange reaction of [Pt(Me)(DMSO)(2,9-Me$_2$phen)](OTf) with di-*n*-butylsulfoxide leads to [Pt(Me)(*n*-Bu$_2$SO)(2,9-Me$_2$phen)](OTf). Substitution of DMSO by triphenylphosphine yields the series [Pt(Me)(PPh$_3$)(2,9-Me$_2$phen)]X (X = PF$_6$, BF$_4$, OTf, ClO$_4$). The PF$_6^-$ ion in the latter was exchanged for BPh$_4^-$ and B{(3,5-(CF$_3$)$_2$C$_6$H$_3$)$_4$}$^-$. The PF$_6$-complex also reacted with AsPh$_4$Cl to afford [Pt(Me)(PPh$_3$)(2,9-Me$_2$phen)]Cl. The latter with AsPh$_4$NO$_2$ gives [Pt(Me)(PPh$_3$)(2,9-Me$_2$phen)](NO$_2$). [Pt(Me)(DMSO)(2,9-Me$_2$phen)](OTf) also reacted with a number of ligands including cyclohexylamine, *i*-propylamine, 2,6-dimethylpyridine, ethylamine, triphenylarsine, N,N'-diemthylthiourea, and the products of the ligand-exchange reactions are [Pt(Me)(L)(2,9-Me$_2$phen)](OTf) with an appropriate set of ligands. The reaction of [Pt(Me)(DMSO)(2,9-Me$_2$phen)](OTf) with *n*-butylthiocyanate and potassium selenocyanate gives neutral complexes [Pt(Me)(X)(2,9-Me$_2$phen)] (X = SCN, SeCN). Related is [Pt(Me)(phen)(Me$_2$SO)](PF$_6$) (96AX(C)827). [Pt(Me)(2,9-Me$_2$phen)(P(*o*-Tol)$_3$)]X (X = PF$_6$, SbF$_6$, OTf, BF$_4$, B B(3,5-(CF$_3$)$_2$C$_6$H$_3$)$_4$) undergo ready cyclometalation to yield [Pt(2,9-Me$_2$phen){CH$_2$C$_6$H$_4$ P(*o*-tolyl)$_2$-η^2 (C, P)}]X (07IC10681).

[Ni(TMEDA)Me$_2$] with 2,2'-bipyridine gives [Ni(bipy)Me$_2$] (71JA3350), but with 2,2':6',2''-terpyridine it gives a unique nickel(I) derivative [Ni(η^3-terpy)Me] (04JA8100). Nickel acetylacetonate [Ni(acac)$_2$] reacts with [AlEt$_2$(OEt)] in the presence of 2,2'-bipyridine to yield [NiEt$_2$(bipy)] (65JA4652, 66JA5198, 66JOM(6)572, 99JCS(D)1027). The product reacts with acrolene to give [(η^2-olefin)NiEt$_2$(bipy)] (67JA5989), which at elevated temperatures reductively eliminates butane and forms [Ni(olefin)$_n$(bipy)] (71JA3350, 71JA3360). [NiEt$_2$(bipy)] with chlorobenzene gives [Ni(Ph)Cl(bipy)] and *n*-butane (70JOM(24)C63, 76JOM(84)93). Reductive elimination of R-R from [Ni(bipy)R$_2$] and Ar-Ar from [(bipy)NiAr$_2$] is promoted by acrylonitrile and protic acids (79JA5876, 84JA8181, 97IC5682, 99CL419, 02BCJ1997).

[(η^4-cod)Pd(Me)Cl] with 2,2'-bipyridine yields [Pd(Me)Cl(bipy)] (96JOM(508)109). The product reacts with silver tetrafluoroborate in acetonitrile to give the ionic [Pd(Me)(AN)(bipy)](BF$_4$). Both products under carbon monoxide give the acetyl complexes [Pd(COMe)Cl(bipy)] and [Pd(COMe)(AN)(bipy)](BF$_4$). [Pd(COMe)(CO)(phen)](B(3,5-(CF$_3$)$_2$ C$_6$H$_3$)$_4$) · CH$_2$Cl$_2$ can be mentioned as well (96JA4746). Acylpalladium(II) [Pd(I)(COMe)(bipy)] can be prepared by two routes: insertion of carbon monoxide into the palladium–carbon bond of [PdIMe(bipy)] and ligand exchange from [Pd(I)(COMe)(TMEDA)] (92JOM(424)C12). Complex [(phen)PdC(O)ON(Ar)C(O)] prepared from [Pd(phen)(OAc)$_2$] and nitrobenzene under carbon monoxide, is regarded as an intermediate in the catalytic reductive carbonylation of nitrobenzene and other nitro compounds (90CC1616, 97JOM(545)89, 98OM1052, 98OM2199). The reaction of [Pd(phen)(p-benzoquinone)] with nitrobenzene gives [Pd$_2$ (phen)$_2$(PhNO)$_3$] with an uncertain structure (99JOM(586)190). The product reacts with methanol under carbon monoxide to yield [Pd(phen) (COMe)$_2$]. The structures of this complex (98JOM(566)37) and the 2,2'-bipyridine analog (93OM568) were determined.

[Pd(LL)(Me)Cl] (LL = 1,10-phenanthroline and its 4,7-dimethyl and 3,4,7,8-tetramethyl derivatives) with silver triflate in methylene chloride gives cationic [Pd(LL)(Me)(AN)](OTf) (02ICA(327)188). In excess polypyridine ligand in chloroform complexes [Pd(η^2-LL)(η^1-LL)(Me)](OTf) resulted. [Pd(LL)(Me)(AN)](OTf) with the monodentate ligands (L) pyridine in methylene chloride and 2-phenylpyridine (LL = phen) or 7,8-benzoquinoline (LL = 3,4,7,8-Me$_4$phen) in chloroform yields [Pd(LL)(L)(Me)](OTf). Methylene chloride solutions of [Pd(η^2-LL) (η^1-LL)(Me)](OTf) under carbon monoxide produce [Pd(η^2-LL)(η^1-LL) (COMe)](OTf) and small amounts of [Pd(η^2-LL)(CO)(COMe)](OTf). Similar complexes are [Pd(bipy)(4-Mepy)(Me)](BF$_4$) (90JOM(393)299), [Pd(phen)(Me)(Et$_2$O)]$^+$ (96JA2436), [Pd(LL)$_2$(CH$_2$NO$_2$)](PF$_6$) (99EJI2085). [Ni(Me)(O-p-C$_6$H$_4$CN)(bipy)] inserts carbon monoxide into the nickel–methyl bond (85OM1130). According to theoretical modeling, the reaction proceeds by a five-coordinate intermediate (04OM891).

Treatment of the complexes of the type [MMe$_2$(bipy)] (M = Ni, Pd) and [PdMe$_2$(phen)] (89OM2907) with [H(OEt$_2$)$_2$]$^+$[B(C$_6$H$_3$(CF$_3$)$_2$-3,5)$_4$]$^-$ gives cationic complexes (92JA5895, 92OM3920, 95JA1137, 95JA6414). [PtClMe(4,4'-t-Bu$_2$bipy)] (95OM1030) reacts with silver triflate to yield [Pt(OTf)Me(4,4'-t-Bu$_2$bipy)] (96JCS(D)1809). [PtMe$_2$(4,4'-t-Bu$_2$bipy)] with B(C$_6$F$_5$)$_3$ in the presence of ethylene or carbon monoxide forms [Pt(Me)L(4,4'-t-Bu$_2$bipy)][MeB(C$_6$F$_5$)$_3$] (L = CO, C$_2$H$_4$) (96JCS(D)1809, 00CRV1391). [PtMe$_2$(4,4'-t-Bu$_2$bipy)] (93OM4592) with CF$_3$COOH forms [PtMe(OOCCF$_3$)(4,4'-t-Bu$_2$bipy)]. [Pt(OTf)Me(4,4'-t-Bu$_2$bipy)] reacts with carbon monoxide to yield [Pt(CO)Me(4,4'-t-Bu$_2$bipy)](OTf). The latter on treatment with diethylamine leads to [Pt(CONEt$_2$)Me(4,4'-t-Bu$_2$bipy)], the result of nucleophilic attack of the amine on the carbonyl ligand. [PtMe$_2$(4,4'-t-Bu$_2$bipy)] with B(C$_6$F$_5$)$_3$ in the presence of carbon monoxide produces [Pt(CO)Me(4,4'-t-Bu$_2$bipy)]{BMe(C$_6$F$_5$)$_3$}. In the presence of ethylene, the product is [Pt(η^2-C$_2$H$_4$)Me(4,4'-t-Bu$_2$bipy)]{BMe(C$_6$F$_5$)$_3$}. 2,2'-Bipyridine, 1,10-phenanthroline, 2,9-dimethyl-, 3,4-7,8-tetramethyl-, 5-nitro-, and 4,7-diphenyl-1,10-phenanthroline (LL) with [Pt(DMSO)$_2$Me (Cl)] and silver hexafluorophosphate afford [Pt(LL)(DMSO)Me](PF$_6$) (96IC5087).

Palladium(II) complexes containing methyl or acyl ligands and terdentately coordinated 2,2':6',2''-terpyridine react with carbon monoxide and diene ligands (96OM668). 2,6-Bis(pyrimidin-2-yl)pyridine with [(η^4-cod)Pd(Me)Cl] form the terdentately coordinated **40** (R = Me, X = Cl) (98JCS(D)113). The product reacts with silver triflate to give **40** (R = Me, X = OTf) and sodium tetrakis(3,5-bis(trifluoromethyl)phenyl)borate to give **40** (R = Me, X = B{3,5-(CF$_3$)$_2$C$_6$H$_3$}$_4$). Under carbon monoxide, **40** (R = Me, X = Cl, OTf, B{3,5-(CF$_3$)$_2$C$_6$H$_3$}$_4$) give rise to acylated **40** (R = COMe, X = Cl, OTf, B{3,5-(CF$_3$)$_2$C$_6$H$_3$}$_4$). $trans$-[Pt(Cl)Me(DMSO)$_2$] reacts with 2,2':6',2''-terpyridine and sodium tetraphenylborate in methanol to yield the terdentately coordinated [Pt(terpy)(Me)](BPh$_4$) (95IC2994, 98IS153, 00JOM(593)403). Similar C-coordinated complexes are [Pt(terpy) (CH$_2$NO$_2$)](ClO$_4$) (98ICC61) and **41** (99IC4262).

40   41

2-Ferrocenyl-1,10-phenanthroline reacts with [(η^4-cod)Pd(Me)(Cl)] to yield **42** (07OM810). With silver tetrafluoroborate in acetonitrile, the product forms cationic **43** and with sodium acetate – the neutral cyclometalated **44** (R = Me). The starting ligand also reacts with palladium(II) acetate in the presence of trifluoroacetic acid to yield cyclometalated **44** (R = COOCF$_3$). On interaction with sodium hexafluorophosphate in acetonitrile, the latter gives rise to cationic **45** (L = MeCN, X = PF$_6$) and with triphenylphosphine cationic **45** (L = PPh$_3$, X = CF$_3$COO). Diphenyl-phosphinoethane or –propane give rise to the dinuclear cyclometalated complexes with bridging diphosphine ligands, **46** (n = 1, 2).

[M(Me)R(bipy)] (M = Pd, R = Me, Ph; M = Pt, R = Me) and [Pd{(CH$_2$)$_4$}(bipy)] react with 8-(bromomethyl)quinoline to yield the derivatives of type **47** (99OM2660). M = Pt and R = Me react with silver tetrafluoroborate to yield the product of metathesis, the cationic complex with a BF$_4^-$ ion. [PtMe$_2$(bipy)] reacts with [Cd(cyclen)(MeOH)$_2$](ClO$_4$)$_2$

to yield [{PtMe$_2$(bipy)}{Cd(cyclen)}](ClO$_4$)$_2$ containing a platinum(II)–cadmium(II) dative bond (99JA7405).

47

Platinum(II) complexes such as [PtMe$_2$(bipy)] or [PtClMe(bipy)] (73JOM(59)411) being electron-rich react as metallonucleophiles (76JOM(117)297, 77IC2171, 77JCS(D)1466, 85MI1, 87OM2548, 88JCS(D)595, 91OM2672, 95OM2188). [PtClMe(bipy)] can be prepared from 2,2'-bipyridine and [(η^4-cod)Pt(Me)Cl] in methylene chloride (73JOM(59)411). 6-Substituted 2,2'-bipyridines (H, Me, Et, i-Pr, CH$_2$Ph, Ph) with $trans$-[Pt(Cl)(Me)(SMe$_2$)$_2$] gives adducts [Pt(Cl)(Me)(6-R-2,2'-bipy)] (00EJI2555). Similar palladium(II) complexes are [PdMe$_2$(LL)] (LL = bipy, phen) (90OM210, 98IS162), [Pd(Me)Cl(phen)] (96JOM (508)109), [Pd(Me)Cl(2,9-Me$_2$phen)] (91JOM(463)269), and [Pd(CH$_2$)$_4$(bipy)] (80JCS(D)1633, 93IS167). A platinum analog of the latter was described (98JOM(568)53) as well as the complex [Pt{(CH$_2$)$_4$}I(Me)(bipy)] (98OM2046). [Pt{(CH$_2$)$_4$}(LL)] (LL = bipy, phen) react with dibromoalkanes Br(CH$_2$)$_n$Br (n = 3–6) to give platinacyclopentane complexes [Pt{(CH$_2$)}$_4$Br{(CH$_2$)$_n$Br}(LL)] (99JOM(574)286).

1,2,4,5-Tetrakis(1-N-7-azaindolyl)benzene with [Pt$_2$Me$_4$(SMe$_2$)$_2$] yields species **48** (02OM4978). The product interacts with [H(Et$_2$O)$_2$][B(C$_6$H$_3$(CF$_3$)$_2$-3,5)$_4$], then with Me$_2$S in benzene to yield cationic **49** (04OM1194). 1,2-Bis(N-7-azaindolyl)benzene (LL) with [Pt$_2$Me$_4$(μ-SMe$_2$)$_2$] gives [PtMe$_2$(LL)] (05OM3290). The product reacts with (H(Et$_2$O)$_2$)(B{3,5-(CF$_3$)$_2$C$_6$H$_3$}$_4$) in benzene to yield cationic [Pt(LL)Me(solvent)](B{3,5-(CF$_3$)$_2$C$_6$H$_3$}$_4$). After standing and adding dimethylsulfide, the benzene C–H activation product [Pt(LL)Ph(SMe$_2$)](B{3,5-(CF$_3$)$_2$C$_6$H$_3$}$_4$) resulted. A similar reaction series occurs in toluene. Bis(N-7-azaindolyl)methane forms **50**, which in the presence of (H(Et$_2$O)$_2$)(B{3,5-(CF$_3$)$_2$C$_6$H$_3$}$_4$) and benzene readily activates a benzene C–H bond and then yields **51** (03OM2187, 04CIC1) with dimethylsulfide. Both 1,2,4,5-tetrakis(1-N-7-azaindolyl)benzene and $tris$(N-7-azaindolyl)methane platinum(II) dimethyl complexes were found as efficient agents in the C–H activation of ethylbenzene (06OM5979).

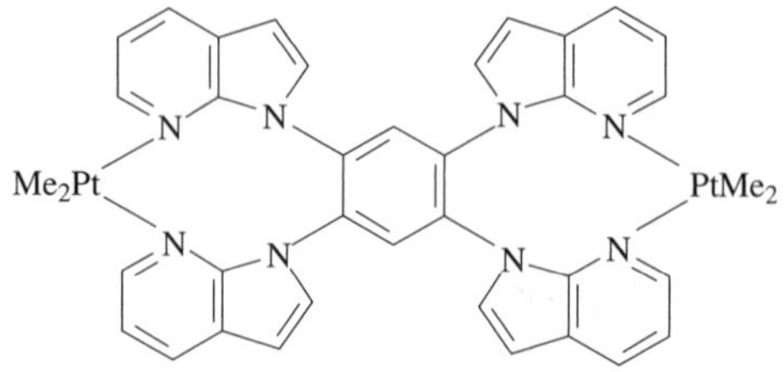

48

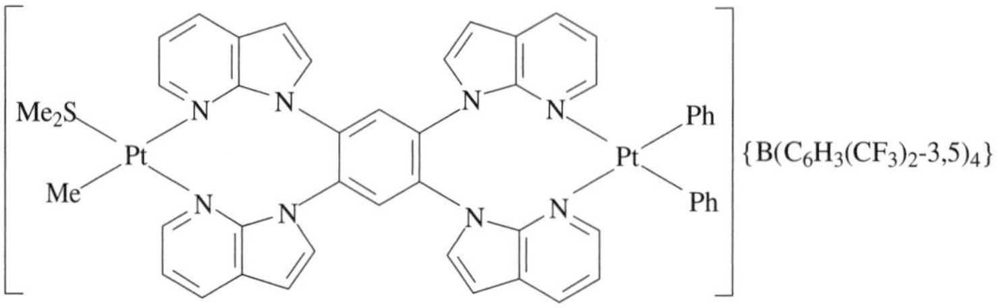

49

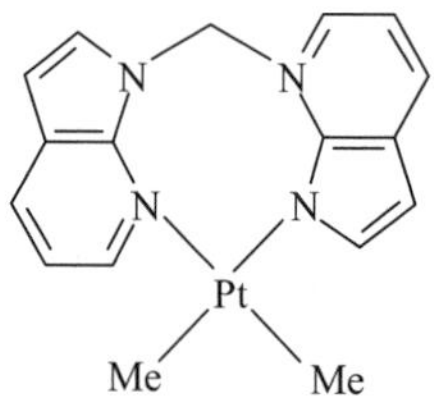

50 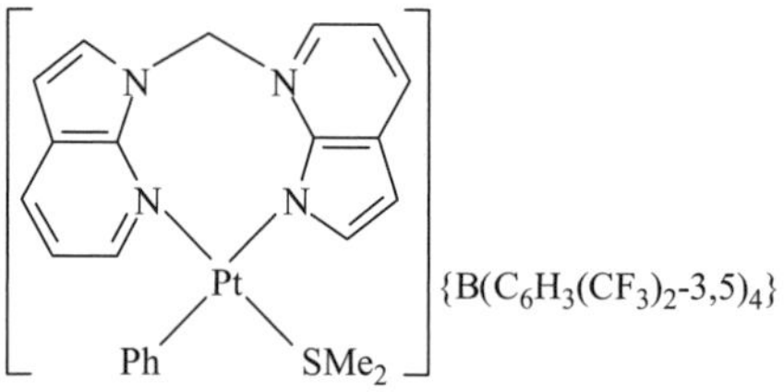   **51**

Palladacyclo-2,4-pentadiene compounds containing a 2,2'-bipyridine ligand [Pd{C(COOMe) = C(COOMe)-C(COOMe) = C(COOMe)}(bipy)] have been prepared from [Pd(dba)$_2$], 2,2'-bipyridine, and dimethyl 2-butynedioate or hexafluorobutyne (98OM1812). The product oxidatively adds methyl iodide, benzyl bromide, or iodobenzene. 2,2'-Bipyridine or 1,10-phenanthroline reacts with [Pd(dba)$_2$] and methyl phenylpropynoate to yield palladacyclopentadiene derivatives **52** (06ICA1773).

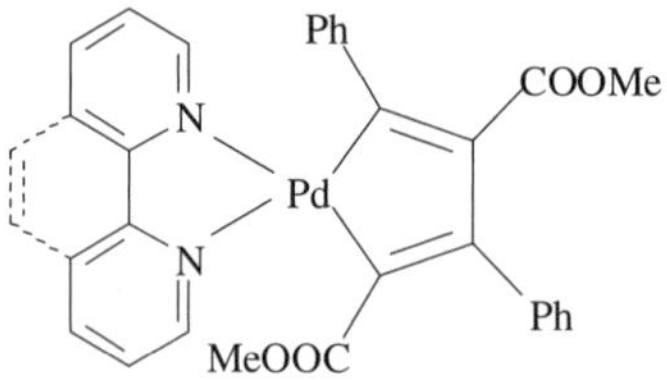

52

3.2.2 Insertion reactions

[Pt{(CH$_2$)$_4$}(LL)] (LL = bipy, phen) reacts with 2,3-epoxypropylphenyl ether in acetone in the presence of carbon dioxide to yield **53** (R = CH$_2$OH) (04JCS(D)619). [Pt{(CH$_2$)$_4$}(phen)] with styrene oxide in the presence of CO$_2$ gives **53** (R = phen). In a similar manner, [PtMe$_2$(LL)] (LL = bipy, phen) reacts with styrene oxide and carbon dioxide (89AG(E)767, 90JA2464). The dialkyl nickel(II) complexes [Ni(bipy)R$_2$] (R = Et, i-Bu) and nitrous oxide generate nickel(II) alkoxy alkyl species [Ni(bipy)(OR)(R)] (R = Et, i-Bu) (93JA2075, 95POL175). Nickelacyclopentane derivatives [Ni{(CH$_2$)$_4$}(LL)] (LL = bipy, phen) with nitrous oxide give the oxanickelacycles [Ni{O(CH$_2$)$_3$CH$_2$} (LL)]. Nickelacycle [Ni{(CH$_2$)$_3$CHCH$_3$}(bipy)] reacts with N$_2$O to give [Ni{OCH(CH$_3$)CH$_2$CH$_2$CH$_2$}(bipy)]. [Ni{CH$_2$CH(CH$_3$)CH(CH$_3$)CH$_2$} (bipy)] reacts with N$_2$O to produce the nickelacycle [Ni{OCH$_2$ CH(CH$_3$)CH(CH$_3$)CH$_2$}(bipy)]. [Ni(bipy)R$_2$] (R = Me, Et) with p-tolyl azide gives [Ni{N(p-Tol)R}R(bipy)] (94JA3665). Nickelacyclopentane [Ni{(CH$_2$)$_4$}(bipy)] yields the azanickelacycle [Ni{(CH$_2$)$_4$}{N(p-Tol) (bipy)].

53

Ionic palladium(II) complexes containing 2,2'-bipyridine or 1,10-phenanthroline catalyze co-polymerization of carbon monoxide and alkenes (91AG(E)989, 93MM911, 95MR9). Ionic complex [(η^2-LL)Pd(AN)-Me](OTf) (LL = bis(anisylimino)acenaphthene) reacts with 1,10-phenanthroline to afford [(η^2-phen)Pd(η^1-LL)Me](OTf) and with 2,9-dimethyl-1,10-phenanthroline to yield [(η^2-LL)Pd(η^1-2,9-Me$_2$phen)Me](OTf) (99JOM(573)3). These complexes insert carbon monoxide in methylene chloride to give [(η^2-phen)Pd(η^1-LL)(COMe)](OTf) and [(η^2-LL)Pd(η^1-2,9-Me$_2$phen)(COMe)](OTf), respectively. Norbornene is inserted into the acetyl complexes followed by the formation of complexes of type **54** where the η^1-coordinated ligand is removed from the coordination sphere. $Ortho$-aromatic electrophilic substitution on phenol by [Pt(Cl) (η^1-C$_2$H$_4$OR)(phen)] (R = Me, Ph) followed by cyclization gives platina-chromane complex **55** (07JCS(D)5720). [Pd(Me)(Cl)(6-Mebipy)] inserts carbon monoxide in methylene chloride to yield the acetyl complex

[Pd(COMe)(Cl)(6-Mebipy)], which on interaction with $NaB(C_6H_3(CF_3)_2$-$3,5)_4$ yields [Pd(Me)(CO)(6-Mebipy)]{$B(C_6H_3(CF_3)_2$-$3,5)_4$} (01OM4111). The product inserts ethylene to yield **56**.

54   55

56

[Pd(Me)Cl(LL)] (LL = bipy, phen) inserts 2,6-dimethylisocyanide, *t*-butylisocyanide, and tosylmethylisocyanide to yield [Pd(C(Me)= NR)Cl(LL)] (R = 2,6-Me$_2$C$_6$H$_3$, *t*-Bu, CH$_2$Tos) (97OM2948). [Pd(Me)-Cl(4,4'-*t*-Bu$_2$bipy)] inserts 2,6-Me$_2$C$_6$H$_3$NC to yield the iminoacyl **57** (03OM3025). The product with PhCH=NMe in the presence of silver tetrafluoroborate gives cationic **58**. In the process of standing, two imidoyl groups of **58** enter a C–C coupling reaction and form the *bis*-imino-ligand in dinuclear complex **59**.

57   58

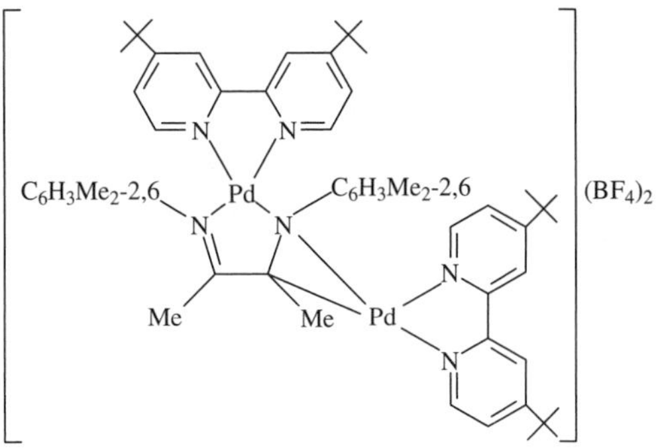

59

[Pd(Me)(phen)(OEt$_2$)]{B(3,5-(CF$_3$)$_2$C$_6$H$_3$)$_4$} serves as a catalyst for the cyclization/hydrosilylation reactions of 1,6- (98JA3805, 99JOC8681, 99TL8499, 02ACR905) and 1,7-dienes (99TL1451) as well as 1-vinyl-1-(3-butenyl)cycloalkanes (01JOC1755). The cationic complexes [Pd(phen)(Me)(L)]{B(3,5-(CF$_3$)$_2$C$_6$H$_3$)$_4$} (L = Et$_2$O, Me$_3$SiC≡CSiMe$_3$) are catalysts for the hydrosilylation and dehydrogenative silylation of alkenes (97JA906). In an effort to elucidate the mechanism of the catalytic reactions, [Pd(Me)(phen)(N≡CC$_6$H$_3$(CF$_3$)$_2$)]{B(3,5-(CF$_3$)$_2$C$_6$H$_3$)$_4$} was reacted with triethylsilane to yield [Pd(SiEt$_3$)(phen)(N≡CC$_6$H$_3$(CF$_3$)$_2$)]{B(3,5-(CF$_3$)$_2$C$_6$H$_3$)$_4$} and then with dimethyl diallylmalonate (01OM5251, 04JA6332). The product **60**, after warming in the presence of N≡CC$_6$H$_3$(CF$_3$)$_2$ undergoes migratory insertion to yield **61** and on further warming it rearranges to **62**.

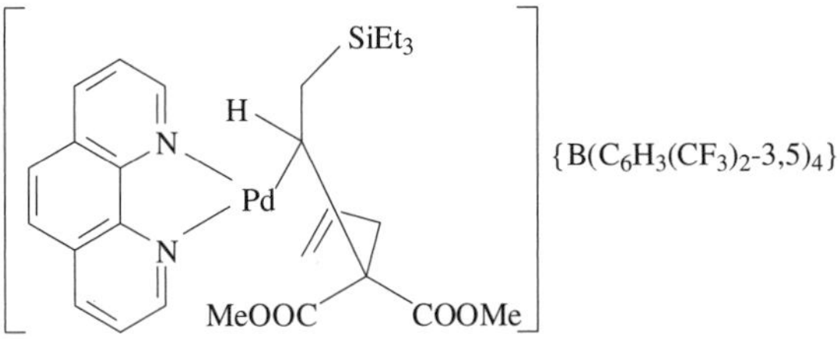

60

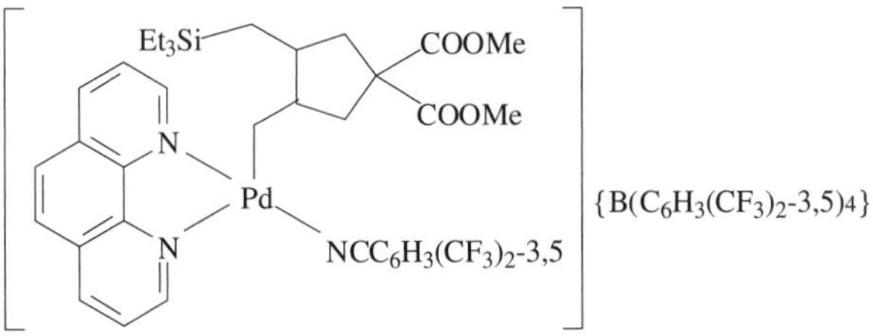

61

62

3.2.3 Oxidative addition reactions

i-Propyl iodide oxidatively adds to [PtMe$_2$(phen)] (75JOM(84)105) to yield [PtMe$_2$(phen)(i-Pr)I] (83CC267, 85OM1669). In the presence of O$_2$ or CH$_2$=CHCN, the latter insert into the i-propyl radical to yield [PtMe$_2$(phen)(A-B-i-Pr)I] (A = B = O; A = CH$_2$, B = CHCN) (83OM1698, 85JA1218). [PtMe$_2$(2,9-Me$_2$phen)] oxidatively adds a wide variety of halides RX to yield [PtMe$_2$(R)(2,9-Me$_2$phen)]X] (R = Me, Et, CH$_2$=CHCH$_2$, CH$_2$=CH, PhCH$_2$, X = Br, I; R = n-Pr, i-Pr, X = I) (00JOM(593)445). [PtMe$_2$(LL)] (LL = t-Bu$_2$bipy, bipy) with Me$_3$SiX (X = Cl, Br, I) and water (formally with HX) gives a new series of alkylhydrido complexes [PtMe$_2$(H)(X)(LL)] (95OM4966). Related to these are alkylhydrido platinum(IV) complexes [Pt(H)(X)I$_2$(2,9-Me$_2$phen)] (95JOM(488)C13). The products readily undergo reductive elimination of methane to yield [Pt(Me)(X)(LL)]. Other oxidative addition products include [Pt(phen)IMe$_2$(CH$_2$CH$_2$COOMe)] (88JOM(342)399) and [Pt(Me)(CF$_3$)$_2$I(bipy)] (88OM1454).

[PdIMe$_3$(bipy)] (72JOM(38)403) in acetonitrile is in equilibrium with [PdMe$_3$(bipy)(AN)]I (90OM826). This complex reductively eliminates ethane with the evolution of iodide (89OM1518, 94OM2412). It can add triphenylphosphine in dichloromethane to yield [PdMe$_3$(bipy)(PPh$_3$)]I (00JOM(595)296), which also reductively eliminates ethane. [PdIMe$_3$(bipy)] reacts with silver acetate in acetone to yield [PtMe$_3$(OAc)(bipy)] (06ICA4326). The same product follows from 2,2'-bipyridine and [PtMe$_3$(OAc)(Me$_2$CO)$_x$] (x = 1, 2). In the presence of silver acetate, it reacts with pyrimidine-2-thione, pyridine-2-thione, thiazoline-2-thione, and thiophene-2-thiol (LH) to yield the products of substitution of the acyl group [PtMe$_3$(L)(bipy)] with S-coordination of the thione or thiol group of the incoming ligand. The platinum(IV) complexes [PtIMe$_3$(LL)] (LL = bipy, phen) react with silver tetrafluoroborate and then pyridazine (L) in THF to yield [PtMe$_3$(LL)(L)](BF$_4$), where the azine ligand is monodentately coordinated (95JCS(D)3165). Complex [PtIMe$_3$(bipy)] is stable, but the palladium analog decomposes by reductive elimination (88OM1363, 94OM2412).

[PtMe$_2$(bipy)] reacts with (IPh$_2$)(OTf) to yield [Pt(OTf)Me$_2$(Ph)(bipy)], which on interaction with sodium iodide gives the oxidative addition product [PtMe$_2$(bipy)(Ph)(I)] (00ICC575). [M{(CH$_2$)$_4$}(bipy)] (M = Pd, Pt) containing pallada- or platinacyclopentane rings reacts with (IPh$_2$)(OTf)/ NaI or LiCl similarly yielding [M{(CH$_2$)$_4$}(Ph)(bipy)X] (M = Pd, Pt; X = Cl, I) (03ICC1382, 04OM3466, 07IC1924). [PtMe$_2$(bipy)] reacts with (IPh(C$_6$H$_4$I-4))(OTf)/NaI to form a mixture of products [PtMe$_2$(Ph)(I) (bipy)] and [Me$_2$Pt(C$_6$H$_4$I-4)(I)(bipy)]. In the reaction series with (IPh(C$_6$H$_4$-4-IPh))(OTf)$_2$, the product is the dinuclear complex [Me$_2$(I) (bipy)Pt(μ-C$_6$H$_4$)Pt(bipy)IMe$_2$]. [PtPh$_2$(4,4'-t-Bu$_2$bipy)] reacts with (IPh$_2$)(OTf)/NaI in a normal way affording [PtPh$_3$(4,4'-t-Bu$_2$bipy)I]. [NiMe$_2$(4,4'-Me$_2$bipy)] (88JOM(355)525) oxidatively adds 2-iodo-2',4',6'-triphenylbiphenyl (RI) at elevated temperatures in benzene to afford [NiMe$_2$(4,4'-Me$_2$bipy)(R)(I)] (04OM3071).

[PtMe$_2$(bipy)] (76TMC199) is capable of oxidative addition of R$_n$ECl$_{4-n}$ (E = Sn, Ge; R = Me, Ph) to yield the platinum(IV) complexes (77IC2171, 78IC77). Interaction of [Me$_2$Pt(μ-SMe$_2$)$_2$PtMe$_2$] with 2,2'-bipyridine and 4,4'-di-t-butyl-2,2'-bipyridine (LL) gives the platinum(II) complexes [PtMe$_2$(LL)], also capable of further oxidative addition (84OM444, 94OM1559, 96OM2108, 97CRV1735, 97JA10127). Both complexes react with trimethylchlorostannane to yield [PtMe$_2$(LL)(SnMe$_3$)Cl]. The 2,2'-bipyridine derivative gives [PtMe$_2$(bipy)(SnMe$_3$)X] (X = Br, I), whereas the 4,4'-di-t-butyl-2,2'-bipyridine complex under these conditions produces 63 (X = Br, I). Me$_2$SnCl$_2$ forms [PtMe$_2$(4,4'-t-Bu$_2$bipy) Cl(SnMe$_2$Cl)] and with Me$_3$SnGeX [PtMe$_2$(4,4'-t-Bu$_2$bipy)X(GeMe$_3$)] (X = Cl, Br, I). Both complexes give [Me$_2$Pt(LL)X(SiMe$_3$)] with Me$_3$SiX (X = Br, I) arises. [PtMe$_2$(4,4'-t-Bu$_2$py)] with Me$_3$SnCl in acetone is described as an equilibrium of the starting complex, solvated species [PtMe$_2$(4,4'-t-Bu$_2$py)(Me$_2$Sn)(acetone)]Cl, and the oxidative addition product [PtMe$_2$(4,4'-t-Bu$_2$py)(Me$_3$Sn)Cl] (95CC2115). With trimethyltin bromide and iodide, the equilibrium is predominantly between [PtMe$_2$ (4,4'-t-Bu$_2$py)] and [PtMe$_2$(4,4'-t-Bu$_2$py)(Me$_3$Sn)X] (X = Br, I). The product of oxidative addition of Me$_2$SnCl$_2$ to [(η^2-MeOOCCH═CHCOO-Me)Pt(2,9-Me$_2$phen)] is [(η^2-MeOOCCH═CHCOOMe)Pt(2,9-Me$_2$phen) (Cl)(SnMe$_2$Cl)] (96OM4012). [PtMe$_2$(4,4'-t-Bu$_2$bipy)] oxidatively adds (R$_2$SnS)$_3$ (R = Me, Ph) to yield 64 (R = Me, Ph) (96OM1750). Similar reactions with (Me$_2$GeTe)$_3$ and (R$_2$SnE)$_3$ (R = Me, Ph; E = Se, Te) give 64 containing five-membered metallacycles (99IC2123).

63   **64**

Platinum(IV) thiolate and selenolate complexes [PtMe$_2$(EPh)$_2$(phen)] (E = S, Se) are also stable (93JCS(D)1835). [PdMe$_2$(bipy)] reacts with dibenzoylperoxide to yield [PdMe$_2$(OOCPh)$_2$(bipy), which decomposes into [Pd(OOCPh)$_2$(bipy)] and other products (98IC3975). Similarly with (EPh)$_2$, the products [PdMe$_2$(EPh)$_2$(bipy)] (E = S, Se) are formed: they readily decompose to various compounds. [PdMe$_2$(bipy)$_2$] and [Pd(Me)(4-Tol)(bipy)$_2$] interact with dibenzoyl- and bis(4-trifluoromethyl-benzoyl)peroxide in a stepwise manner as a function of temperature (01ICC648, 04OM1122). In the range from -70°C to -30°C, the oxidative addition products [Pd(OOCAr)$_2$Me(R)(bipy)] (Ar = Ph, p-C$_6$H$_4$CF$_3$; R = Me, 4-Tol) result. Above -30°C, the products enter an aroate exchange forming palladium(II) [Pd(OOCAr)(R)(bipy)] and palladium(IV) [Pd(OOCAr)Me$_2$(R)(bipy)] species. At elevated temperatures, various reductive elimination processes occur. Oxidative addition of bis(4-chlorophenyl)diselenide to [Pd(Me)(Ar)(bipy)] (Ar = 4-Tol, 4-MeOC$_6$H$_4$) gives octahedral palladium(IV) compounds [Pd(SeC$_6$H$_4$Cl-4)$_2$(Me)(Ar)(bipy)] (04JOM672).

[PtMe$_2$(LL)] (LL = bipy, 2,9-Me$_2$-4,7-Ph$_2$phen) reacts with Hg(OOCCF$_3$)$_2$ to yield various products, among them [(LL)Me$_2$(OOCCF$_3$)Pt-Hg-Pt(OOCCF$_3$)Me$_2$(LL)], [(LL)Me$_2$(OOCCF$_3$)Pt-Pt(HgOOCCF$_3$)Me$_2$(LL)], [(LL)Me$_2$(OOCCF$_3$)Pt-PtMe$_2$(LL)-PtMe$_2$(LL)-Pt(HgOOCCF$_3$)Me$_2$(LL)] (78IC1458). [PtMe$_2$(4,4'-t-Bu$_2$bipy)] with HgX$_2$ (X = Cl, Br, OOCCF$_3$, OOCMe) gives the products of an oxidative addition [PtMe$_2$X(HgX)(t-Bu$_2$bipy)] (01IC1728, 05ICA1614). In excess of the platinum(II) complex, the products are [PtMe$_2$X(4,4'-t-Bu$_2$bipy)(μ-HgX)PtMe$_2$X(4,4'-t-Bu$_2$bipy)].

Complexes of platinum(IV) with 2,2':6',2''-terpyridine contain the bidentately coordinated ligand, which possesses fluxional behavior (92CC303). 2,2':6',2''-Terpyridine reacts with [PtXMe$_3$]$_4$ (X = Cl, Br, I) in benzene to yield fac-[PtClMe$_3$(terpy)], where the polypyridine ligand is also bidentate and flexible (93JCS(D)291). The same type of reaction is observed for (6-pyrazol-1-yl)-2,2'-bipyridine, 6-(4-methylpyrazol-1-yl)-2,2'-bipyridine, or 6-(3,5-diemthylpyrazol-1-yl)-2,2'-bipyridine. In solution, there is an equilibrium of complexes with bidentate coordination through two nitrogens of the 2,2'-bipyridine moiety and one nitrogen of this moiety and one nitrogen of the pyrazolyl group, 65 (96POL3203). Fluxionality occurs in the complexes cis-[M(C$_6$F$_5$)$_2$(LL)] (M = Pd, Pt; LL = terpy, 4-Me-4'-(4-ClC$_6$H$_4$)-terpy) (96CC2329, 96JCS(D)253). The dinuclear complex [(PtMe$_3$I)$_2$(LL)] is formed with 2,2:6',2'':6'',2'''-quater-pyridine, where each PtMe$_3$I moiety is bidentately coordinated to half of the ligand (99POL1285).

65

Other palladium(IV) complexes include [PdMe$_2$Br(CH$_2$Ph)(bipy)] (90OM3080, 92ACR83, 95MI1), [Pd(Br)(Me)(Ph)(CH$_2$Ph)(bipy)] (88OM2046, 94OM2053, 97JOM(532)235), [PtHMe$_3$(4,4'-t-Bu$_2$bipy)] (96JA8745), [PtMe$_2$(PR$_3$)$_2$(bipy)]$^{2+}$ (72IC2749). [Pt(LL)Me$_2$] (LL = bipy, phen) are oxidized by molecular oxygen in methanol to the alkoxoplatinum(IV) complexes [Pt(LL)PtMe$_2$(OH)(OMe)] (98OM4530, 07OM4860).

Potassium salts of di(2-pyridyl)methanesulfonate and its methyl analog (LL) react with [Pt$_2$Me$_4$(SMe$_2$)$_2$] in methanol to yield the anionic platinum(II) K[Pt(η^2(N,N)-LL)Me$_2$], which can be protonated at the platinum site in methanol-water to yield the neutral complexes (04JA11160). Platinum complexes of [2.1.1]-(2,6)-pyridinophane were utilized in C–H bond cleavage reactions, for example, **66** (02AG(E)4102). Complexes of this ligand (LL) having composition [Pt(Me)(LL)]$^+$ cleave C–H bonds of alkanes (03CC358, 03NJC665). Interaction of [Pt(Me$_2$)(LL)] with Na{B(C$_6$H$_3$(CF$_3$)$_2$-3,5)$_4$} in methylene chloride/pentafluoro-benzene and further with triflic acid gives [Pt(HL$^+$)(C$_6$F$_5$)$_2$] {B(C$_6$H$_3$(CF$_3$)$_2$-3,5)$_4$} (04IC3642). Of interest is the platinum(II) complex of dimethyl(2-pyridyl)borate **67**, which on treatment with methanol or ethanol in the presence of oxygen experiences methyl group transfer from the boron to the platinum center and formation of the B-O-Pt framework in **68** (R = Me, Et) (07AG(E)6309).

66   **67**   **68**

Oxidative additions leading to dendrimers are of interest (94AG(E)847, 94CC1895, 96OM43, 04JOM4016). [PtMe$_2$(4,4'-t-Bu$_2$bipy)] oxidatively adds 4,4'-bromomethyl-2,2'-bipyridine to yield the dinuclear **69**. Further reaction with [Me$_2$Pt(μ-SMe$_2$)PtMe$_2$] gives trinuclear **70**, and the sequence can be continued to dendrimer formation.

69   **70**

3.3 Metal–aryl complexes

Oxidative addition of RX (R = Ph, X = Br, I; PhCH$_2$, X = Cl) to [Ni(bipy)]$^{2+}$ gives [Ni(R)(X)(bipy)] (88OM2203, 91JEAC125). [Ni^0(bipy)] formed in the process of electrochemical reduction of [NiBr$_2$(bipy)] oxidatively adds p-tolylbromide to yield [Ni(Br)(p-Tol)(bipy)] (01JOM(630)185, 02RCB796, 03RCB567, 03RJE1261). [Ni(Mes)$_2$(bipy)] is implicated in the electrochemical reduction of [NiBr$_2$(bipy)] in the presence of mesityl bromide (02RJGC168) on the way to [Ni(Br)(Mes)(bipy)] (06POL1607). Neutral [NiBr(Mes)(LL)] complexes (LL = bipy, phen) in the presence of thallium(I) tetrafluoroborate and 3,5-lutidine or acetonitrile give [Ni(Mes)(3,5-Me$_2$py)(LL)](BF$_4$) or [Ni(Mes) {NH=C(Me)(2,4,6-Me$_3$C$_6$H$_2$)}(LL)](BF$_4$), respectively (07JCS(D)83). In the latter case, the nitrile molecule initially inserts into the nickel–carbon bond. 2,2'-Bipyridine reacts with [(η^4-cod)$_2$Ni] in THF and further with 1-chloronaphthalene to yield [Ni(1-naphthyl)(bipy)Cl] (02ICA(334)149). The other method of preparation involves the reaction of trans-[Ni (1-naphthyl)Ni(PPh$_3$)$_2$Cl] with 2,2'-bipyridine in THF. [PtMes$_2$(3,4,7,8-Me$_4$phen)] follows from [PtMes$_2$(DMSO)$_2$] (02CCR(230)193, 03EJI1917). A series of mesityl complexes of 2,2'-bipyridine and various

1,10-phenanthrolines (parent, 2,9-Me$_2$, 3,8-Me$_2$-, 4,7-Me$_2$, 5,6-Me$_2$-) can be prepared similarly (92JOM(440)207, 95OM1176, 02JCS(D)2371). 2,2'-Bipyridine, 4,4'-dimethyl-2,2'-bipyridine, and 3,4,7,8-tetramethyl-1,10-phenanthroline (LL) enter the ligand-substitution reaction with [Ni(PPh$_3$)$_2$(Mes)Br] to yield [Ni(LL)(Mes)Br] (03EJI839, 04EJI2784). [Ni(Mes$_2$)(bipy)] can be prepared in several ways (85ZC411, 01ZAAC645). 4,4'-Dimethyl-2,2'-bipyridine and 2,2'-biquinoline (LL) react with [Pd(2,4,6-(CF$_3$)$_3$C$_6$H$_2$)$_2$(Me$_2$S)$_2$] to yield [Pd(2,4,6-(CF$_3$)$_3$C$_6$H$_2$)$_2$(LL)] (04EJI2326). 2,2'-Bipyridine or 1,10-phenanthroline (LL) react with [Pt(η^1-biphenyl monoanion)$_2$(DMSO)$_2$] in methylene chloride to yield [Pt(η^1-biphenyl monoanion)$_2$(LL)] (04JA6470). 2,2'-Bipyridine and 4,4'-di-t-butyl-2,2'-bipyridine (LL) react with [Pd(dba)$_2$] and 1,3,5-triiodomesy-tylene in toluene to yield [I(LL)Pd(μ-C$_6$Me$_3$)Pd(LL)I] (01OM4695).

Complexes [Pd(Ph)I(LL)] (LL = bipy, phen) can be prepared by the ligand-exchange reaction from [Pd(Ph)I(TMEDA)], but 2,9-dimethyl-1,10-phenthroline does not enter this reaction (94JOM(482)191, 95JOM(493)1). Complex [Pd(dba)$_2$] oxidatively adds aryl iodides in the presence of 2,2'-bipyridine or 4,4'-dimethyl-2,2'-bipyridine (LL) to yield [Pd(LL)(Ar)I] (Ar = Ph, C$_6$H$_3$(CF$_3$)$_2$-3,5, C$_6$H$_3$Me$_2$-3,5, LL = bipy; Ar = C$_6$H$_4$OMe-4, C$_6$H$_3$(CF$_3$)$_2$-3,5, C$_6$H$_3$Me$_2$-3,5, C$_6$H$_4$CF$_3$-4, LL = 4,4'-Me$_2$bipy) (01OM1087). [Pd(LL)(C$_6$H$_3$(CF$_3$)$_2$-3,5)I] with silver tetrafluoroborate forms cationic [Pd(LL)(C$_6$H$_3$(CF$_3$)$_2$-3,5)(solvent)](BF$_4$) (LL = bipy, solvent = AN, acetone, THF; LL = 4,4'-Me$_2$bipy, solvent = AN). [Pd$_2$(dba)$_3$] reacts with aryl iodides and 4,4'-di-t-butyl-2,2'-bipyridine to yield [Pd(4,4'-t-Bu$_2$bipy) (Ar)I] (Ar = Ph, 4-NH$_2$C$_6$H$_4$, 4-MeC$_6$H$_4$, 2-MeC$_6$H$_4$, 4-MeOC$_6$H$_4$, 2-MeOC$_6$H$_4$, 2,6-Me$_2$C$_6$H$_3$, 4-CNC$_6$H$_4$, 4-NO$_2$C$_6$H$_4$, 4-ClC$_6$H$_4$) (06OM5746). Further reaction of the products with Ag(P(O)(OR)$_2$) (R = Et, Ph) gives arylpalladium phosphonate [Pd(4,4'-t-Bubipy)(Ar)(P(O)(OR)$_2$)]. [Pd(LL) (Me)(P(O)(OPh)$_2$) (LL = bipy, phen, 4,4'-(nonyl)$_2$bipy) are intermediates in the P(O)-C(sp^3) bond-forming reactions (02OM3278).

[NiEt$_2$(bipy)] reacts with C$_6$F$_6$ to give {Ni(C$_6$F$_5$)$_2$(bipy)] (97JOM(535)209). cis-[Pt(C$_6$F$_5$)$_2$L$_2$] (L = Me$_2$SO, Et$_2$S, THT) reacts with 2,2'-bipyridine to yield [Pt(C$_6$F$_5$)$_2$(bipy)] (82TMC97, 93JCS(D)1343). [Pd(C$_6$F$_5$)(bipy)(CO)](ClO$_4$) is known (76JOM(112)105). [Pd(C$_6$F$_5$)(Me$_2$ CO)(LL)](ClO$_4$) (LL = bipy, 4,4'-Me$_2$bipy, phen) reacts with potassium hydroxide in methanol to yield [Pd(C$_6$F$_5$)(OH)(LL)] (03IC3650). Under carbon monoxide in methanol–water, they are converted into alkox-ycarbonyl complexes [Pd(C$_6$F$_5$)(COOMe)(LL)]. Related is [Pd(bipy) (COOMe)$_2$] (95JOM(495)185). [Pd(C$_6$F$_5$)(OH)(LL)] (LL = bipy, 4,4'-Me$_2$bipy) under carbon dioxide in THF or dichloromethane yield dinuclear [(LL)(C$_6$F$_5$)Pd(μ-OC(O)OPd(C$_6$F$_5$)(LL)] with the bridging car-bonato group (03IC3650). The reaction may proceed by the insertion of carbon dioxide into the palladium–hydroxyl bonds. [Pd(C$_6$F$_5$)(OH)(LL)] (LL = bipy, 4,4'-Me$_2$bipy, phen) reacts with PhNCS in methanol, ethanol, or n-propanol to afford the thiocarbamate [Pd(C$_6$F$_5$)(SC(=NPh) OR)(LL)] (R = Me, Ph, n-Pr), the products of insertion of PhNCS into the palladium–alkoxy bond. In the presence of diethylamine, PhNCS gives thioureide [Pd(C$_6$F$_5$)(NPhCSNEt$_2$)(LL)], and its course is interpreted as a sequence of amide complexes [Pd(C$_6$F$_5$)(Et$_2$N)(LL)] and insertion of PhNCS into the Pd–NEt$_2$ bond. [Pd(LL)(C$_6$F$_5$)(OH)] (LL = bipy, 4,4'-Me$_2$bipy) with sulfur dioxide in methanol, ethanol, or n-propanol gives alkyl sulfites [Pd(LL)(C$_6$F$_5$)(SO$_2$OR)] (LL = bipy, R = Me, Et, n-Pr; L = 4,4'-Me$_2$bipy, R = Me, Et) (04JCS(D)929). [Pt(bipy)(C$_6$F$_5$)(OH)] with sulfur dioxide in THF gives the hydrogen sulfite complex [Pd(bipy)(C$_6$F$_5$)(SO$_2$OH)]. [Pd(LL)(C$_6$F$_5$)(OH)] (LL = bipy, 4,4'-Me$_2$bipy) reacts with silanols in toluene to yield [Pd(LL)(C$_6$F$_5$)(OSiR$_3$)] (R$_3$ = Ph$_3$, Et$_3$, Me$_2$Bu-t) (04JOM1872). [Pd(C$_6$F$_5$)(OH)(LL)] (LL = bipy, 4,4'-Me$_2$bipy, phen) reacts with acetone or methyl-i-butylketone to yield [Pd(C$_6$F$_5$)(CH$_2$COR)(LL)] (R = Me, i-Bu) (04JCS(D)3521). [Pd(C$_6$F$_5$) (OH)(LL)] (LL = bipy, phen) reacts with diethylmalonate to give [Pd(C$_6$F$_5$)(COOEt)(LL)], and with malonitrile to afford [Pd(C$_6$F$_5$) (CH(CN)$_2$)(LL)], with nitromethane to form [Pd(C$_6$F$_5$)(CH$_2$NO$_2$)(LL)].

1,8-Naphthyridine reacts with [N(n-Bu)$_4$]$_2$[Pt$_2$(μ-C$_6$F$_5$)$_2$(C$_6$F$_5$)$_4$] to yield mononuclear [N(n-Bu)$_4$][Pt(C$_6$F$_5$)$_3$(napy)] (96IC7345). The product is stable in contrast to the 1,10-phenanthroline analog, which undergoes rearrangement into [Pt(C$_6$F$_5$)$_2$(phen)] and [N(n-Bu)$_4$]$_2$[Pt(C$_6$F$_5$)$_4$] (95ICA(235)51). (N(n-Bu)$_4$)[Pt(C$_6$F$_5$)$_3$(napy)] reacts with cis-[Pt(C$_6$F$_5$)$_2$ (THF)$_2$] to yield dinuclear (N(n-Bu)$_4$)[Pt$_2$(μ-C$_6$F$_5$)(μ-napy)(C$_6$F$_5$)$_4$] (96IC7345). The reaction between (N(n-Bu)$_4$)[Pt(C$_6$F$_5$)$_3$(bipy)] and cis-[Pt(C$_6$F$_5$)$_2$(THF)$_2$] gives cis-[Pt(C$_6$F$_5$)$_2$(bipy)] and (N(n-Bu)$_4$)[Pt$_2$ (μ-C$_6$F$_5$)$_2$(C$_6$F$_5$)$_4$]. The product with the bridging 1,8-naphthyridine ligand reacts with water and HX (X = Cl, Br, I, SPh) to yield the products of substitution of the pentafluorophenyl bridge, [N(n-Bu)$_4$] [Pt$_2$(μ-X)(μ-napy)(C$_6$F$_5$)$_4$] (X = OH, Cl, Br, I, SPh). 1,8-Naphthyridine with [Pt(C$_6$F$_5$)$_2$(THT)$_2$] yields mononuclear [Pt(C$_6$F$_5$)$_2$(η^2-napy)] with a strained four-membered chelate ring (04JCS(D)2733). Under carbon monoxide, the product forms [Pt(C$_6$F$_5$)$_2$(η^1-napy)(CO)] with loss of chelation. [Pt(C$_6$F$_5$)$_2$(η^2-napy)] inserts diazomethane into the platinum–nitrogen bond in methylene chloride to yield **71**. Similar complex [Pt(C$_6$F$_5$)$_2$(η^2-bipy)] containing a five-membered chelate ring does not possess this reactivity pattern (74JOM(81)115).

71

[Pd$_2$(dba)$_3$]·dba reacts with 2-ROC$_6$H$_4$I and 2,2'-bipyridine to form arylpalladium [Pd(C$_6$H$_4$OR-2)I(bipy)] (R = H, C(O)Me) (01OM2704). Carbonylation gives the insertion compounds [Pd{C(O)C$_6$H$_4$OR-2}I(bipy)] (R = H, C(O)Me). Isonitriles R'NC also insert and lead to [Pd{C(=NR')C$_6$H$_4$OR-2}I(bipy)] (R = H, R' = 2,6-Me$_2$C$_6$H$_3$, t-Bu; R = C(O)Me, R' = 2,6-Me$_2$C$_6$H$_3$). [Pd(C$_6$H$_4$OH-2)I(bipy)] reacts with (p-Tol) N=C=N(Tol-p) and thallium(I) triflate by insertion of one of the C=N groups into the palladium–carbon bond and protonation of the second nitrogen atom to yield **72** (R = p-Tol) (06OM1851). The cyclohexyl analog, CyN=C=NCy, gives two products, one of insertion, **72** (R = Cy), and another of addition of the O–H group to one C=N, **73**.

72   **73**

A complex with a palladium–aryl bond [Pd(I)(C$_6$H$_3$Me$_2$-3,5)(bipy)], prepared from [Pd(dba)$_2$], 2-2'-bipyridine, and 3,5-Me$_2$C$_6$H$_3$I, enters the insertion reaction with dimethylacetylene dicarboxylate and silver tetrafluoroborate in acetone to yield **74** (00OM2125). In the absence of silver tetrafluoroborate, the insertion reaction leads to **75**. It reacts with silver tetrafluoroborate to yield cationic **76** (01OM1087). If dimethylacetylene dicarboxylate is taken in excess, then in the presence of silver tetrafluoroborate, **77** is the product. [Pd(I)(C$_6$H$_3$Me$_2$-3,5)(bipy)] also reacts with phenylallene and silver tetrafluoroborate in acetone, and the reaction is accompanied by the insertion of the C=C bond into the palladium–aryl bond to yield palladium allyl **78**.

74   **75**

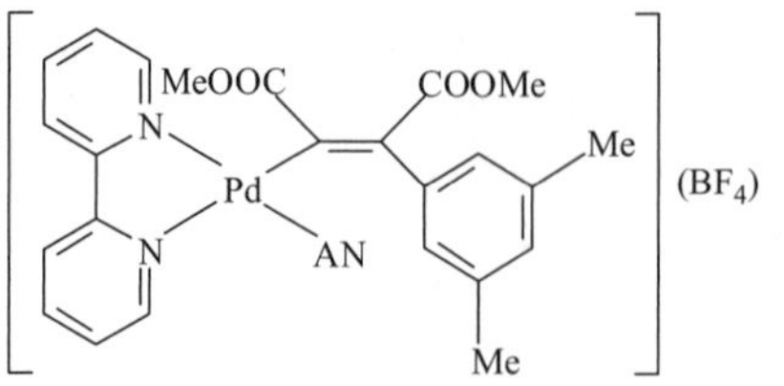

76

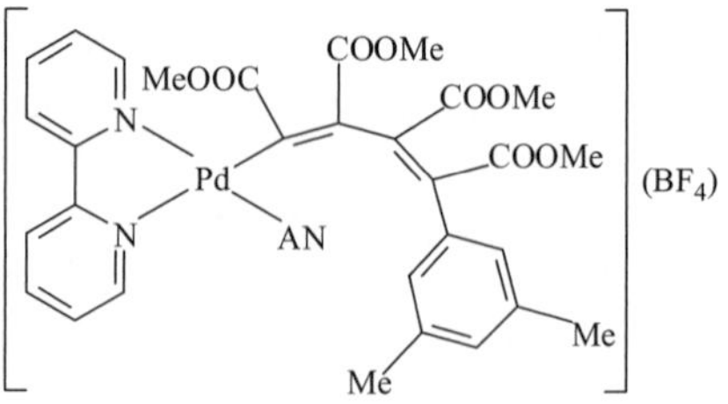

77

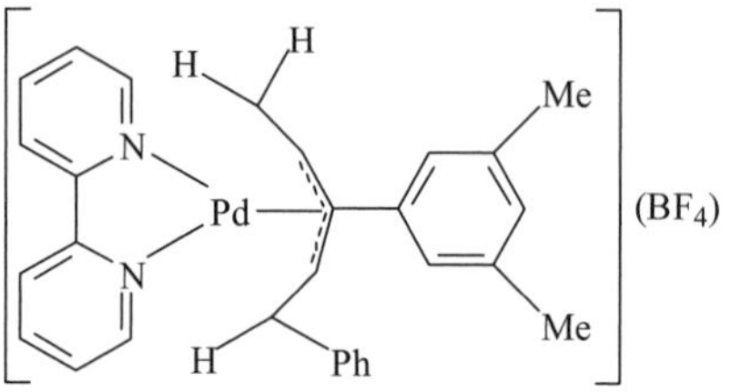

78

[PtCl(R)(2,9-Me$_2$phen)] (R = Me, Et, 4-MeOC$_6$H$_4$) can be carbonylated to yield the five-coordinate [PtCl(R)(2,9-Me$_2$phen)(CO)] (94GCI117, 96JOM(513)97). 2-Methyl-1,10-phenanthroline and 1,10-phenanthroline (LL) also produce the four-coordinate cationic complexes [Pt(R)(LL) (CO)]Cl. [PtPh$_2$(bipy)] undergoes two successive one-electron reductions (92IC222) and oxidatively adds methyl iodide to yield [Pt(Ph$_2$) (Me)(I)(bipy)] (90JOM(406)261). [Pt(p-MeC$_6$H$_4$)$_2$(LL)] (LL = bipy, 4,4'-Me$_2$bipy, 4,4'-t-Bu$_2$bipy) oxidatively add methyl iodide and yield [Pt(p-MeC$_6$H$_4$)$_2$(LL)(Me)I] (05OM2528). [PtAr$_2$(LL)] (Ar = Ph, p-MeC$_6$H$_4$, m-MeC$_6$H$_4$, and p-MeOC$_6$H$_4$; LL = bipy, phen) oxidatively add R$_2$O$_2$ (R = H, COPh) to yield [PtAr$_2$(LL)(OH)$_2$] or [PtAr$_2$(LL)(OCOPh)$_2$] (01JCS(D)3430).

3.4 Alkene, alkyne, and allyl complexes

1,1'-Bis(2-pyridyl)ferrocene and [(η^4-cod)Pd(Me)Cl] and Zeise's salt give **79** and **80**, respectively (96JOM(514)125). Octamethyl-5,5'-di(2-pyridyl)-ferrocene (LL) with K[PtCl$_3$(C$_2$H$_4$)] gives [{PtCl$_2$(C$_2$H$_4$)}$_2$(LL)] (97CC1723, 97JCS(D)4705).

79   80

2,2'-Bipyridine reacts with $[(\eta^2\text{-}C_{10}H_{12}R)PtCl]_2$ (R = OMe, NHPh) to yield ionic $[(\eta^2\text{-}C_{10}H_{12}R)Pt(bipy)]Cl$ (76JCS(D)2355). Cationic methylpalladium $[Pd(phen)(Me)(OEt_2)]\{B(3,5\text{-}(CF_3)_2C_6H_3)_4\}$ reacts with $RC\equiv CR'$ to form the η^2-alkyne $[(\eta^2\text{-}RC\equiv CR)Pd(phen)(Me)]\{B(3,5\text{-}(CF_3)_2C_6H_3)_4\}$ (R = R' = SiMe$_3$; R = Ph, R' = H) (98OM1530).

2,2'-Bipyridine with $[(\eta^4\text{-cod})_2Ni]$ in THF and further with diphenylacetylene gives $[(\eta^2\text{-PhC}\equiv CPh)Ni(bipy)]$, which on the basis of structural data was described as the nickelacyclopropene **81** (01EJI77). Protonolysis with phosphoric or acetic acid gives *cis*-stilbene, whereas hydrochloric acid gives **82**. Complex **81** catalyzes cyclotrimerization of alkynes. $[(\eta^4\text{-cod})Ni(bipy)]$ reacts with dimethyl 3-methyl-3-*t*-butyl-1-cyclopropene-1,2-dicarboxylate to yield first the isolable intermediate **83**, which was structurally characterized, and then complex **84** as the product (77JOM(135)405, 01OM1713).

81   82

83   **84**

[PtPh$_2$(bipy)] reacts with HBF$_4$ in acetonitrile to give cationic [Pt(Ph)(bipy)(AN)](BF$_4$) (02OM2088). [Pt(Ph)I(bipy)] reacts with silver tetrafluoroborate in acetone or acetonitrile to yield [Pt(Ph)(bipy)(solvent)](BF$_4$) (solvent = acetone, AN). Under carbon monoxide, the latter transform into the carbonyl complex [Pt(Ph)(CO)(bipy)](BF$_4$). The acetone complex inserts phenylallene into the platinum–phenyl bond to yield **85**. Under carbon monoxide, **85** readily rearranges into σ-allyl **86**. [Pt(Ph)(bipy)(AN)](BF$_4$) oxidatively adds methyl iodide to yield [Pt(Ph)(bipy)(AN)(Me)(I)](BF$_4$).

85   **86**

Complexes of C-metalated dimethyl malonate containing polypyridine ligands are known (89OM2513). [(η^2-Olefin)Pt(LL)] oxidatively add REER (LL is 2,9-dimethyl-1,10-phenanthroline; R = Me, Ph; E = S, Se; olefin = R^{1}OOCCH = CHCOOR^1, R^1 = Me, Ph) to yield [(η^2-olefin) Pt(ER)$_2$(LL)] (01JA4352). [(η^2-Olefin)Pt(2,9-Me$_2$phen)] where olefin is dimethyl maleate or dimethyl fumarate oxidatively add hydrogen peroxide, diacetyl peroxide, and dibenzoylperoxide to yield [(η^2-olefin) Pt(2,9-Me$_2$phen)(RO)$_2$] (R = H, COMe, COPh) (03OM1223). [(η^2-Dimethyl fumarate)Pt(2,9-Me$_2$phen)(HO)$_2$] is readily acetylated by acetic anhydride to give [(η^2-dimethylfumarate)Pt(2,9-Me$_2$phen)(AcO)$_2$]. [(η^2-RCOOCH = CHCOOR)Pt(EMe)$_2$(2,9-Me$_2$phen)] (R = Me, Ph; E = S, Se) are readily methylated by the E atom of Me$_3$OBF$_4$ to afford cationic [(η^2-RCOOCH = CHCOOR)Pt(EMe$_2$)$_2$(2,9-Me$_2$phen)](BF$_4$)$_2$. [Pt(4,4′-t-Bu$_2$bipy)(R^1)(R^2)] (R^1 = 4-MeOC$_6$H$_4$, R^2 = CH(COOMe)$_2$, CH(COOEt)$_2$, CH(COOPr-i)$_2$; R^1 = Me, R^2 = CH(COOMe)$_2$) can be prepared from [Pt(4,4′-t-Bu$_2$bipy)(R^1)(I)] with

HR^2 in the presence of potassium carbonate (02OM3503). They oxidatively add RSeSeR (R = Me, Ph) to yield [Pt(4,4'-t-Bu$_2$bipy)(R^1) (R^2)(SeR)$_2$]. [PtMe$_2$(4,4'-t-Bu$_2$bpy)] reacts with [Hg$_4$(EPh)$_6$(L)$_4$][ClO$_4$]$_2$ (E = S, Se; L = PEt$_3$, PPh$_3$) to give cationic platinum(IV) [PtMe$_2$(EPh) L(4,4'-t-Bu$_2$bpy)](ClO$_4$)   (02CJC41).   [Pt(Me)(CH(COOMe)$_2$)(4,4'-t-Bu$_2$ (bipy)], [Pt(CH(COOMe)$_2$)$_2$(4,4'-t-Bu$_2$(bipy)], and [Pt(CN)(CH(COOMe)$_2$) (4,4'-t-Bu$_2$(bipy)] oxidatively add PhSeX (X = Cl, Br) to yield [Pt(Me)(X) (SePh)(CH(COOMe)$_2$)(4,4'-t-Bu$_2$(bipy)],   [Pt(X)(SePh)(CH(COOMe)$_2$)$_2$ (4,4'-t-Bu$_2$(bipy)], and [Pt(CN)(X)(SePh) (CH(COOMe)$_2$)(4,4'-t-Bu$_2$(bipy)], respectively (03ICC1282).

Complexes of the general formula [(η^2-olefin)Pt(X)(Y)(LL)] are coordinatively saturated and stable both to heating and to nucleophilic attack (93CIC349, 94CCR(133)67). Starting materials may be [(η^2-ethylene)Pt(2,9-Me$_2$phen)] (91JCS(D)1007, 91JOM(403)243, 93GCI65, 98IS158) and [(η^2-dimethylmaleate)Pt(2,9-Me$_2$phen)] (94ICA(219)169). These platinum(0) complexes oxidatively add (CN)$_2$ (the ethylene complex) and (SCN)$_2$ (the dimethylmaleate complex) to yield the platinum(II) derivatives [(η^2-ethylene)Pt(2,9-Me$_2$phen)(CN)$_2$] and [(η^2-dimethylmaleate)Pt(2,9-Me$_2$phen)(SCN)$_2$], respectively (95ICA(239)61). The reaction of [(η^2-ethylene)Pt(2,9-Me$_2$phen)] with dimethyl oxalate (COOMe)$_2$ involves splitting of the C–C bond in the electrophilic substrate, evolution of carbon dioxide and formation of the platinum–methyl bond in [(η^2-ethylene)Pt(2,9-Me$_2$phen)Me(COOMe)]. Vinylchloroacetate and vinylchloroformate oxidatively add to [(η^2-dimethylmaleate) Pt(2,9-Me$_2$phen)] to yield [(η^2-dimethylmaleate)Pt(Cl)(2,9-Me$_2$phen)(CH$_2$)$_n$ COOCH=CH$_2$] (n = 0, 1). Similar complexes involve [(η^2-dimethylmaleate)Pt(Cl)(2,9-Me$_2$phen)H] (94JOM(469)237), [(η^2-dimethylmaleate)Pt(Cl) (2,9-Me$_2$phen)HgMe] (93JCS(D)3421), [(η^2-ethylene)Pt(Cl)(2,9-Me$_2$phen) (SnPh$_2$Cl)] (92JOM(425)177), [(η^2-dimethylfumarate)Pt(Cl)(2,9-Me$_2$phen)] (BF$_4$) (95ACDD43), [(η^2-ethylene)Pt(Cl)(2,9-Me$_2$phen)(py)]$^+$ (92OM3665), [(η^2-olefin)Pt(Cl)(Me)(2,9-Me$_2$phen)](BF$_4$)   (92ICA(197)51)   [(η^2-olefin) Pt(Me)L(NN)]$^+$ (90JOM(388)243),   [(η^2-dimethylmaleate)Pt(Me)(H$_2$O) (2,9-Me$_2$phen)](BF$_4$) (96JOM(519)75), [(η^2-dimethylmaleate)Pt(Me)(phen)] (97OM5981),   [(η^2-CH$_2$=CHOH)Pt(X)(Y)(2,9-Me$_2$phen)]   (X = Cl, Br, I; Y = Cl, Me) (97ICA(264)231), [(η^2-NCCH=CHCN)Pt(Me)(CH(COOMe)$_2$) (2,9-Me$_2$phen)] (97ICA(265)35). [(η^2-Olefin)Pt(2,9-Me$_2$phen)] oxidatively add the iodonium salt (I(py)$_2$)(NO$_3$) to yield five coordinate cationic [(η^2-olefin)Pt(2,9-Me$_2$phen)(I)(py)](NO$_3$)   (olefin = dimethyl   maleate, dimethyl fumarate, maleic anhydride, methyl acrylate, dibenzoylethylene, diethyl maleate, diethyl fumarate) (00EJI1717). Five-coordinate platinum (II) complexes [(η^2-ethylene)PtCl$_2$(LL)] (LL = 4,7-diphenyl-1,10-phenanthroline disulfonic acid disodium salt, 2,9-dimethyl-4,7-diphenyl-1,10-phenanthroline disulfonic acid disodium salt) have cytotoxic activity (06JCS(D)5077).

Oxidative addition of chloromethylenedimethylammonium chloride [(ClCH=NMe$_2$)Cl] to [(η^2-olefin)Pt(LL)] (LL = 2,9-Me$_2$phen, olefin = dimethylmaleate, dimethylfumarate, fumarodinitrile; LL = 6,6'-Me$_2$bipy, olefin = dimethylmaleate) gives the carbene [(η^2-olefin)Pt(LL)Cl(=CHNMe$_2$)]Cl (99OM3482). [(η^2-Dimethylmaleate)Pt(R)(2,9-Me$_2$phen)(H$_2$O)](BF$_4$) (R = Me, Ph) with diazoacetato derivatives N$_2$CHX (X = COOEt, CONMe$_2$, CN), however, does not lead to carbene products, but to complexes **87**.

87

[(η^2-ROOCCH=CHCOOR)Pt(2,9-Me$_2$phen)] (R = Me, t-Bu) oxidatively add R'HgCl (R' = Me, t-Bu) to yield [(η^2-ROOCCH=CHCOOR)Pt(2,9-Me$_2$phen)(HgR')Cl] (R = R' = Me; R = Me, R' = t-Bu; R = t-Bu, R' = Me) (95JOM(503)251) that react with silver tetrafluoroborate in acetonitrile-aqueous medium to produce cationic [(η^2-ROOCCH=CHCOOR)Pt(2,9-Me$_2$phen)(HgR')(H$_2$O)](BF$_4$). When PhHgCl or HgCl$_2$ was tried, the reaction course was different and lead to dinuclear [(η^2-ROOCCH=CHCOOR)(2,9-Me$_2$phen)ClPtHgPtCl(η^2-ROOCCH=CHCOOR)(2,9-Me$_2$phen)] and [(η^2-ROOCCH=CHCOOR)(2,9-Me$_2$phen)(H$_2$O)PtHgPt(OH$_2$)(η^2-ROOCCH=CHCOOR)(2,9-Me$_2$phen)](BF$_4$)$_2$.

[(η^2-Alkene)Pt(LL)(Cl)R], [(η^2-alkyne)Pt(LL)(Cl)R], and [(η^2-alkyne)Pt(LL)(X)Me] where LL are substituted polypyridine ligands were studied and X = Cl, Br, I, and R is the hydrocarbyl radical (81IC467, 87GCI445, 87OM517, 89OM1180, 90OM1269, 92OM3669). [Pt(Cl)Me(2,9-Me$_2$phen)] reacts with alkynes R^{1}C≡CR^2 (R^1 = R^2 = CF$_3$, COOMe, CH$_2$Cl, Me; R^1 = H, R^2 = COOMe) to yield [(η^2-alkyne)Pt(2,9-Me$_2$phen)(Cl)Me] (92CC333, 93JCS(D)1927, 94IC3331, 98ICA(275)500, 00JOM(600)37). With but-2-yne or phenylacetylene, the products are [(2,9-Me$_2$phen)(Me)Pt(μ-Cl)(μ, η^2-alkyne)Pt(Me)Cl]. With silver tetrafluoroborate and acetonitrile, the mononuclear neutral complexes are converted to the mononuclear cationic complexes followed by the formation of [(η^2-alkyne)Pt(2,9-Me$_2$phen)(AN)Me](BF$_4$). Structures of similar complexes were described (89GCI235, 92JOM(455)177).

[(η^2-Dimethylmaleate)Pd(2,9-Me$_2$phen)] reacts with PbPh$_2$Cl$_2$ to yield [Pd(Ph)Cl(2,9-Me$_2$phen)] (95OM5410) that then reacts with ethylene to afford $[(\eta^2\text{-}C_2H_4)Pd(Ph)Cl(2,9\text{-}Me_2phen)]$. Silver tetrafluoroborate in acetonitrile forms cationic $[Pd(Ph)(2,9\text{-}Me_2phen)(AN)](BF_4)$. $[(\eta^2\text{-Olefin})Pt(2,9\text{-}Me_2phen)]$ (olefin $= C_2H_4$, $CH_2=CHCOOMe$, $CH_2=CHCN$, $MeOOCCH=CHCOOMe$) oxidatively add R_2PbCl_2 (R = Me, Ph) to yield $[(\eta^2\text{-olefin})Pt(PbR_2Cl)Cl(2,9\text{-}Me_2phen)]$ (95OM4213). The tin analogs have composition $[(\eta^2\text{-olefin})Pt(SnR_2Cl)Cl(2,9\text{-}Me_2phen)]$ (92ACS499). Less sterically hindered 2-methyl-1,10-phenanthroline and 3,4,7,8-tetramethyl-phenanthroline (LL) give $[Pt(R)Cl(LL)]$ (R = Me, Ph) (95OM4213). $[(\eta^2\text{-}C_2H_4)Pt(2,9\text{-}Me_2phen)]$ oxidatively adds Me_nGeCl_{4-n} ($n = 1$–3) or Ph_2GeCl_2 to yield $[(\eta^2\text{-}C_2H_4)Pt(2,9\text{-}Me_2phen)Cl(GeR_nCl_{3-n})]$ (R = Me, $n = 0$–2; R = Ph, $n = 1$) (99ICA(285)70).

Polymers **88** (R = Me, t-Bu) and **89** were prepared by reacting the corresponding polymer with $[(\eta^2\text{-olefin})Pt(LL)]$ (LL = phen, 2,9-Me_2phen) (98ICA(281)141). The polyvinylpyridine derivatives **90** and **91** follow from the polymer and $[(\eta^2\text{-dimethylmaleate})Pt(Me)(H_2O)(2,9\text{-}Me_2phen)](BF_4)$. The second type of polymers, **92**, follows from the polymer and $[Pt(Me)Cl(2,9\text{-}Me_2phen)]$. The third type of polymers, **93**, is prepared first by incorporating 2-methylphenanthroline into the backbone of the polymer, and then reacting it with $[Pt(Me)Cl(SMe_2)_2]$ and ethylene.

92   93

1,10-Phenanthroline reacts with [(η^3-C$_3$H$_5$)PdCl]$_2$ in chloroform to yield [(η^3-C$_3$H$_5$)Pd(phen)Cl] (77JA8413). The η^3-allyl chloride complexes of palladium(II) react with 2,9-dimethyl-1,10-phenanthroline and silver tetrafluoroborate to yield [(η^3-CH$_2$CHCRR^1)Pd(2,9-Me$_2$phen)](BF$_4$) (R = H, R^1 = Me, Et, n-Pr, i-Pr; R = Me, R^1 = CH$_2$CH$_2$CH=CMe$_2$) (90JA4587). The η^3-allyl palladium complexes of 2,2'-bipyridine and 1,10-phenanthroline are characterized by an apparent rotation of the π-allyl groups (69ICA(R)109, 69JA518, 70JA3034, 75MI1, 90OM1826, 91OM1800, 92OM3954, 93JOC5445, 93MRC954, 93OM4940, 94JA3631). 2,2':6',2''-Terpyridine with [(η^3-C$_3$H$_5$)PdCl]$_2$ and silver tetrafluoroborate in methylene chloride gives [(η^3-C$_3$H$_5$)Pd(η^3-terpy)](BF$_4$) in the solid state, which becomes [(η^3-C$_3$H$_5$)Pd(η^2-terpy)](BF$_4$) and flexible in solution (96OM5442).

3.5 Diene complexes

[(η^4-cod)Pd(Me)(OTf)] reacts with 2,9-diaryl-1,10-phenthroline to yield cationic [(η^4-cod)Pd(Me)(2,9-Ar$_2$phen)](OTf) (in the 2,4,6-trisubstitued aryl groups 2-R and 4-R are H, and 6-R is t-Bu, or 2-R and 4-R are OMe and 6-R is H) (99JOM(575)214).

Oxidative addition of benzonitriles to [(η^4-cod)Ni(bipy)] gives **94** and **95** (97JOM(532)267). Acrylonitrile also promotes reductive elimination from the products to yield uncomplexed benzonitriles (03JOM(671)179). [(η^4-cod)Ni(bipy)] (76ZAAC422) catalyzes the reaction of carbon dioxide and acetaldehyde and forms nickela-5-methyl-2,4-dioxilan-3-one (2,2'-bipyridine) (82JOM(224)81). [(η^4-cod)Ni(bipy)] obtainable from 2,2'-bipyridine and [Ni(η^4-cod)$_2$] in THF is used as a reductant for various transformations of organopalladium compounds (02ZAAC20, 06JOM4868). [(η^4-cod)Ni(bipy)] with isoprene and carbon dioxide forms the ((3-5-η^3)-3-methyl-3-pentenylato) nickel bipyridine product (98OM4400). 2,2'-Bipyridine with [(η^3-cyclododecatriene)Ni] gives [Ni(bipy)$_2$] which reacts *in situ* with carbon dioxide and ethylene to yield the cyclic nickelacroxylate **96**, which further interacts with hydrochloric acid (83JOM(251)C51, 84JOM(266)203, 87AG(E)771, 04OM5252).

94   **95**   **96**

2,2'-Bipyridine reacts with [(η^4-C$_4$Me$_2$(t-Bu)$_2$)PdCl$_2$]$_2$ in methylene chloride to yield [(η^4-C$_4$Me$_2$(t-Bu)$_2$)Pd(Cl)(bipy)]Cl (05OM5136). Subsequent treatment with triethylamine and silver tetrafluoroborate gives [(η^3-C$_4$(=CH$_2$)Me(t-Bu)$_2$)Pd(bipy)](BF$_4$), where activation of the cyclobutadiene ring takes place. [(η^1,η^2-C$_8$H$_{12}$OMe)Pd(Cl)]$_2$ reacts with 2,2'-bipyridine in methanol in the presence of a corresponding salt to yield cationic [(η^1,η^2-C$_8$H$_{12}$OMe)Pd(bipy)]X (X = BPh$_4$, OTf, BF$_4$, PF$_6$, SbF$_6$, B(3,5-(CF$_3$)$_2$C$_6$H$_3$)$_4$ (78JOM(155)122, 00JCS(D)3055). These complexes promote copolymerization reactions (99OM3061, 00CRV1169, 02AG(E)544).

3.6 Monoacetylide complexes

Interaction of [(η^3-terpy)Pt(AN)](OTf)$_2$ with terminal alkynes RC≡CH in methanol with sodium hydroxide or triethylamine and ammonium hexafluorophosphate gives series [(η^3-terpy)Pt-C≡CR](PF$_6$) (R = Ph, C$_6$H$_4$Cl-4, C$_6$H$_4$Me-4, C$_6$H$_4$OMe-4, C$_6$H$_4$NO$_2$-4, benzo-15-crown-5, C$_6$H$_3$(OMe)$_2$-3,4) (01OM4476, 04JOM1393). They are prone to self-aggregation and interact with nucleic acids (97CC1451, 98IC2763, 98ICA(275)242). [Pt(4,4',4''-t-Bu$_3$tpy)(C≡CR)]$^+$ (R = Alk, Ar) are prepared from [Pt(4,4',4''-t-Bu$_3$tpy)Cl]$^+$ (07IC3038). Similar complexes are [(η^3-4-R-terpy)Pt(C≡CR^1)](ClO$_4$) (R = p-Tol, R^1 = CH$_2$OH, n-Pr, C$_6$H$_4$Cl-4, Ph, C$_6$H$_4$Me-4; R = H, R^1 = Ph) (02IC5653). The one-electron reduction of [(4'-p-Tolterpy)Pt(C≡CC$_6$H$_4$Cl)](ClO$_4$) and a series of amine donors occurs in acetonitrile (06IC4319). Similarly, complexes with R = NMe$_2$ and N-15-monoazacrown-5-ether were prepared and by metathesis reactions were converted to the tetrafluoroborate and perchlorate species (04EJI1948).

[Pt(4,4',4''-*t*-Bu$_3$terpy)(AN)](OTf)$_2$ reacts with 4-aminophenylacetylene and sodium hydroxide in methanol to yield [Pt(4,4',4''-*t*-Bu$_3$terpy)(C≡CC$_6$H$_4$NH$_2$-4)](OTf) (04OM3459). The product reacts with SCCl$_2$ and iodoacetic anhydride to afford [Pt(4,4',4''-*t*-Bu$_3$terpy)(C≡CC$_6$H$_4$X-4)](OTf) (X = NCS, NHCOCH$_2$I). [Pt(PhCN)Cl$_2$] reacts with silver hexafluoroantimonate in acetonitrile followed by 4'-(1-naphthyl)-2,2:6',2''-terpyridine (LL), and then by AgC≡CR (R = Ph, *n*-Bu) in pyridine to yield [Pt(η^3-LL)(C≡CR)](SbF$_6$) (05ICA4567). 4'-Tolylterpyridyl platinum(II) complexes with phenylacetylide, 4-bromophenylacetylide, 4-nitrophenylacetylide, 4-methoxyphenylacetylide, 4-dimethylaminophenylacetylide, 1-naphthylacetylide, and 3-quinolinylacetylide find application in non-linear optics (03APL850, 05IC4055). These and similar systems were used for the preparation of photosensitizers (03OL3221, 04JOC4788) and photocatalytic hydrogen production (04JA3440). A cyclometalated platinum(II) bipyridyl acetylide complex containing a monoazacrown moiety in the acetylide ligand serves as a chemosensor with respect to Mg^{2+} ions (04IC5195). The cytotoxicity of the platinum terpyridine complexes with glycosylated acetylide and arylacetylide ligands was evaluated for possible use for medicinal purposes (05CC4675). Another similar class of complexes is prepared when a zinc or magnesium porphyrin and a platinum 4-R-terpyridine (R = OC$_7$H$_{15}$-*n*, PO$_3$Et$_2$, H) acetylide complex are assembled through a *para*-phenylene *bis*(acetylene) spacer (05IC4806). [Pt(4'-(4-pyridin-1-ylmethylphenyl)terpy)(*p*-C≡C-C$_6$H$_4$-NH-CO-C$_6$H$_2$(OMe)$_3$)](PF$_6$)$_2$, [Pt((4'-*p*-Tol)terpy)(*p*-C≡C-C$_6$H$_4$-NH-CO-C$_6$H$_2$(OMe)$_3$)](PF$_6$) are prepared by a copper(I) iodide-catalyzed coupling of the acetylene with the terpyridine chloride cationic complex (05IC6284). Platinum(II) terpyridyl complexes containing phenolic ethynyl ligands 97 reveal interesting photophysical properties in the presence of anions such as fluoride, acetate, and dihydrogenphosphate (07JCS(D)3885).

97

1,3-*Bis*(4'-phenyl-2'-quinolinyl)benzene reacts with K$_2$[PtCl$_4$] in acetic acid and further with terminal arylacetylenes to yield alkynyl comlexes

98 (R = Ph, $C_6H_4NMe_2$, $C_6H_4NO_2$) (06JOM5900). 4-*Tert*-butyl-(1,3-*bis* (4-phenyl-2-quinolyl)pyridine reacts with $[Pt(DMSO)_2Cl_2]$ and silver tetrafluoroborate in acetonitrile, further with terminal arylacetylene, and then with ammonium hexafluorophosphate to give **99** (R = Ph, $C_6H_4NMe_2$, $C_6H_4NO_2$). $[Pt(terpy)(C\equiv C-C_6H_4-NR_2-4)]X$ (X = OTf, Cl; R = Me, CH_2CH_2OMe, H) are protonated on the amino nitrogen atom (05IC1492). Other alkynyl complexes are $[Pt(terpy)(C\equiv CC\equiv CH)]X$, (X = OTf, PF_6, ClO_4, BF_4, BPh_4), $[Pt(terpy)(C\equiv CPh)]X$ (X = OTf, PF_6, ClO_4^-, BF_4), $[Pt(terpy)(C\equiv CC_6H_4OMe-4)](OTf)$, and $[Pt(4'-CH_3O-terpy) (C\equiv CPh)](OTf)$ (05CEJ4535).

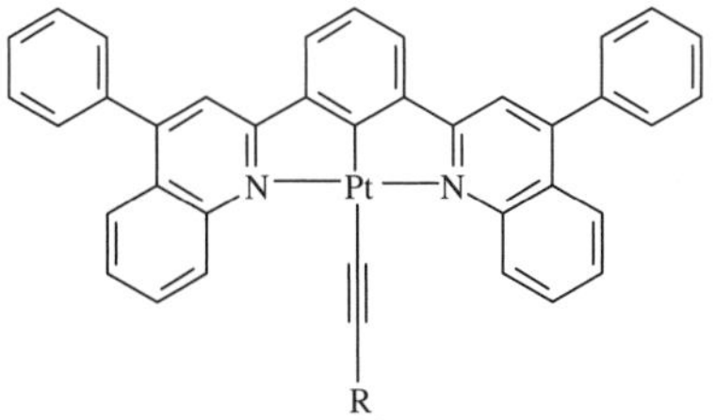

98 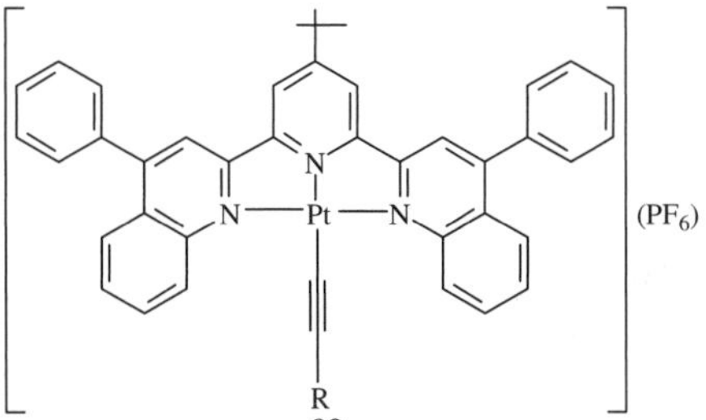   99   (PF$_6$)

$[Pt(4,4',4''-t-Bu_3terpy)(AN)](OTf)_2$ reacts with $Me_3SiC\equiv CH$ or $Me_3SiC\equiv CC\equiv CSiMe_3$ and potassium fluoride in methanol to yield **100** (n = 1, 2; X = OTf) (03AG(E)1400). The oxidative coupling of $[Pt(4,4',4''-t-Bu_3terpy)(C\equiv CC\equiv H)](OTf)$ (02JA6506) in the presence of copper(II) acetate in pyridine followed by metathesis with ammonium hexafluorophosphate gives **100** (n = 4, X = PF_6) (03AG(E)1400). $(n-Bu_4N)_2[Pt(C\equiv CC_5H_4N-4)_4]$ reacts with $[Pt(4,4',4''-t-Bu_3terpy)(AN)] (OTf)_2$ in methanol and potassium hexafluorophosphate to yield **101** (04IC812). A dinuclear platinum(II) complex containing 4,4',4''-tri-*t*-butyl-2,2':6',2''-terpyridine and bridged by the diethynylcalix[4]arene moiety is characterized by strong Pt–Pt interactions in the crystalline state (07JCS(D)4386). Platinum(II) bearing alkyne-4'-terpyridine can be assembled in step-by-step copper(I) iodide catalyzed cross-coupling reactions and form **102** (M = Fe, Zn) (06JCS(D)3285).

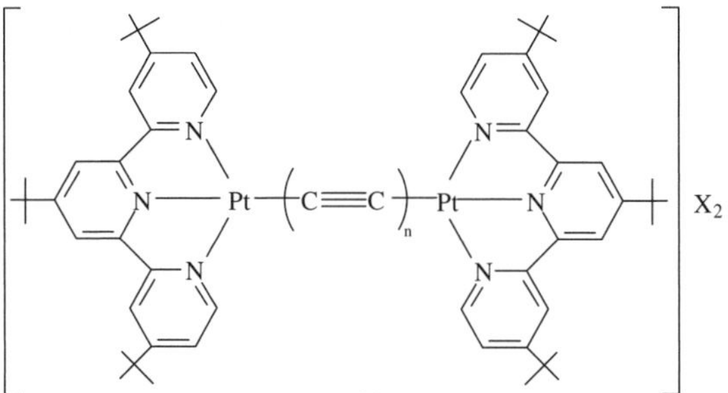

100

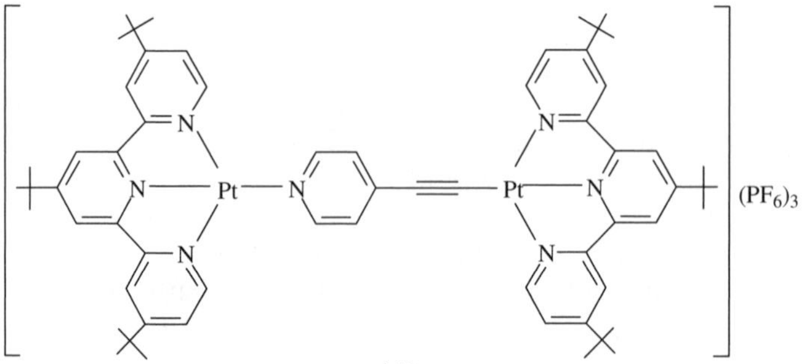

101

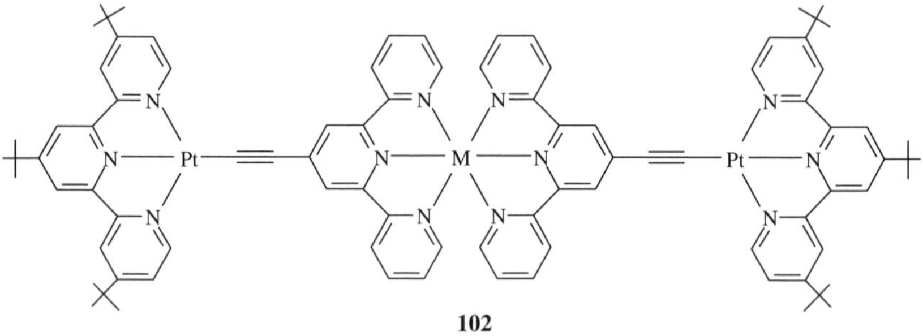

102

Self-assembly of an alkynylplatinum(II) terpyridyl is popular due to the interesting photochemical properties of the products (05AG(E)791). Hexanuclear $[Pt_2(\mu\text{-dppm})_2(C\equiv CC_5H_4N)_4\{Pt(terpy)\}_4]$ (OTf)$_8$ can be synthesized from $[Pt_2(\mu\text{-dppm})_2(C\equiv CC_5H_4N)_4]$ (02IC6178). Diethynylcarbazole-bridged platinum(II) and $[Pt(4,4',4''\text{-}t\text{-}Bu_3terpy)(C\equiv CC_6H_4C\equiv CH)]OTf$ were used to assemble tetranuclear **103** (03NJC150). $[Pt(4,4',4''\text{-}R_3terpy)Cl](OTf)$ (R = H, t-Bu) reacts with $[Re(CO)_3(LL)(C\equiv C(C_6H_4)_nC\equiv CH)]$ (LL = bipy, 4,4'-t-Bu$_2$bipy, 4,4'-(CF$_3$)$_2$bipy, 5-NO$_2$phen, n = 1; LL = 4,4'-t-Bu$_2$bipy, n = 0) in DMF with catalytic amounts of copper(I) iodide and triethylamine to yield a series of heterodinuclear complexes $[(LL)Re(CO)_3(C\equiv C(C_6H_4)_nC\equiv C)Pt(4,4',4''\text{-}R_3terpy)](OTf)$ (LL = bipy, 4,4'-t-Bu$_2$bipy, 4,4'-(CF$_3$)$_2$bipy, 5-NO$_2$phen, R = t-Bu, n = 1; LL = 4,4'-t-Bu$_2$bipy, R = H, n = 1; LL = 4,4'-t-Bu$_2$bipy, R = t-Bu, n = 0) (05OM4298). $[Pt(terpy)Cl]$ with iridium-stabilized p-dithiobenzoquinone $[(\eta^5\text{-}Cp^*)Ir(\eta^4\text{-}C_6H_4S_2)]$ produced the supramolecular assembly $\{[Pt(terpy)((\eta^5\text{-}Cp^*)Ir(p\text{-}(\eta^4\text{-}C_6H_4S_2)))Pt(terpy)](OTf)_4\}_n$ with π–π and Pt–Pt interactions within a one-dimensional chain (07JCS(D)3526).

103

5-(2-(Trimethylsilyl-1-ethynyl))-2,2'-bipyridine reacts with [PtCl$_2$ (η^2-dppm)] in the presence of copper(I) iodide and potassium fluoride to yield dinuclear **104**, which serves as a source of new heteronuclear complexes (07ICA163). This ligand contains four donor sites and was used to prepare different hexanuclear complexes **105** (ML$_n$ = Eu (hfacac)$_3$, Nd(hfacac)$_3$, Yb(hfacac)$_3$) from [Ln(hfacac)$_3$(H$_2$O)$_2$] (Ln = Eu, Nd, Yb) (06CC1601), **105** (ML$_n$ = Re(CO)$_3$Cl) from Re(CO)$_5$Cl, **105** (ML$_n$ = Ru(bipy)$_2$) from [Ru(bipy)$_2$Cl$_2$], and **105** (ML$_n$ = Gd(hfacac)$_3$) from [Gd(hfacac)$_3$(H$_2$O)$_2$] (07ICA163). 4'-(2-propyn-1-oxy)-2,2':6',2''-ter-pyridine (05JCS(D)234) and 4'-(4,7,10-trioxadec-1-yn-10-yl)-2,2':6',2''-terpyridine (07ICA4069) are used to assemble macromolecules. The latter reacts with *trans*-[PtI$_2$(PEt$_3$)$_2$] in the presence of copper(I) iodide and triethylamine in methylene chloride to yield **106**. The same reaction run in THF/DMF gives mononuclear **107**. For 4'-(2-propyn-1-oxy)-2,2':6',2''-terpyridine, similar reactions occur, but the analog of **106** with iron(II) chloride and ammonium hexafluorophosphate in methanol gave metallacycle **108**. 4'-(4-Ethynylphenyl)-2,2':6',2''-terpyridine reacts with [Pt(dppe)Cl$_2$] or [Pt(dppp)Cl$_2$] in the presence of copper(I) iodide and tri-*i*-propylamine in methylene chloride to yield **109** (*n* = 0, 1) (07OM4483). The products with [Ln(hfacac)$_3$(H$_2$O)$_2$] (Ln = Eu, Nd, Yb) give the heterotrinuclear **110**.

104

105

106

N
N
O
O
O
N
PEt3
Pt
I
PEt3

107

PEt3
Pt
PEt3
O
O
N
N
N
Fe
N
N
N
PEt3
Pt
PEt3
(PF6)4

108

Ph2
P
(CH2)n
Pt
P
Ph2
N
N
N

109

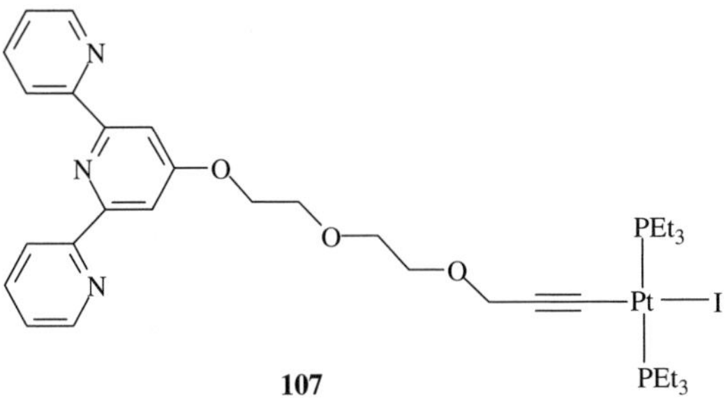

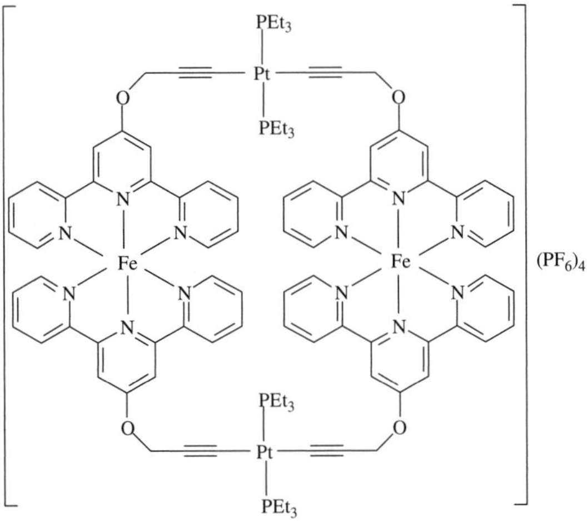

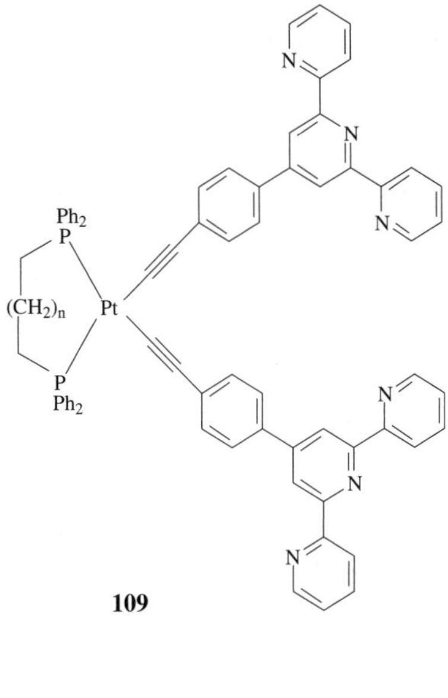

110

3.7 Diacetylide complexes

[Pt(4,4'-Me$_2$bipy)Cl$_2$] reacts with HC≡CAr in methylene chloride containing HNPr$_2^i$ and copper(I) iodide to yield [Pt(4,4'-Me$_2$bipy)(C≡CAr)$_2$] (Ar = Ph, C$_6$H$_4$-4-Me, C$_6$H$_4$-4-NO$_2$) (97JOM(543)233, 00JCS(D)63). [Pt(t-Bu$_2$bipy)(C≡CAr)$_2$] (Ar = 4-pyridyl, 3-pyridyl, 2-pyridyl, 4-ethynylpyridyl, 2-thienyl, pentafluorophenyl) (03CEJ6155), [Pt(4,4'-t-Bu$_2$bipy)(C≡CR)$_2$] (R = SiMe$_3$, C≡C-SiMe$_3$, t-Bu) (01CC789, 05IC471), [Pt(LL)(C≡C$_6$H$_4$R)$_2$] (LL = bipy, phen; R = N-7-azaindolyl, 2,2'-dipyridylamino) (03JCS(D)3493) are interesting due to their luminescent properties.   3,8-Di-n-pentyl-4,7-di(phenylethynyl)-1,10-phenanthroline (LL) was the basis for [Pt(LL)(C≡C–Ph)$_2$], [Pt(LL)(C≡C–Fc)$_2$], and [Pt(LL)(C≡C–p-C$_6$H$_4$–C≡C–Fc)$_2$] (05JCS(D)2365). Luminescent [Pt(4,7-Ph$_2$phen)bis(4-ethynylbenzaldehyde)] can be used in the synthesis of chromophores through Schiff base condensation of anilines to give imine-linked species and through imine reduction with sodium tetrahydroborate from amine-linked species (05IC2628). [Pt(4,4'-t-Bu$_2$bipy)(C≡CR)$_2$] (R = C$_6$H$_4$Me, SiMe$_3$) with MSCN (M = Cu, Ag) gives heterodinuclear [Pt(4,4'-t-Bu$_2$bipy)(C≡CR)$_2$MSCN], where copper or silver are coordinated through the triple bonds of the alkynyl (99JOM(578)178). Complex **111** using the bidentate diacetylide ligand tolan-2,2'-diacetylide contains a planar platinacycle (06IC4304, 07IC8771).

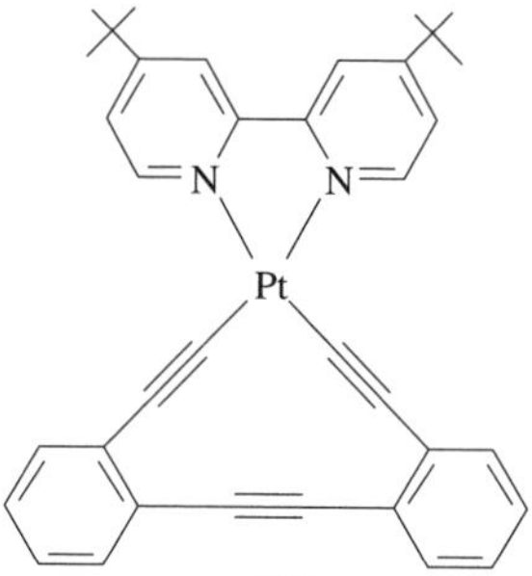

111

[Pt(phen)(C≡CPh)$_2$] (99IC3264), [Pt(5-Rphen)(C≡CPh)$_2$] (R = H, Me, Cl, Br, NO$_2$, C≡CPh), [Pt(4,4'-t-Bu$_2$bipy)(C≡CC$_6$H$_4$(p-X))$_2$] (X = H, Me, CF$_3$, F, NO$_2$, OMe, NMe$_2$), [Pt(phen)(C≡CC$_6$H$_4$CHO)$_2$] (03IC3772), [Pt(4,4'-t-Bu$_2$bipy)(C≡C-pyrene)$_2$] (03IC1394, 04JPC(A)3485, 06CCR1819), [Pt(5,6-dithienyl)phen(C≡CPh)$_2$] (07OM12) posses luminescent properties (00IC447, 01IC4053). On the way to a molecular photochemical device, **112** was constructed (01IC4510). This group of complexes is regarded as possible materials for organic light-emitting diodes (01CEJ4180). Another preparation of this kind involves the reaction of [PtCl$_2$(4,4'-X$_2$bipy)] (X = t-Bu, COOEt) with p-ethynylbenzyl-N-phenothiazine or p-ethynylbenzyl-N-(2-trifluoromethyl)phenothiazine containing copper(I) iodide and triethylamine in methylene chloride to yield **113** (R = t-Bu, COOEt; R^1 = H, CF$_3$) (03IC4355).

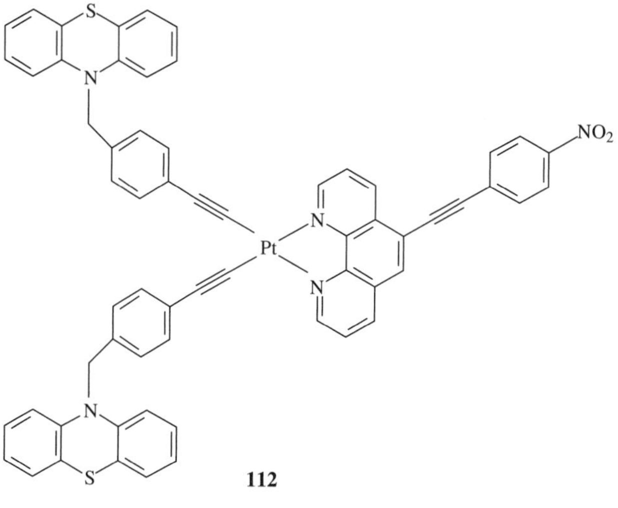

112

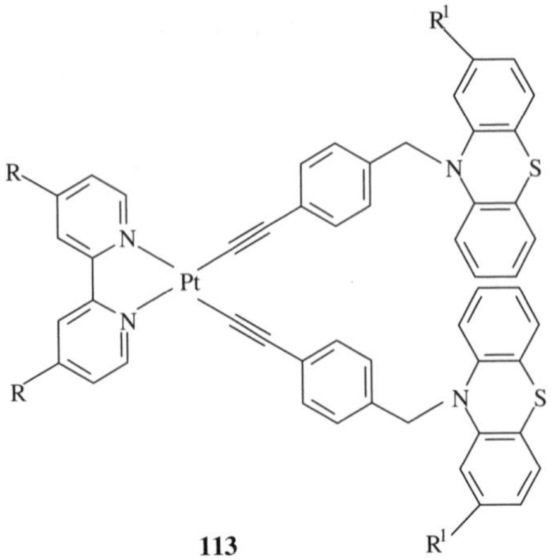

113

(2,2'-Bipyridine-5-ylethynyl)ferrocene reacts with $[PtCl_2(Me_2S)_2]$ and further with trimethylsilylacetylene in the presence of copper(I) iodide and di-i-propylamine to yield *bis*-acetylide **114**, which forms heterodinuclear **115** on interaction with $[Cu(AN)_4](BF_4)$ (06OM4579). Clusters $[Pt_2Cu_4(C_6F_5)_4(C\equiv CBu\text{-}t)_4(acetone)_2]$ or $[PtCu_2(C_6F_5)_2(C\equiv CPh)_2]_x$ react with 2,2'-bipyridine to yield **116** (R = t-Bu, Ph) with interesting luminescent properties (03JOM(670)221).

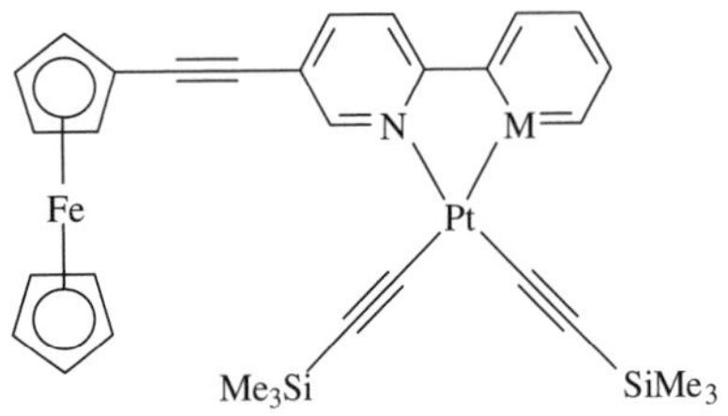

114

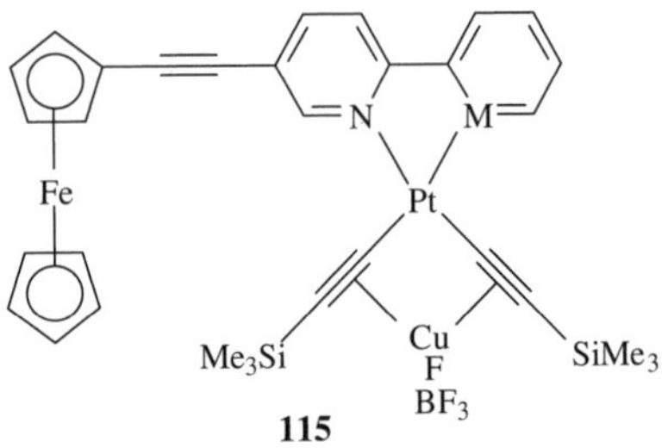

115

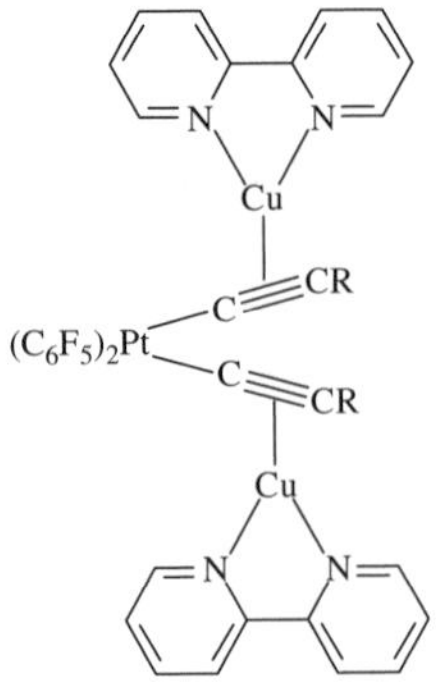

116

3.8 Cyclometalation

2,2'-Bipyridine and 1,10-phenanthroline (LL) react with [Pd(L)(C$_6$F$_5$)(AN)] (L = ((dimethylamino)methyl)phenyl-C^2,N) to form the mononuclear derivatives [Pd(L)(C$_6$F$_5$)(η^2-LL)] (LL = bipy, phen) (90OM2422, 93IC3675). 1,10-Phenanthroline reacts in a similar manner with [Pd(L)(C$_6$F$_5$)(AN)] (L = 8-quinolylmethyl-C,N) to yield [Pd(L) (C$_6$F$_5$)(phen)]. However, 2,2'-bipyridine reacts with this complex differently to yield the dinuclear complex with an extremely rare bridging function of the 2,2'-bipyridine ligand, [(C$_6$F$_5$)(L)Pd(μ-bipy)Pd(L)(C$_6$F$_5$)]. Such a function was also discovered in (PPN)$_2$[(C$_6$F$_5$)$_3$Pt(μ-bipy) Pt(C$_6$F$_5$)$_3$] (89POL2209). Another example with a cyclometalated ligand and 2,2'-bipyridine co-ligand is the product of interaction of [Pd$_2$L(AN)$_2$(ClO$_4$)$_2$] (H$_2$L = N,N'-diethyl-2,6-dialdiminobenzene) with 2,2'-bipyridine in acetonitrile, whose composition is [Pd$_2$L(bipy)$_2$](ClO$_4$)$_2$ (91POL1513). Some illustrations include [Pt(2-Phpy)(bipy)]Cl (89IC309, 90CCR(97)193) and other complexes (93OM4151, 96OM24, 97GCI119, 00OM752).

Derivatives of polypyridine ligands can form cyclometalated complexes. Thus, 6-(1-methylbenzyl)-2,2'-bipyridine (LL) forms **117** (R^1 = H, R^2 = Me, X = Cl) (86JOM(307)107, 90IC5137, 92GCI455, 94JOM(481)195, 95JCS(D)777). This complex on reaction with silver tetrafluoroborate and two-electron ligands L forms the derivatives [(LL)Pt(L)](BF$_4$) (L = Ph$_3$P, py, AN) (89JOM(363)401, 93JOM(452)257). By exchange of the chloride ligands, a series of the neutral complexes **117** (R^1 = R^2 = H; R^1 = H, R^2 = Me, R^1 = R^2 = Me; X = I, CN) and cationic chloride complexes **117** (R^1 = R^2 = H; R^1 = H, R^2 = Me, R^1 = R^2 = Me; X = CO, PPh$_3$, py, AN) can be prepared (00EJI2555, 00ICA(305)189). From the chlorides with NaBH$_4$, the cyclometalated hydrides follow (07OM5621). Complexes of 6-phenyl-2,2'-bipyridine **118** (X = Cl) possess interesting photochemical properties (90CC513, 90JCS(D)443, 96JCS(D)1645, 99CEJ2845). They react with isonitriles in the presence of lithium perchlorate to yield cationic **119** (R = t-Bu, n-Pr, i-Pr, Cy, 2,6-Me$_2$C$_6$H$_3$) (99OM3327, 02OM226). The latter react with amines or hydrazine to form the diamino carbene **120** (R = t-Bu, 2,6-Me$_2$C$_6$H$_3$; R^1 = Me, NH$_2$, CH$_2$Ph). With 3,3'-diamino-N-methyl-di-n-propyl amine the dinuclear carbene complex **121** results. Photochemical reaction of **120** (R = t-Bu, R^1 = Me) with methyl iodide gives platinum(IV) **122**. The photochemical properties of the following mono- and dinuclear complexes [Pt(4-Ar-6-Ph-bipy)Cl] (Ar = H, Ph, 4-ClC$_6$H$_4$, 4-Tol, 4-MeOC$_6$H$_4$, 3,4,5-(MeO)$_3$C$_6$H$_2$), [Pt(6-Phbipy)L]$^+$ (L = py, PPh$_3$), [Pt$_2$(4-Ar-6-Phbipy)$_2$(μ-dppm)]$^{2+}$ (Ar = H, Ph, 4-ClC$_6$H$_4$, 4-Tol, 4-MeOC$_6$H$_4$, 3,4,5-(MeO)$_3$C$_6$H$_2$), [Pt$_2$(6-Phbipy)(μ-pz)]$^+$, and [Pt$_2$(6-Phbipy)(μ-L)]$^{2+}$ (L = diphenylphosphinopropane, diphenylphosphinopentane) are of interest (98CC1127, 98CC2295, 99IC4046, 00IC255). 2,9-Diphenyl-1,10-phenanthroline (LL) and [Pt(AN)$_4$](ClO$_4$)$_2$ form the cyclometalated complex [Pt(LL)(AN)](ClO$_4$), where one of the phenyl groups of the phenanthroline ligand is engaged in coordination along with both nitrogen atoms (93JA11245, 94CCR(132)87, 00CCR(208)115). An interesting class of photochemical materials contains [(6-Arbipy)Pt(C≡C)$_n$R] (Ar = Ph, thienyl; n = 1–4, R = Ar, SiMe$_3$) (02CC206, 02CC900). 6-Thienyl-2,2'-bipyridine (LL) with [MCl$_4$]$^{2-}$ (M = Pd, Pt) forms cyclometalated [M(LL)Cl], where the thienyl group is coordinated through the C^3 center (92JCS(D)2251). The product reacts with trimethyl phosphite to yield [M(LL)(P(=O)(OMe)$_2$] and with nickel acetylacetonate to give [M(LL)(acac)] where the acetylacetonate group is C-coordinated. 6-Aryl-2,2'-bipyridines under reflux with K$_2$[PtCl$_4$] in acetonitrile/water and on further interaction with various terminal alkynes in methylene chloride in the presence of copper(I) iodide and di-i-propylamine produce a wide variety of platinum alkynyl complexes: [Pt(6-Arbipy)(C≡CR)] (Ar = Ph, R = Ph, C$_6$H$_4$Me-4, C$_6$H$_4$OMe-4, C$_6$H$_4$Cl-4, C$_6$H$_4$F-4, C$_6$H$_4$NO$_2$-4, C$_6$F$_5$, C≡CPh, t-Bu, SiMe$_3$, C≡CSiMe$_3$, C≡CC≡CSiMe$_3$, C≡CC≡CC≡CSiMe$_3$; Ar = 5-MeC$_6$H$_4$, 6-MeC$_6$H$_4$, 6-CF$_3$C$_6$H$_4$, 4,6-F$_2$C$_6$H$_3$, R = Ph); [Pt(4-Arbipy)(C≡CPh)] (Ar = Ph, 4-MeC$_6$H$_4$, 4-MeOC$_6$H$_4$, 4-ClC$_6$H$_4$, 4-NO$_2$C$_6$H$_4$); [Pt(4,4'-t-Bu$_2$bipy)(C≡CR)] (R = Ph, C$_6$F$_5$, pyrenyl-1); [Pt(4-EtOOCbipy)(C≡CR)] (R = Ph, C$_6$F$_5$); [(6-hetaryl)Pt(C≡CR)] (hetaryl = furyl, thienyl, R = Ph, p-Tol, C$_6$F$_5$) (04JA4958). Another cationic complex of interest is [Pt(C,N,N-2,9-Me$_2$phen)(AN)]$^+$ (96CC1039).

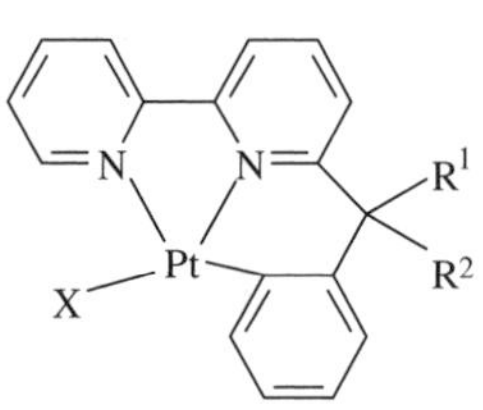

117

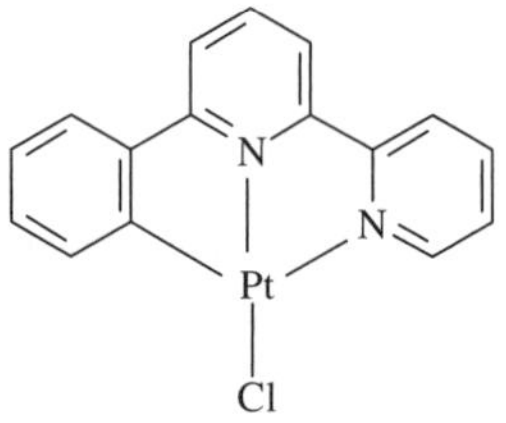

118

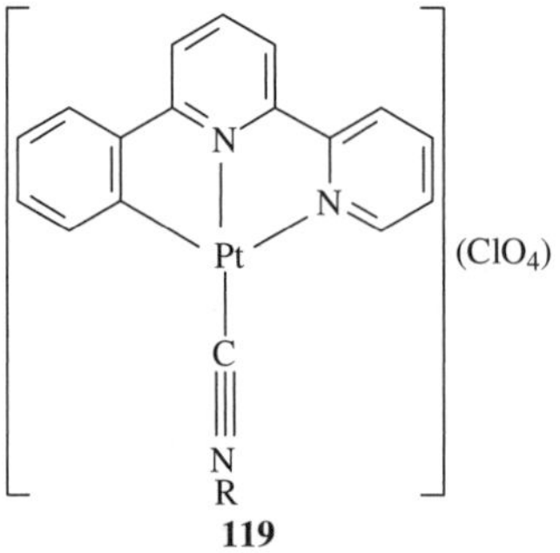

119

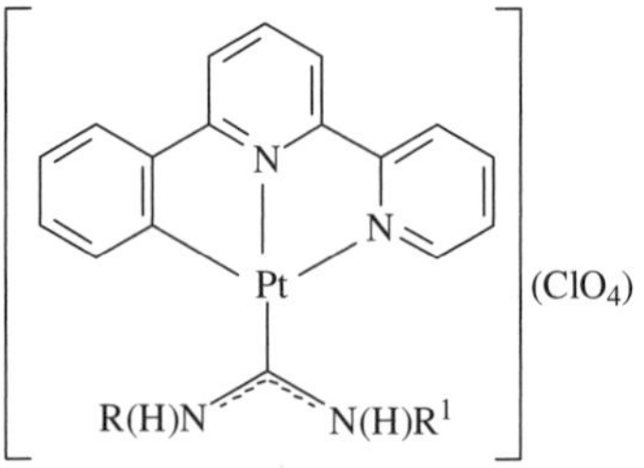

120

121   122

6-(2-Tolyl)-2,2'-bipyridine with Na$_2$[PdCl$_6$] in H$_2$O/HCl at elevated temperatures forms cyclometalated **123** (X = Cl), whereas 6-(2,6-xylyl)-2,2'-bipyridine (LL) similarly gives adduct [PdCl$_2$(LL)] (03JOM(679)1). With lithium iodide, **123** (X = Cl) gives **123** (X = I) and with triphenylphosphine and sodium tetrafluoroborate, cationic **124** results. 6-(2-Tolyl)-2,2'-bipyridine with palladium(II) acetate generates the cyclometalated derivative **123** (X = OAc) and further with lithium chloride **123** (X = Cl) can be prepared. Both ligands in adducts [Pd(LL)(Me)Cl] react with NaB(3,5-(CF$_3$)$_2$C$_6$H$_3$)$_4$ and subsequently triphenylphosphine to give **125** differing by the mode of cyclometalation. Cyclometalated [Pt(LL)Cl] complexes where LL originate from deprotonated alkyl-, phenyl-, and benzyl-6-substituted 2,2'-bipyridines react with triphenylphosphine to yield [Pt(N-N-C)(PPh$_3$)]Cl (00IC4749). [Pt(LL)Cl] (LLH = 6-phenyl-2,2'-bipyridine, 1,3-di(2-pyridyl)benzene) enter substitution reactions with Br$^-$, I$^-$, thiourea, N,N-dimethylthiourea, and N,N,N',N'-tetramethylthiourea (03IC6528). 1,3-Bis(1-(pyridin-2-yl)ethyl)benzene and 1,3-bis(2-pyridinecarbonyl)benzene react with Pd(MeCOO)$_2$ to form the cyclopalladated [Pd{(NC$_5$H$_4$CHMe)$_2$C$_6$H$_3$}(O$_2$CMe)] and [Pd{(NC$_5$H$_4$CO)$_2$C$_6$H$_3$}(O$_2$CMe)], respectively, possessing tridentate NCN co-ordination involving two six-membered palladacycle rings (87JCS(D)1477).

123   124

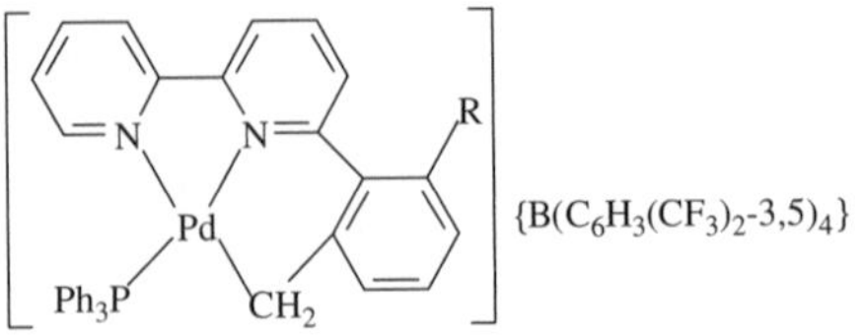

125

[Pd(6-Phbipy)Cl] enters a metathesis reaction with sodium azide to produce [Pd(6-Phbipy)N$_3$] (04JCS(D)3699). The product with aryl isocyanides leads to carbodiimide complexes [Pd(6-Phbipy)(N=C= NAr)] (Ar = C$_6$H$_3$Et$_2$-2,6, C$_6$H$_3$(i-Pr)$_2$-2,6). In contrast, with 2,6-dimethyl-phenylisocyanide the carbodiimide complex containing an imidoyl group follows [Pd(6-Phbipy)(N=C=NC$_6$H$_3$Me$_2$-2,6)(-C=NC$_6$H$_3$Me$_2$ 2,6)]. The imidoyl group is the result of insertion into the palladium–carbon bond. Alkyl isocyanides n-BuNC and i-PrNC react differently and lead to the C-coordinated tetrazolato complexes **126** (R = n-Bu, i-Pr) by dipolar cycloaddition of an isocyanide to the Pd-N$_3$ bond. The same pattern is observed in allyl isothiocyanate when the product is the S-coordinated tetrazole thiolate **127**.

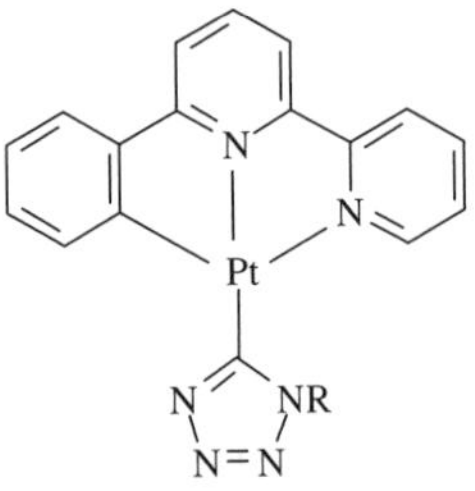

126 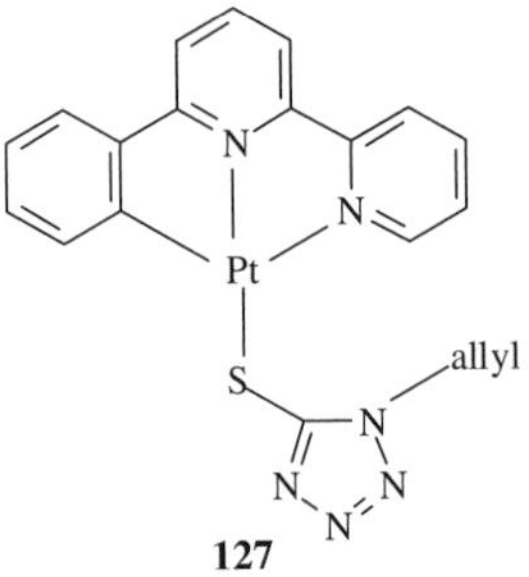   **127**

1,3-*Bis*(7-azaindolyl)benzene, 1-bromo-3,5-*bis*(7-azaindolyl)benzene, and 1,3,5-*tris*(7-azaindolyl)benzene with K$_2$[MCl$_4$] (M = Pd, Pt) give the cyclometalated **128** (01OM4683). 1,3,5-*Tris*(di-2-pyridylamino)benzene with palladium acetate or chloride produces derivative **129** (03OM2358).

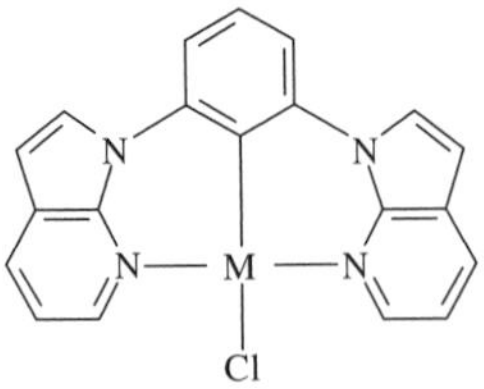

128 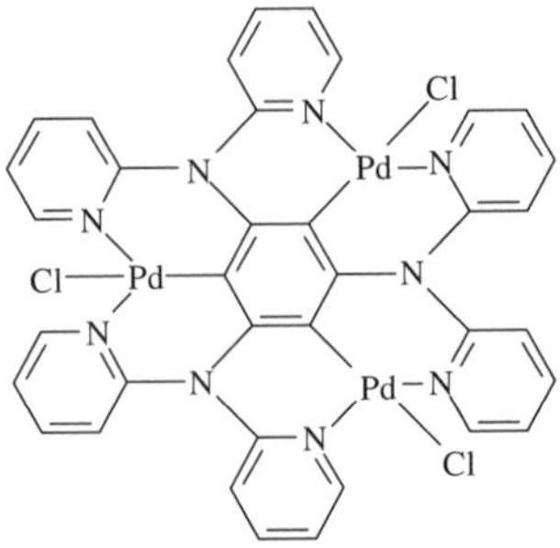   **129**

2,2′:6′,2″-Terpyridine reacts with [PtMe$_2$(DMSO)$_2$] in toluene to give dinuclear cyclometalated **130** (L = DMSO) with the unique N, C∩C, N coordination mode (01OM1148). The ligand substitution reactions form **130** (L = AN, CO, PPh$_3$, PCy$_3$). In excess 3,5-lutidine and prolonged heating **130** (L = 3,5-lutidine) can be prepared (06JOM4135). Complex **130** (L = DMSO) reacts with tetracyanoethylene to yield η^2-coordinated **131**. Complex **130** (L = PPh$_3$) oxidatively adds methyl iodide to yield **132**. Complex **130** (L = DMSO) with hydrochloric acid gives dinuclear **133** and tetranuclear **134**. The DMSO ligands in **133** can be readily substituted by triphenylphosphine and with more difficulty by 3,5-lutidine. The triphenylphosphine-substitution product of **133** under carbon monoxide allows substitution of one dimethyl sulfoxide or one triphenylphosphine ligand, respectively, to yield the monocarbonyl complexes.

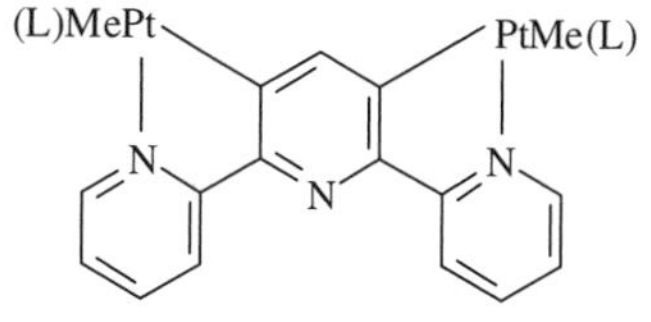

130

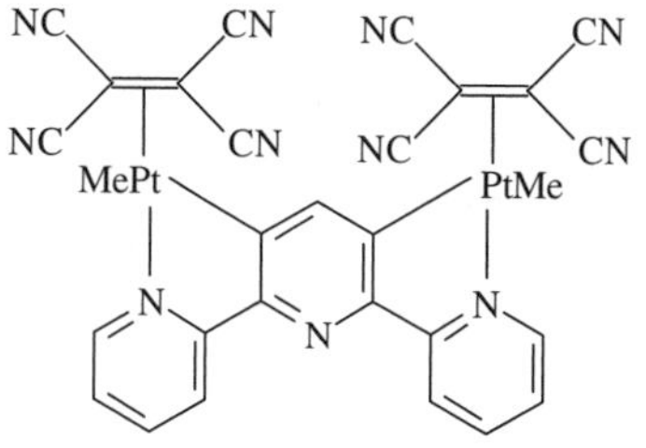

131

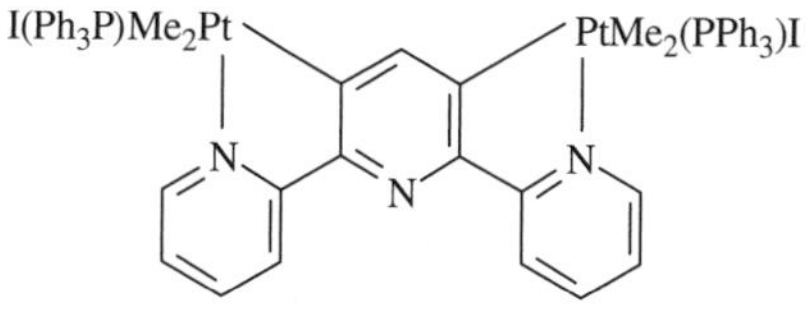

132

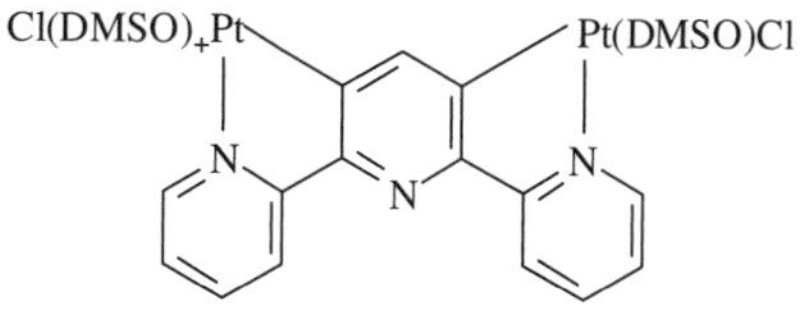

133

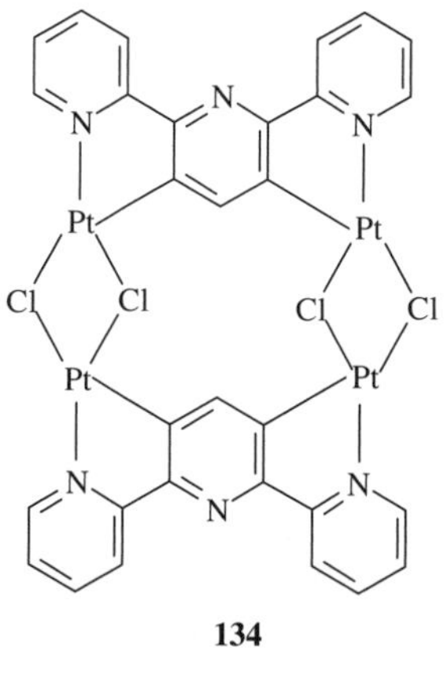

134

6'-Phenyl-2,2'-bipyridine and its 4'-functionalized derivatives (LL) form cyclometalated palladium(II) complexes [Pd(LL)Cl] (96CC2463, 97IC6150, 98LC673). [Pt(6-Phbipy)(L)](ClO$_4$) (L = py, 2-NH$_2$py, 4-NH$_2$py, 2,6-(NH)$_2$)$_2$py contains the pyridine ligands coordinated through the nitrogen heteroatom (00IC3537). 4-Carboxy-6-phenyl-2,2'-bipyridine, 4-carboxyphenyl-6-phenyl-2,2'-bipyridine, or 4-hydroxyphenyl-6-phenyl-2,2'-bipyridine (LL) react with [Pd(PhCN)$_2$Cl$_2$] in benzene to yield the cyclometalated derivatives [Pd(LL)Cl] (02OM3511). 1,3-Di(2-pyridyl)-benzene on interaction with K$_2$[PtCl$_4$] gives classical cyclometalated **135** (M = Pt, X = Cl), whereas cyclometalation with palladium(II) acetate leads to tetranuclear **136** (99OM3337, 01AG(E)3750). The C-coordinated mercury salt of 1,3-*bis*(2-pyridyl)benzene (LL), [Hg(η^1(C)-LL)Cl] enters a transmetalation reaction with palladium(II) acetate and further with a lithium halide (chloride, bromide, iodide) to yield cyclometalated **135** (M = Pd, X = Cl, Br, I) (05OM53). Complex **135** (M = Pd, X = Cl) with silver tetrafluoroborate in dichloromethane/acetone gives respective cationic [Pd(η^3(N,C,N)-LL)(H$_2$O)](BF$_4$). In turn, **135** (M = Pd, X = Cl, Br, I) with NaB{3,5-(CF$_3$)$_2$C$_6$H$_3$}$_4$ in dichloromethane gives dinuclear [(η^3(N,C,N)-LL)Pd(μ-X)Pd(η^3(N,C,N)-LL)](B(3,5-(CF$_3$)$_2$C$_6$H$_3$)$_4$). Chloride **135** (M = Pd, X = Cl) also reacts with silver acetate to yield the acetate complex [Pd(η^3(N,C,N)-LL)(OAc)], which in a similar fashion generates the dinuclear species [(η^3(N,C,N)-LL)Pd(μ-OAc)Pd(η^3(N,C,N)-LL)](B(3,5-(CF$_3$)$_2$C$_6$H$_3$)$_4$).

135   136

6-C(R)(R^1)Me-2,2'-bipyridine (LL; R = R^1 = H; R^1 = H, R^2 = Me) and 6-CH$_2$CMe$_3$-2,2'-bipyridine (LL) with Na$_2$[PdCl$_6$] in aqueous hydrochloric acid leads to adducts [PdCl$_2$(LL)], while that of 6-*t*-butyl-2,2'-bipyridine gives the cyclometalated species **137** (R = R^1 = Me) with C–H activation of the *t*-butyl group (93JOM(450)C15, 99CHC992, 00OM4295). 6-Ethyl and 6-*i*-propyl-2,2'-bipyridine with palladium acetate in benzene at elevated temperatures followed by treatment with lithium chloride in acetone/water give cyclometalated **137** ($R = R^1 = H$, $R = H$, $R^1 = Me$). In the case of the 6-i-propyl derivative, the other product with rare C-H activation of one of the pyridine rings is formed, **138** ($R = i$-Pr) along with **137** ($R = R^1 = Me$). In the case of 6-neopentyl-2,2′-bipyridine, the chelate complex **138** ($R = CH_2Bu$-t) is the sole product. The same reaction but conducted in acetic acid at room temperature followed by lithium chloride in acetone/water gives **139** for 6-i-propyl-2,2′-bipyridine, **137** ($R = CH_2OC(O)Me$, $R^1 = H$) for 6-neopentyl-2,2′-bipyridine, and complexes **140** and **137** ($R = Ph$, $R^1 = Me$) for 6-PhCMe$_2$-2,2′-bipyridine. C^3, N'-Coordination was postulated to interpret the thermal rearrangement of the diaryl(2,2′-bipyridyl)platinum(II) complexes (85CC609). 6-Phenyl-2,2′-bipyridine with cis-[Me$_2$Pt(DMSO)$_2$] or cis-[Ph$_2$Pt(DMSO)$_2$] in toluene gives products with C^3,N^1-coordination, **141** ($R = Me$, Ph) (02OM783, 03OM4770). With an excess of the starting platinum(II) complexes, unique dinuclear products **142** ($R = Me$, Ph; $L = L' = DMSO$) are formed. Ligand substitution reactions allowed the preparation of a series of dinuclear complexes **142** ($R = Me$, Ph, $L = L' = PPh_3$, CO; $R = Me$, $L = 3,5$-Me$_2$py, $L' = DMSO$) (04EJI4484). Ligand substitution with 1,3-bis(diphenylphosphino)propane gives tetra-nuclear **143**. With hydrochloric acid, **142** ($R = Me$, Ph; $L = L' = DMSO$) form **142** where R is the chloride ligand. In contrast, **142** ($R = Me$, Ph; $L = L' = PPh_3$) with hydrochloric acid retain the Pt–R bonds and rearrange into **144** ($R = Me$, Ph). 6-Alkyl-2,2′-bipyridines with K$_2$[PtCl$_4$] form the cyclometalated complexes [Pt(6-Alkbipy)Cl] (Alk = CH$_2$Me, CHMe$_2$, and CH$_2$CMe$_3$) having five- and six-membered chelate rings (02EJI3336).

137

138

139

140

141

142

143   **144**

6,6'-Diphenyl-2,2'-bipyridine (LL) with $[Pt(Me)_2(DMSO)_2]$ gives dinuclear **145** (L = DMSO) with a bridging cyclometalated polypyridine ligand (06OM2253). With Ph_3P, Cy_3P, 3,5-dimethylpyridine, quinoline the products are **145** (L = Ph_3P, Cy_3P, 3,5-diemthylpyridine, quinoline), and with $Na_3(4-SO_3C_6H_4)_3P$ the substitution product is $[Pt_2(\mu\text{-}LL)(P(4-SO_3C_6H_4Na)_3)_2]$. Substitution products are also formed with dimethylphosphinomethane, -ethane, and -propane, whose structure was assumed to match with the formation of at least a tetranuclear species. Complex **145** (L = DMSO) also reacts with carbon monoxide in methylene chloride to yield substituted **145** (L = CO). In excess platinum(II) precursor, mononuclear **146** can also be prepared. Rare examples of cyclometalation occur in *N*-substituted polypyridines, in particular in the *N*-methyl-2,2'-bipyridinium (bipyMe) complexes $[M(bipyMe)X_3]$ (M = Pd, Pt; X = Cl, Br, I) (83ICA(69)179).

145   **146**

3.9 Clusters and polynuclear complexes

2,6-*Bis*(2,2'-carbalkoxy)ethylpyridines (alkoxy = methoxy, ethoxy) with $K_2[PdCl_4]$ and 4,4'-bipyridine give dinuclear **147** (R = Me, Et) with the cyclometalated pyridine ligand and bridging polypyridine ligand along with minor amounts of mononuclear species **148** (R = Me, Et) (82JA994). Dinuclear complexes predominate when instead of 4,4'-bipyridine, 1,2-*bis*(4-pyridyl)ethylene-acetylene and ethane bridges were used. [Pd(C_6H_4I)(PPh_3)$_2$(OTf)] with 4,4'-bipyridine toluene/methylene chloride gives the dinuclear complex [(C_6H_4I)(PPh_3)$_2$Pd(μ-4,4'-bipy)Pd (C_6H_4I)(PPh_3)$_2$](OTf)$_2$ (06JOM3834).

147   148

Iodobenzoic acids oxidatively add to [Pt(PPh_3)$_4$] and the product further interacts with silver triflate and 1,1-*bis*(4-pyridyl)ethylene, 4,4'-bipyridine, or 4,7-phenanthroline to yield dinuclear complexes with a bridging polypyridine ligand, shown for 4,7-phenanthroline **149** below (L = PPh_3, R = R^1 = 3-COOH, 4-COOH) (01OM3373). [Pt(I)(PPh_3)$_2$Ar] (Ar = 3-COOSi(Bu-*t*)Ph$_2$C$_6$H$_4$, 4-COOSi(Bu-*t*)Ph$_2$C$_6$H$_4$, 3-CONHMe, 4-CONHMe) reacts with silver triflate and then 4,7-phenanthroline in methylene chloride to form mononuclear cationic **150** where R is the 3- or 4-substituted carboxylate (04JOM1288). The products **150** react further with [Pt(Ar')L$_2$(OTf)] (Ar' = Ph, L = PEt$_3$; Ar' = 3-COOSi(Bu-*t*)Ph$_2$C$_6$H$_4$, 4-COOSi(Bu-*t*)Ph$_2$C$_6$H$_4$, 3-CONHMe, L = PPh_3) to yield dinuclear **149** (L = PEt$_3$, R = 4-COOSi(Bu-*t*)Ph$_2$, R^1 = H; L = PPh_3, R = 4-CONHMe, R^1 = 4-COOSi(Bu-*t*)Ph$_2$; L = PPh_3, R = 4-COOSi(Bu-*t*)Ph$_2$, R = 3-CXONHMe; L = PPh_3, R = 3-CONHMe, R^1 = 3-COOSi(Bu-*t*)Ph$_2$).

149   **150**

[Pt(1,3-*bis*(piperidylmethyl)benzene)Cl] reacts with 4,4′-bipyridine and silver triflate to yield dinuclear [Cl(1,3-*bis*(piperidylmethyl)benzene) Pt(μ-4,4′-bipy)Pt(1,3-*bis*(piperidylmethyl)benzene)Cl] (04IC725). [Pd (PPh$_3$)$_2$(C$_6$H$_4$-4-SC(O)Me)(I)] with silver triflate and then 4,4′-bipyridine gives binuclear [(Me(O)CS-4-C$_6$H$_4$)(Ph$_3$P)$_2$Pd(μ-4,4′-bipy)Pd(PPh$_3$)$_2$ (C$_6$H$_4$-4-SC(O)Me)](OTf)$_2$ with the bridging 4,4′-bipyridine ligand (06ICA1899).

2,7-*Bis*(diphenylphosphino)-1,8-naphthyridine reacts with [Pt(2,6-xylylisocyanide)$_4$](PF$_6$)$_2$ and sodium tetrahydroborate to generate [Pt$_2$Na(μ-2,7-(Ph$_2$P)$_2$napy)$_3$(2,6-xylylisocyanide)$_2$](PF$_6$), a cryptate containing an encapsulated sodium ion (04OM6042). The polypyridine ligand is coordinated by its nitrogen heteroatoms to the sodium ion and by its phosphorus atoms to platinum centers. A similar palladium complex was prepared as well. Terminal isocyanide ligands can be readily substituted by the carbonyl ligands under carbon monoxide.

Complex [Pd(bipy)(COOMe)$_2$] (93OM568) reacts with [Os$_3$(CO)$_{10}$ (AN)$_2$] in dichloromethane to give a heteropolynuclear cluster [Os$_6$Pd (CO)$_{18}$(bipy)] (94JCS(D)1605, 96JOM(510)219). The same palladium precursor with [H$_2$Os$_3$(CO)$_{10}$] gives [{(bipy)Pd}$_2$Os$_3$(CO)$_{12}$] (95JOM(489)C78, 96JOM(510)219). With the chloride salt, the cluster is destroyed, and one of the products is dinuclear [{(bipy)Pd}$_2$(μ-H)$_2$(μ-CO)]$^+$.

Oxidative addition of [Co(COMe)(CO)$_4$] on [Pd(dba)$_2$] in the presence of 2,2′-bipyridine or 1,10-phenanthroline (LL) gives [(LL)(COMe)Pd-Co(CO)$_4$], which can also be obtained by metathesis of [Pd(LL)(Me)X] (X = NO$_3$, Cl) with Na[Co(CO)$_4$] followed by carbon monoxide (00OM5251).

4. CONCLUSIONS

1. Polypyridine ligands form a wide variety of organometallic species with representatives of cobalt and nickel group. They include carbonyl, alkene, alkyne, diene, cyclopentadienyl, alkyl, aryl,

monoacetylide, and diacetylide complexes. Different cyclometalated species are formed.

2. Along with the classical polypyridine ligands, various new classes
    are considered, including pyridinophanes and pyridylborate
    ligands. When they play the traditional role of spectator ligands,
    new reactivity patterns arise, especially insertion and oxidative
    addition reactions, dendrimer, assembly, and cluster formation.
    Compounds perform new catalytic functions and are the basis of the
    new chemical materials.

3. Polypyridine ligands themselves in organometallic compounds of
    these metal groups often perform an exotic coordination function,
    including bridging of the 2,2'-bipyridine ligand, roll-over C, N,
    and even C, C-coordination functions. Derivatized polypyridine
    ligands form various cyclometalated structures, including those
    with double and multiple coordination.

LIST OF ABBREVIATIONS

Ac	Acetyl
acac	Acetylacetonate
Alk	Alkyl
AN	Acetonitrile
bipy	2,2'-bipyridine
Bu	Butyl
cod	1,5-cyclooctadiene
COE	Cyclooctene
COT	Cyclooctatriene
Cp	Cyclopentadienyl
Cp*	Pentamethylcyclopentadienyl
Cy	Cyclohexyl
cyclen	1,4,7,10-tetraazacyclododecane
dba	Dibenzylidene acetonate
DMF	Dimethylformamide
DMSO	Dimethylsulfoxide
dppe	Diphenylphosphinoethane
dppm	Diphenylphosphinomethane
dppp	Diphenylphosphinopropane
Et	Ethyl
Fc	Ferrocenyl
hfacac	Hexafluoroacetylacetonate
Me	Methyl
Mes	Mesityl
napy	1,8-naphthyridine

nbd	Norbornadiene
OTf	Triflate
Ph	Phenyl
phen	1,10-phenanthroline
Pr	Propyl
py	Pyridine
pz	Pyrazol-1-yl
terpy	2,2':6',2'-terpyridine
tfb	Tetrafluorobenzobarrelene
THF	Tetrahydrofuran
THT	Tetrahydrothiophene
TMEDA	N,N,N',N'-tetramethyl-1,2-ethanediamine
Tol	Tolyl
Tos	Tosyl

REFERENCES

57JCS4735	J. Chat, and L. M. Venanzi, *J. Chem. Soc.*, 4735 (1957).
65IC161	L. M. Vallarino, *Inorg. Chem.*, **4**, 161 (1965).
65JA4652	A. Yamamoto, K. Morifuji, S. Ikeda, T. Saito, Y. Uchida, and A. Misono, *J. Am. Chem. Soc.*, **87**, 4652 (1965).
66JA5198	T. Saito, Y. Uchida, A. Misono, A. Yamamoto, K. Morifuji, and S. Ikeda, *J. Am. Chem. Soc.*, **88**, 5198 (1966).
66JOM(6)572	T. Saito, Y. Uchida, A. Misono, A. Yamamoto, K. Morifuji, and S. Ikeda, *J. Organomet. Chem.*, **6**, 572 (1966).
67JA5989	A. Yamamoto and S. Ikeda, *J. Am. Chem. Soc.*, **89**, 5989 (1967).
68JCS(A)583	J. Powell and B. L. Shaw, *J. Chem. Soc.*, A, 583 (1968).
68JOM(15)209	F. Guerrieri and G. P. Chiusoli, *J. Organomet. Chem.*, **15**, 209 (1968).
68JPC1853	D. M. Classen and G. A. Crosby, *J. Phys. Chem.*, **48**, 1853 (1968).
69IC298	V. G. Albano, P. L. Bellon, and M. Sansoni, *Inorg. Chem.*, **8**, 298 (1969).
69ICA(R)109	K. Vrieze, H. C. Volger, and P. W. N. M. van Leeuwen, *Inorg. Chim. Acta Rev.*, 109 (1969).
69JA518	J. W. Faller, M. J. Incorvia, and M. E. Thomsen, *J. Am. Chem. Soc.*, **91**, 518 (1969).
70CC54	G. K. N. Reddy and G. H. Susheelanma, *J. Chem. Soc., Chem. Commun.*, 54 (1970).
70JA3034	D. L. Tibbetts and T. L. Brown, *J. Am. Chem. Soc.*, **92**, 3034 (1970).
70JOM(24)C63	M. Uchino, A. Yamamoto, and S. Ikeda, *J. Organomet. Chem.*, **24**, C63 (1970).
71JA3350	T. Yamamoto, A. Yamamoto, and S. Ikeda, *J. Am. Chem. Soc.*, **93**, 3350 (1971).
71JA3360	T. Yamamoto, A. Yamamoto, and S. Ikeda, *J. Am. Chem. Soc.*, **93**, 3360 (1971).
71JCS(A)2334	M. Green, T. A. Kuc, and S. H. Taylor, *J. Chem. Soc.*, A, 2334 (1971).
72IC2749	H. C. Clark and L. E. Manzer, *Inorg. Chem.*, **11**, 2749 (1972).
72JOM(35)389	C. Cosevar, G. Mestroni, and A. Camus, *J. Organomet. Chem.*, **35**, 389 (1972).

72JOM(38)403	D. E. Clegg, J. R. Hall, and G. A. Swile, *J. Organomet. Chem.*, **38**, 403 (1972).
73JOM(59)411	H. C. Clark and L. E. Manzer, *J. Organomet. Chem.*, **59**, 411 (1973).
74JOM(65)119	G. Mestroni, A. Camus, and G. Zassinovich, *J. Organomet. Chem.*, **65**, 119 (1974).
74JOM(67)443	G. K. N. Reddy and B. R. Ramesh, *J. Organomet. Chem.*, **67**, 443 (1974).
74JOM(81)115	R. Uson, J. Fornies, J. Gimeno, P. Espinet, and R. Navarro, *J. Organomet. Chem.*, **81**, 115 (1974).
75INCL359	G. Mestroni, G. Zassinovich, and A. Camus, *Inorg. Nucl. Chem. Lett.*, **11**, 359 (1975).
75JCS(D)133	R. Guillard, K. Harrison, and I. H. Mather, *J. Chem. Soc., Dalton Trans.*, 133 (1975).
75JOM(84)105	N. Chaudhury and R. J. Puddephatt, *J. Organomet. Chem.*, **84**, 105 (1975).
75JOM(91)379	G. Zassinovich, A. Camus, and G. Mestroni, *J. Organomet. Chem.*, **91**, 379 (1975).
75MI1	K. Vrieze, in *"Dynamic Nuclear Magnetic Resonance Spectroscopy"* (L. M. Jackman and F. A. Cotton, eds.), Academic Press, New York (1975).
76CL1217	T. Yamamoto, T. Kohara, and A. Yamamoto, *Chem. Lett.*, 1217 (1976).
76JA4115	P. E. Garrou and R. F. Heck, *J. Am. Chem. Soc.*, **98**, 4115 (1976).
76JCS(D)762	N. Bresciani-Pahor, M. Calligaris, G. Nardin, and P. Delise, *J. Chem. Soc., Dalton Trans.*, 762 (1976).
76JCS(D)2355	R. N. Hazeldine, R. V. Parish, and D. W. Robins, *J. Chem. Soc., Dalton Trans.*, 2355 (1976).
76JOM(84)93	M. Uchino, K. Asagi, A. Yamamoto, and S. Ikeda, *J. Organomet. Chem.*, **84**, 93 (1976).
76JOM(105)365	R. Uson, L. A. Oro, C. Claver, and M. A. Garralda, *J. Organomet. Chem.*, **105**, 365 (1976).
76JOM(112)105	R. Uson, J. Fornies, and F. Martinez, *J. Organomet. Chem.*, **112**, 105 (1976).
76JOM(116)C35	R. Uson, L. A. Oro, J. A. Cuchi, and M. A. Garralda, *J. Organomet. Chem.*, **116**, C35 (1976).
76JOM(117)297	J. Jawad and R. J. Puddephatt, *J. Organomet. Chem.*, **117**, 297 (1976).
76TMC199	J. Kuyper, R. van der Laan, F. Jenneans, and K. Vrieze, *Transit. Met. Chem.*, **1**, 199 (1976).
76ZAAC422	E. Dinjus, I. Gorsky, E. Uhlig, and H. Walther, *Z. Anorg. Allg. Chem.*, **75**, 422 (1976).
77IC2171	J. Kuyper, *Inorg. Chem.*, **16**, 2171 (1977).
77JA8413	C. P. Cheng, B. Plankey, J. V. Rund, and T. L. Brown, *J. Am. Chem. Soc.*, **99**, 8413 (1977).
77JCS(D)1466	J. Jawad and R. J. Puddephatt, *J. Chem. Soc., Dalton Trans.*, **1466** (1977).
77JOM(135)405	P. Binger, M. J. Doyle, J. McMeeking, C. Kruger, and Y. H. Tsay, *J. Organomet. Chem.*, **135**, 405 (1977).
77JOM(140)63	G. Mestroni, G. Zassinovich, and A. Camus, *J. Organomet. Chem.*, **140**, 63 (1977).
78IC77	J. Kuyper, *Inorg. Chem.*, **17**, 77 (1978).
78IC1458	J. Kuyper, *Inorg. Chem.*, **17**, 1458 (1978).

78JOM(148)81	L. A. Oro, E. Pinilla, and M. L. Tenajas, *J. Organomet. Chem.*, **148**, 81 (1978).
78JOM(155)122	G. Pietropaolo, F. Cusmano, E. Rotondo, and A. Spadaro, *J. Organomet. Chem.*, **155**, 122 (1978).
78JOM(157)345	G. Mestroni, R. Spogliarich, A. Camus, F. Martinelli, and G. Zassinovich, *J. Organomet. Chem.*, **157**, 345 (1978).
79CL1513	T. Kohara, S. Komiya, T. Yamamoto, and A. Yamamoto, *Chem. Lett.*, 1513 (1979).
79JA5876	B. Akermark, H. Johansen, B. Ross, and U. Wahlgren, *J. Am. Chem. Soc.*, **101**, 5876 (1979).
79JCS(D)1569	P. N. A. Seth, A. E. Underhill, and D. M. Watkins, *J. Chem. Soc., Dalton Trans.*, 1569 (1979).
79TMC55	R. Uson, L. A. Oro, M. A. Garralda, M. C. Claver, and P. Lahuerta, *Transit. Met. Chem.*, **4**, 55 (1979).
80IC7	W. J. Louw and C. E. Hepner, *Inorg. Chem.*, **19**, 7 (1980).
80JCS(D)1633	P. Diversi, G. Igrosso, A. Lucherini, and S. Murtas, *J. Chem. Soc., Dalton Trans.*, 1633 (1980).
80JOM(197)87	R. Uson, L. A. Oro, R. Sariego, M. Valderrama, and C. Rebullida, *J. Organomet. Chem.*, **197**, 87 (1980).
81BCJ2161	T. Yamamoto, T. Kohara, and A. Yamamoto, *Bull. Chem. Soc. Japan*, **54**, 2161 (1981).
81IC467	N. Chaudhury and R. J. Puddephatt, *Inorg. Chem.*, **20**, 467 (1981).
81ICA(51)241	H. van der Poel, G. van Koten, and K. Vrieze, *Inorg. Chim. Acta*, **51**, 241 (1981).
81NJC543	P. A. Marnot, R. R. Ruppert, and J. P. Sauvage, *Nouv. J. Chim.*, **5**, 543 (1981).
81TMC45	J. V. Heras, E. Pinilla, and L. A. Oro, *Transit. Met. Chem.*, **6**, 45 (1981).
82IC1023	W. A. Fordyce and G. A. Crosby, *Inorg. Chem.*, **21**, 1023 (1982).
82IC1027	W. A. Fordyce, K. H. Pool, and G. A. Crosby, *Inorg. Chem.*, **21**, 1027 (1982).
82JA994	G. R. Newkome, D. K. Kohli, and F. R. Fronczek, *J. Am. Chem. Soc.*, **104**, 994 (1982).
82JOM(224)81	J. Kaiser, J. Sieler, U. Braun, L. Golic, E. Dinjus, and D. Walther, *J. Organomet. Chem.*, **224**, 81 (1982).
82TMC97	G. B. Deacon and I. L. Grayson, *Transit. Met. Chem.*, **7**, 97 (1982).
83CC267	A. Ferguson, M. Parvez, P. K. Monaghan, and R. J. Puddephatt, *J. Chem. Soc., Chem. Commun.*, 267 (1983).
83IC3429	G. Nord, A. C. Hazell, R. G. Hazell, and O. Farver, *Inorg. Chem.*, **22**, 3429 (1983).
83IC4060	P. J. Spellane, R. J. Watts, and C. J. Curtis, *Inorg. Chem.*, **22**, 4060 (1983).
83ICA(69)179	S. Dholakia, R. D. Gillard, and F. L. Wimmer, *Inorg. Chim. Acta*, **69**, 179 (1983).
83JOM(251)C51	H. Hoberg and D. Schafer, *J. Organomet. Chem.*, **251**, C51 (1983).
83OM1698	P. K. Monaghan and R. J. Puddephatt, *Organometallics*, **2**, 1698 (1983).
84IC3425	P. S. Braterman, G. H. Heat, A. J. Mackenzie, B. C. Noble, R. D. Peacock, and L. J. Yellowlees, *Inorg. Chem.*, **23**, 3425 (1984).

| 84JA6647 | S. Sprouse, K. A. King, P. J. Spellane, and R. J. Watts, *J. Am. Chem. Soc.*, **106**, 6647 (1984). |

84JA6647 · S. Sprouse, K. A. King, P. J. Spellane, and R. J. Watts, *J. Am. Chem. Soc.*, **106**, 6647 (1984).

84JA8181 · K. Tatsumi, A. Nakamura, S. Komiya, T. Yamamoto, and A. Yamamoto, *J. Am. Chem. Soc.*, **106**, 8181 (1984).

84JOM(266)203 · H. Hoberg, D. Schafer, G. Burkhart, C. Kruger, and M. J. Romao, *J. Organomet. Chem.*, **266**, 203 (1984).

84OM444 · P. K. Monahan and R. J. Puddephatt, *Organometallics*, **3**, 444 (1984).

84POL799 · E. Delgado-Laita and E. Sanchez-Munoyerro, *Polyhedron*, **3**, 799 (1984).

85CC609 · A. C. Skapski, V. F. Sutcliffe, and G. B. Young, *J. Chem. Soc., Chem. Commun.*, 609 (1985).

85JA1218 · R. H. Hill and R. J. Puddephatt, *J. Am. Chem. Soc.*, **107**, 1218 (1985).

85JOM(282)123 · S. Morton and J. F. Nixon, *J. Organomet. Chem.*, **282**, 123 (1985).

85MI1 · J. K. Stille, in *"The Nature of the Metal – Carbon Bond"* (F. R. Hartley and S. Patai, eds.), Vol. 2, p. 625, Wiley, New York (1985).

85NJC225 · A. A. Bahsoun and R. Ziessel, *Nouv. J. Chim.*, **9**, 225 (1985).

85OM1130 · S. Komiya, Y. Akai, K. Tanaka, T. Yamamoto, and A. Yamamoto, *Organometallics*, **4**, 1130 (1985).

85OM1669 · G. Ferguson, P. K. Monaghan, M. Parvez, and R. J. Puddephatt, *Organometallics*, **4**, 1669 (1985).

85POL325 · L. A. Oro, M. T. Pinillos, and M. P. Jarauta, *Polyhedron*, **4**, 325 (1985).

85TMC288 · V. Garcia, M. A. Garralda, and L. Ibarlucea, *Transit. Met. Chem*, **10**, 288 (1985).

85ZC411 · W. Siedel, *Z. Chem.*, **25**, 411 (1985).

86JOM(307)107 · G. Minghetti, M. A. Cinellu, G. Chelucci, S. Gladiali, F. Demartin, and M. Manassero, *J. Organomet. Chem.*, **307**, 107 (1986).

86OM980 · U. Kolle, B. Fuss, M. Belting, and E. Raabe, *Organometallics*, **5**, 980 (1986).

86ZNK413 · Y. N. Kukushkin, A. V. Iretskii, L. I. Danilina, and R. O. Abdurakhmanov, *Zh. Neorg. Khim.*, **31**, 413 (1986).

87AG(E)568 · U. Kolle and M. Gratzel, *Angew. Chem., Int. Ed. Engl.*, **26**, 568 (1987).

87AG(E)771 · H. Hoberg, Y. Peres, C. Kruger, and Y. Tsay, *Angew. Chem., Int. Ed. Engl.*, **26**, 771 (1987).

87CC246 · N. G. Connelly and G. Garcia, *J. Chem. Soc., Chem. Commun.*, 246 (1987).

87GCI445 · A. De Renzi, G. Morelli, A. Panunzi, and A. Vitagliano, *Gazz. Chim. Ital.*, **117**, 445 (1987).

87JCS(D)1477 · A. J. Canty, N. J. Minchin, B. W. Skelton, and A. H. White, *J. Chem. Soc., Dalton Trans.*, 1477 (1987).

87JPC1047 · Y. Ohsawa, S. Sprouse, K. A. King, M. K. DeArmond, K. W. Hanck, and R. J. Watts, *J. Phys. Chem.*, **91**, 1047 (1987).

87OM517 · V. G. Albano, D. Braga, V. De Felice, A. Panunzi, and A. Vitagliano, *Organometallics*, **6**, 517 (1987).

87OM2548 · M. Crespo and R. J. Puddephatt, *Organometallics*, **6**, 2548 (1987).

88CC16 · R. Ziessel, *J. Chem. Soc., Chem. Commun.*, 16 (1988).

88CC1150 · R. Ruppert, S. Herrmann, and E. Steckhan, *J. Chem. Soc., Chem. Commun.*, 1150 (1988).

88IC4582	C. M. Bolinger, N. Story, B. P. Sullivan, and T. J. Meyer, *Inorg. Chem.*, **27**, 4582 (1988).
88JCS(D)595	P. K. Monaghan and R. J. Puddephatt, *J. Chem. Soc., Dalton Trans.*, 595 (1988).
88JOM(342)399	T. G. Appleton, R. D. Berry, J. D. Hall, and D. W. Neale, *J. Organomet. Chem.*, **342**, 399 (1988).
88JOM(355)525	W. Kascube, K. R. Poerschke, and G. Wilke, *J. Organomet. Chem.*, **355**, 525 (1988).
88OM1363	P. K. Byers, A. J. Canty, M. Crespo, R. J. Puddephatt, and J. D. Scott, *Organometallics*, **7**, 1363 (1988).
88OM1454	K. Aye, D. Colpitts, and R. J. Puddephatt, *Organometallics*, **7**, 1454 (1988).
88OM2046	A. J. Canty, J. L. Hoare, N. W. Davies, and P. R. Traill, *Organometallics*, **7**, 2046 (1988).
88OM2203	C. Amatore and A. Jutand, *Organometallics*, **7**, 2203 (1988).
89AG(E)767	K. T. Aye, G. Ferguson, A. J. Lough, and R. J. Puddephatt, *Angew. Chem., Int. Ed. Engl.*, **28**, 767 (1989).
89CB1869	U. Kolle, B. S. Kang, P. Infelta, P. Compte, and M. Gratzel, *Chem. Ber.*, **122**, 1869 (1989).
89CC1259	S. Cosnier, A. Deronzier, and N. Vlachopoulos, *J. Chem. Soc., Chem. Commun.*, 1259 (1989).
89GCI235	V. Albano, C. Castellari, G. Morelli, and A. Vitagliano, *Gazz. Chim. Ital.*, **119**, 235 (1989).
89IC309	C. A. Craig and R. J. Watts, *Inorg. Chem.*, **28**, 309 (1989).
89IC2097	J. J. Robertson, A. Kadziola, R. A. Krause, and S. Larsen, *Inorg. Chem.*, **28**, 2097 (1989).
89JCS(D)2049	T. Brauns, C. Carriedo, J. S. Cockayne, N. G. Connelly, G. G. Herbosa, and A. G. Orpen, *J. Chem. Soc., Dalton Trans.*, 2049 (1989).
89JOM(363)197	M. T. Youino and R. Ziessel, *J. Organomet. Chem.*, **363**, 197 (1989).
89JOM(363)401	M. A. Cinellu, S. Gladiali, and G. Minghetti, *J. Organomet. Chem.*, **363**, 401 (1989).
89JOM(367)347	L. Garnier, Y. Rollin, and J. Perichon, *J. Organomet. Chem.*, **367**, 347 (1989).
89NJC53	L. Garnier, Y. Rollin, and J. Perichon, *New J. Chem.*, **13**, 53 (1989).
89OM1180	E. Cucciolito, V. De Felice, A. Panunzi, and A. Vitagliano, *Organometallics*, **8**, 1180 (1989).
89OM1518	K. T. Aye, A. J. Canty, M. Crespo, R. J. Puddephatt, J. D. Scott, and A. A. Watson, *Organometallics*, **8**, 1518 (1989).
89OM2513	G. R. Newkome, K. J. Theriot, F. R. Fronzek, and B. Villar, *Organometallics*, **8**, 2513 (1989).
89OM2907	W. de Graaf, J. Boersma, W. J. J. Smeets, A. L. Spek, and G. van Koten, *Organometallics*, **11**, 2907 (1989).
89POL2209	R. Uson, J. Fornies, M. Tomas, J. M. Casas, and C. Fortuno, *Polyhedron*, **8**, 2209 (1989).
90AG(E)388	E. Steckhan, S. Herrmann, R. Ruppert, J. Thommes, and C. Wandrey, *Angew. Chem., Int. Ed. Engl.*, **29**, 388 (1990).
90CC513	E. C. Constable, R. P. G. Henney, T. A. Leese, and D. A. Tocher, *J. Chem. Soc., Chem. Commun.*, 513 (1990).
90CC1616	P. Leconte, F. Metz, A. Mortreux, J. A. Osborn, F. Paul, F. Petit, and A. Pillot, *J. Chem. Soc., Chem. Commun.*, 1616 (1990).

90CCR(97)193	C. A. Craig, F. O. Garces, R. J. Watts, R. Palmans, and A. J. Frank, *Coord. Chem. Rev.*, **97**, 193 (1990).
90IC5137	G. Minghetti, M. A. Cinellu, S. Stoccoro, G. Chelucci, and A. Zacca, *Inorg. Chem.*, **29**, 5137 (1990).
90JA2464	K. T. Aye, L. Gelmini, N. C. Payne, J. J. Vittal, and R. J. Puddephatt, *J. Am. Chem. Soc.*, 2464 (1990).
90JA4587	B. Akermark, S. Hansson, and A. Vitagliano, *J. Am. Chem. Soc.*, **112**, 4587 (1990).
90JCS(D)443	E. C. Constable, R. P. G. Henney, T. A. Leese, and D. A. Tocher, *J. Chem. Soc., Dalton Trans.*, 443 (1990).
90JOM(388)243	C. Castellari, V. De Felice, A. Panunzi, A. Sanchez, and A. Vitagliano, *J. Organomet. Chem.*, **388**, 243 (1990).
90JOM(393)299	P. K. Byers, A. J. Canty, B. W. Skelton, and A. H. White, *J. Organomet. Chem.*, **393**, 299 (1990).
90JOM(406)261	M. Rashidi, Z. Fakhroeian, and R. J. Puddephatt, *J. Organomet. Chem.*, **406**, 261 (1990).
90OM210	P. K. Byers and A. J. Canty, *Organometallics*, **9**, 210 (1990).
90OM826	P. K. Byers, A. J. Canty, B. W. Skelton, and A. J. White, *Organometallics*, **9**, 826 (1990).
90OM1269	V. G. Albano, C. Castellari, M. E. Cucciolito, A. Panunzi, and A. Vitagliano, *Organometallics*, **9**, 1269 (1990).
90OM1826	A. Albinati, C. J. Ammann, P. S. Pregosin, and H. Ruegger, *Organometallics*, **9**, 1826 (1990).
90OM2422	J. Fornies, R. Navarro, V. Sicilia, and M. Tomas, *Organometallics*, **9**, 2422 (1990).
90OM3080	P. K. Byers, A. J. Canty, B. W. Skelton, P. R. Traill, A. A. Watson, and A. H. White, *Organometallics*, **9**, 3080 (1990).
91AG(E)844	R. Ziessel, *Angew. Chem., Int. Ed. Engl.*, **30**, 844 (1991).
91AG(E)989	M. Barsacchi, G. Consiglio, L. Medici, G. Petrucci, and U. W. Suter, *Angew. Chem., Int. Ed. Engl.*, **30**, 989 (1991).
91ICA(184)73	P. Imhoff, R. van Asselt, C. J. Elsevier, and M. C. Zouthberg, *Inorg. Chim. Acta*, **184**, 73 (1991).
91JCS(D)1007	F. P. Panizi, F. P. Intini, L. Maresca, G. Natile, M. Lanfranchi, and A. Tiripicchio, *J. Chem. Soc., Dalton Trans.*, 1007 (1991).
91JEAC125	C. Amatore, A. Jutand, and L. Mottier, *J. Electroanal. Chem.*, **306**, 125 (1991).
91JOM(403)243	V. De Felice, M. Funicello, A. Panunzi, and F. Ruffo, *J. Organomet. Chem.*, **403**, 243 (1991).
91JOM(463)269	V. De Felice, V. G. Albano, C. Castellari, M. E. Cucciolito, and A. De Renzi, *J. Organomet. Chem.*, **403**, 269 (1991).
91JOM(419)233	M. Ladwig and W. Kaim, *J. Organomet. Chem.*, **419**, 233 (1991).
91OM1568	E. Steckhan, S. Herrmann, R. Ruppert, E. Dietz, M. Frede, and E. Spika, *Organometallics*, **10**, 1568 (1991).
91OM1800	A. Albinati, R. W. Kunz, C. J. Ammann, and P. S. Pregosin, *Organometallics*, **10**, 1800 (1991).
91OM2672	C. M. Anderson, M. Crespo, M. C. Jennings, A. J. Lough, and R. J. Puddephatt, *Organometallics*, **10**, 2672 (1991).
91POL1513	S. Chakladar, P. Paul, and K. Nag, *Polyhedron*, **10**, 1513 (1991).
91RKCL185	V. R. Parameswaran and S. Vancheesan, *Reac. Kinet. Catal. Lett.*, **44**, 185 (1991).
92ACR83	A. J. Canty, *Acc. Chem. Res.*, **25**, 83 (1992).

92ACS499   V. De Felice, A. Panunzi, F. Ruffo, and B. Akermark, *Acta Chem. Scand.*, **46**, 499 (1992).

92AG(E)1529   D. Westerhausen, S. Herrmann, W. Hummel, and E. Steckhan, *Angew. Chem., Int. Ed. Engl.*, **31**, 1529 (1992).

92AX(B)515   R. Ziessel, S. Chardon-Noblat, A. Deronzier, D. Matt, L. Toupet, F. Balgroune, and D. Grandjean, *Acta Crystallogr.*, **B49**, 515 (1992).

92CC143   N. G. Connelly, T. Eining, G. G. Herbosa, P. M. Hopkins, C. Mealli, A. G. Orpen, and G. M. Rosair, *J. Chem. Soc., Chem. Commun.*, 143 (1992).

92CC303   E. W. Abel, N. J. Long, K. G. Orrell, A. G. Osborne, H. M. Pain, and V. Sik, *J. Chem. Soc., Chem. Commun.*, 303 (1992).

92CC333   F. P. Fanizzi, L. Maresca, G. Natile, M. Lanfranchi, A. Tiripicchio, and G. Pacchioni, *J. Chem. Soc., Chem. Commun.*, 333 (1992).

92CRV1051   G. Zassinovich, G. Mestroni, and S. Gladiali, *Chem. Rev.*, **92**, 1051 (1992).

92GCI455   G. Minghetti, M. A. Cinellu, S. Stoccoro, and A. Zucca, *Gazz. Chim. Ital.*, **122**, 455 (1992).

92IC222   P. C. Braterman, J. I. Song, C. Vogler, and W. Kaim, *Inorg. Chem.*, **31**, 222 (1992).

92ICA(197)51   S. Bartolucci, P. Carpinelli, V. De Felice, F. Giovannitti, and A. De Renzi, *Inorg. Chim. Acta*, **197**, 51 (1997).

92JA5895   M. Brookhart, F. C. Rix, and J. M. DeSimone, *J. Am. Chem. Soc.*, **114**, 5895 (1992).

92JCS(D)2251   E. C. Constable, R. P. G. Henney, P. R. Raithby, and L. R. Sousa, *J. Chem. Soc., Dalton Trans.*, 2251 (1992).

92JCS(D)2907   N. G. Connelly, P. M. Hopkins, A. G. Orpen, G. M. Rosair, and F. Viguri, *J. Chem. Soc., Dalton Trans.*, 2907 (1992).

92JOM(424)C12   B. A. Markies, M. H. P. Rietveld, J. Boersma, A. L. Spek, and G. van Koten, *J. Organomet. Chem.*, **424**, C12 (1992).

92JOM(425)177   V. G. Albano, C. Castellari, V. De Felice, A. Panunzi, and F. Ruffo, *J. Organomet. Chem.*, **425**, 177 (1992).

92JOM(439)79   M. Ladwig and W. Kaim, *J. Organomet. Chem.*, **439**, 79 (1992).

92JOM(440)207   A. Klein, H. D. Hausen, and W. Kaim, *J. Organomet. Chem.*, **440**, 207 (1992).

92JOM(455)177   V. G. Albano, C. Castellari, V. De Felice, A. Panunzi, and F. Ruffo, *J. Organomet. Chem.*, **455**, 177 (1992).

92OM3665   V. G. Albano, C. Castellari, V. De Felice, M. Monari, A. Panunzi, and F. Ruffo, *Organometallics*, **11**, 3665 (1992).

92OM3669   V. De Felice, A. De Renzi, D. Tesauro, and A. Vitagliano, *Organometallics*, **11**, 3669 (1992).

92OM3920   M. Brookhart, B. Grant, and A. F. Volpe, *Organometallics*, **11**, 3920 (1992).

92OM3954   M. Sjogren, S. Hansson, P. O. Norrby, B. Akermark, M. E. Cucciolito, F. Giordano, and A. Vitagliano, *Organometallics*, **11**, 3954 (1992).

92ZOB821   M. A. Mitryaikina, L. S. Gracheva, V. K. Polovnyak, A. E. Usachev, and Y. V. Yablokov, *J. Gen. Chem. USSR*, **62**, 821 (1992).

93CIC349   L. Maresca and G. Natile, *Comments Inorg. Chem.*, **14**, 349 (1993).

93GCI65   V. De Felice, B. Giovanniti, A. Panunzi, and F. Ruffo, *Gazz. Chim. Ital.*, **123**, 65 (1993).

93IC3081   M. G. Colombo and H. U. Gudel, *Inorg. Chem.*, **32**, 3081 (1993).

93IC3088	M. G. Colombo, A. Hauser, and H. U. Gudel, *Inorg. Chem.*, **32**, 3088 (1993).
93IC3675	J. Fornies, R. Navarro, V. Sicilia, and M. Tomas, *Inorg. Chem.*, **32**, 3675 (1993).
93IS167	P. Diversi, G. Igrosso, and A. Lucherini, *Inorg. Synth.*, **22**, 167 (1993).
93JA118	R. Ziessel, *J. Am. Chem. Soc.*, **115**, 118 (1993).
93JA2075	P. T. Matsanuga, G. L. Hillhouse, and A. L. Rheingold, *J. Am. Chem. Soc.*, **115**, 2075 (1993).
93JA11245	C. W. Chan, T. F. Lai, C. M. Che, and S. M. Peng, *J. Am. Chem. Soc.*, **115**, 11245 (1993).
93JCS(D)291	E. W. Abel, V. S. Dimitrov, N. J. Long, K. G. Orrell, A. G. Osborne, V. Sik, M. B. Hursthouse, and M. A. Mazid, *J. Chem. Soc., Dalton Trans.*, 291 (1993).
93JCS(D)1343	D. Minniti, *J. Chem. Soc., Dalton Trans.*, 1343 (1993).
93JCS(D)1835	K. T. Aye, J. J. Vittal, and R. J. Puddephatt, *J. Chem. Soc., Dalton Trans.*, 1835 (1993).
93JCS(D)1927	V. De Felice, A. De Renzi, F. Giordano, and D. Tesauro, *J. Chem. Soc., Dalton Trans.*, 1927 (1993).
93JCS(D)3421	M. E. Cucciolito, V. De Felice, F. Giordano, A. Panunzi, and F. Ruffo, *J. Chem. Soc., Dalton Trans.*, 3421 (1993).
93JOC5445	J. E. Backvall, K. L. Granberg, P. G. Anderson, R. Gatti, and A. Gogoll, *J. Org. Chem.*, **58**, 5445 (1993).
93JEAC213	S. Chardon-Noblat, S. Cosnier, A. Deronzier, and N. Vlachopoulos, *J. Electroanal. Chem.*, **352**, 213 (1993).
93JOM(450)C15	S. Stoccoro, G. Chelucci, M. A. Cinellu, G. Minghetti, and M. Manassero, *J. Organomet. Chem.*, **450**, C15 (1994).
93JOM(452)257	G. Minghetti, M. I. Pilo, G. Sanna, R. Seeber, S. Stoccoro, and F. Laschi, *J. Organomet. Chem.*, **452**, 257 (1993).
93JOM(463)215	M. Aresta and E. Quaranta, *J. Organomet. Chem.*, **463**, 215 (1993).
93MM911	A. Sen and Z. Jiang, *Macromolecules*, **26**, 911 (1993).
93MRC954	A. Gogoll and H. Greenberg, *Magn. Reson. Chem.*, **31**, 954 (1993).
93OM568	G. D. Smith, B. E. Hanson, J. S. Merola, and F. J. Walles, *Organometallics*, **12**, 568 (1993).
93OM4151	J. Vicente, J. A. Abad, J. Gil-Rubio, P. G. Jones, and E. Bembenek, *Organometallics*, **12**, 4151 (1993).
93OM4592	S. Achar, J. D. Scott, J. J. Vittal, and R. J. Puddephatt, *Organometallics*, **12**, 4592 (1993).
93OM4940	S. Hansson, P. O. Norrby, M. Sjogren, B. Akermark, M. E. Cucciolito, F. Giordano, and A. Vitagliano, *Organometallics*, **12**, 4940 (1993).
93ZAAC1998	R. Reinhardt and W. Kaim, *Z. Anorg. Allg. Chem.*, **691**, 1988 (1993).
94AG(E)497	A. Togni and L. M. Venanzi, *Angew. Chem., Int. Ed. Engl.*, **33**, 497 (1994).
94AG(E)847	S. Achar and R. J. Puddephatt, *Angew. Chem., Int. Ed. Engl.*, **33**, 847 (1994).
94CC1895	S. Achar and R. J. Puddephatt, *J. Chem. Soc., Chem. Commun.*, 1895 (1994).
94CCR(132)87	C. W. Chan, L. K. Cheng, and C. M. Che, *Coord. Chem. Rev.*, **132**, 87 (1994).
94CCR(133)67	V. G. Albano, G. Natile, and A. Panunzi, *Coord. Chem. Rev.*, **133**, 67 (1994).

94GCI117	V. De Felice, M. L. Ferrara, F. Giordano, and F. Ruffo, *Gazz. Chim. Ital.*, **124**, 117 (1994).
94IC3331	F. P. Fanizzi, M. Lanfranchi, G. Natile, and A. Tiripicchio, *Inorg. Chem.*, **33**, 3331 (1994).
94IC4453	W. Kaim, R. Reinhardt, and M. Sieger, *Inorg. Chem.*, **33**, 4453 (1994).
94ICA(219)169	V. De Felice, A. De Renzi, F. Ruffo, and D. Tesauro, *Inorg. Chim. Acta*, **219**, 169 (1994).
94JA3631	A. Gogoll, J. Ornebro, H. Greenberg, and J. E. Backvall, *J. Am. Chem. Soc.*, **116**, 3631 (1994).
94JA3665	P. T. Matsanuga, C. R. Hess, and G. L. Hillhouse, *J. Am. Chem. Soc.*, **116**, 3665 (1994).
94JCS(D)1605	S. Chan and W. T. Wong, *J. Chem. Soc., Dalton Trans.*, 1605 (1994).
94JOM(469)237	V. G. Albano, C. Castellari, M. L. Ferrara, A. Panunzi, and F. Ruffo, *J. Organomet. Chem.*, **469**, 237 (1994).
94JOM(481)195	G. Minghetti, A. Zucca, S. Stoccoro, M. A. Cinellu, M. Manassero, and M. Sansoni, *J. Organomet. Chem.*, **481**, 195 (1995).
94JOM(482)191	B. A. Markies, A. J. Canty, W. De Graaf, J. Boersma, M. D. Janssen, M. P. Hogerheide, W. J. J. Smeets, A. L. Spek, and G. van Koten, *J. Organomet. Chem.*, **482**, 191 (1994).
94OM1559	C. J. Levy, R. J. Puddephatt, and J. J. Vittal, *Organometallics*, **13**, 1559 (1994).
94OM2053	B. A. Markies, A. J. Canty, J. Boersma, and G. van Koten, *Organometallics*, **13**, 2053 (1994).
94OM2412	C. Ducker-Benfer, R. van Eldik, and A. J. Canty, *Organometallics*, **13**, 2412 (1994).
95ACDD43	S. Bartolucci, P. Carpinelli, and F. Ruffo, *Anti Cancer Drug Design*, **10**, 43 (1995).
95CC2115	C. J. Levy and R. J. Puddephatt, *J. Chem. Soc., Chem. Commun.*, 2115 (1995).
95CC2331	M. Ocafrain, M. Devaud, M. Toupet, and J. Perichon, *J. Chem. Soc., Chem. Commun.*, 2331 (1995).
95IC2994	G. Arena, L. M. Scolaro, R. F. Pasternack, and R. Romeo, *Inorg. Chem.*, **34**, 2994 (1995).
95ICA(235)51	R. Uson, J. Fornies, M. Tomas, J. M. Casas, and C. Fortuno, *Inorg. Chim. Acta*, **235**, 51 (1995).
95ICA(239)61	F. Giordano, B. Panunzi, A. Roviello, and F. Ruffo, *Inorg. Chim. Acta*, **239**, 61 (1995).
95JA1137	F. C. Rix and M. Brookhart, *J. Am. Chem. Soc.*, **117**, 1137 (1995).
95JA6414	L. K. Johnson, C. M. Killian, and M. Brookhart, *J. Am. Chem. Soc.*, **117**, 6414 (1995).
95JCS(D)777	G. Minghetti, M. A. Cinellu, S. Stoccoro, A. Zucca, and M. Manassero, *J. Chem. Soc., Dalton Trans.*, 777 (1995).
95JCS(D)3165	E. W. Abel, P. J. Heard, K. G. Orrell, M. B. Hursthouse, and K. M. A. Malik, *J. Chem. Soc., Dalton Trans.*, 3165 (1995).
95JEAC195	P. A. Christensen, A. Hamnett, S. J. Higgins, and J. A. Timney, *J. Electroanal. Chem.*, **395**, 195 (1995).
95JOM(488)C13	V. De Felice, A. De Renzi, A. Panunzi, and D. Tesauro, *J. Organomet. Chem.*, **488**, C13 (1995).
95JOM(489)C78	S. Chan and W. T. Wong, *J. Organomet. Chem.*, **489**, C78 (1995).
95JOM(493)1	V. De Felice, M. E. Cucciolito, A. De Renzi, F. Ruffo, and D. Tesauro, *J. Organomet. Chem.*, **493**, 1 (1995).

95JOM(495)185   J. Fornies, F. Martinez, R. Navarro, and E. P. Urriolabeitia, *J. Organomet. Chem.*, **495**, 185 (1995).

95JOM(503)251   M. E. Cucciolito, F. Giordano, F. Ruffo, and V. De Felice, *J. Organomet. Chem.*, **503**, 251 (1995).

95MI1   A. J. Canty, in *"Comprehensive Organometallic Chemistry"* 2nd edn. (R. J. Puddephatt, ed.), Vol. 9, p. 225, Pergamon Press, New York (1995) .Ch. 5.

95MR9   S. Bartolini, C. Carfagna, and A. Musco, *Macromol. Rapid Commun.*, **16**, 9 (1995).

95OM1030   L. M. Rendina, J. J. Vittal, and R. J. Puddephatt, *Organometallics*, **14**, 1030 (1995).

95OM1176   A. Klein and W. Kaim, *Organometallics*, **14**, 1176 (1995).

95OM2188   L. M. Rendina, J. J. Vittal, and R. J. Puddephatt, *Organometallics*, **14**, 2188 (1995).

95OM4213   V. G. Albano, C. Castellari, M. Monari, V. De Felice, M. L. Ferrara, and F. Ruffo, *Organometallics*, **14**, 4213 (1995).

95OM4966   G. S. Hill, L. M. Rendina, and R. J. Puddephatt, *Organometallics*, **14**, 4966 (1995).

95OM5410   M. E. Cucciolito, A. De Renzi, F. Giordano, and F. Ruffo, *Organometallics*, **14**, 5410 (1995).

95POL175   P. T. Matsanuga, J. C. Mavrapoulos, C. R. Hess, and G. L. Hillhouse, *Polyhedron*, **14**, 175 (1995).

96AX(C)827   G. Bruno, F. Niccolo, R. Scopelliti, and G. Arena, *Acta Crystallogr.*, **C52**, 827 (1996).

96CB319   J. Forstner, R. Kettenbach, R. Goddard, and H. Butenschon, *Chem. Ber.*, **129**, 319 (1996).

96CC2329   E. W. Abel, A. Gelling, K. G. Orrell, A. G. Osborne, and V. Sik, *Chem. Commun.*, 2329 (1996).

96CC1039   H. Q. Liu, T. C. Cheung, and C. M. Che, *Chem. Commun.*, 1039 (1996).

96CC2463   F. Neve, M. Ghedini, and A. Crispini, *Chem. Commun.*, 2463 (1996).

96IC5087   R. Romeo, L. M. Scolaro, N. Nastasi, and G. Arena, *Inorg. Chem.*, **35**, 5087 (1996).

96IC7345   I. Ara, J. M. Casas, J. Fornies, and A. J. Rueda, *Inorg. Chem.*, **35**, 7345 (1996).

96ICA(244)11   R. Ros, R. Bertrani, A. Tassan, D. Braga, F. Grepioni, and E. Tedesco, *Inorg. Chim. Acta*, **244**, 11 (1996).

96JA2436   F. C. Rix, M. Brookhart, and P. S. White, *J. Am. Chem. Soc.*, **118**, 2436 (1996).

96JA4746   F. C. Rix, M. Brookhart, and P. S. White, *J. Am. Chem. Soc.*, **118**, 4746 (1996).

96JA8745   G. S. Hill and R. J. Puddephatt, *J. Am. Chem. Soc.*, **118**, 8745 (1996).

96JCS(D)253   E. Rotondo, G. Giordano, and D. Minniti, *J. Chem. Soc., Dalton Trans.*, 253 (1996).

96JCS(D)1645   T. C. Cheung, K. K. Cheung, S. M. Peng, and C. M. Che, *J. Chem. Soc., Dalton Trans.*, 1645 (1999).

96JCS(D)1809   G. S. Hill, L. M. Rendina, and R. J. Puddephatt, *J. Chem. Soc., Dalton Trans.*, 1809 (1996).

96JCS(D)2503   M. N. Collomb-Dunand-Sauthier, A. Deronzier, J. C. Moutet, and S. Tingry, *J. Chem. Soc., Dalton Trans.*, 2503 (1996).

96JEAC189   C. Caix, S. Chardon-Noblat, A. Deronzier, and R. Ziessel, *J. Electroanal. Chem.*, **403**, 189 (1996).

96JOM(508)109   R. E. Rulke, J. G. P. Delis, A. M. Groot, C. J. Elsevier, P. W. N. M. van Leeuwen, K. Vrieze, K. Goubitz, and H. Schlenk, *J. Organomet. Chem.*, **508**, 109 (1996).

96JOM(510)219   S. Chan, S. M. Lee, Z. Lin, and W. T. Wong, *J. Organomet. Chem.*, **510**, 219 (1996).

96JOM(513)97   V. De Felice, A. De Renzi, M. L. Ferrara, and A. Panunzi, *J. Organomet. Chem.*, **513**, 97 (1996).

96JOM(514)125   J. G. P. Delis, P. W. N. van Leeuwen, K. Vrieze, N. Veldman, A. L. Spek, J. Fraanje, and K. Goubitz, *J. Organomet. Chem.*, **514**, 125 (1996).

96JOM(519)75   M. L. Ferrara, I. Orabona, F. Ruffo, and V. De Felice, *J. Organomet. Chem.*, **519**, 75 (1996).

96JOM(524)195   W. Kaim, R. Reinhardt, E. Waldhoer, and J. Fiedler, *J. Organomet. Chem.*, **524**, 195 (1996).

96NJC659   M. Durandetti, M. Devaud, and J. Perichon, *New J. Chem.*, **20**, 659 (1996).

96OM24   J. Vicente, J. A. Abad, R. F. de Bobadilla, P. G. Jones, and M. C. R. de Arellano, *Organometallics*, **15**, 24 (1996).

96OM43   S. Achar, J. J. Vittal, and R. J. Puddephatt, *Organometallics*, **15**, 43 (1996).

96OM668   R. E. Rulke, V. E. Kaasjager, D. Kliphuis, C. J. Elsevier, P. W. N. M. van Leeuwen, and K. Goubitz, *Organometallics*, **15**, 668 (1996).

96OM1750   L. M. Rendina, J. J. Vittal, and R. J. Puddephatt, *Organometallics*, **15**, 1750 (1996).

96OM2108   C. J. Levy, J. J. Vittal, and R. J. Puddephatt, *Organometallics*, **15**, 2108 (1996).

96OM4012   V. G. Albano, C. Castellari, M. Monari, V. De Felice, A. Panunzi, and F. Ruffo, *Organometallics*, **15**, 4012 (1996).

96OM5442   S. Ramdeehul, L. Barloy, J. A. Osborn, A. De Cian, and J. Fischer, *Organometallics*, **15**, 5442 (1996).

96POL1823   S. A. Moya, R. Pastene, R. Sariego, R. Sartori, and H. Le Bozec, *Polyhedron*, **15**, 1823 (1996).

96POL3203   A. Gelling, K. G. Orrell, A. G. Osborne, V. Sik, M. B. Hursthouse, and S. J. Coles, *Polyhedron*, **15**, 3203 (1996).

96TMC305   H. Pasternak and F. P. Pruchnik, *Transit. Met. Chem.*, **21**, 305 (1996).

97CC1451   L. M. Scolaro, A. Romeo, and A. Terracina, *Chem. Commun.*, 1451 (1997).

97CC1723   U. Siemeling, U. Vorfeld, B. Neumann, and H. G. Stammler, *Chem. Commun.*, 1723 (1997).

97CRV1735   R. J. Puddephatt and L. M. Rendina, *Chem. Rev.*, **97**, 1735 (1997).

97JA10127   C. J. Levy and R. J. Puddephatt, *J. Am. Chem. Rev.*, **119**, 10127 (1997).

97GCI119   J. A. Abad, *Gazz. Chim. Ital.*, **127**, 119 (1997).

97IC5682   Y. Murakami and T. Yamamoto, *Inorg. Chem.*, **36**, 5682 (1997).

97IC6150   F. Neve, A. Crispini, and S. Campagna, *Inorg. Chem.*, **36**, 6150 (1997).

97ICA(255)351   R. Dhillon, A. Elduque, L. A. Oro, and M. T. Pinillos, *Inorg. Chim. Acta*, **255**, 351 (1997).

97ICA(264)231   F. Giordano, F. Ruffo, A. Saporito, and A. Panunzi, *Inorg. Chim. Acta*, **264**, 231 (1997).

97ICA(265)35   V. G. Albano, M. Monari, I. Orabona, F. Ruffo, and A. Vitagliano, *Inorg. Chim. Acta*, **265**, 35 (1997).

97JA906   A. M. LaPointe, F. C. Rix, and M. Brookhart, *J. Am. Chem. Soc.*, **119**, 906 (1997).

97JCS(D)3777   R. Ziessel, L. Toupet, S. Chardon-Noblat, A. Deronzier, and D. Matt, *J. Chem. Soc., Dalton Trans.*, 3777 (1997).

97JCS(D)4705   B. Neumann, U. Siemeling, H. G. Stammler, U. Vorfeld, J. G. P. Delis, P. W. N. M. van Leeuwen, K. Vrieze, J. Fraanje, K. Goubitz, F. F. de Biani, and P. Zanello, *J. Chem. Soc., Dalton Trans.*, 4705 (1997).

97JOM(532)31   S. Hamar-Thibault, J. C. Moutet, and S. Tingry, *J. Organomet. Chem.*, **532**, 31 (1997).

97JOM(532)235   D. Kruis, B. A. Markies, A. J. Canty, J. Boersma, and G. van Koten, *J. Organomet. Chem.*, **532**, 235 (1997).

97JOM(532)267   M. Abla and T. Yamamoto, *J. Organomet. Chem.*, **532**, 267 (1997).

97JOM(535)209   T. Yamamoto and M. Abla, *J. Organomet. Chem.*, **535**, 209 (1997).

97JOM(543)233   S. L. James, M. Younus, P. R. Raithby, and J. Lewis, *J. Organomet. Chem.*, **543**, 233 (1997).

97JOM(545)89   A. S. Santi, B. Milani, E. Mestroni, G. Zangrando, and L. Randaccio, *J. Organomet. Chem.*, **545**, 89 (1997).

97OM2948   J. G. P. Delis, P. G. Aubel, K. Vrieze, P. W. N. M. van Leeuwen, N. Veldmen, A. L. Spek, and F. J. R. van Neer, *Organometallics*, **16**, 2948 (1997).

97OM5981   M. Fusto, F. Giordano, I. Orabona, F. Ruffo, and A. Panunzi, *Organometallics*, **16**, 5981 (1997).

97SCI2100   K. Toellner, R. Popovitz-Biro, M. Lahav, and D. Milstein, *Science*, **278**, 2100 (1997).

97TL17089   E. Dolhem, M. Ocafrain, J. Y. Nedelec, and M. Toupet, *Tetrahedron Lett.*, **53**, 17089 (1997).

98CC1127   L. Z. Wu, T. C. Cheung, and C. M. Che, *Chem. Commun.*, 1127 (1998).

98CC2295   M. C. Tse, K. K. Cheung, M. K. W. Chan, and C. M. Che, *Chem. Commun.*, 2295 (1998).

98IC2763   G. Arena, G. Calogero, S. Campagna, L. M. Scolaro, V. Ricevuto, and R. Romeo, *Inorg. Chem.*, **37**, 2763 (1998).

98IC3975   A. J. Canty, H. Jin, B. W. Skelton, and A. H. White, *Inorg. Chem.*, **37**, 3975 (1998).

98ICA(275)242   M. Casamento, G. Arena, C. L. Passo, I. Pernice, A. Romeo, and L. M. Scolaro, *Inorg. Chim. Acta*, **275–276**, 242 (1998).

98ICA(275)500   F. P. Fanizzi, G. Natile, M. Lanfranchi, A. Tiripicchio, and G. Pacchioni, *Inorg. Chim. Acta*, **275–276**, 500 (1998).

98ICA(281)141   B. Panunzi, A. Roviello, F. Ruffo, and A. Vivo, *Inorg. Chim. Acta*, **281**, 141 (1998).

98ICC61   M. Akita and Y. Sasaki, *Inorg. Chem. Commun.*, **1**, 61 (1998).

98IS153   R. Romeo and L. M. Scolaro, *Inorg. Synth.*, **32**, 153 (1998).

98IS158   A. De Renzi, A. Panunzi, and F. Ruffo, *Inorg. Synth.*, **32**, 158 (1998).

98IS162   P. K. Byers, A. J. Canty, H. Jin, D. Kruis, B. A. Markies, J. Boersma, and G. van Koten, *Inorg. Synth.*, **32**, 162 (1998).

98JA3805	R. A. Widenhoefer and M. A. DeCarli, *J. Am. Chem. Soc.*, **120**, 3805 (1998).
98JCS(D)113	J. H. Groen, P. W. N. M. van Leeuwen, and K. Vrieze, *J. Chem. Soc., Dalton Trans.*, 113 (1998).
98JOM(560)103	M. Ocafrain, M. Devaud, J. Y. Nedelec, and M. Toupet, *J. Organomet. Chem.*, **560**, 103 (1998).
98JOM(566)37	R. Santi, A. M. Romano, R. Garrone, and R. Millini, *J. Organomet. Chem.*, **566**, 37 (1998).
98JOM(568)53	M. Rashidi, A. R. Esmaeilbeig, N. Shahabadi, S. Tangestaninejad, and R. J. Puddephatt, *J. Organomet. Chem.*, **568**, 53 (1998).
98JOM(571)37	M. Ocafrain, M. Devaud, J. Y. Nedelec, and M. Toupet, *J. Organomet. Chem.*, **571**, 37 (1998).
98LC673	F. Neve, M. Ghedini, O. Francescangeli, and S. Campagna, *Liq. Cryst.*, **24**, 673 (1998).
98OM1052	N. Masciocchi, F. Ragaini, S. Cenini, and A. Seroni, *Organometallics*, **17**, 1052 (1998).
98OM1530	A. M. La Pointe and M. Brookhart, *Organometallics*, **17**, 1530 (1998).
98OM1812	R. van Belzen, R. A. Klein, H. Koojman, N. Veldman, A. L. Spek, and C. J. Elsevier, *Organometallics*, **17**, 1812 (1998).
98OM2046	A. J. Canty, J. L. Hoare, N. W. Davies, and P. R. Traill, *Organometallics*, **17**, 2046 (1998).
98OM2199	F. Paul, J. Fischer, P. Ochsenbein, and J. Osborn, *Organometallics*, **17**, 2199 (1998).
98OM4400	C. Geyer and S. Schindler, *Organometallics*, **17**, 4400 (1998).
98OM4530	V. V. Rostovtsev, J. A. Labinger, J. E. Bercaw, T. L. Lasseter, and K. I. Goldberg, *Organometallics*, **17**, 4530 (1998).
98POL299	H. Aneetha, P. S. Zacharias, B. Srinivas, G. H. Lee, and Y. Wang, *Polyhedron*, **18**, 299 (1998).
99CEJ2845	K. H. Wong, M. C. W. Chan, and C. M. Che, *Chem. Eur. J.*, **5**, 2845 (1999).
99CHC992	G. Minghetti, A. Doppiu, A. Zucca, S. Stoccoro, M. A. Cinellu, M. Manassero, and M. Sansoni, *Chem. Heterocycl. Comp.*, **35**(8), 992 (1999).
99CL419	T. Yamamoto, T. Murakami, and M. Abla, *Chem. Lett.*, 419 (1999).
99EJI27	J. G. Donkervoort, M. Buhl, J. M. Ernsting, and C. J. Elsevier, *Eur. J. Inorg. Chem.*, 27 (1999).
99EJI2085	B. Milani, G. Corso, E. Zangrando, L. Randaccio, and G. Mestroni, *Eur. J. Inorg. Chem.*, 2085 (1999).
99IC2123	M. C. Janzen, H. A. Jenkins, L. M. Rendina, J. J. Vittal, and R. J. Puddephatt, *Inorg. Chem.*, **38**, 2123 (1999).
99IC2250	F. Neve, A. Crispini, S. Campagna, and S. Serroni, *Inorg. Chem.*, **38**, 2250 (1999).
99IC3264	W. B. Connick, D. Geiger, and R. Eisenberg, *Inorg. Chem.*, **38**, 3264 (1999).
99IC4046	S. W. Lai, M. C. W. Chan, T. C. Cheung, S. M. Peng, and C. M. Che, *Inorg. Chem.*, **38**, 4046 (1999).
99IC4262	S. W. Lai, M. C. W. Chan, K. K. Cheung, and C. M. Che, *Inorg. Chem.*, **38**, 4262 (1999).
99ICA(285)70	V. G. Albano, M. L. Ferrara, M. Monari, A. Panunzi, and F. Ruffo, *Inorg. Chim. Acta*, **285**, 70 (1999).

99JA7405	T. Yamaguchi, F. Yamazaki, and T. Ito, *J. Am. Chem. Soc.*, **121**, 7405 (1999).
99JCS(D)1027	A. Yamamoto, *J. Chem. Soc., Dalton Trans.*, 1027 (1999).
99JOC8681	R. A. Widenhoefer and C. N. Stengone, *J. Org. Chem.*, **64**, 8681 (1999).
99JOM(573)3	J. H. Groen, B. J. de Jong, J. M. Ernsting, P. W. N. M. van Leeuwen, K. Vrieze, J. J. Smeets, and A. L. Spek, *J. Organomet. Chem.*, **573**, 3 (1999).
99JOM(574)286	M. Rashidi and B. Z. Momeni, *J. Organomet. Chem.*, **574**, 286 (1999).
99JOM(575)214	P. Burger and J. M. Baumeister, *J. Organomet. Chem.*, **575**, 214 (1999).
99JOM(578)178	C. J. Adams and P. R. Raithby, *J. Organomet. Chem.*, **578**, 178 (1999).
99JOM(586)190	E. Gallo, F. Ragaini, S. Cenini, and F. Demartin, *J. Organomet. Chem.*, **586**, 190 (1999).
99OM2660	A. J. Canty, J. L. Hoare, J. Patel, M. Pfeffer, B. W. Skelton, and A. H. White, *Organometallics*, **18**, 2660 (1999).
99OM3061	A. Macchioni, G. Bellachioma, G. Cardaci, M. Travaglia, C. Zuccaccia, B. Milani, G. Corso, E. Zangrando, G. Mestroni, C. Carfagna, and M. Formica, *Organometallics*, **18**, 3061 (1999).
99OM3327	S. W. Lai, C. W. Chan, K. K. Cheung, and C. M. Che, *Organometallics*, **18**, 3327 (1999).
99OM3337	D. J. Cardenas, A. M. Echavarren, and R. M. C. de Arellano, *Organometallics*, **18**, 3337 (1999).
99OM3482	M. E. Cucciolito, A. Panunzi, F. Ruffo, V. G. Albano, and M. Monari, *Organometallics*, **18**, 3482 (1999).
99POL1285	A. Gelling, K. G. Orrell, A. G. Osborne, and V. Sik, *Polyhedron*, **18**, 1285 (1999).
99TL1451	C. N. Stengone and R. A. Widenfhoefer, *Tetrahedron Lett.*, **40**, 1451 (1999).
99TL8499	R. A. Widenhoefer and A. Vadehra, *Tetrahedron Lett.*, **40**, 8499 (1999).
99ZK463	A. Felix, A. R. Guadalupe, S. D. Huang, and Z. Krist, *New Cryst. Struct.*, **214**, 463 (1999).
00CCR(208)115	M. Hissler, J. E. McGarrah, W. B. Connick, D. K. Geiger, S. D. Cummings, and R. Eisenberg, *Coord. Chem. Rev.*, **208**, 115 (2000).
00CRV1169	S. D. Ittel, L. K. Johnson, and M. Brookhart, *Chem. Rev.*, **100**, 1169 (2000).
00CRV1391	E. Y. X. Chen and T. J. Marks, *Chem. Rev.*, **100**, 1391 (2000).
00CRV1527	H. Butenschon, *Chem. Rev.*, **100**, 1527 (2000).
00EJI1039	F. Neve and A. Crispini, *Eur. J. Inorg. Chem.*, 1039 (2000).
00EJI1717	M. Bigioni, P. Ganis, A. Panunzi, F. Ruffo, C. Salvatore, and A. Vito, *Eur. J. Inorg. Chem.*, 1717 (2000).
00EJI2555	A. Doppiu, M. A. Cinellu, G. Minghetti, S. Stoccoro, A. Zucca, M. Manassero, and M. Sansoni, *Eur. J. Inorg. Chem.*, 2555 (2000).
00IC255	S. W. Lai, T. C. Cheung, M. C. W. Chan, K. K. Cheung, S. M. Peng, and C. M. Che, *Inorg. Chem.*, **39**, 255 (2000).
00IC447	M. Hissler, W. B. Connick, D. K. Geiger, J. E. McGarrah, D. Lipa, R. Lachicotte, and R. Eisenberg, *Inorg. Chem.*, **39**, 447 (2000).

00IC3537	J. H. K. Yip, Suwarno, and J. J. Vittal, *Inorg. Chem.*, **39**, 3537 (2000).
00IC4749	R. Romeo, M. R. Plutino, L. M. Scolaro, S. Stoccoro, and G. Minghetti, *Inorg. Chem.*, **39**, 4749 (2000).
00ICA(305)189	G. Sanna, G. Minghetti, A. Zucca, M. I. Pilo, R. Seeber, and F. Laschi, *Inorg. Chim. Acta*, **305**, 189 (2000).
00ICC575	A. Bayler, A. J. Canty, J. H. Ryan, B. W. Skelton, and A. H. White, *Inorg. Chem. Commun.*, **3**, 575 (2000).
00ICC620	J. C. Moutet and L. Y. Cho, *Inorg. Chem. Commun.*, **3**, 620 (2000).
00JCS(D)63	C. J. Adams, S. L. James, X. Liu, P. R. Raithby, and L. J. Yellowlees, *J. Chem. Soc., Dalton Trans.*, 63 (2000).
00JCS(D)3055	B. Milani, F. Paronetto, and E. Zangrando, *J. Chem. Soc., Dalton Trans.*, 3055 (2000).
00JOM(593)403	R. Romeo, L. M. Scolaro, M. R. Plutino, and A. Albinati, *J. Organomet. Chem.*, **593–594**, 403 (2000).
00JOM(593)445	V. De Felice, B. Giovanitti, A. De Renzi, D. Tesauro, and A. Panunzi, *J. Organomet. Chem.*, **593–594**, 445 (2000).
00JOM(595)296	A. Bayler, A. J. Canty, B. W. Skelton, and A. H. White, *J. Organomet. Chem.*, **595**, 296 (2000).
00JOM(600)37	U. Belluco, R. Bertrani, R. A. Michelin, and M. Mozzon, *J. Organomet. Chem.*, **600**, 37 (2000).
00OM622	H. Yang, H. Gao, and R. J. Angelici, *Organometallics*, **19**, 622 (2000).
00OM752	J. Vicente, J. A. Abad, E. Martinez-Viviente, M. C. R. de Arellano, and P. G. Jones, *Organometallics*, **19**, 752 (2000).
00OM1247	C. P. Lenges, P. S. White, W. J. Marshall, and M. Brookhart, *Organometallics*, **19**, 1247 (2000).
00OM2125	T. Yagyu, K. Osakada, and M. Brookhart, *Organometallics*, **19**, 2125 (2000).
00OM4295	A. Zucca, M. A. Cinellu, M. V. Pinna, S. Stoccoro, G. Minghetti, M. Manassero, and M. Sansoni, *Organometallics*, **19**, 4295 (2000).
00OM5251	N. Komine, H. Hoh, M. Hirano, and S. Komiya, *Organometallics*, **19**, 5251 (2000).
01AG(E)3750	M. A. Albrecht and G. van Koten, *Angew. Chem., Int. Ed. Engl.*, **40**, 3750 (2001).
01CC789	V. W. W. Yam, *Chem. Commun.*, 789 (2001).
01CC1200	M. Stradiotto, K. L. Fujdala, and T. D. Tilley, *Chem. Commun.*, 1200 (2001).
01CCC207	S. Chardon-Noblat, A. Deronzier, and R. Ziessel, *Collect. Czech. Chem. Commun.*, **66**, 207 (2001).
01CEJ4180	S. C. Chan, M. C. W. Chan, C. M. Che, Y. Wang, K. K. Cheung, and N. Y. Zhu, *Chem. Eur. J.*, **7**, 4180 (2001).
01EJI69	J. Romero, M. N. Collomb, A. Deronzier, A. Llobet, E. Perret, J. Pecaut, L. Le Pape, and J. M. Latour, *Eur. J. Inorg. Chem.*, 69 (2001).
01EJI77	J. J. Eisch, X. Ma, K. I. Han, J. N. Gitua, and C. Kruger, *Eur. J. Inorg. Chem.*, 77 (2001).
01EJI613	S. Chardon-Noblat, A. Deronzier, F. Hartl, J. van Slageren, and T. Mahabiersing, *Eur. J. Inorg. Chem.*, 613 (2001).
01IC1093	F. Neve, A. Crispini, S. Serroni, F. Loiseau, and S. Campagna, *Inorg. Chem.*, **40**, 1093 (2001).

01IC1728	M. C. Janzen, M. C. Jennings, and R. J. Puddephatt, *Inorg. Chem.*, **40**, 1728 (2001).
01IC3293	R. Romeo, L. Fenech, L. M. Scolaro, A. Albinati, A. Macchioni, and C. Zuccaccia, *Inorg. Chem.*, **40**, 3293 (2001).
01IC4053	C. E. Whittle, J. A. Weinstein, M. W. Grorge, and K. S. Schanze, *Inorg. Chem.*, **40**, 4053 (2001).
01IC4150	M. Rodriguez, I. Romero, A. Llobet, A. Deronzier, M. Biner, T. Parella, and H. Stoeckli-Evans, *Inorg. Chem.*, **40**, 4150 (2001).
01IC4510	J. E. McGarrah, Y. J. Kim, M. Hissler, and R. Eisenberg, *Inorg. Chem.*, **40**, 4510 (2001).
01ICC648	A. J. Canty, M. C. Done, B. W. Skelton, and A. H. White, *Inorg. Chem. Commun.*, **4**, 648 (2001).
01JA4352	V. G. Albano, M. Monari, I. Orabona, A. Panunzi, and F. Ruffo, *J. Am. Chem. Soc.*, **123**, 4352 (2001).
01JCS(D)3430	M. Rashidi, M. Nabavizadeh, R. Hakimelahi, and S. Jamali, *J. Chem. Soc., Dalton Trans.*, 3430 (2001).
01JOC1755	X. Wang, H. Chakrapani, C. N. Stengone, and R. A. Widenhoefer, *J. Org. Chem.*, **66**, 1755 (2001).
01JOM(626)118	J. van Slageren, A. L. Verner, D. J. Stufkens, M. Lutz, and A. L. Spek, *J. Organomet. Chem.*, **626**, 118 (2001).
01JOM(630)185	Y. H. Budnikova, J. Perichon, D. G. Yakharov, Y. M. Kargin, and O. G. Sinyashin, *J. Organomet. Chem.*, **630**, 185 (2001).
01JPC(B)4801	H. Laguitton-Pasquier, A. Martre, and A. Deronzier, *J. Phys. Chem.*, **B105**, 4801 (2001).
01OM1087	T. Yagyu, M. Hamada, K. Osakada, and T. Yamamoto, *Organometallics*, **20**, 1087 (2001).
01OM1148	A. Doppiu, G. Minghetti, M. A. Cinellu, S. Stoccoro, A. Zucca, and M. Manassero, *Organometallics*, **20**, 1148 (2001).
01OM1668	S. Chardon-Noblat, G. H. Cripps, A. Deronzier, J. S. Field, S. Gouws, R. Haines, and F. Southway, *Organometallics*, **20**, 1668 (2001).
01OM1713	H. Weiss, F. Hampel, W. Donaubauer, M. A. Grundl, J. W. Bats, A. S. K. Hashmi, and S. Schnidler, *Organometallics*, **20**, 1713 (2001).
01OM2704	J. Vicente, J. A. Abad, W. Fortsch, P. G. Jones, and A. K. Fischer, *Organometallics*, **20**, 2704 (2001).
01OM3373	D. P. Gallasch, E. R. T. Tiekink, and L. M. Rendina, *Organometallics*, **20**, 3373 (2001).
01OM4111	S. Stoccoro, G. Minghetti, M. A. Cinellu, A. Zucca, and M. Manassero, *Organometallics*, **20**, 4111 (2001).
01OM4476	V. W. W. Yam, R. P. L. Tang, K. M. C. Wong, and K. K. Cheung, *Organometallics*, **20**, 4476 (2001).
01OM4683	D. Song, Q. Wu, A. Hook, I. Korin, and S. Wang, *Organometallics*, **20**, 4683 (2001).
01OM4695	J. Vicente, M. Lyakhovych, D. Bautista, and P. G. Jones, *Organometallics*, **20**, 4695 (2001).
01OM4903	S. Ogo, N. Makihara, Y. Kaneko, and Y. Watanabe, *Organometallics*, **20**, 4903 (2001).
01OM5251	N. S. Perch and R. A. Widenhoefer, *Organometallics*, **20**, 5251 (2001).
01ZAAC645	A. Klein, *Z. Anorg. Allg. Chem.*, **627**, 645 (2001).
01ZAAC1146	T. Marx, L. Wessemann, and S. Dehnen, *Z. Anorg. Allg. Chem.*, **627**, 1146 (2001).

02ACR905	R. A. Wiedenhoefer, *Acc. Chem. Res.*, **35**, 905 (2002).
02AG(E)544	S. Mecking, A. Held, and F. M. Bauers, *Angew. Chem., Int. Ed. Engl.*, **41**, 544 (2002).
02AG(E)3056	T. Ishiyama, J. Takagi, J. F. Hartwig, and N. Miyara, *Angew. Chem., Int. Ed. Engl.*, **41**, 3056 (2002).
02AG(E)4102	A. N. Vedernikov and K. G. Caulton, *Angew. Chem., Int. Ed. Engl.*, **41**, 4102 (2002).
02BCJ1997	T. Yamamoto, M. Abla, and Y. Murakami, *Bull. Chem. Soc. Japan*, **75**, 1997 (2002).
02CC206	W. Lu, B. X. Mi, M. C. W. Chan, Z. Hui, N. Zhu, S. T. Lee, and C. M. Che, *Chem. Commun.*, 206 (2002).
02CC900	W. Lu, N. Zhu, and C. M. Che, *Chem. Commun.*, 900 (2002).
02CCR(230)193	J. van Slageren, A. Klein, and S. Zalis, *Coord. Chem. Rev.*, **230**, 193 (2002).
02CEJ372	Y. Yamamoto, H. Suzuzki, N. Tajima, and K. Tatsumi, *Chem. Eur. J.*, **8**, 372 (2002).
02CJC41	M. C. Janzen, M. C. Jennings, and R. J. Puddephatt, *Can. J. Chem.*, **80**, 41 (2002).
02EJI1827	R. Dorta, L. J. W. Shimon, H. Rozenberg, and D. Milstein, *Eur. J. Inorg. Chem.*, 1827 (2002).
02EJI3336	A. Zucca, S. Stoccoro, M. A. Cinellu, G. Minghetti, M. Manassero, and M. Sansoni, *Eur. J. Inorg. Chem.*, 3336 (2002).
02EJO1685	C. Maillet, P. Janvier, M. J. Bertrand, T. Praveen, and B. Bujoli, *Eur. J. Org. Chem.*, 1685 (2002).
02IC5653	Q. Z. Yang, L. Z. Wu, Z. X. Wu, L. P. Zhang, and C. H. Tung, *Inorg. Chem.*, **41**, 5653 (2002).
02IC6178	C. K. Hui, B. W. K. Chu, N. Zhu, and V. W. W. Yam, *Inorg. Chem.*, **41**, 6178 (2002).
02IC6521	H. Hadadzadeh, M. C. DeRosa, G. P. A. Yap, A. R. Rezvani, and R. J. Crutchley, *Inorg. Chem.*, **41**, 6521 (2001).
02ICA(327)188	B. Milani, A. Marson, E. Zangrando, G. Mestroni, J. M. Ernsting, and C. J. Elsevier, *Inorg. Chim. Acta*, **327**, 188 (2002).
02ICA(334)149	D. Gelman, S. Dechert, H. Schumann, and J. Blum, *Inorg. Chim. Acta*, **334**, 149 (2002).
02ICA(330)128	C. Dragonetti, M. Pizotti, D. Roberto, and S. Galli, *Inorg. Chim. Acta*, **330**, 128 (2002).
02JA390	T. Ishiyama, J. Takagi, K. Ishida, N. Miyaura, N. R. Anastasi, and J. F. Hartwig, *J. Am. Chem. Soc.*, **124**, 390 (2002).
02JA6506	V. W. W. Yam, K. M. C. Wong, and N. Zhu, *J. Am. Chem. Soc.*, **124**, 6506 (2002).
02JCS(D)2371	A. Klein, E. J. L. McInnes, and W. Kaim, *J. Chem. Soc., Dalton Trans.*, 2371 (2002).
02OM226	S. W. Lai, H. W. Lam, W. Lu, K. K. Cheung, and C. M. Che, *Organometallics*, **21**, 226 (2002).
02OM783	A. Zucca, A. Doppiu, M. A. Cinellu, S. Stoccoro, G. Minghetti, and M. Manassero, *Organometallics*, **21**, 783 (2002).
02OM2088	T. Yagyu, Y. Suzaki, and K. Osakada, *Organometallics*, **21**, 2088 (2002).
02OM3278	A. M. Levine, R. A. Stockland, R. Clark, and I. Guzei, *Organometallics*, **21**, 3278 (2002).
02OM3503	A. Panunzi, G. Roviello, and F. Ruffo, *Organometallics*, **21**, 3503 (2002).

02OM3511	F. Neve, A. Crispini, C. Di Pietro, and S. Campagna, *Organometallics*, **21**, 3511 (2002).
02OM4978	D. Song, K. Silowski, J. Pang, and S. Wang, *Organometallics*, **21**, 4978 (2002).
02OM5830	M. Pizzotti, R. Ugo, D. Roberto, S. Bruni, P. Fantucci, and C. Rovizzi, *Organometallics*, **21**, 5830 (2002).
02RCB796	D. G. Yakharov, E. G. Samieva, D. I. Tazeev, and Y. G. Budnikova, *Russ. Chem. Bull.*, **51**, 796 (2002).
02RJGC168	Y. G. Budnikova, D. G. Yakharov, V. I. Morozov, Y. M. Kargin, A. V. Ilyasov, Y. N. Vyakhireva, and O. G. Sinyashin, *Russ. J. Gen. Chem.*, **72**, 168 (2002).
02ZAAC20	D. Walther, K. Heubach, L. Bottcher, H. Schreer, and H. Gorls, *Z. Anorg. Allg. Chem.*, **628**, 20 (2002).
03AG(E)1400	V. V. W. Yam, K. M. C. Wong, and N. Zhu, *Angew. Chem., Int. Ed. Engl.*, **42**, 1400 (2003).
03APL850	W. Sun, Z. X. Wu, Q. Z. Yang, L. Z. Wu, and C. H. Tung, *Appl. Phys. Lett.*, **82**, 850 (2003).
03CC358	A. N. Vedernikov, J. C. Huffman, and K. G. Caulton, *Chem. Commun.*, 358 (2003).
03CEJ475	K. K. W. Lo, C. K. Chung, and N. Zhu, *Chem. Eur. J.*, **9**, 475 (2003).
03CEJ6155	W. Lu, M. C. W. Chan, N. Zhu, C. M. Che, Z. He, and K. Y. Wong, *Chem. Eur. J.*, **9**, 6155 (2003).
03EJI839	M. P. Feth, A. Klein, and H. Bertagnolli, *Eur. J. Inorg. Chem.*, 839 (2003).
03EJI1917	A. Klein, J. van Slageren, and S. Zalis, *Eur. J. Inorg. Chem.*, 1917 (2003).
03IC686	K. K. W. Lo, C. K. Chung, T. K. M. Lee, L. H. Lui, K. H. K. Tsang, and N. Zhu, *Inorg. Chem.*, **42**, 686 (2003).
03IC1394	I. E. Pomestchenko, C. R. Luman, M. Hissler, R. Ziessel, and F. N. Castellano, *Inorg. Chem.*, **42**, 1394 (2003).
03IC3160	R. Dorta, L. J. W. Shimon, H. Rozenberg, Y. Ben-David, and D. Milstein, *Inorg. Chem.*, **42**, 3160 (2003).
03IC3650	J. Ruiz, M. T. Martinez, F. Florenciano, V. Rodriguez, G. Lopez, J. Perez, P. A. Chaloner, and P. B. Hitchcock, *Inorg. Chem.*, **42**, 3650 (2003).
03IC3772	T. J. Wadas, R. J. Lachicotte, and R. Eisenberg, *Inorg. Chem.*, **42**, 3772 (2003).
03IC4355	J. E. McGarrah and R. Eisenberg, *Inorg. Chem.*, **42**, 4355 (2003).
03IC5185	S. Zalis, M. Sieger, S. Greulich, H. Stoll, and W. Kaim, *Inorg. Chem.*, **42**, 5185 (2003).
03IC6528	A. Hofmann, L. Dahlenburg, and R. van Eldik, *Inorg. Chem.*, **42**, 6528 (2003).
03ICC1282	A. Panunzi, G. Roviello, and F. Ruffo, *Inorg. Chem. Commun.*, **3**, 1282 (2003).
03ICC1382	A. J. Canty and T. Rodemann, *Inorg. Chem. Commun.*, **6**, 1382 (2003).
03JA11430	M. A. Iron, A. Sundermann, and J. M. L. Martin, *J. Am. Chem. Soc.*, **125**, 11430 (2003).
03JCS(D)2080	E. A. Plummer, J. W. Hofstraat, and L. De Cola, *Dalton Trans.*, 2080 (2003).
03JCS(D)3493	Y. Kang, J. Lee, D. Song, and S. Wang, *Dalton Trans.*, 3493 (2003).

03JMC(A)95   Y. Himeda, N. Onozawa-Komatsuzaki, H. Sugihara, H. Arakawa, and K. Kasuga, *J. Mol. Catal.*, **A195**, 95 (2003).

03JOM(670)221   I. Ara, J. R. Berenguer, E. Eguizabal, J. Fornies, J. Gomez, and E. Lalinde, *J. Organomet. Chem.*, **670**, 221 (2003).

03JOM(671)179   T. Yamamoto, Y. Yamaguchi, and M. Abla, *J. Organomet. Chem.*, **671**, 179 (2003).

03JOM(679)1   S. Stoccoro, B. Soro, G. Minghetti, A. Zucca, and M. A. Cinellu, *J. Organomet. Chem.*, **679**, 1 (2003).

03NJC150   C. H. Tao, K. M. C. Wong, N. Zhu, and V. W. W. Yam, *New J. Chem.*, **27**, 150 (2003).

03NJC665   A. N. Vedernikov, J. C. Huffman, and K. G. Caulton, *New J. Chem.*, **27**, 665 (2003).

03OL3221   D. Zhang, L. Z. Wu, Q. Z. Zhang, X. H. Li, L. P. Zhang, and C. H. Tung, *Org. Lett.*, **5**, 3221 (2003).

03OM1047   L. Gonsalvi, J. A. Gaunt, H. Adams, A. Castro, G. J. Sunley, and A. Haynes, *Organometallics*, **22**, 1047 (2003).

03OM1223   V. G. Albano, M. Monari, I. Orabona, A. Panunzi, G. Roviello, and F. Ruffo, *Organometallics*, **22**, 1223 (2003).

03OM2187   D. Song and S. Wang, *Organometallics*, **22**, 2187 (2003).

03OM2358   C. J. Sumby and P. J. Steel, *Organometallics*, **22**, 2358 (2003).

03OM3025   G. R. Owen, R. Vilar, A. J. P. White, and D. J. Williams, *Organometallics*, **22**, 3025 (2003).

03OM4770   G. Minghetti, S. Stoccoro, M. A. Cinellu, B. Soro, and A. Zucca, *Organometallics*, **22**, 4770 (2003).

03RCB567   D. G. Yakharov, Y. G. Budnikova, and O. G. Sinyashin, *Russ. Chem. Bull.*, **52**, 567 (2003).

03RJE1261   D. G. Yakharov, Y. G. Budnikova, and O. G. Sinyashin, *Russ. J. Electrochem.*, **39**, 1261 (2003).

04CIC1   D. Song and S. Wang, *Comments Inorg. Chem.*, **25**, 1 (2004).

04EJI1948   Q. Z. Yang, Q. X. Tong, L. Z. Wu, Z. X. Wu, L. P. Zhang, and C. H. Tung, *Eur. J. Inorg. Chem.*, 1948 (2004).

04EJI2326   C. Bartolome, P. Espinet, J. M. Martin-Alvarez, and F. Villafane, *Eur. J. Inorg. Chem.*, 2326 (2004).

04EJI2784   A. Klein, M. P. Feth, H. Bertagnolli, and S. Zalis, *Eur. J. Inorg. Chem.*, 2784 (2004).

04EJI4484   A. Zucca, M. A. Cinellu, G. Minghetti, S. Stoccoro, and M. Manassero, *Eur. J. Inorg. Chem.*, 4484 (2004).

04IC725   H. Jude, J. A. K. Bauer, and W. B. Connick, *Inorg. Chem.*, **43**, 725 (2004).

04IC812   V. W. W. Yam, C. K. Hui, and N. Zhu, *Inorg. Chem.*, **43**, 812 (2004).

04IC3642   A. N. Vedernikov, M. Pink, and K. G. Caulton, *Inorg. Chem.*, **43**, 3642 (2004).

04IC5195   Q. Z. Zhang, L. Z. Wu, H. Zhang, B. Chen, Z. X. Wu, L. P. Zhang, and C. H. Tung, *Inorg. Chem.*, **43**, 5195 (2004).

04IC7180   R. Dorta, R. Dorta, L. J. W. Shimon, and D. Milstein, *Inorg. Chem.*, **43**, 7180 (2004).

04JA2763   J. D. Slinker, A. A. Gorodetsky, M. S. Lowry, J. Wang, S. Parker, R. Rohl, S. Bemhard, and G. G. Malliaras, *J. Am. Chem. Soc.*, **126**, 2763 (2004).

04JA3440   D. Zhang, L. Z. Wu, L. Zhou, X. Han, Q. Z. Yang, L. P. Zhang, and C. H. Tung, *J. Am. Chem. Soc.*, **126**, 3440 (2004).

04JA4958	W. Lu, B. X. Mi, M. C. W. Chan, Z. Hui, C. M. Che, N. Zhu, and S. T. Lee, *J. Am. Chem. Soc.*, **126**, 4958 (2004).
04JA6332	N. S. Perch and R. A. Wiedenhoefer, *J. Am. Chem. Soc.*, **126**, 6332 (2004).
04JA6470	M. R. Plutino, L. M. Scolaro, A. Albinati, and R. Romeo, *J. Am. Chem. Soc.*, **126**, 6470 (2004).
04JA8100	T. J. Anderson, G. D. Jones, and D. A. Vicic, *J. Am. Chem. Soc.*, **126**, 8100 (2004).
04JA11160	A. N. Vedernikov, J. C. Fettinger, and F. Mohr, *J. Am. Chem. Soc.*, **126**, 11160 (2004).
04JCS(D)619	M. Rashidi, N. Shahabadi, and S. M. Nabavizadeh, *Dalton Trans.*, 619 (2004).
04JCS(D)929	J. Ruiz, M. T. Martinez, E. Florenciano, V. Rodriguez, G. Lopez, J. Perez, P. A. Chaloner, and P. B. Hitchcock, *Dalton Trans.*, 929 (2004).
04JCS(D)2733	J. M. Casas, B. E. Diosdano, L. R. Falvello, J. Fornies, A. Martin, and A. J. Rueda, *Dalton Trans.*, 2733 (2004).
04JCS(D)3521	J. Ruiz, M. T. Martinez, V. Rodriguez, G. Lopez, J. Perez, P. A. Chaloner, and P. B. Hitchcock, *Dalton Trans.*, 35277 (2004).
04JCS(D)3699	Y. J. Kim, X. Chang, J. T. Han, M. S. Lim, and S. W. Lee, *Dalton Trans.*, 3699 (2004).
04JOC4788	Y. Yang, D. Zhang, L. Z. Wu, B. Chen, L. P. Zhang, and C. H. Tung, *J. Org. Chem.*, **69**, 4788 (2004).
04JOM672	A. J. Canty, M. C. Denney, J. Patel, H. Sun, B. W. Skelton, and A. J. White, *J. Organomet. Chem.*, **689**, 672 (2004).
04JOM1288	D. P. Gallasch, S. L. Woodhouse, and L. M. Rendina, *J. Organomet. Chem.*, **689**, 1288 (2004).
04JOM1393	V. W. W. Yam, *J. Organomet. Chem.*, **689**, 1393 (2004).
04JOM1872	J. Ruiz, C. Vicente, V. Rodriguez, N. Cutillas, G. Lopez, and C. R. de Arellano, *J. Organomet. Chem.*, **689**, 1872 (2004).
04JOM4016	P. A. Chase, R. J. M. K. Gebbink, and G. van Koten, *J. Organomet. Chem.*, **689**, 4016 (2004).
04JPC(A)3485	I. E. Pomestchenko and F. N. Castellano, *J. Phys. Chem.*, **A108**, 3485 (2004).
04MRC1491	E. Holder, V. Marin, M. A. R. Meier, and U. S. Schubert, *Macromol. Rapid Commun.*, **25**, 1491 (2004).
04OM891	S. A. Mcgregor and G. W. Neave, *Organometallics*, **23**, 891 (2004).
04OM1122	A. J. Canty, M. C. Denney, B. W. Skelton, and A. H. White, *Organometallics*, **23**, 1122 (2004).
04OM1194	D. Song, W. L. Jia, and S. Wang, *Organometallics*, **23**, 1194 (2004).
04OM1480	Y. Himeda, N. Onozawa-Komatsuzaki, H. Sugihara, H. Arakawa, and K. Kasuga, *Organometallics*, **23**, 1480 (2004).
04OM3071	G. D. Jones, T. J. Anderson, N. Chang, R. J. Brandon, G. L. Ong, and D. A. Vicic, *Organometallics*, **23**, 3071 (2004).
04OM3459	K. M. C. Wong, W. S. Tang, B. W. K. Chu, N. Zhu, and V. W. W. Yam, *Organometallics*, **23**, 3459 (2004).
04OM3466	A. J. Canty, J. Patel, T. Rodemann, J. H. Ryan, B. W. Skelton, and A. H. White, *Organometallics*, **23**, 3466 (2004).
04OM5252	I. Papai, G. Schubert, I. Mayer, G. Besenyei, and M. Aresta, *Organometallics*, **23**, 5252 (2004).
04OM5856	F. Neve, M. La Deda, A. Crispini, A. Bellusci, F. Puntoriero, and S. Campagna, *Organometallics*, **23**, 5856 (2004).

04OM6042   E. Goto, M. Usuki, H. Takenaka, K. Sakai, and T. Tanase, *Organometallics*, **23**, 6042 (2004).

05AG(E)791   C. Yu, K. M. C. Wong, K. H. Y. Chan, and V. W. W. Yam, *Angew. Chem., Int. Ed. Engl.*, **44**, 791 (2005).

05CC230   K. J. Arm and J. A. G. Williams, *Chem. Commun.*, 230 (2005).

05CC4675   D. L. Ma, T. Y. T. Shum, F. Zhang, C. M. Che, and M. Yang, *Chem. Commun.*, 4675 (2005).

05CCR1085   P. Thanasekaran, T. T. Liao, Y. H. Liu, T. Rajendran, S. Rajagopal, and K. L. Liu, *Coord. Chem. Rev.*, **249**, 1085 (2005).

05CEJ4535   V. W. W. Yam, K. H. Y. Chan, K. M. C. Wong, and N. Zhu, *Chem. Eur. J.*, **11**, 4535 (2005).

05EJI110   M. Lepeltier, T. K. M. Lee, K. K. W. Lo, L. Toupet, H. Le Bozec, and V. Guerchais, *Eur. J. Inorg. Chem.*, 110 (2005).

05IC471   F. Hua, S. Kinayyigit, J. R. Cable, and F. N. Castellano, *Inorg. Chem.*, **44**, 471 (2005).

05IC1492   K. M. C. Wong, W. S. Tang, X. X. Lu, N. Zhu, and V. W. W. Yam, *Inorg. Chem.*, **44**, 1492 (2005).

05IC2628   T. J. Wadas, R. J. Lachicotte, and R. Eisenberg, *Inorg. Chem.*, **44**, 2628 (2005).

05IC4055   F. Guo, W. Sun, Y. Liu, and K. Schanze, *Inorg. Chem.*, **44**, 4055 (2005).

05IC4806   C. Monnereau, J. Gomez, E. Blart, F. Odobel, S. Wallin, A. Fallberg, and L. Hammarstroem, *Inorg. Chem.*, **44**, 4806 (2005).

05IC6284   S. Chakraborty, T. J. Wadas, H. Hester, C. Flaschenreim, R. Schmehl, and R. Eisenberg, *Inorg. Chem.*, **44**, 6284 (2005).

05IC8723   A. B. Tamayo, S. Garon, T. Sajoto, P. I. Djurovich, I. M. Tsyba, R. Bau, and M. E. Thompson, *Inorg. Chem.*, **44**, 8723 (2005).

05ICA1614   M. C. Janzen, M. C. Jennings, and R. J. Puddephatt, *Inorg. Chim. Acta*, **358**, 1614 (2005).

05ICA4567   J. S. Field, R. J. Haines, D. R. McMillin, O. Q. Munro, and G. C. Summerton, *Inorg. Chim. Acta*, **358**, 4567 (2005).

05ICC94   P. Frediani, A. Salvini, M. Bessi, L. Rosi, and C. Giannelli, *Inorg. Chem. Commun.*, **8**, 94 (2005).

05JCS(D)234   E. C. Constable, C. E. Housecroft, M. Neuburger, S. Schaffner, and E. J. Shardlow, *Dalton Trans.*, 234 (2005).

05JCS(D)2365   U. Siemeling, S. Chakraborty, K. Bausch, H. Fink, C. Bruhn, M. Baldus, B. Angerstein, R. Plessow, and A. Brockhinke, *Dalton Trans.*, 2365 (2007).

05JMTC2820   E. Lafolet, S. Welter, Z. Popovich, and L. De Cola, *J. Mater. Chem.*, **15**, 2820 (2005).

05OM53   B. Soro, S. Stoccoro, G. Minghetti, A. Zucca, M. A. Cinellu, S. Gladiali, M. Manassero, and M. Sansoni, *Organometallics*, **24**, 53 (2005).

05OM2528   M. Rashidi, S. M. Nabavizadeh, A. Akbari, and S. Habibzadeh, *Organometallics*, **24**, 2528 (2005).

05OM3290   S. B. Zhao, D. Song, W. L. Jia, and S. Wang, *Organometallics*, **24**, 3290 (2005).

05OM3516   I. O. Koshevoy, M. Haukka, T. A. Pakkanen, S. P. Tunik, and P. Vainiotalo, *Organometallics*, **24**, 3516 (2005).

05OM4298   S. C. F. Lam, V. W. W. Yam, K. M. C. Wong, E. C. C. Cheng, and N. Zhu, *Organometallics*, **24**, 4298 (2005).

05OM5136	K. Mashima, D. Shimizu, T. Yamagata, and K. Tani, *Organometallics*, **21**, 5136 (2005).
05ZOB705	M. V. Kulikova, N. McClenaghan, and K. P. Balashev, *Russ. J. Gen. Chem.*, **75**, 705 (2005).
06CC1601	H. B. Xu, L. X. Shi, E. Ma, L. Y. Zhang, Q. H. Wei, and Z. N. Chen, *Chem. Commun.*, 1601 (2006).
06CCR1819	F. N. Castellano, I. E. Pomestchenko, E. Shikhova, F. Hua, M. L. Muro, and N. Rajapakse, *Coord. Chem. Rev.*, **250**, 1819 (2006).
06EJI3274	J. Q. Wang, C. X. Ren, and G. X. Jin, *Eur. J. Inorg. Chem.*, 3274 (2006).
06IC4304	F. Hua, S. Kinayyigit, J. R. Cable, and F. N. Castellano, *Inorg. Chem.*, **45**(11), 4304 (2006).
06IC4319	R. Narayana-Prabhu and R. H. Schmehl, *Inorg. Chem.*, **45**, 4319 (2007).
06IC5653	M. Clemente-Leon, E. Coronado, C. J. Gomez-Garcia, and A. Soriano-Portillo, *Inorg. Chem.*, **45**, 5653 (2006).
06IC6152	Q. Zhao, S. Liu, M. Shi, C. Wang, M. Yu, L. Li, F. Li, T. Yi, and C. Huang, *Inorg. Chem.*, **45**, 6152 (2006).
06IC8685	A. J. Wilkinson, H. Puschmann, J. A. K. Howard, C. E. Foster, and J. A. G. Williams, *Inorg. Chem.*, **45**, 8685 (2006).
06IC10990	A. Auffrant, A. Barbieri, F. Barigelletti, J. P. Collin, L. Flamigni, C. Sabatini, and J. P. Sauvage, *Inorg. Chem.*, **45**, 10990 (2006).
06ICA1666	F. Neve, M. La Deda, F. Puntoriero, and S. Campagna, *Inorg. Chim. Acta*, **359**, 1666 (2006).
06ICA1773	A. Holuigue, C. Sirlin, M. Pfeffer, K. Goubitz, J. Fraanje, and C. J. Elsevier, *Inorg. Chim. Acta*, **359**, 1773 (2006).
06ICA1899	D. Taher, B. Walfort, and H. Lang, *Inorg. Chim. Acta*, **359**, 1899 (2006).
06ICA2431	T. Suzuki, *Inorg. Chim. Acta*, **359**, 2431 (2006).
06ICA4326	C. Vetter, C. Wagner, J. Schmidt, and D. Steinborn, *Inorg. Chim. Acta*, **359**, 4326 (2006).
06JCS(D)2468	A. S. Ionkin, W. J. Marshall, D. C. Roe, and Y. Wang, *Dalton Trans.*, 2468 (2006).
06JCS(D)3285	R. Ziessel, S. Diring, and P. Retailleau, *Dalton Trans.*, 3285 (2006).
06JCS(D)4657	S. Ogo, R. Kabe, H. Hayashi, R. Harada, and S. Fukuzumi, *Dalton Trans.*, 4657 (2006).
06JCS(D)5077	S. A. De Pascali, D. Migoni, P. Papadia, A. Muscella, S. Marsigliante, A. Ciccarese, and F. P. Fanizzi, *Dalton Trans.*, 5077 (2006).
06JCS(D)5225	S. Liu, J. Zhang, X. Wang, and G. X. Jin, *Dalton Trans.*, 5225 (2006).
06JOM2037	P. E. A. Ribeiro, C. L. Donnici, and E. N. dos Santos, *J. Organomet. Chem.*, **691**, 2037 (2006).
06JOM3834	H. Lang, D. Taber, B. Walfort, and H. Pritzkow, *J. Organomet. Chem.*, **691**, 3834 (2006).
06JOM4135	S. Stoccoro, A. Zucca, G. L. Petretto, M. A. Cinellu, G. Minghetti, and M. Manassero, *J. Organomet. Chem.*, **691**, 4135 (2006).
06JOM4573	J. Rajput, A. T. Hutton, J. R. Moss, H. Su, and C. Imrie, *J. Organomet. Chem.*, **691**, 4573 (2006).
06JOM4868	M. Schwalbe, D. Walther, H. Schreer, J. Langer, and H. Gorls, *J. Organomet. Chem.*, **691**, 4868 (2006).
06JOM5900	C. Baik, W. S. Han, Y. Kang, S. O. Kang, and J. Ko, *J. Organomet. Chem.*, **691**, 5900 (2006).

06OM74	J. Q. Wang, C. X. Ren, and G. X. Jin, *Organometallics*, **25**, 74 (2006).
06OM1607	P. Sangtrirutnugul, M. Stradiotto, and T. D. Tilley, *Organometallics*, **25**, 1607 (2006).
06OM1851	J. Vicente, J. A. Abad, M. J. Lopez-Saez, and P. G. Jones, *Organometallics*, **25**, 1851 (2006).
06OM2253	A. Zucca, G. L. Petretto, S. Stoccoro, M. A. Cinellu, G. Minghetti, M. Manassero, C. Manassero, L. Male, and A. Albinati, *Organometallics*, **25**, 2253 (2006).
06OM4579	R. Packheiser, B. Walfort, and H. Lang, *Organometallics*, **25**, 4579 (2006).
06OM5746	M. C. Kohler, R. A. Stockland, and N. P. Rath, *Organometallics*, **25**, 5746 (2006).
06OM5979	S. B. Zhao, G. Wu, and S. Wang, *Organometallics*, **25**, 5979 (2006).
06POL1607	D. G. Yakharov, D. I. Tazeev, O. G. Sinyashin, G. Giambastiani, C. Bianchini, A. M. Segarra, P. Lonnecke, and E. Hey-Hawkins, *Polyhedron*, **25**, 1607 (2006).
07AG(E)6309	E. Khaskin, P. Y. Zavalij, and A. N. Vedernikov, *Angew. Chem. Int. Ed. Engl.*, **46**, 6309 (2007).
07AHC(93)179	A. P. Sadimenko, *Adv. Heterocycl. Chem.*, **93**, 179 (2007).
07AHC(94)109	A. P. Sadimenko, *Adv. Heterocycl. Chem.*, **94**, 109 (2007).
07CC4116	C. Dragonetti, S. Righetto, D. Roberto, R. Ugo, A. Valore, S. Fantacci, A. Sgamellotti, and F. De Angelis, *Chem. Commun.*, 4116 (2007).
07IC700	K. K. W. Lo and J. S. Y. Lau, *Inorg. Chem.*, **46**, 700 (2007).
07IC1924	N. R. Deprez and M. S. Sanford, *Inorg. Chem.*, **46**, 1924 (2007).
07IC3038	E. Shikhova, E. O. Danilov, S. Kinayyigit, I. E. Pomestchenko, A. D. Tregubov, F. Camerel, P. Retailleau, R. Ziessel, and F. N. Castellano, *Inorg. Chem.*, **46**, 3038 (2007).
07IC6911	A. Auffrant, A. Barbieri, F. Barigelletti, J. Lacour, P. Mobian, J. P. Collin, J. P. Sauvage, and B. Ventura, *Inorg. Chem.*, **46**, 6911 (2007).
07IC8533	C. Dragonetti, L. Falciola, P. Mussini, S. Righetto, D. Roberto, R. Ugo, A. Valore, F. De Angelis, S. Fantacci, A. Sgamellotti, M. Ramon, and M. Muccini, *Inorg. Chem.*, **46**, 8533 (2007).
07IC8771	F. Hua, S. Kinayyigit, A. A. Rachford, E. A. Shikhova, S. Goeb, J. R. Cable, C. J. Adams, K. Kirschbaum, A. A. Pinkerton, and F. N. Castellano, *Inorg. Chem.*, **46**, 8771 (2007).
07IC9139	M. Schmittel and H. Lin, *Inorg. Chem.*, **46**, 9139 (2007).
07IC10187	F. Shao, B. Elias, W. Lu, and J. K. Barton, *Inorg. Chem.*, **46**, 10187 (2007).
07IC10681	R. Romeo, G. D'Amico, E. Guido, A. Albinati, and S. Rizzato, *Inorg. Chem.*, **46**, 10681 (2007).
07ICA163	H. B. Xu, L. Y. Zhang, and Z. N. Chen, *Inorg. Chim. Acta*, **360**, 163 (2007).
07ICA4069	E. C. Constable, C. E. Housecroft, M. Neuburger, S. Schaffner, and E. J. Shardlow, *Inorg. Chim. Acta*, **360**, 4069 (2007).
07JA14733	F. Shao and J. K. Barton, *J. Am. Chem. Soc.*, **129**, 14733 (2007).
07JCS(D)83	R. M. Ceder, G. Muller, M. Ordinas, and J. I. Ordinas, *Dalton Trans.*, 83 (2007).
07JCS(D)133	C. Crotti, E. Farnetti, S. Filipuzzi, M. Stener, E. Zangrando, and P. Moras, *Dalton Trans.*, 133 (2007).

07JCS(D)3526	J. Moussa, K. M. C. Wong, L. M. Chamoreau, H. Amouri, and V. W. W. Yam, *Dalton Trans.*, 3526 (2007).
07JCS(D)3885	Y. Fan, Y. M. Zhu, F. R. Dai, L. Y. Zhang, and Z. N. Chen, *Dalton Trans.*, 3885 (2007).
07JCS(D)4386	H. S. Lo, S. K. Yip, N. Zhu, and V. W. W. Yam, *Dalton Trans.*, 4386 (2007).
07JCS(D)4457	P. Govindaswamy, D. Linder, J. Lacour, G. Suss-Fink, and B. Therrien, *Dalton Trans.*, 4457 (2007).
07JCS(D)5720	V. M. Vecchio, M. Benedetti, D. Migoni, S. A. De Pascali, A. Ciccarese, S. Marsigliante, F. Capitelli, and F. P. Fanizzi, *Dalton Trans.*, 5720 (2007).
07JOM3810	G. Mansouri, A. R. Rezvani, H. Hadadzadeh, H. R. Khavasi, and H. Saravani, *J. Organomet. Chem.*, **692**, 3810 (2007).
07JOM4545	Y. F. Han, J. S. Zhang, Y. J. Lin, J. Dai, and G. X. Jin, *J. Organomet. Chem.*, **692**, 4545 (2007).
07OM12	J. K. W. Lee, C. C. Ko, K. M. C. Wong, N. Zhu, and V. W. W. Yam, *Organometallics*, **26**, 12 (2007).
07OM702	Y. Himeda, N. Onozawa-Komatsuzaki, H. Sugihara, and K. Kasuga, *Organometallics*, **26**, 702 (2007).
07OM810	J. Durand, S. Gladiali, G. Erre, E. Zangrando, and B. Milani, *Organometallics*, **26**, 810 (2007).
07OM2137	K. J. H. Younf, O. A. Mironov, and R. A. Periana, *Organometallics*, **26**, 2137 (2007).
07OM4483	X. L. Li, F. R. Dai, L. Y. Zhang, Y. M. Zhu, Q. Peng, and Z. N. Chen, *Organometallics*, **26**, 4483 (2007).
07OM4860	J. Chen, W. Zhang, Z. Dong, G. F. Swiegers, and G. G. Wallace, *Organometallics*, **26**, 4860 (2007).
07OM5557	P. Sangtrirutnugul and T. D. Tilley, *Organometallics*, **26**, 5557 (2007).
07OM5621	A. Zucca, S. Stoccoro, M. A. Cinellu, G. L. Petretto, and G. Minghetti, *Organometallics*, **26**, 5621 (2007).
07OM5922	Q. Zhao, S. Liu, M. Shi, F. Li, H. Jing, T. Yi, and C. Huang, *Organometallics*, **26**, 5922 (2007).
08AHC(95)219	A. P. Sadimenko, *Adv. Heterocycl. Chem.*, **95**, 219 (2008).

CHAPTER 3

Azulenes Fused to Heterocycles [☆]

Gunther Fischer

[☆] Dedicated to the memory of my highly regarded teacher Wilhelm Treibs (1890–1978), one of the Masters of azulene chemistry, on the 30th anniversary of his death.

Geibelstraße 15, D-04129 Leipzig, Germany

Advances in Heterocyclic Chemistry, Volume 97
ISSN 0065-2725, DOI 10.1016/S0065-2725(08)00203-1

1. INTRODUCTION

1.1 General survey

The fascinating chemistry of deep blue *azulene* (**1**; Scheme 1) and its derivatives began to develop in the 1930s and has in quick succession been recorded in early reviews (e.g., 48FOR(5)40, 50AG281, 52CRV127, 55FCF(3)334, 58AG419, 59MI1, 59MI2).

Together with tropoids (cf. 95AHC(64)81, 96AHC(66)285) azulenes have proven to be one of the most important classes of non-benzenoid aromatic substances (66MI1, 84MI1). They were later classified as cross-conjugated compounds and zero-bridged annulenes (00MI1). Azulenes have continued to attract attention in synthetic and theoretical chemistry (77RCR530), especially by Japanese groups (71MI1, 73MI1, 76MI1, 96MI2, 04EJO899).

Nearly from the beginning of the azulene chemistry many *heterocyclic*-fused derivatives have also been described, for instance, linderazulene (**2**) that even is a natural product (cf. Section 2.2). They have been treated in small special chapters of several important azulene reviews (61FOR(19)32, 82RCR1089, 85HOU(5/2c)127).

1.2 Scope and limitation

This chapter reviews relevant literature from the beginning through 2007 together with some work published in 2008. Patents are included provided they reveal essential synthetic aspects or applications.

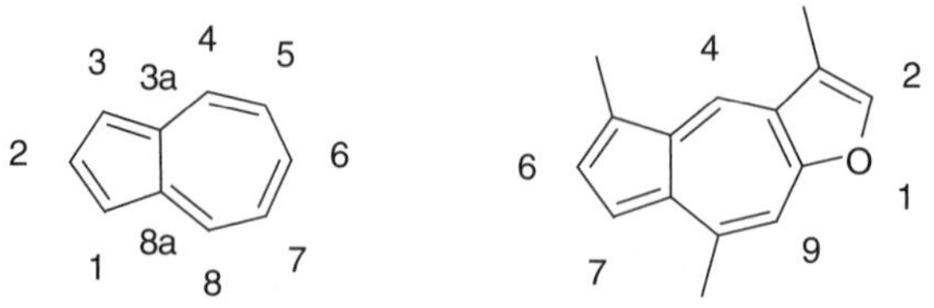

(**1**)   (**2**)

Scheme 1

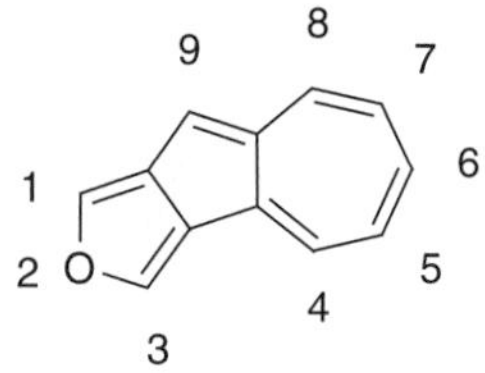

Scheme 2

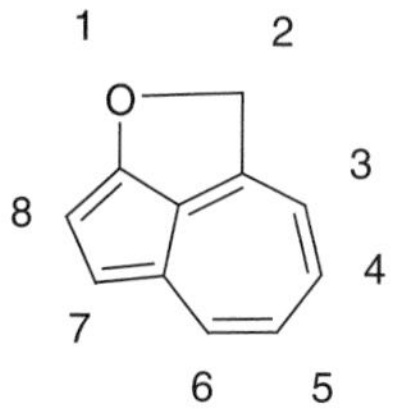

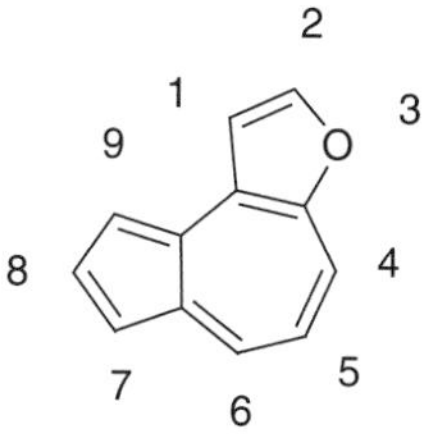

Coverage is restricted to azulenes directly fused to heterorings showing maximum unsaturation or to the corresponding tautomers.

1.3 Nomenclature

In this review, for consistency, the bicyclic azulene moiety is generally drawn and numbered as shown in formula **1**. Given this numbering, the heterorings are attached in a way that enables the attribution of the lowest possible numbers to the fusing C-C bond of the parent azulene system, for instance, "[6,5-*b*]" in linderazulene (**2**).

Linderazulene, when numbered according to the rules of Chemical Abstracts Service (CAS) (98MI1), is named 3,5,8-trimethylazuleno[6,5-*b*] furan. Other examples of azulenofurans are, in the CAS numbering, those of formulas

- **3**: Azuleno[1,2-*c*]furan (*o*-quinonoid),
- **4**: Azuleno[1,8-*bc*]furan (*peri*-annulated),
- **5**: Azuleno[5,4-*b*]furan (numbering not beginning on the heteroatom) (Scheme 2).

Several authors, however, prefer non-CAS nomenclature and numbering.

2. OCCURRENCE AND SYNTHESIS

2.1 Survey

After a view of the natural occurrence (Section 2.2) general synthetic methods useful for azulenes fused to *different* heterocyclic rings will be

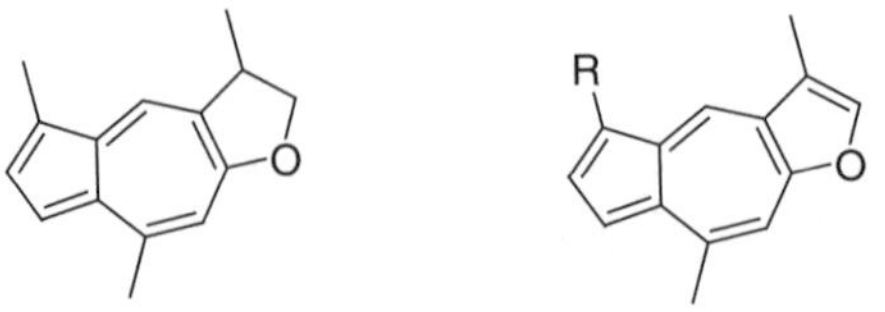

(6)   (7a, b)   (R = CHO,COOMe)

Scheme 3

treated in Section 2.3. The subsequent Section 2.4 will contain special synthetic paths leading to *individual*-fused heterocycles. Most syntheses under review are based on azulene, cycloheptatriene or cyclopentadiene derivatives and consist in cyclizations to form the heterorings. Detailed descriptions of modern heterocycle synthesis have been covered in recent handbooks and textbooks by Katritzky et al. (96CHEC2(1–10), 08CHEC3(1–15)), Eicher and Hauptmann (03MI1), and other authors. The final Section 2.5 will summarize all the relevant syntheses arranged according to the heterocyclic systems (Table 2).

2.2 Natural occurrence

The formation of deep-colored heterocycloazulenes from colorless natural precursors has been known for a long time (see Section 2.3.1). It was surprising, however, when in the 1980s linderazulene (2) itself was detected to be a genuine constituent of certain gorgonian corals, especially deep-sea corals (81E442, 82PAC1907, 84TL2109). It is partly accompanied by 2,3-dihydrolinderazulene (6; Scheme 3) (87E624, 93CL2003) or linderazulene metabolites 7a and 7b (05JNP248).

Furthermore, linderazulene was even found in the essential oils of higher plants as in those of Kwangsi turmeric or *Curcuma kwangsiensis* (83MI1) and of the deep violet fruits (seeds) of Perfoliate Alexanders or *Smyrnium perfoliatum* (98P1079), a native Mediterranean plant invading Western and Central Europe and found in Leipzig, too (Figure 1).

2.3 General syntheses

2.3.1 From natural proazulenes

A first indication of a novel structural principle "azulene" had been a blue coloration in certain essential oils that was observed after simple operations of isolation that favor spontaneous dehydrogenation (52CRV127). Thus, the essential oil of wormwood (*Artemisia absinthium*), according to the conditions of its isolation, turns more or less blue (52N571) and was later found to contain artemazulene (8; Scheme 4) among other azulenes (06MI1).

Figure 1 *Smyrnium perfoliatum*: (a) in the flowering state; (b) in the fruit-bearing state (photo by Silvia Fischer) (for color pictures, see back cover of this book).

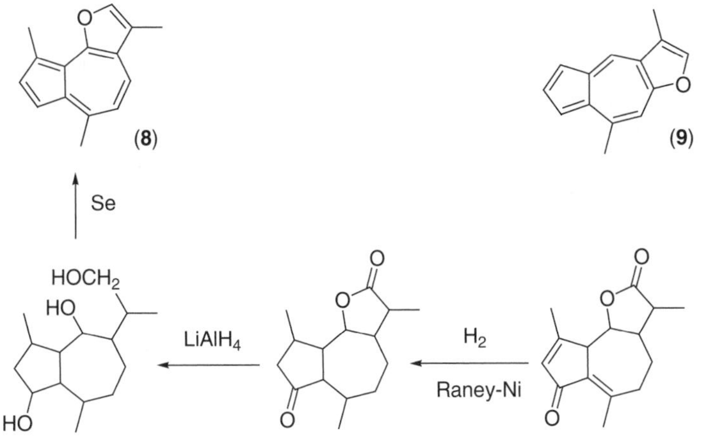

Scheme 4

Deliberate dehydrogenation of natural hydroazulenes and other sesquiterpene derivatives, sometimes after preceding dehydration, hydrogenation, and reduction, has been the first synthetic path to azulenes (55FCF(3)334, p. 355; 59MI2, p. 301). Similar reactions yielded azulenes fused to heterorings, such as furan and thiophene (Table 1). Their precursors were essentially guaianolides, that is sesquiterpenic lactones of the perhydroazulene type (79FOR(38)47, p. 166).

In this way, linderazulene (2) and artemazulene (8) have been obtained from numerous precursors (62MI1), partly together with ujacazulene (5-norlinderazulene, 9) and possibly 9-norartemezulene (64JCS3577).

Typical reactions are exemplified by the transformation of leucomysine (10a) to artemazulene (63ZOB2734, 68MI1). Reductive lactone ring cleavage leads to 1,4-glycol 10b that when dehydrogenated suffers recyclization.

Whether linderazulene (2) or artemazulene (8) forms from a given guaianolide mainly depends on how the lactone ring is positioned in the precursor. When, however, this lactone ring is flanked by a hydroxyl or derivative group, both isomeric azulenofuranes (2 and 8) are formed as shown by the reaction of globicin (11; Scheme 5) (63CCC1202).

Takeda et al. (71CPB676, cf. 70PAC181), on resuming earlier work published in 1939–1953 (Table 1, first entries), studied thoroughly the dehydrogenation of lindenenyl acetate by palladium and found the reaction to proceed partly by a direct route leading to linderazulene (2) and partly through intermediate 12.

Most of the reactions listed in Table 1 were meant less for synthetic purpose than for elucidation of the precursor structure. Dehydrogenation as the crucial step, owing to thermal stress, usually gives small yields and sometimes mixtures of azulenes (48MI1; 52CRV127, p. 166; 55FCF(3)334, p. 362; 59MI2; 85HOU(5/2c)127; cf. 65CPB717). Thanks to their intense color, their basicity and adsorptivity to alumina, however, even small amounts of azulenes are easily detected, separated, and purified (cf. 59MI2, p. 322), for instance, in the case of linder- and artemazulene (53CPB164, 60CCC1702).

An example of the technical dehydrogenation by sulfur is that producing guaiazulene from guaiene (14), in turn made from the non-lactonic sesquiterpene guaiol (13). Among the minor by-products are two isomeric azulenothiophenes 15 and 16 (Scheme 6) that are analogs of linder- and artemazulene, respectively (79BCJ1549). Azulene 15 was also obtained, possibly by a secondary reaction, among the dehydrogenation products of the sesquiterpene hydrocarbons from camphor oil (68BCJ2182).

2.3.2 From bicyclo[5,3,0]decanones

Bicyclo[5,3,0]decanones (perhydroazulenones), preferentially made from natural precursors (52CRV127, p. 163; 59MI2, p. 302), are versatile intermediates in azulene syntheses, the final steps again consisting in dehydrogenation. Thus, semisynthetic ketodiol 17 (made from

Table 1 Azulenofurans and azulenothiophenes from natural precursors

Product	Precursor	Reaction[a]	Reference
Linderazulene (2)	Linderene	D (Pd-asbestos)	39YZ162
		R/D	44YZ(is.9A)32
		Zn dust distillation	53CPB164
	Tenulin	H/R (LiAlH$_4$)/D (Pd-C)	56JCS142
	Pseudoivalin	R (LiAlH$_4$)/D (Pd-C)	65JOC118
	Zederone	R (LiAlH$_4$)/D (Pd-C)	66CPB550
	Cumanin	R (LiAlH$_4$)/D (Pd-C)	66T1499
	Virginolid	H (H$_2$/PtO$_2$)/R (LiAlH$_4$)/D (Pd-C)	67JOC507
	Cumambrin B	H (H$_2$/Pd-C)/R (LiAlH$_4$)/D (Pd-C)	68T5625
	Pulchellidine	R (LiAlH$_4$)/D (Pd-C)	69TL2073
	2,3-Dihydrolinderazulene (6)	D (Pd-C)	87E624
Artemazulene (8)	Prochamazulenogen	R (LiAlH$_4$)/D (S)	54CCC792
	Arborescine	R (LiAlH$_4$)/TsCl/D (Pd-C)	56CIL492
	Ambrosin	H (H$_2$/Pd-C)/D (Se)	59CCC1548
	Cnicin	R (LiAlH$_4$)/D (Se)	59CIL517
	Absinthin, anabsinthin	R (LiAlH$_4$)/D (Se)	60CCC1492
	Parthenin	R (LiAlH$_4$)/D (Pd-C)	62JA2601
	Estafiatin	H (H$_2$/PtO$_2$)/R (LiAlH$_4$)/D (Se)	63T1285
	Leucomysine (leucodin, 10a)	H (H$_2$/Ni)/R (LiAlH$_4$)/D (Se)	63ZOB2734
		H (H$_2$/Pt)/R (LiAlH$_4$)/D (Pd-C)	68MI1
	Laserolide	R (LiAlH$_4$)/D (Se)	64CCC938
	Grosshemine	H (H$_2$/PtO$_2$)/R (LiAlH$_4$)/D (Se)	65ZOB580

Table 1 *(Continued)*

Product	Precursor	Reaction[a]	Reference
	Mokko lactone	H (H$_2$/PtO$_2$)/R (LiAlH$_4$)/D (Se)	67YZ70
	Saupirin		71KPS727
	Gnididione	R (LiAlH$_4$)/D (Pd-C)	77JOC348
Linderazulene (**2**) and Artemazulene (**8**)	Matricin	H (H$_2$/PtO$_2$)/R (LiAlH$_4$)/D (Se)	56CIL1234, 57CCC1921
	Lactucin	H (H$_2$/Pd-SrCO$_3$)/R (LiAlH$_4$)/D (Se)	58CCC2195
	Matricarin	R (LiAlH$_4$)/D (Se)	59CCC1554
	Cynaropicrin	H (H$_2$/PtO$_2$)/R (LiAlH$_4$)/D (Se)	60CCC507
	Globicin (**11**)	H (H$_2$/PtO$_2$)/R/D	63CCC1202
	Amberboin	H (H$_2$/Pd)/R (LiAlH$_4$)/D (Se)	69AQ285
Ujacazulene (**9**)	Mexicanin E	H (H$_2$/PtO$_2$)/R (NaBH$_4$)/D (Pd-C)	61JA2326, 63T2317
	Linderane	D (Pd-C)	64JCS3577
	Linderalactone	D (Pd-C)	64JCS4578
Thienoazulenes **15** and **16**	Guaiene (**14**)	D (S)	79BCJ1549

[a]D, dehydrogenation; H, hydrogenation; R, reduction.

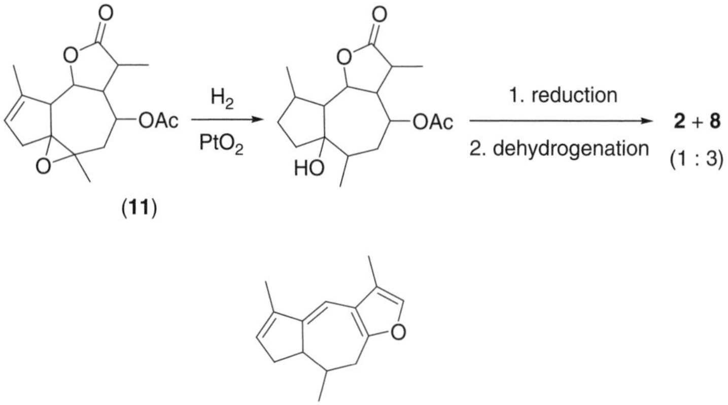

(11)

(12)

Scheme 5

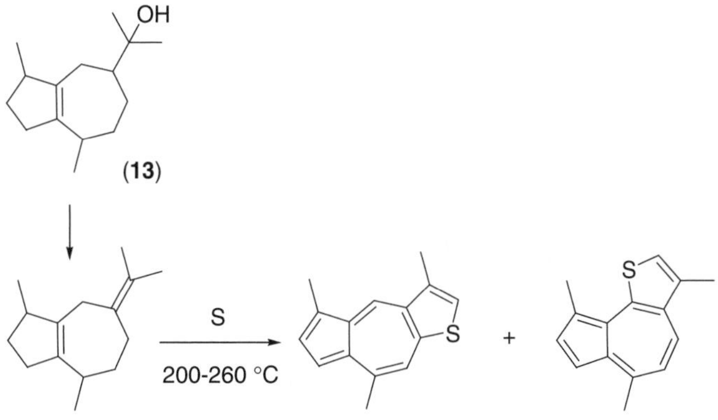

(13)

(14)   (15) (1%)   (16) (0.3%)

Scheme 6

guaianolide mexicanin E) by a Grignard reaction is transformed to linderazulene (**2**; Scheme 7) (63T2317).

In other cases, the keto group is the starting point of a furan ring annulation by an initial Reformatzky reaction or α-alkylation, respectively. In multi-step syntheses, linderazulene (**2**; Scheme 8) is obtained from ketone **18** (made from guaiol, **13**), finally by dehydrogenation of octahydrolinderazulene (**19**) (63JCS2591), and ujacazulene (**9**) is formed by a similar route (64JCS3577). Using microtechnics of dehydrogenating intermediate **19**, the yield of linderazulene under optimum conditions could be raised to 18% (65CPB717).

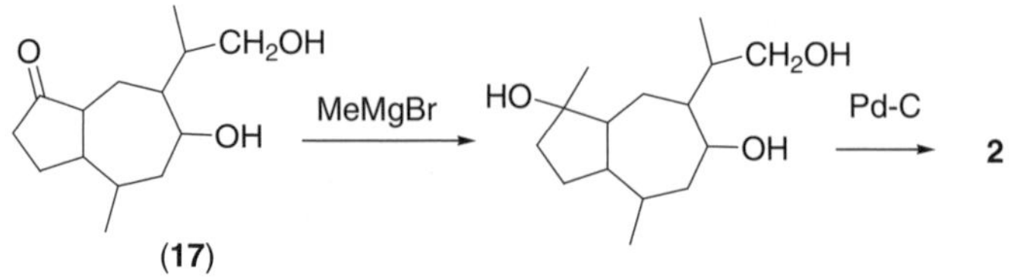

Scheme 7

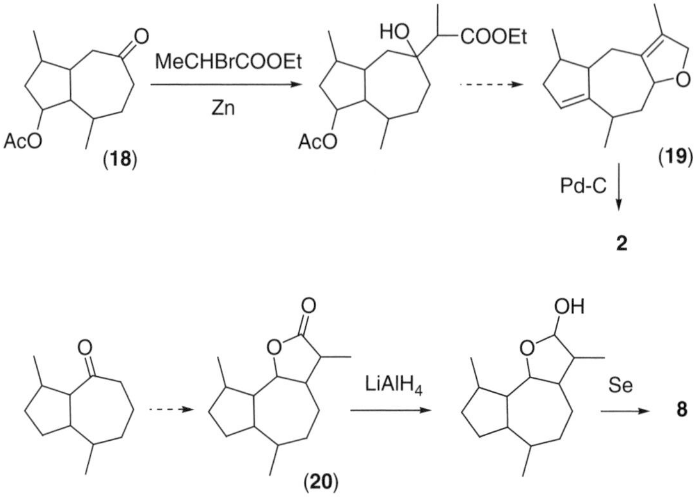

Scheme 8

On the other hand, artemazulene (**8**) is formed from completely synthetic guaianolide **20** (60G322) by reduction and dehydrogenation (63G395).

Besides a number of azulenofuran syntheses just mentioned, Hantzsch's thiazole synthesis (Scheme 9) was applied to bicyclo[5,3,0]-decanones **21** and **24** (the latter being made from guaiol, **13**, by oxidation). After dehydrogenation, these reactions yielded azulenothiazole **23** and a mixture of isomers **25** and **26**, respectively (63ZC26, 64UP1). These isomers are easily separated by column chromatography and can structurally be characterized by comparing their visual spectra with those of linder- and artemazulene (**2** and **8**, respectively). Among the isomers, presumably for steric reasons, [5,6]-fused compound **25** predominates just as in the case of azulenothiophene couple **15** and **16** (Scheme 6).

Dehydrogenation to yield thiazoles **25** and **26** is preferentially managed by means of palladium-on-charcoal at 300 °C and better in the gas than in the liquid phase; in this way yields could be increased to

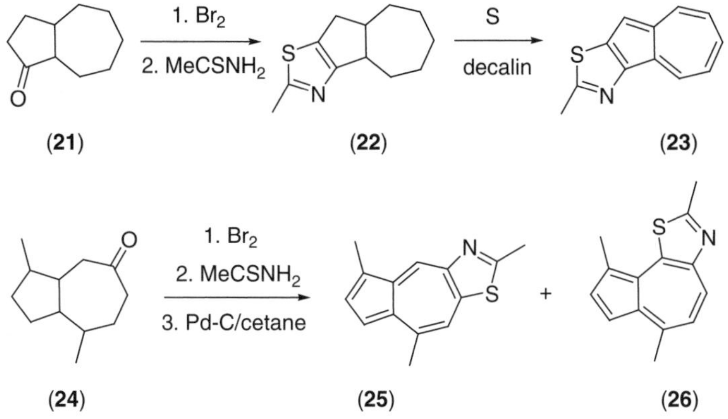

(21)   **(22)**   **(23)**

(24)   **(25)**   +   **(26)**

Scheme 9

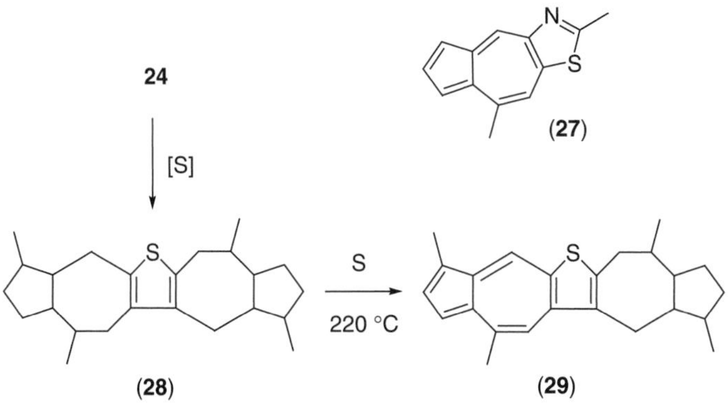

24

[S]

(28)   **(29)**

Scheme 10

24% (**25**) and 12% (**26**) (64UP1). Nor-compound **27** (Scheme 10) was isolated as a minor by-product of the dehydrogenation.

The application of the Hantzsch synthesis to ketone **24** was accompanied by the formation of crystalline by-product **28**. The synthesis of thiophenes by the action of sulfur (or sulfur donors) on symmetrical or unsymmetrical ketones has been reviewed (74AHC(16)181, p. 238; 94HOU(E6a)186, p. 235, respectively). Thiophene **28** could only partially be dehydrogenated to yield octahydro compound **29** or its isomer.

2.3.3 From 1-oxa-2-azulenones

2.3.3.1 With active methylene compounds. 1-Oxa-2-azulenones (2H-cyclohepta[*b*]furan-2-ones) are attacked by carbanions (that are produced from active methylene compounds) at C-8a to yield, after recyclization, bicyclic azulene derivatives (85HOU(5/2c)127, p. 223). In addition, the

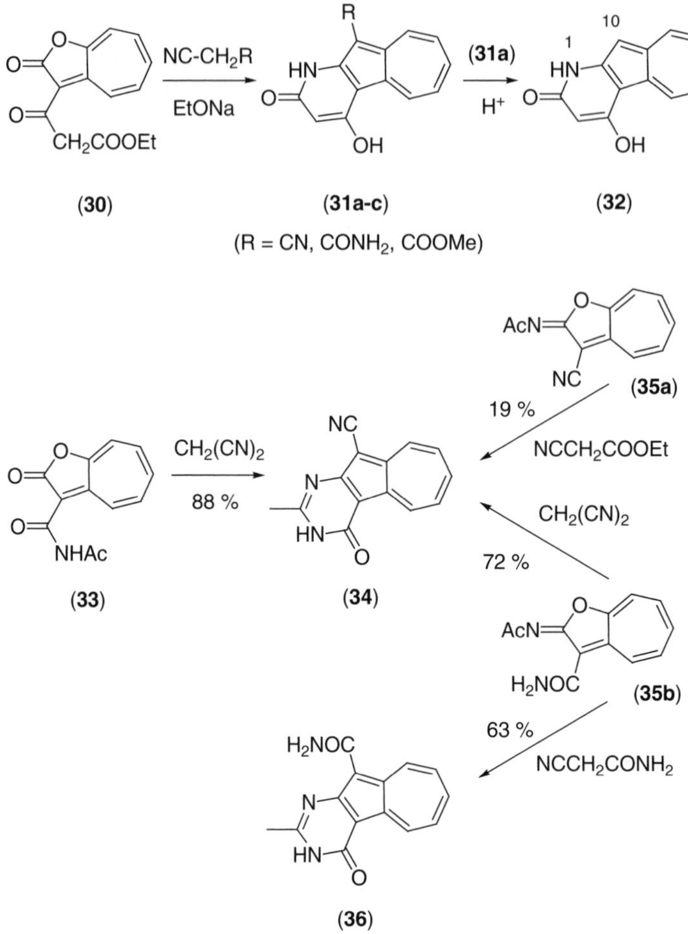

Scheme 11

use of suitably substituted compounds (e.g., **30** or **33**) can result in [1,2]-fused tricyclic systems (Scheme 11), such as pyridones **31a–31c**[1] (74BCJ1750, 79MI1) and pyrimidone **34** (81H(15)839) that can be degraded to yield parent compounds (e.g., **32**). A similar route leads from acetylimino derivatives **35a** and **35b** to pyrimidones **34** and **36**.

2.3.3.2 With furan derivatives. The reaction of furans with 1-oxa-2-azulenones (e.g., **37**; Scheme 12) yields functionalized azulene **38** and

---

[1] In the captions beneath the formulas, substituents (R) or chain and ring atoms (Z) in parentheses refer, in the order given, to substructures **a**, **b**, **c**, etc. of the respective formula or all formulas of the reaction.

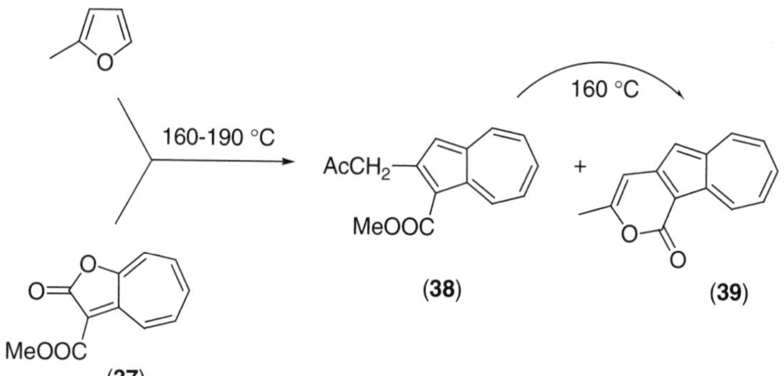

Scheme 12

azulenopyranone **39** (92H(34)429). This synthesis is believed to involve [8+2] cycloaddition of educt **37** and an α,β-unsaturated ether.

2.3.3.3 With enamines. 1-Oxa-2-azulenones (as tethered heptafulvenes) with enamines undergo [8+2] cycloaddition as well (85HOU(5/2c)127, p. 218; 98SL950). This way, in the case of heterocyclic enamines (such as **40**), azuleno[1,2]-fused azoles and azines are formed (Schemes 13 and 14): The reactions of oxaazulenone **37** with the isomer mixtures of enamines **40a** or **40b** yield mixtures of isomeric dihydrothiophenes **41** and **42** that are dehydrogenated by 2,3-dichloro-5,6-dicyano-p-benzoquinone (DDQ) to yield thiophenes **43** and **44**, respectively (83CL1721). Whereas the latter [1,2-c]-fused compound is unstable because of its o-quinonoid structure, thiophenes **43a** and **b** are demethoxycarbonylated to yield substances **45a** and **b**.

In a similar manner, the reaction of the 6-isopropyl homolog of oxaazulenone **37** with enamines **40** yields isomeric dihydrothiophenes; again only the [1,2-b]-fused isomer can be transformed to thiophene **46** (03OBC2572). Moreover, lactone **37**, when reacted with the morpholino enamines of 3-oxotetrahydrofuran or 1-ethoxycarbonyl-3-pyrrolidone, after dehydrogenation and dealkoxycarbonylation yields furan **47a** and pyrrole **47b**, respectively (86CL1021). Finally, azulenopyridines **50** and **51** can be made from unsubstituted 1-oxa-2-azulenone (**48**) and the pyrrolidino enamines of 1-ethoxycarbonyl-4-piperidone (enamine **49**) or -3-piperidone, respectively (87SAA1067).

2.3.4 From cycloheptatrienes

Yamamura's group developed a synthetic route to (β-acylvinyl)azuleno[1,2-b]- or -[2,1-b]azoles, which is exemplified in Scheme 15. Stille coupling of thienylcycloheptatriene **52b** with 2-alkyl- or 2-arylfuran derivatives (e.g., **52a**) yields furylthiophenes (e.g., **53**), which after

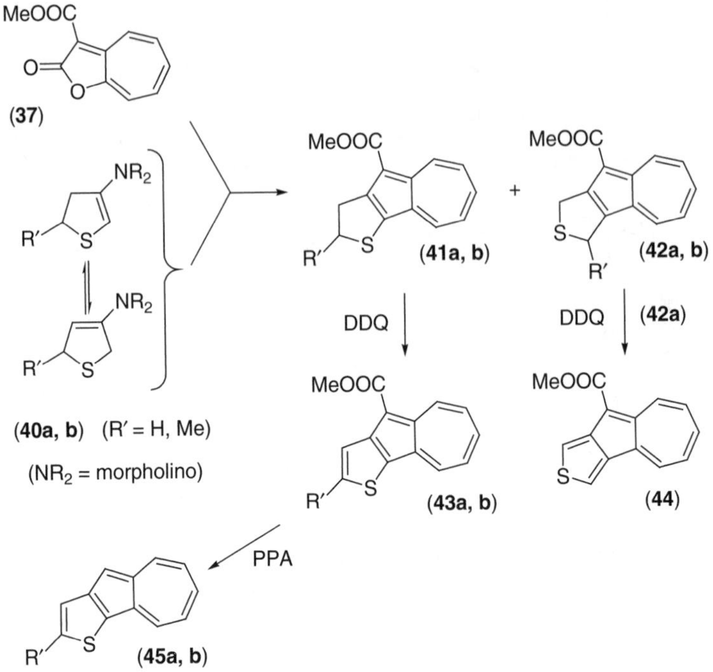

Scheme 13

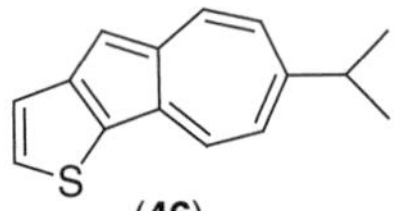

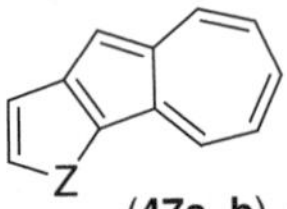

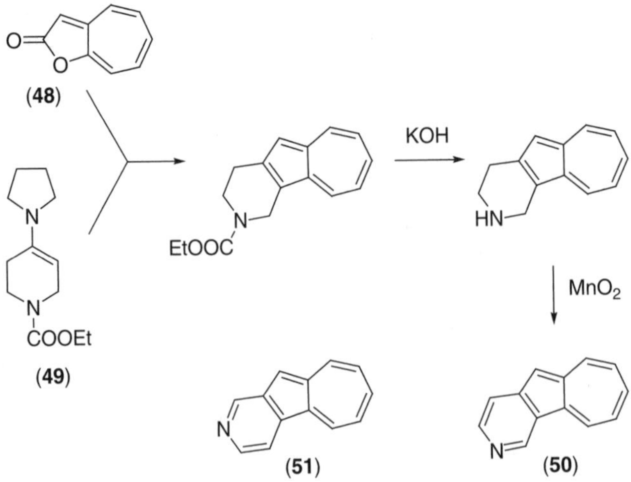

Scheme 14

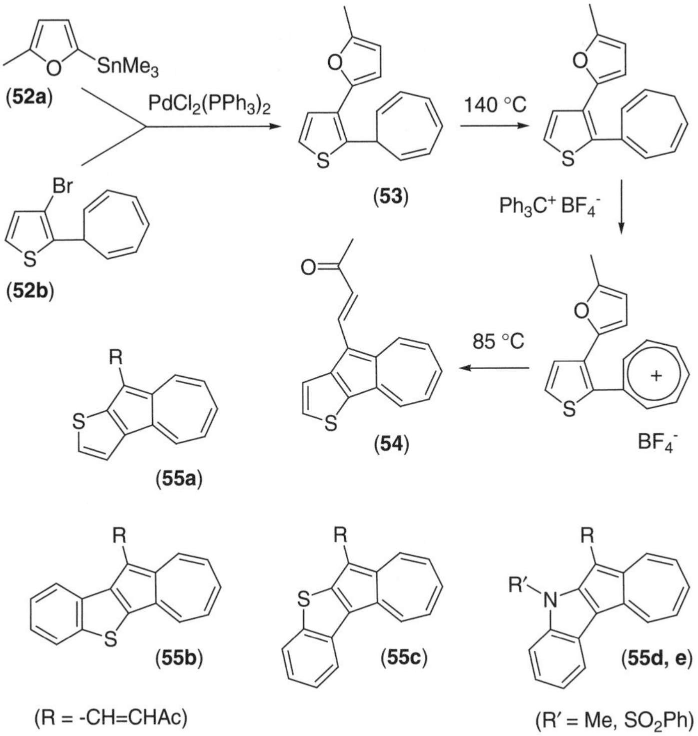

Scheme 15

isomerization (to diminish the steric hindrance), hydride abstraction, and thermal cyclization yield azuleno[1,2-*b*]thiophene derivatives, such as **54** (02T7653). A mechanism involving furan-ring opening by intramolecular attack of the tropylium ion onto the 2-position is postulated.

Analogous reactions lead to acylvinyl derivatives (types **55a–55e**) of azuleno[2,1-*b*]thiophene (02T7653), azuleno[1,2-*b*]- and azuleno[2,1-*b*]-benzothiophene (04OBC1413), and azuleno[2,1-*b*]indole (06TL8535, 07H(74)951), respectively, and the corresponding azuleno[1,2-*b*]indole derivatives (07H(74)951).

2.3.5 From azulenes by annulation onto the five-membered ring

2.3.5.1 From 1-haloazulenes. The use of functionalized azulenes, especially halo and amino derivatives (Sections 2.3.5.1–2.3.5.5), is regarded to be the most universal approach to polycyclic systems based on azulene (82JOU132). Many of the precursors for their part are readily accessible from *active troponoids* (04EJO899).

Thus, azulene derivative **56** (Scheme 16) is transformed by nucleophilic substitution into benzothiazine **57** (91H(32)213). Iodides **58a** and

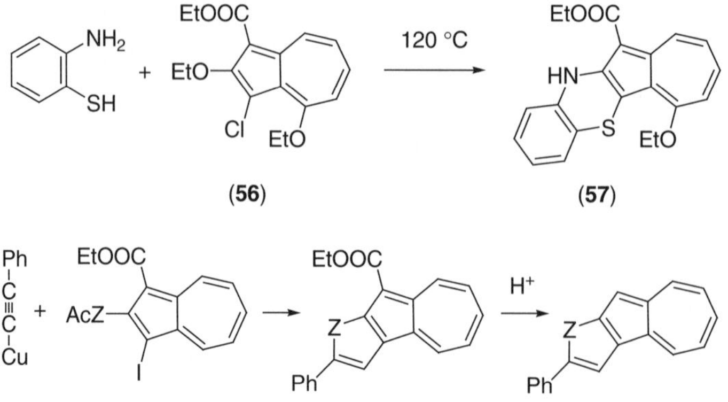

(56)   (57)

(58a, b) (Z = O, NH)   (59a, b)   (60a, b)

Scheme 16

58b with cuprous phenylacetylide, after an initial C-alkynylation, yield furan **59a** and pyrrole **59b**, which after desethoxycarbonylation yield parent substances **60a** and **60b**, respectively (81H(15)835).

2.3.5.2 From 2-chloroazulenes. According to Matsui and co-workers, thioglycolic acid derivative **62** (made from 2-chloroazulene **61**; Scheme 17) and its ester (**65**) by Dieckmann condensation yield azuleno[2,1-*b*]thiophenes **64** and **66**, respectively (60CIL1302, 61NKZ1517). 6-Isopropyl and 5-isopropyl homologs of thioglycolate **65** yield the 6-isopropyl derivative of thiophene **66** (61NKZ1665) or, respectively, a mixture of isomeric 5- and 7-isopropyl derivatives (61NKZ1522). Hydroxyl-free analog **68** is prepared from formyl compound **67** and is degraded to unsubstituted tricyclus **69** (81BCJ2537).

In a different mode (Scheme 18), chloroazulene **61** reacts with malonitrile and the product can be cyclized to yield azulenopyranone **70**, a compound of 1,2-azulenoquinone dimethide type (04BML63).

2.3.5.3 From 1-aminoazulenes. Usual methods of heterocyclization transform, for instance, 1,2-diaminoazulene **71** (Scheme 19) by the action of formic acid, nitrite, diacetyl or 5-nitrosotropolone to imidazole **72**, triazoles **73** and **271** (see Scheme 66), pyrazine **74** (and defunctionalized compound **75**), and pyrazinotroponoxime **76**, respectively (73BCJ3161, 85TL335).

1-Acetylaminoazulene derivative **77** (Scheme 20) cyclizes to yield thiazoloazulene **78** (86BCJ3320), thus providing a second access to defunctionalized compound **23** (cf. Section 2.3.2). 1-Aminoazulene **79**

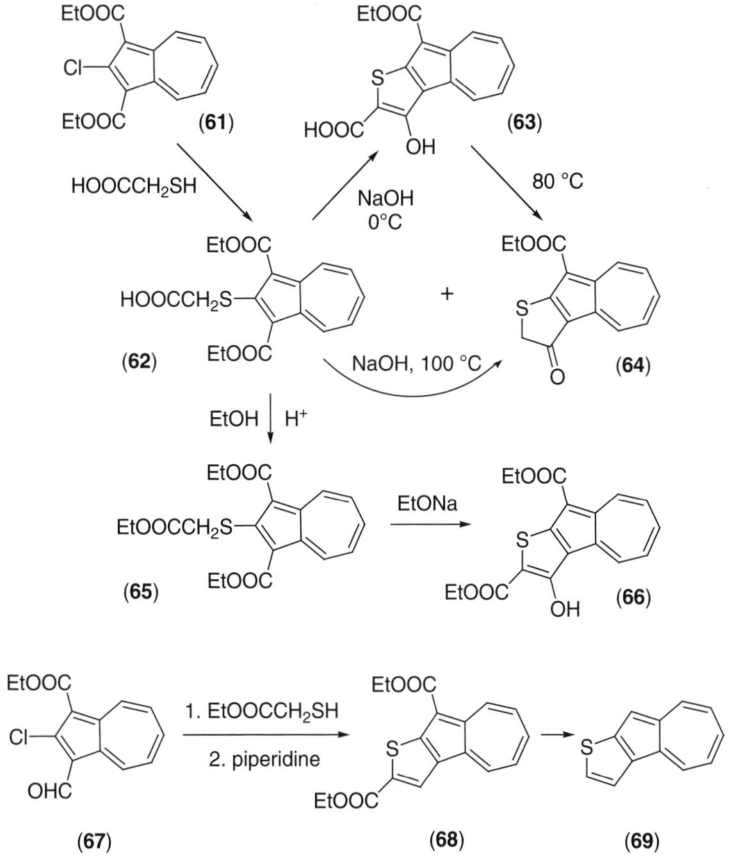

Scheme 17

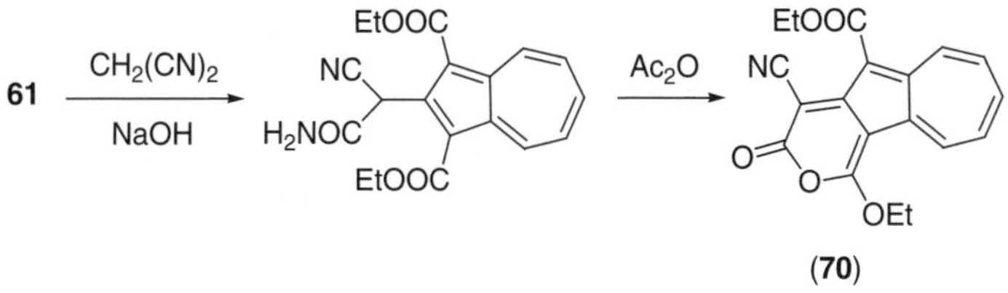

Scheme 18

condenses with a trimethinium salt to yield vinylogous amidine **80** that thermally cyclizes to yield azulenopyridine **81** (74CB2383).

2.3.5.4 From 2-aminoazulenes to form azulenoazoles.

Cyclohexanone 2-azulenylhydrazone (Scheme 21) by the Fischer indole synthesis yields

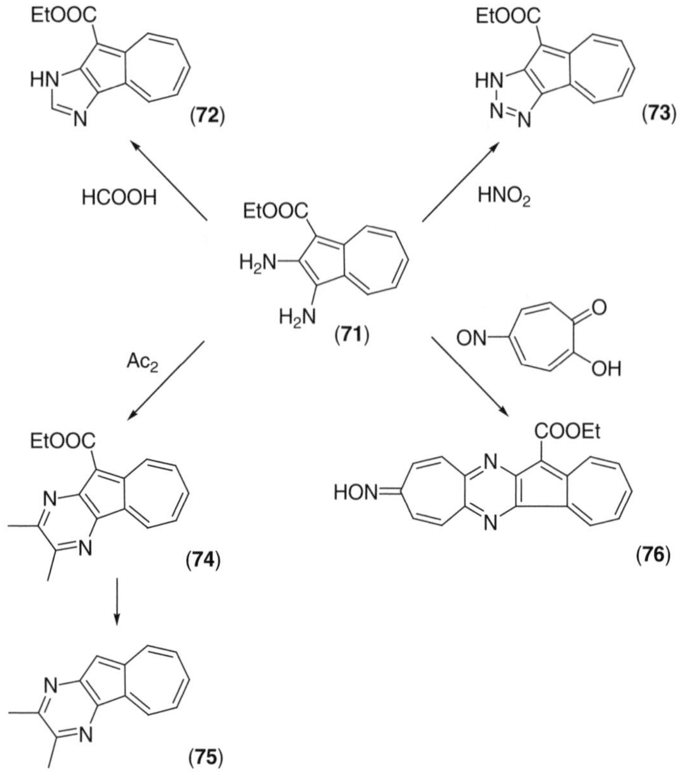

Scheme 19

tetrahydroindole **82** (61FOR(19)32, p. 51). Thiolester **83a** (an isomer of **77**) and thiocyano compound **83b** cyclize to form thiazoles **84a** and **84b**, respectively, which can be degraded to yield compound **85a** (an isomer of **23**) and parent compound **85b**, respectively (83H(20)1263, 82CL707, respectively).

(2-Azulenylimino)phosphoranes (e.g., **86**; Scheme 22) are easily prepared by the Staudinger reaction of 2-azidoazulenes and tertiary phosphines. They are (vinylimino)phosphoranes (equivalents of primary enamines), the vinylene group of which is part of an aromatic system. Nitta et al. reacted phosphorane **86** *in situ* with phenacyl bromide in an enamine-type alkylation and subsequent aza-Wittig reaction to form pyrrole **87** (93MI1), whereas the action of 2-bromotropone yielded azulenoazaazulene **88** (93TL831). The enhanced reactivity of reagent **86** as compared with phenyliminophosphorane is ascribed to the low resonance energy of azulene.

In an alternative synthesis, tetracyclus **88** was obtained from tropylidylazulene derivative **89** by hydride abstraction and subsequent

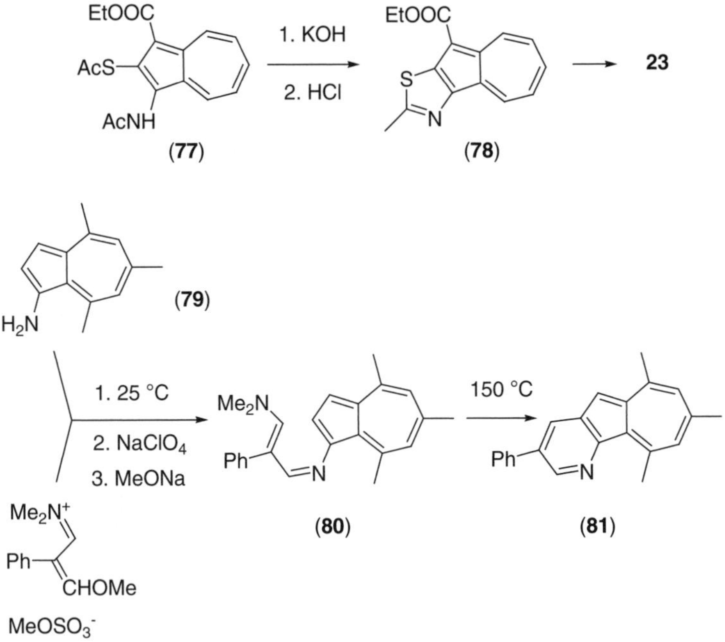

Scheme 20

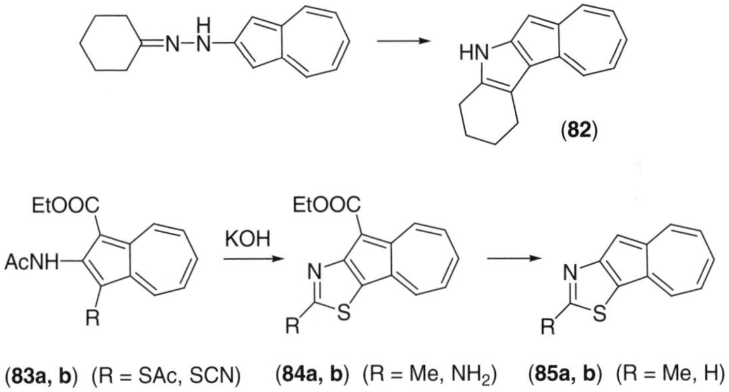

(83a, b) (R = SAc, SCN)   **(84a, b)** (R = Me, NH$_2$)   **(85a, b)** (R = Me, H)

Scheme 21

defunctionalization (93TL831). For the sake of completeness, a similar strategy to prepare derivatives (such as **90**) of other azulenoazaazulenes (isomers of **88**, but heteroannulated onto the seven-membered azulene ring) from tropylidenes deserves mentioning (81H(15)547).

2.3.5.5 *From 2-aminoazulenes to form azulenoazines.* The heterocyclization according to the method of Nitta et al. (see the preceding section),

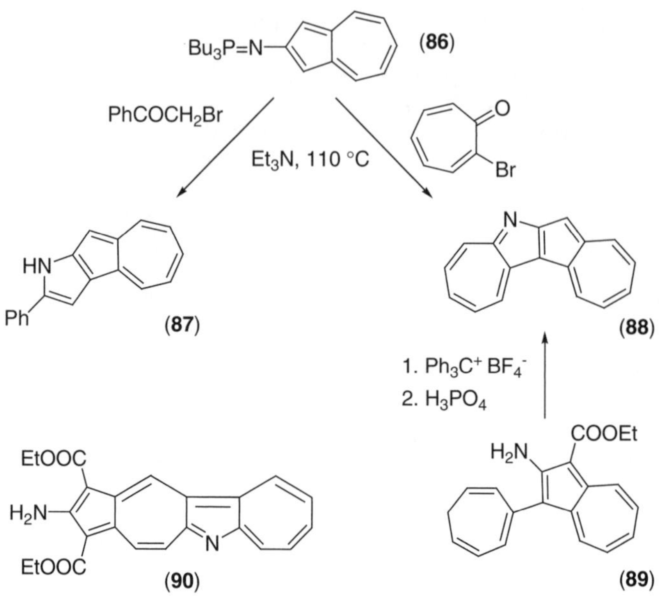

Scheme 22

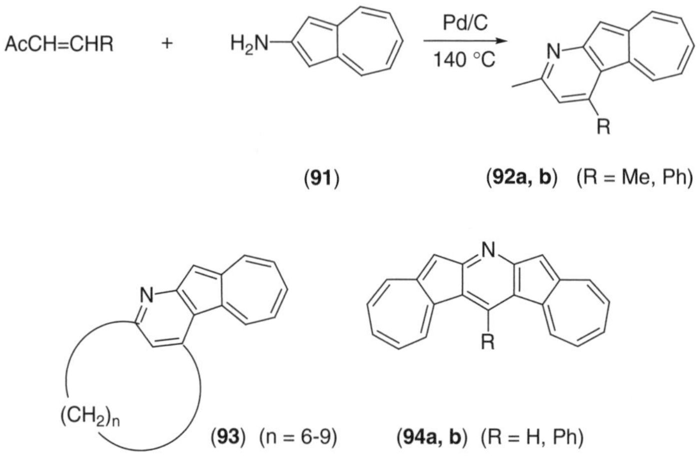

Scheme 23

when accomplished by acyclic α,β-unsaturated carbonyl compounds, is more conveniently applied to 2-aminoazulene (**91**) itself (94JOC1309). Thus, in the presence of palladium-on-charcoal as dehydrogenation catalyst, **azulenopyridines 92a** and **92b** are obtained (Scheme 23). Similar reactions of 2-cycloalkenones (C_{9-12}) with 2-aminoazulene (**91**) afford,

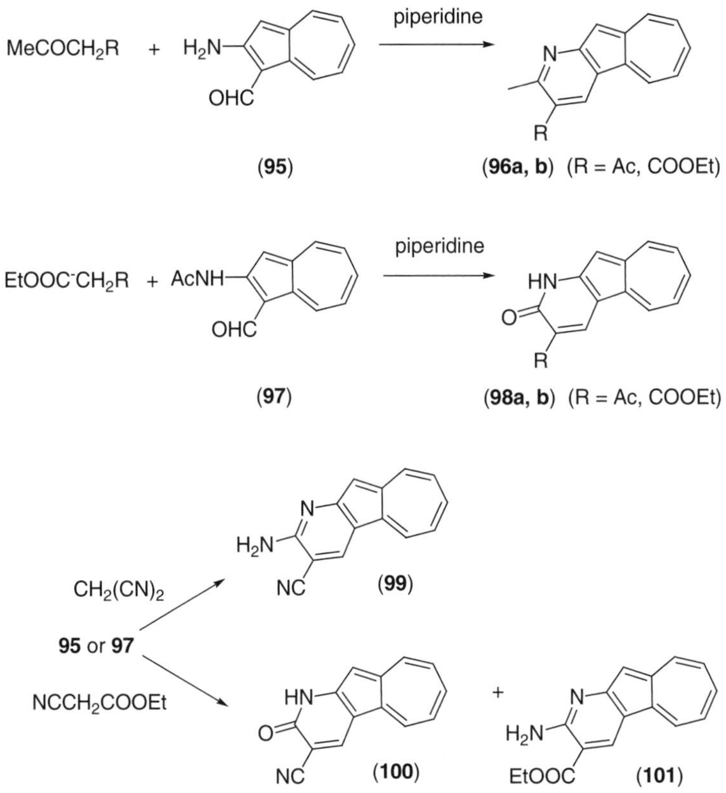

Scheme 24

partly after dehydrogenation by DDQ, azuleno-annulated [*n*](2,4)pyridinophanes (**93**).

The reaction of 2-aminoazulene with formaldehyde and benzaldehyde does not stop at the expected di(1-azulenyl)methane stage but immediately affords diazulenopyridines **94a** and **94b**, respectively (90PAC507, 01H(54)667).

Bifunctional derivatives, such as 1-formyl-2-aminoazulene (**95;** Scheme 24) and 1-formyl-2-acetylaminoazulene (**97**), with 1,3-dicarbonyl compounds or their cyano-group containing analogs undergo Friedländer reactions to yield azuleno[2,1-*b*]pyridines and -2-pyridones (63BCJ633). In this way cyclization with acetylacetone, ethyl acetoacetate or diethyl malonate yields pyridines **96a** and **96b** and pyridones **98a** and **98b**, whereas the condensation of azulene **95** and diethyl acetonedicarboxylate leads to the 2-CH$_2$COOEt analog of product **96b**. Remarkably, azulenes **95** and **97** with ethyl acetoacetate, due to different modes of cyclization, yield different products (**96b** and **98a**, respectively).

Finally, the attack of cyano-containing reagents leads to pyridines **99** and **101** and pyridone **100**.

2-Pyridones (Scheme 25) are also the result of condensing 2-amino-azulene (**91**) or its derivative **104** with electron-deficient acetylenes, such as methyl propiolate or dimethyl acetylenedicarboxylate (91BCJ2393). The Michael addition dominates in these reactions. 2-Aminoazulene yields pyridones **102** and **103**, respectively (the latter forming by incorporation of two mol ester). In the corresponding reactions of ester **104** bicyclic intermediates (**106a** and **106b**) occur as by-products that can also cyclize to yield the target pyridones (**105a** and **105b**).

In related syntheses (Scheme 26), aminoazulenes are reacted with dibenzoylacetylene or diethyl (ethoxymethylene)malonate to afford pyridines **107a** and **107b** and 4-pyridone **108**, respectively.

Classic methods of pyrimidine synthesis enable the preparation of **azulenopyrimidines**, azuleno-4-pyrimidones, and azulenopyrimidine-diones. One type of reactions (Scheme 27) consists in the cyclization of 2-aminoazulenes (**95**, **110**) by means of nitrogen-containing reagents

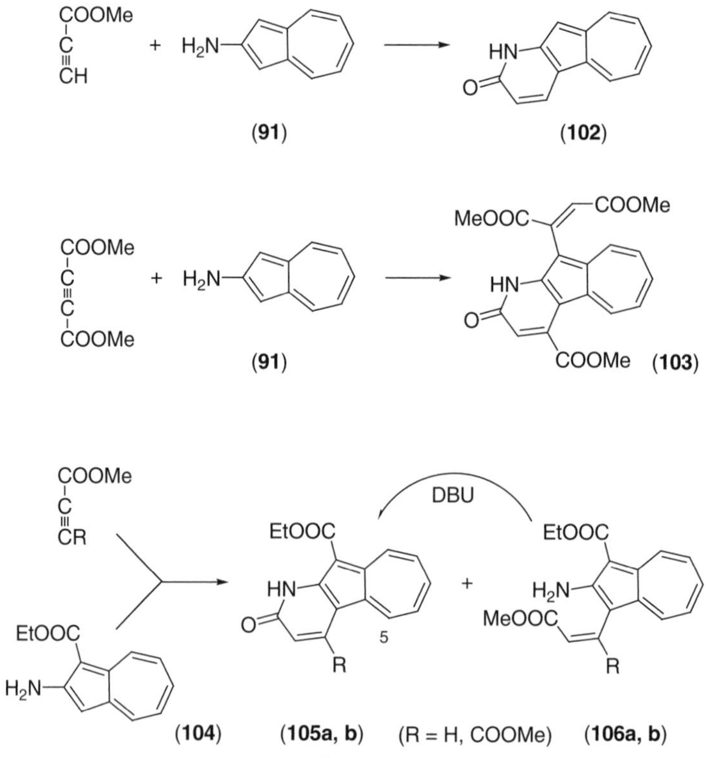

Scheme 25

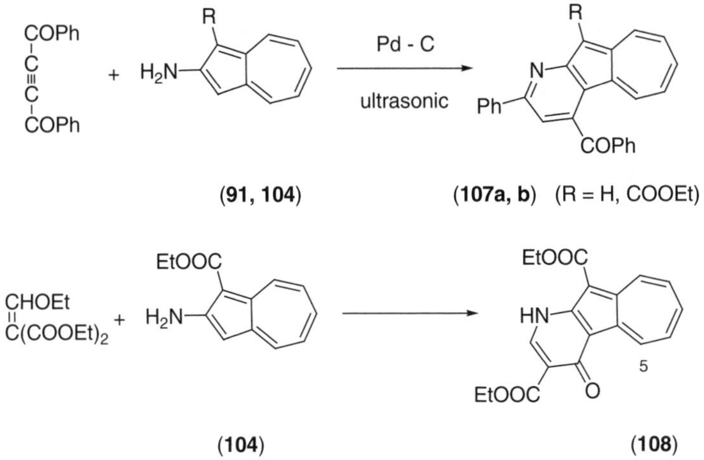

(91, 104)   (107a, b) (R = H, COOEt)

(104)   (108)

Scheme 26

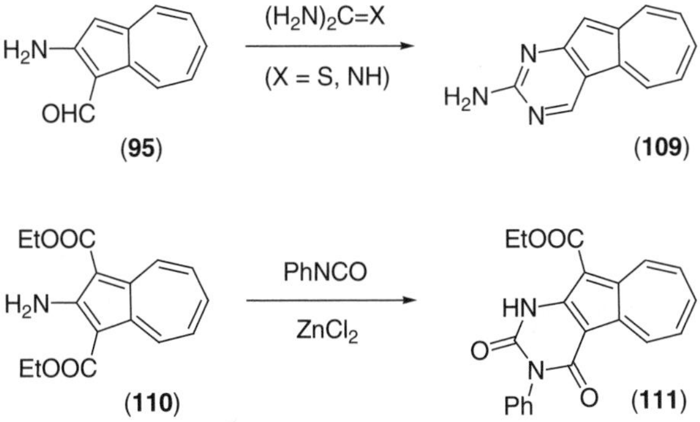

(95)   (109)

(110)   (111)

Scheme 27

(thiourea, guanidine, isocyanate) to yield aminopyrimidine **109** (73BCJ3161) or pyrimidine **111**, the latter of which is also a by-product in the corresponding reaction of monocarboxylic ester **104** (97JRM2434).

Another reaction type (Scheme 28) leads from 2-aminoazulenes having an additional nitrogen functionality (e.g., **112**, **114**) by means of formic acid or acetanhydride to 4-pyrimidones, for instance, substances **113a**, **113b**, and **115** (64BCJ859, 81H(15)839).

Finally, **azulenopyrazines 117a** and **117b** are synthesized from nitrosoazulene **116** and active methylene compounds (68BCJ2095).

2.3.5.6 From 1-acylazulenes. 1-Acetylazulene **118** (Scheme 29) by the action of phenylhydrazines yields azuleno[1,2-*d*]pyrazoles, for instance,

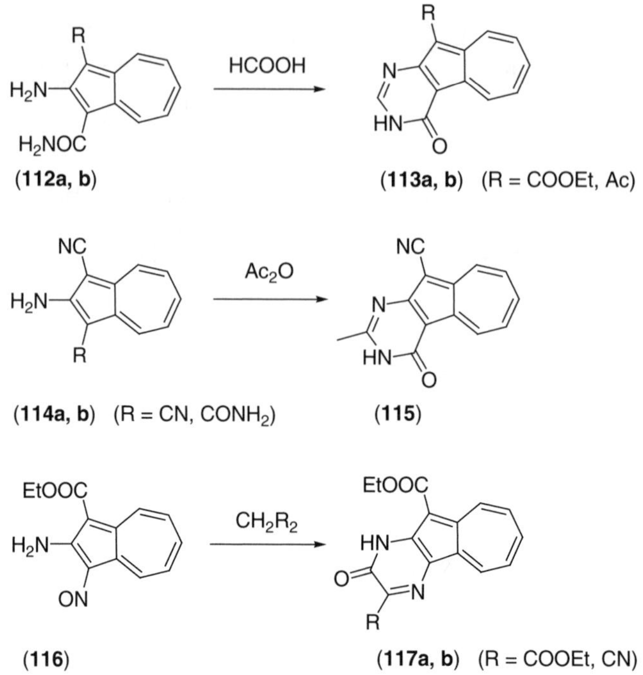

Scheme 28

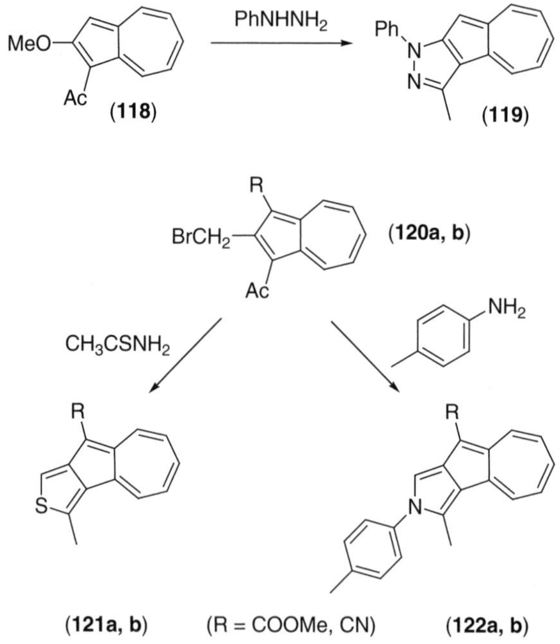

Scheme 29

substance **119** (02H(56)497). Reacting 1-acyl-2-bromomethylazulenes (e.g., **120a** and **120b**) with thioamides or anilines, on the other hand, allows access to *o*-quinonoid azuleno[1,2-*c*]thiophenes (e.g., **121a** and **121b**) (02H(58)405, 06IJB731) and azuleno[1,2-*c*]pyrroles (e.g., **122a** and **122b**) (01H(54)647), respectively.

2.3.5.7 From azulenic dicarbonyl compounds. Condensation of 1,2-azulenequinones **123a–123c** (Scheme 30) with *o*-phenylenediamine leads to azulenoquinoxalines **124a–124c** (80CL197). In an inverse manner, 1,2-diacylazulenes **125a–125c** (Scheme 31) in the presence of hydrazine cyclize to yield azulenopyridazines **126a** (74JOU1320), **126b** (88BCJ1225), and **127** (82JOU132), respectively. Finally, when both the 1,4-functionalities are

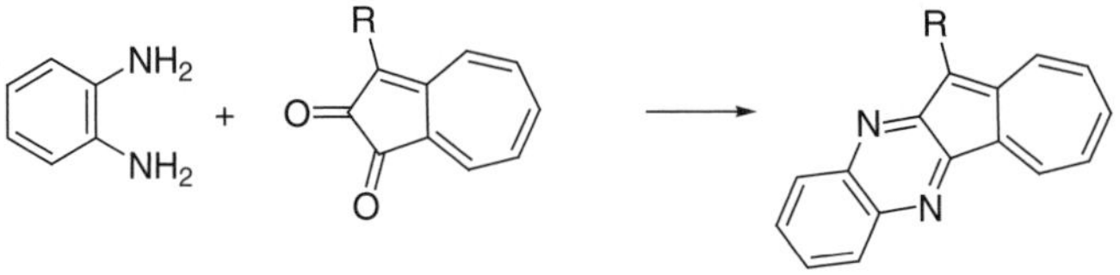

(123a-c)   (R = H, COOEt, CN)   (124a-c)

Scheme 30

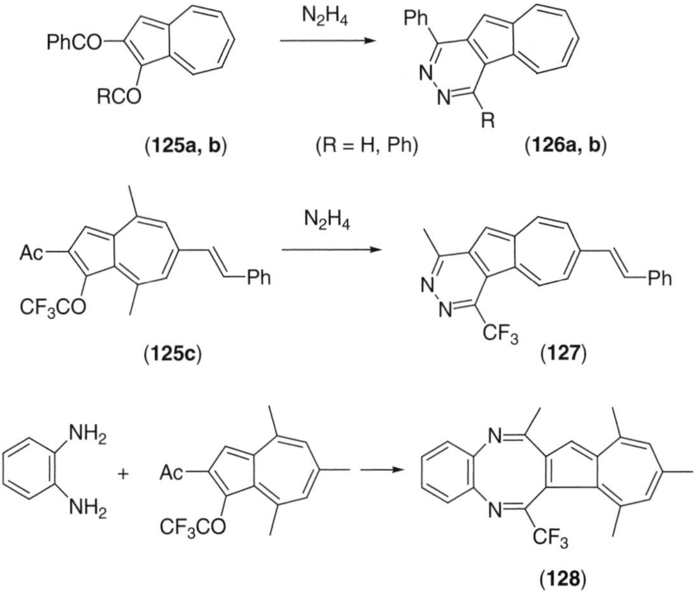

(125a, b)   (R = H, Ph)   (126a, b)

(125c)   (127)

(128)

Scheme 31

combined, the unusual azuleno[1,2-f][1,4]diazocine ring system (as in **128**) can be constructed (82JOU132).

2.3.6 From azulenes by annulation at the *peri*-position

Guaiazulene (**129**; Scheme 32) as starting material is characterized by its active 4-methyl group and the possibility of electrophilic attack at the 3-position. Thus, its sodium salt (**130**) with benzonitriles by imination and hydrolysis yields 4-phenacyl derivatives **131a** and **131b** that in the presence of thiocyanogen yield 3-thiocyano compounds **132a** and **132b** (68JOC823); these are capable of various reactions.

Under the influence of base in the absence of air they release hydrocyanic acid and cyclize to yield azuleno[1,8-*bc*]thiophenes **133a** and **133b**, whereas in the presence of air dimer **134** is obtained (71JA2196). Reduction of compound **132a**, however, results in the cyclization of an intermediate mercaptan to yield an unstable cyclic hemimercaptal (**135**), which is readily dehydrated to form thiopyran **136** (68JOC823). Another thiopyran can be prepared from 4,6,8-trimethylazulene though on a more complex reaction path because the starting azulene offers two reactive sites for thiocyanation.

A by-product of the initial imination (of **130**) is azuleno[1,8-*cd*]azepine **137**, an intramolecular cyclization of the imine salt presumably being involved.

2.3.7 From fulvenes

In a number of syntheses (Schemes 33 and 34) specific fulvene and heterocyclic derivatives contribute the five-membered azulene ring and the heterocyclic moiety, respectively, to the target azuleno[5,6-*b*]heterocycles, the seven-membered ring being newly formed in the reaction. Thus, [6+4] cycloaddition of electron-rich fulveneketene acetals (e.g., **138a**) or 6-aminofulvenes (e.g., **138b**) and electron-deficient 2-pyrones (e.g., **139**) followed by extrusion of carbon dioxide provides azulenoindols **140a** and **140b**, respectively (96CC937, 01BML1981). Microwave assistance raises the yield, for instance, that of indole **140b** from 26% to 65%.

In another reaction type, related to the *Ziegler–Hafner azulene synthesis* (cf. 85HOU(5/2c)127, p. 182), the cyclic heptamethinium salt **141** condenses with acidic three-carbon components (allylic anions). Subsequent thermal cyclization and elimination of dimethylamine yields azulenes, for instance, in the case of 2-methylchromone (**142**) as component, azulenochromone **143** (74CB2956).

Several quinonoid azuleno[*c*]furans can be prepared by a *tandem cycloaddition–cycloreversion strategy* in the course of multi-step syntheses (Scheme 34), the furan ring being introduced at an early stage (03OBC2383). Thus, the cycloaddition of sulfone **144** (generated *in situ*)

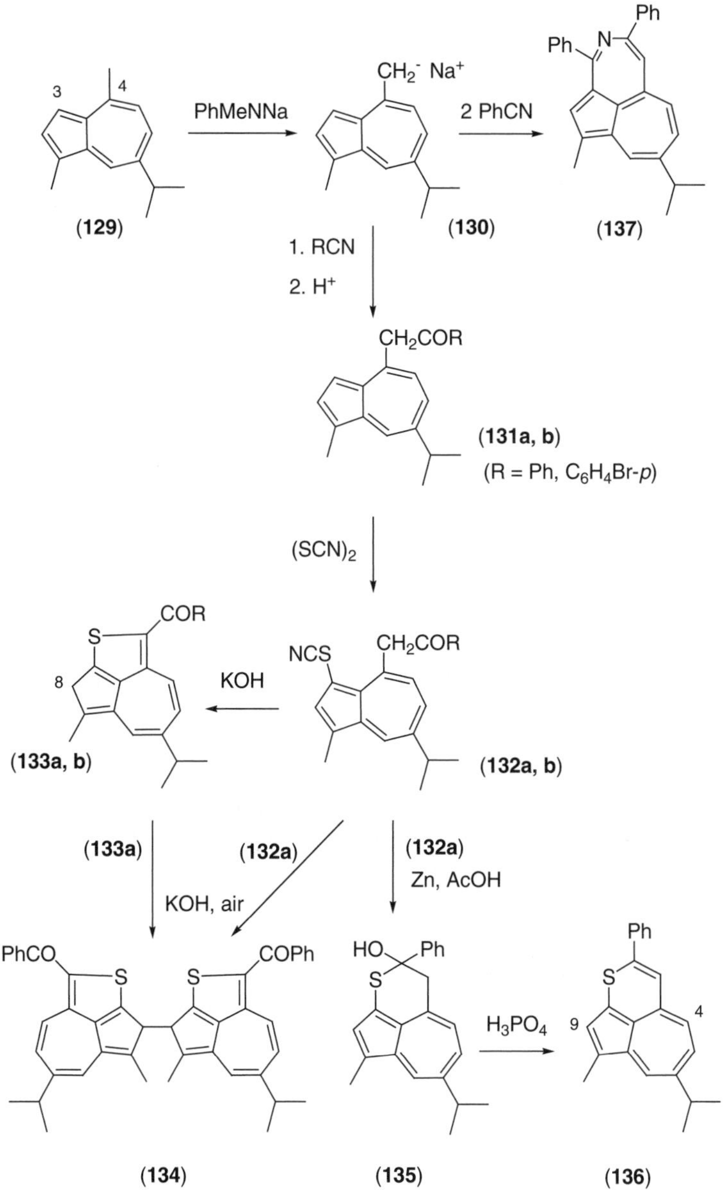

Scheme 32

and furan yields endoxide **145** as a mixture of roughly equal amounts of *endo* and *exo* isomers. The application of the *Houk–Leaver azulene synthesis* (cf. 85HOU(5/2c)127, p. 205) using the Diels–Alder reaction with 6-dimethylaminofulvene (**138b**) in the presence of base, through

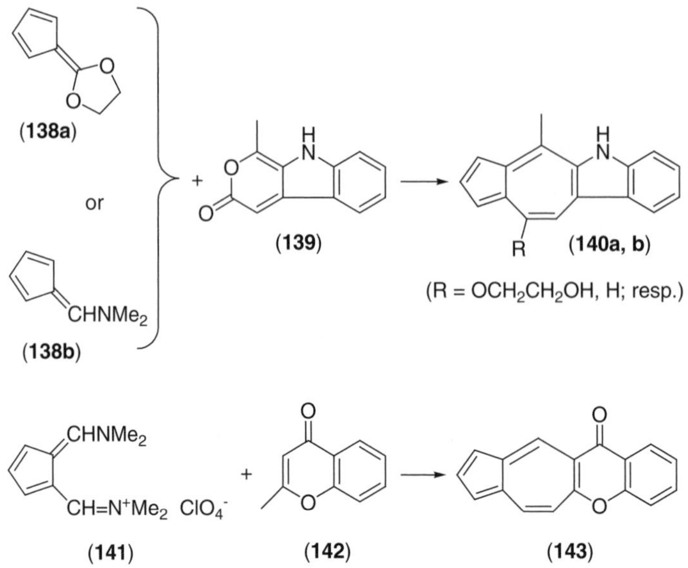

Scheme 33

intermediate **146**, yields endoxybenzazulene **147**. The extrusion of the etheno bridge is readily effected by cycloaddition of 3,6-di(2-pyridyl) [1,2,4,5]tetrazine (**148**) and cycloreversion to yield moderately stable azuleno[4,5-*c*]furan **149**.

On a partially analogous route adduct **151**, obtained by trapping 5,6-didehydroazulene (from **150**) with furan, is transformed by means of tetrazine **148** to stable azuleno[5,6-*c*]furan **152**. Both azulenes **149** and **152** can be characterized as adducts (see Section 4.5.3).

2.3.8 From bis(ethynylphenyl) compounds

The *diyne reaction* can be utilized to obtain pentacyclic azulene derivatives (Scheme 35). Silane **153a**, phosphine oxide **153b**, and sulfoxide **153c**, in the presence of palladium(II) chloride or its complexes, isomerize, presumably passing coordinated cyclobutadiene intermediates **154a–154c**, to yield silepin **155a** (73CZ447), phosphepin oxide **155b** (76ZNB1116), and thiepin oxide **155c** (74CZ41), respectively.

2.3.9 Azuliporphyrins and analogs

Lash's group constructed *azuliporphyrins* (azulene-containing porphyrinoids), such as monoazulene macrocyclus **157** (Scheme 36), by exploiting the *[3+1] methodology*, for instance, from tripyrrane **156** and azulene-1,3-dialdehyde (97AGE839). An inverse variant of this [3+1] method (Scheme 37) is exemplified by the reaction of an azulitripyrrane analog

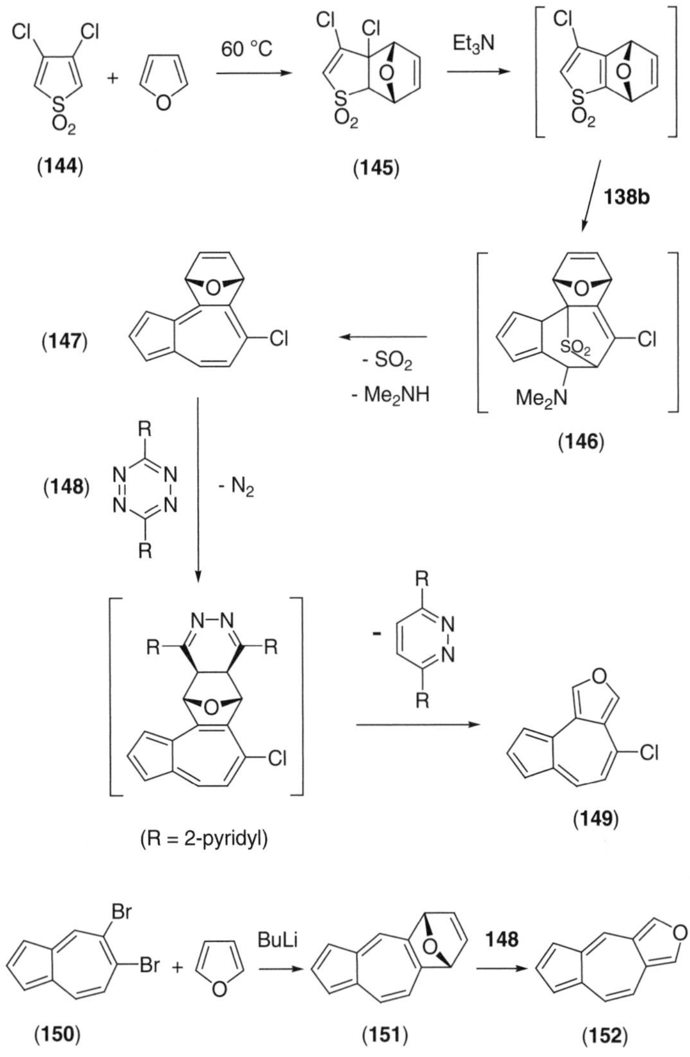

Scheme 34

with the dialdehydes of diethylpyrrole, thiophene or indene under acidic conditions and subsequent oxidation (DDQ or $FeCl_3$) to yield azuliporphyrin **159a** and analogs **159b** and **160** (02AGE1371, 07JOC8402). A one-pot synthetic method is available (07EJO3981).

A third method (Scheme 38) utilizes the suitability of azulene as a substrate for *Rothemund-type condensations*. The action of 2,5-bis (α-hydroxy-p-methylbenzyl)thiophene on azulene in the presence of boron trifluoride yields porphyrinogen **161a**. This intermediate is

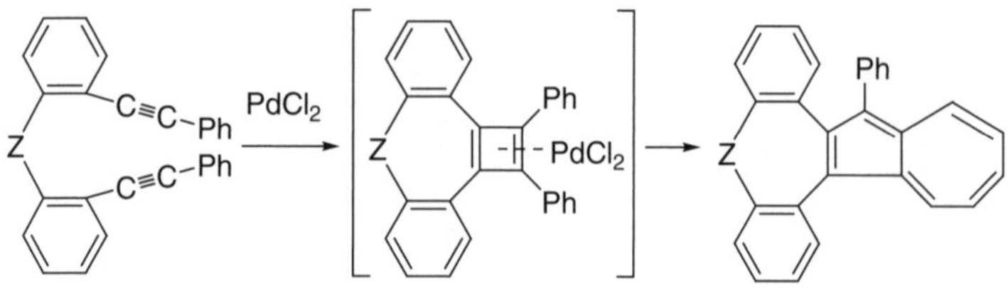

(153a-c)   (154a-c)   (155a-c)

[Z = SiMe$_2$, P(=O)Ph, S=O]

Scheme 35

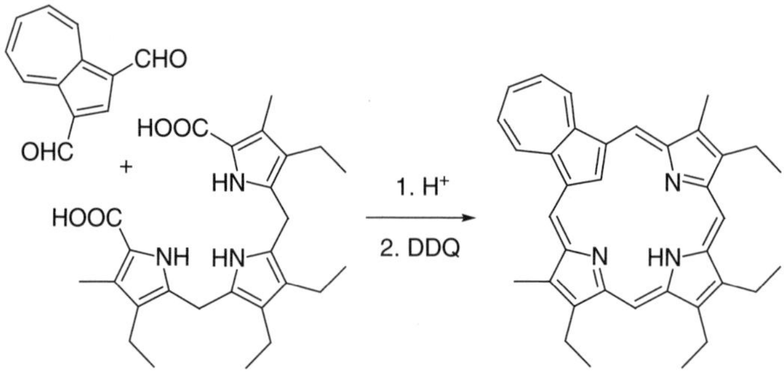

(156)   (157)

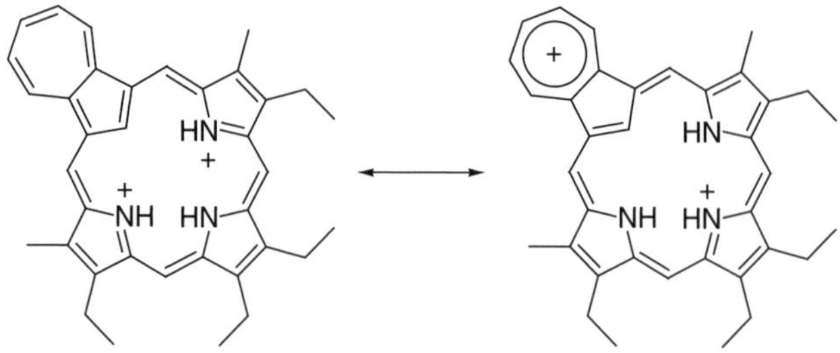

(158a)   (158b)

Scheme 36

oxidized by DDQ to mixtures of dithiadiazuliporphyrin **161b**, its radical cation, and dication, these three species constituting a *multielectron redox system* (05JA13108).

Just recently a [2+2] methodology has been described (07JA13800, 08MI1), which enables the synthesis of *adj*-diazuliporphyrins (dicarbaporphyrinoid systems with two adjacent azulene subunits) by the condensation of dipyrrylmethanes with diazulene dialdehydes.

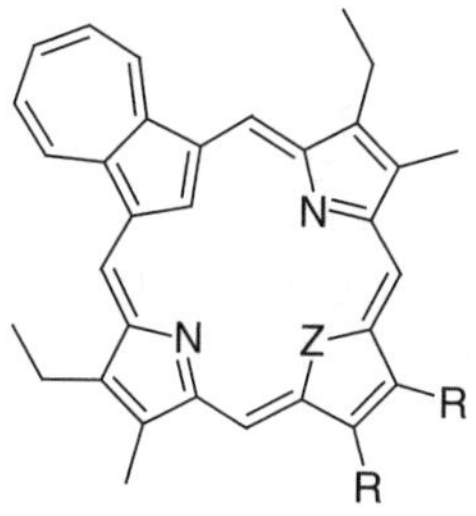

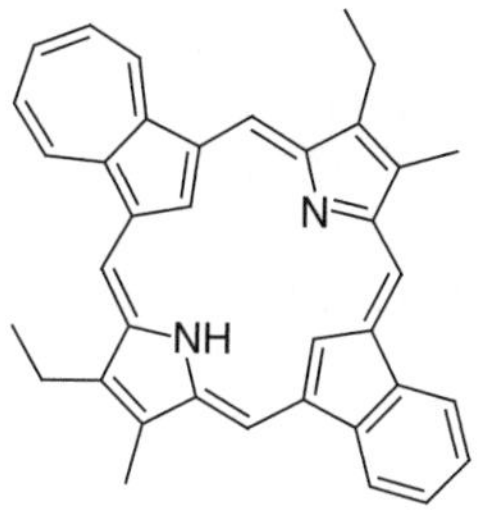

(**159a**: Z = NH, R = Et)

(**159b**: Z = S, R = H)

(**160**)

Scheme 37

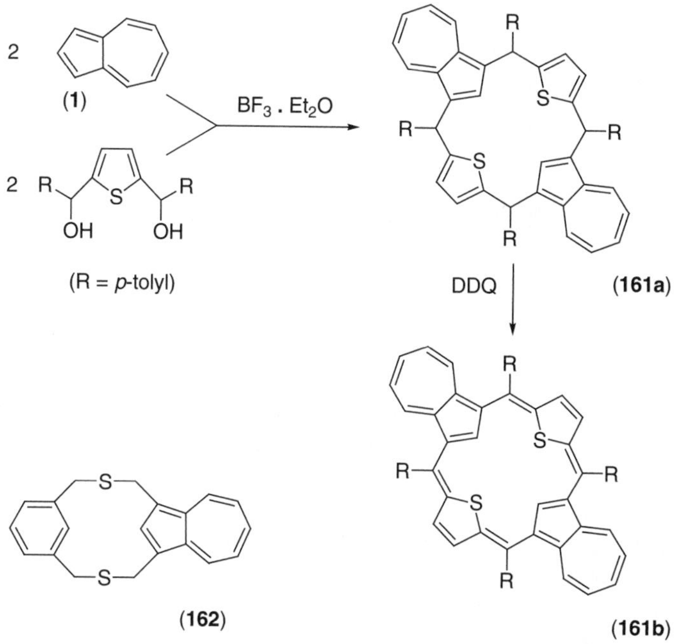

(**162**)

(**161b**)

Scheme 38

The unprotonated and protonated forms of azuliporphyrins mentioned above show differentiated degrees of aromaticity. (General remarks may be found in Sections 3.4.1 and 4.1.1.) Thus, spectroscopic data suggest that compound **157** has borderline porphyrinoid aromaticity, dipolar tropylium forms contributing but not dominating. Its dication **158a**, obtained by addition of trace amounts of trifluoroacetic acid, however, displays typical porphyrin-like nature dominated by the mesomeric tropylium form **158b** (97AGE839). Compounds **159a** and **159b** (quite similar to **157**) show fully aromatic character, whereas indene **160** does not but forms mono- and diprotonated species stabilized by charge delocalization (02AGE1371). X-ray crystallography of compound **161b** finally discloses the azulene geometry to be largely unaltered, suggesting that these rings are not conjugated with the macrocyclus (05JA13108). The azuliporphyrin topic has been reviewed in Ref. (07EJO5461).

Among other macrocyclic azulene derivatives, sulfur-containing metacyclo(1,3)azulenophanes (e.g., **162**) have been described (79MI2).

2.4 Syntheses of azulenes fused to individual heterocyclic rings

2.4.1 Five-membered rings

2.4.1.1 Furans. Aminoazulene **163b** (Scheme 39) can be obtained from β-dolabrin (**163a**) by Nozoe's azulene synthesis. Deamination is accompanied by furan-ring closure to yield azuleno[6,5-*b*]furan **164** and, after defunctionalization, compound **165** (61CIL1715).

In an attempt to generate parent azuleno[4,5-*c*]furan **168** by reducing ester **166**, etherifying hemiacetal **167**, and splitting off methanol the target substance could not be isolated but was trapped by means of maleic anhydride as cycloadduct **169** (83CC1025).

2.4.1.2 Thiophenes. Heating properly substituted azulenes with sulfur yields azuleno[*b*]thiophenes, sulfur serving as both dehydrogenating and thiophene-ring forming agent (Scheme 40). Thus, vetivazulene (**170**) yields azuleno[1,2-*b*]thiophene **171** (69BCJ1404), whereas guaiazulene (**129**) forms azuleno[6,5-*b*]thiophene **15** (67MI1) that is also known as a minor by-product of dehydrogenating several proazulenes with sulfur (Section 2.3.1). 3-Acetylguaiazulene (**172**) when sulfurized yields 2-acetyl- (**173a**) and 3-acetylazulenothiophene derivative **173b**, the former being the result of a thermal acetyl-group migration (70BCJ509). Finally, azuleno[1,2-*b*]thiophenes **175** and **176**, respectively, are formed in the reactions of 2-styrylazulene **174** or of azulene and diphenylacetylene with sulfur in boiling quinoline (82JOU132).

1-Azulenylthioketones, for instance, phenyl derivative **177b** (Scheme 41) or di(1-azulenyl)thioketone, when prepared *in situ* from ketones (e.g., **177a**) by sulfurization, undergo intramolecular pericyclization to yield 3*H*-azuleno[8,1-*bc*]thiophenes (e.g., **178a**) provided they bear alkyl groups in the 3-position (08JOC2256). The action of phosphorus pentasulfide/triethylamine as the sulfur-donating reagent on ketones (such as **177a**) yields isolable thioketones that under thermal or acid-catalyzed conditions pericyclize. The proposed mechanism of this reaction involves a concerted 1,5-hydrogen transfer as demonstrated by NMR spectra (including those obtained in the presence of DCl) and by calculation. Hydride abstraction

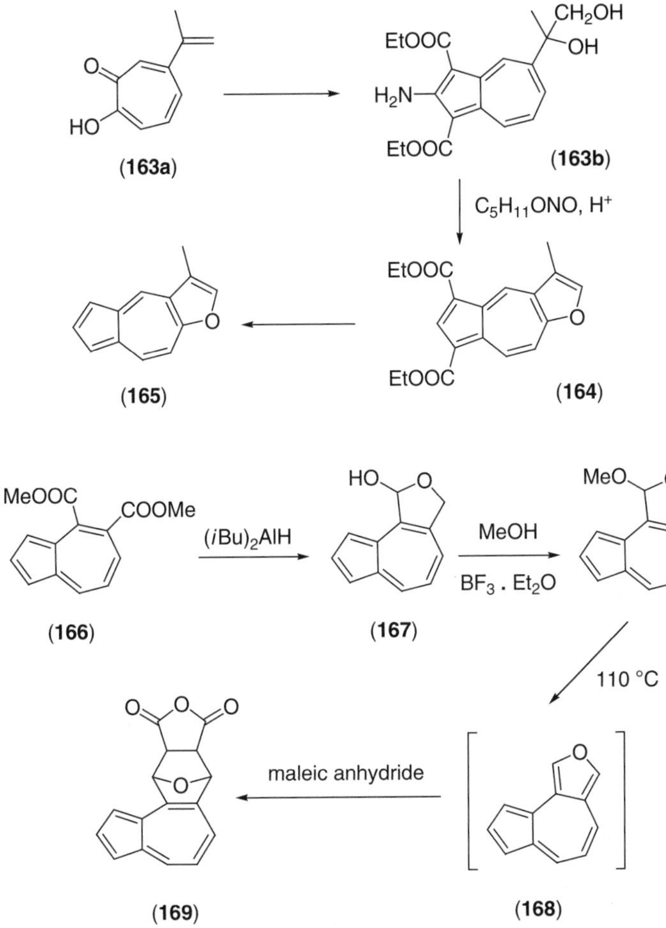

Scheme 39

by means of DDQ forms stable (resonance-stabilized) salts (e.g., **178b**), which have been studied by cyclic voltammetry (relating to the redox behavior), electrochromic and ESR analysis.

In contrast, azuleno[1,8-*bc*]thiophene **179c** is generated by the acid-catalyzed cyclodehydration of sulfine **179a** and demethylation of the resulting sulfenium salt **179b** (71JA2196).

2.4.1.3 [1,3]Dithioles. 5,6-Dichloroazulene is smoothly converted into azulenodithiolethione **180** (98JMC289).

2.4.1.4 Isoxazoles. The diazotization of aminoazulene **181** (Scheme 42) proceeds in an unexpected manner and is dependent on small

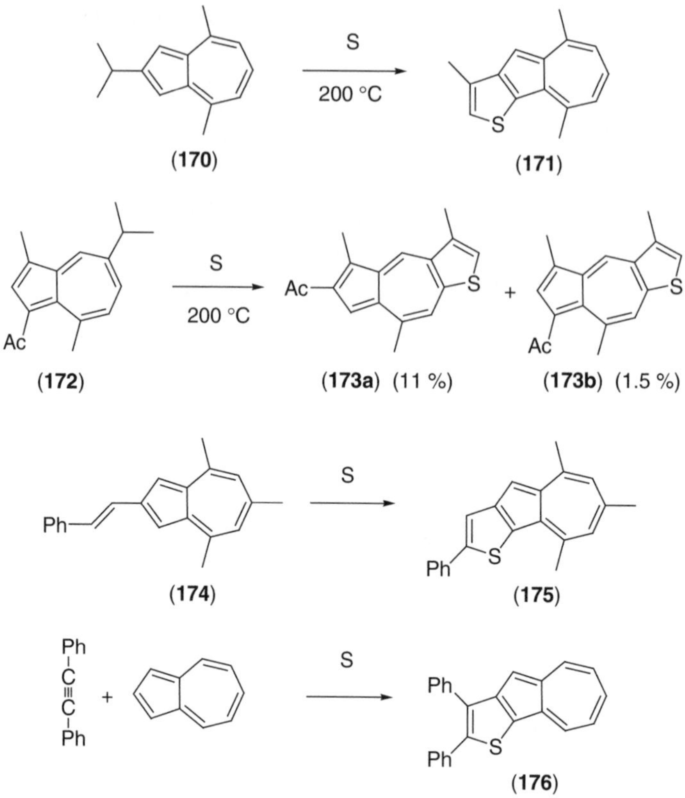

Scheme 40

differences in the conditions (86CB2956). Among the main products are isoxazole **182** and labile diazo compound **183**, respectively, the latter being easily transformed to chlorine-free isoxazole **184**. A third reaction path through oxime **185** yields, again dependent on the conditions of diazotization, isoxazoles **182** and **184**, respectively. A suggested mechanism involves oximation of acetate **181** (to yield **185**), diazotization on C-2, cyclization and oxidation (to form the isoxazole ring), and decomposition of the diazonium salt to yield isoxazoles **182** and **184**, dependent on the presence or absence, respectively, of hydrochloric acid.

2.4.1.5 Thiazoles. 6-Amino-5-thiocyanoazulenes (e.g., **186**; Scheme 43) or 5-bromo-6-thioureidoazulenes (e.g., **192**; Scheme 44) are starting materials to get azuleno[6,5-*d*]thiazoles. Thus, compound **186** cyclizes to yield 2-aminothiazole **187** (degraded step by step to **188a** and **188b**) or 2-methylthiazoles **189a**, **189b** and (after degradation) **190** [96H(43)1049]. The hydroxy analog of acetamide **186** similarly yields 2-methylthiazole

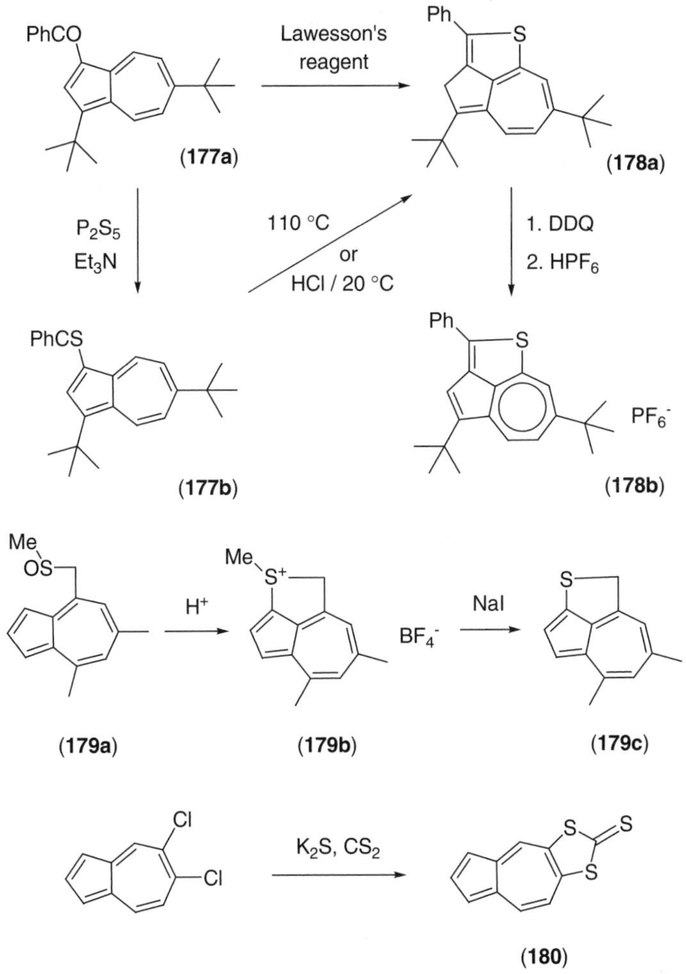

Scheme 41

191 (96MI1). 2-Benzamido analogs (**193a** and **193b**) are available from benzoylthioureido compounds **192a** and **192b** (Scheme 44).

2.4.1.6 Pyrroles. The action of liquid ammonia upon furan **164** initiates heteroatom exchange to afford azulenopyrroles **194** and (after degradation) **195** (61CIL1715).

Azaazulenes are versatile precursors in syntheses of *peri*-fused azulenopyrrole derivatives in the course of the [8+2] cycloaddition by means of electron-deficient acetylenes or of acetylene equivalents (Schemes 45 and 46). Treatment of 1-aza-2-azulenones **196a** and **196b**

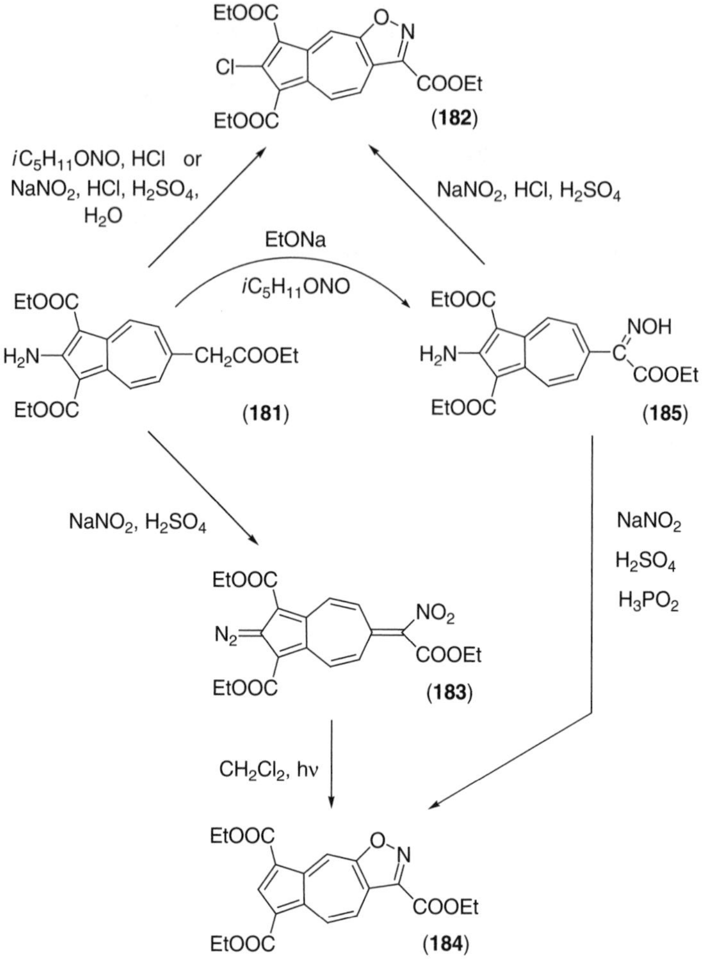

Scheme 42

with dimethyl acetylenedicarboxylate or dibenzoylacetylene yields, among other products, azuleno[8,1-*bc*]pyrrolones **198a**–**198c**, partly together with dihydro compounds (**197a** and **197b**) that can be dehydrogenated to yield target substances **198a** and **198b** (88BCJ1225).

The action of endoxide **199** on azaazulene **196c** under high pressure in a thermal cycloaddition–cycloreversion reaction affords the *N*-acetyl derivative (**198d**) of pyrrolone **198c** (89BCJ1567). On the other hand, benzyne prepared from precursors **200a** or **200b** reacts *in situ* with azaazulenes **201a** and **201b** to yield small amounts of benzazulenopyrroles **202a**, **202b**, and **203** along with other products (02JRM162).

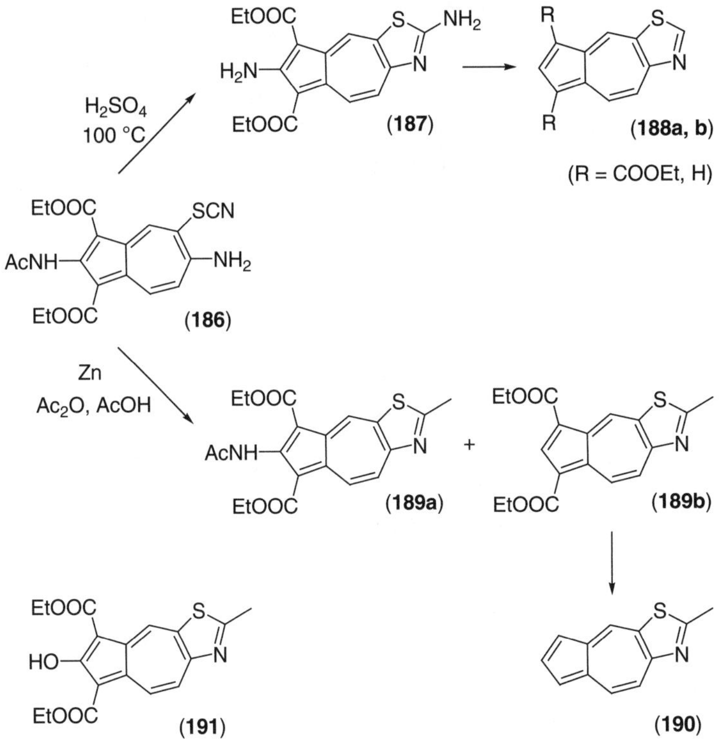

Scheme 43

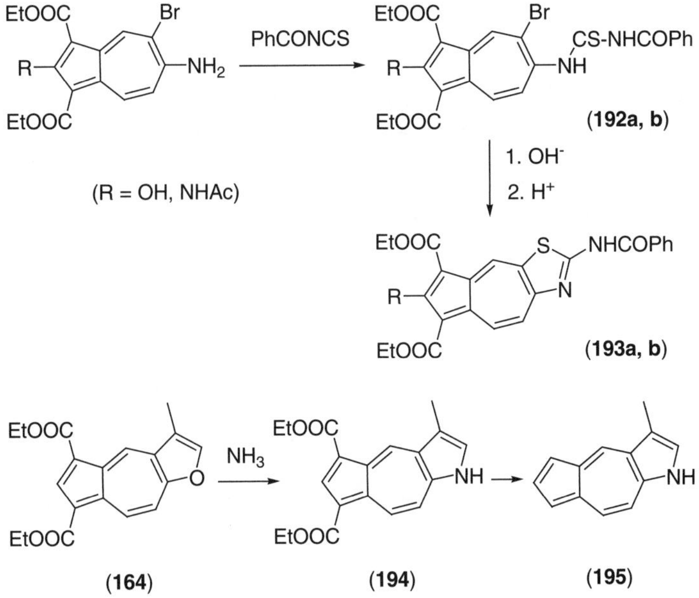

Scheme 44

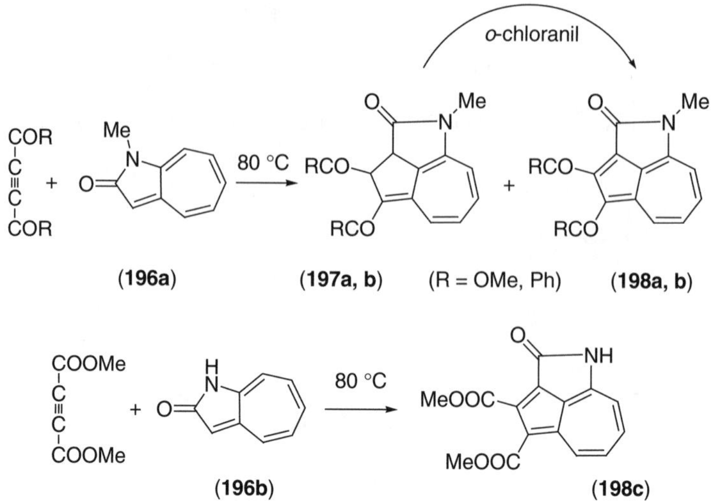

Scheme 45

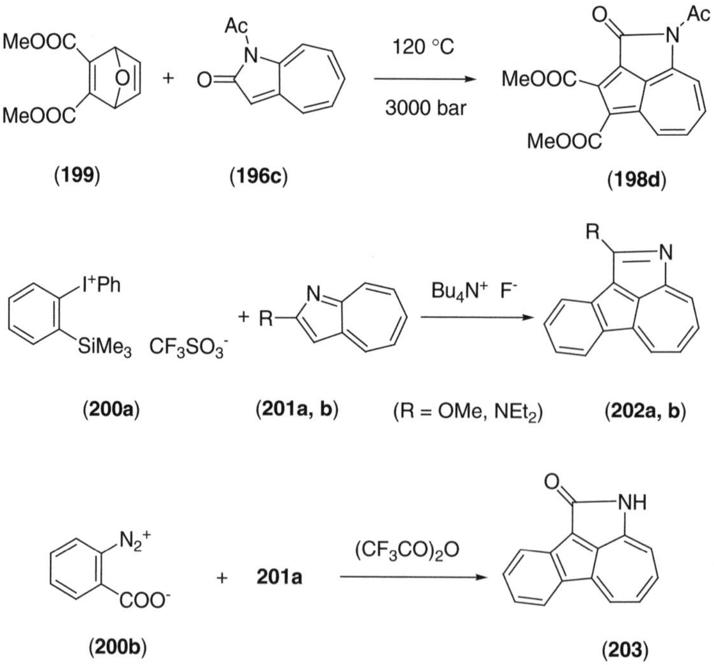

Scheme 46

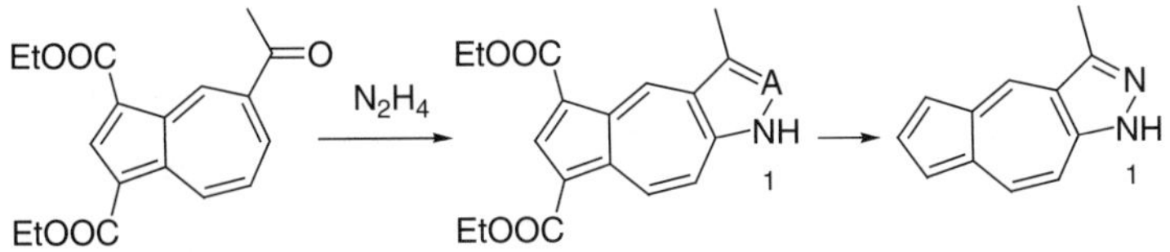

(204)   (205)   (206)

Scheme 47

2.4.1.7 Pyrazoles.   The condensation of 5-acetylazulene **204** (Scheme 47) with unsubstituted hydrazine, contrary to similar reactions, results not only in the normal ketone derivatization but also in a subsequent unusual cyclization and spontaneous dehydrogenation to yield pyrazoles **205** and (after degradation) **206** (63BCJ1016).

2.4.2 Six-membered rings

2.4.2.1 Pyrans.   Nucleophilic   substitution   of   chloroazulene   **207** (Scheme 48) with cyanoacetate and subsequent reaction with DDQ yields *peri*-fused pyran **208**, presumably through dehydrogenation, intramolecular cyclization, and aromatization (97MI1).

The reaction of azulenotropone **209** with sulfur ylide ethyl dimethylsulfuranylidene acetate (EDSA) yields, instead of the expected homotropone, dihydropyran **210** (82CL2027). This compound by hydride abstraction transforms to a cation (in **211a**) or its mesomers that when neutralized yields tetracyclic pyran **212**, the protonation of which regenerates the cation (in **211b**). Pyran **212** is also obtained from dihydro compound **210** by treatment with DDQ and neutralization.

2.4.2.2 Pyridines.   Azuleno[1,2-*b*]pyridone (**214**; Scheme 49) forms in the vacuum pyrolysis of 3-ethynyl-2-phenylpyridine (**213**) as a by-product (02EJO2547). Azuleno[1,8-*cd*]pyridinedione **216** similarly occurs as a minor by-product on irradiating the bridged dihydronaphthalene **215** (a tetraenic propellane) to form tetracyclodecadienes by an electrocyclic reaction (69ISJ435). The formation of azulene **216** involves dehydrogenation.

Other *peri*-fused azulenopyridines (e.g., **218a**) are purposefully synthesized from amides, such as **217** (derived from guaiazulene, **129**), by the intramolecular condensation of an *N*-methylacetamido group in the 1-position and an acidic 8-methyl group (64JA3137). Corresponding amides derived from 4,6,8-trimethylazulene yield analogous products (**218b** and **218c**).

2.4.2.3 Pyridazines.   Hafner   et   al.   (78H(11)387)   used   variants   of their azulene synthesis to construct azuleno[8,1-*cd*]pyridazines by

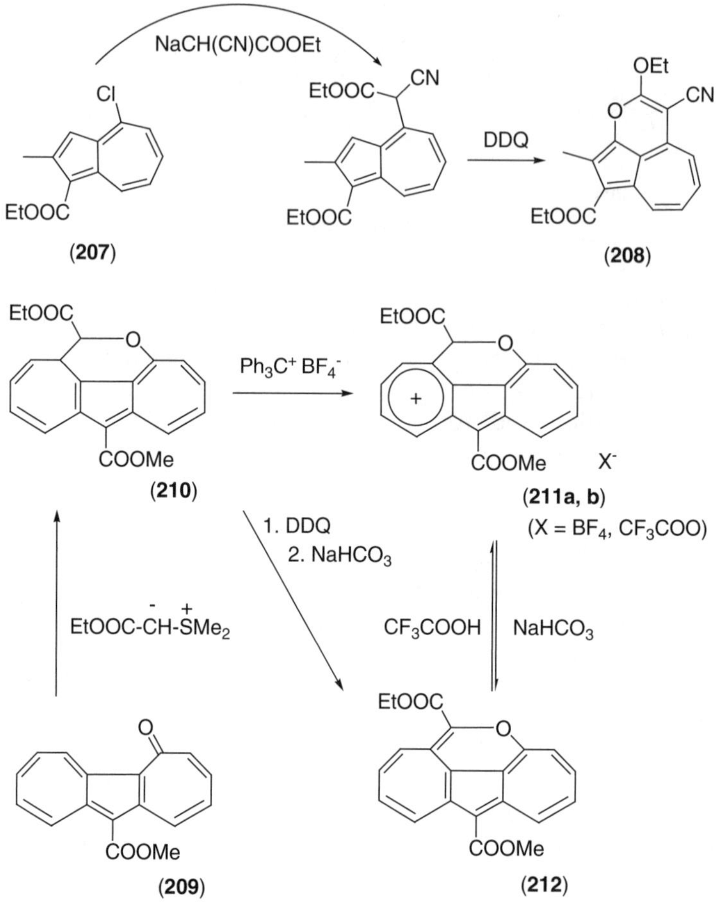

Scheme 48

peri-annulation of a seven-membered ring to cyclopenta[*d*]pyridazine **219** (Scheme 50) that resembles azulene in its reactivity. Its 1-methyl group is easily deprotonated; the anion reacts with several electrophiles to yield substitution products, which condense intramolecularly to afford azulenes **220a** and **220b**.

A two-step [4+2] cycloaddition of azulene or 1-nitroazulene and the extremely electron-deficient diazadiene system of tetrazine **221** (Scheme 51) yields, together with other substances, azuleno[1,2-*d*]pyridazines **222a** and **222b** (93AP29).

2.4.2.4 Pyrazines. Photofragmentation (elimination of nitrogen) of spiro-pyrazole **223** affords a highly strained spiro-cyclopropene that

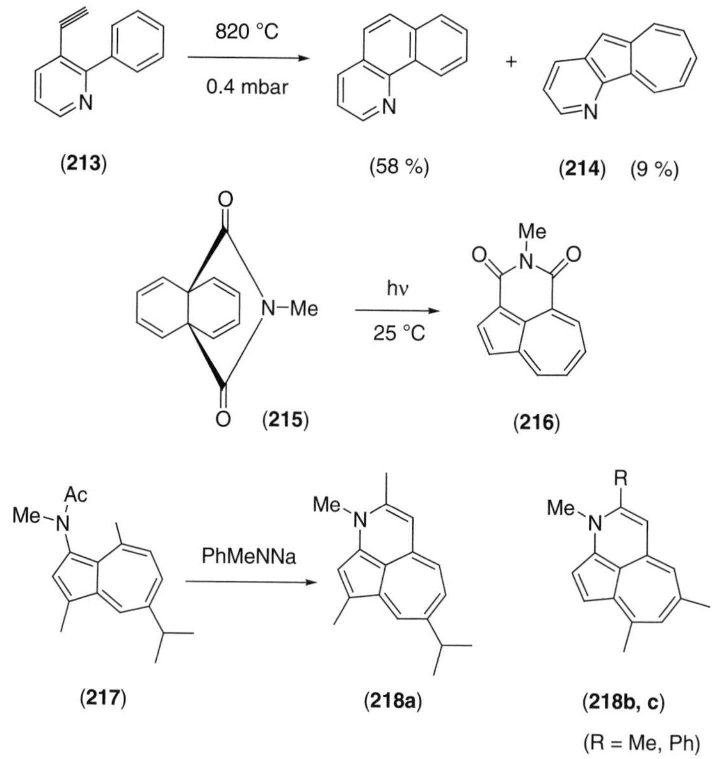

(213)   (58 %)   **(214)**   (9 %)

(215)   **(216)**

PhMeNNa

(217)   **(218a)**   **(218b, c)**

(R = Me, Ph)

Scheme 49

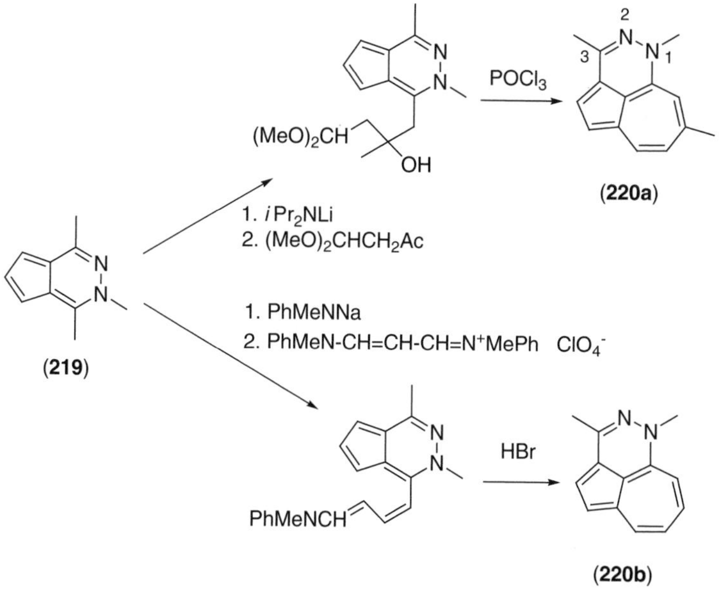

(MeO)$_2$CH

OH

POCl$_3$

(220a)

1. iPr$_2$NLi
2. (MeO)$_2$CHCH$_2$Ac

1. PhMeNNa
2. PhMeN-CH=CH-CH=N$^+$MePh   ClO$_4^-$

(219)

PhMeNCH

HBr

(220b)

Scheme 50

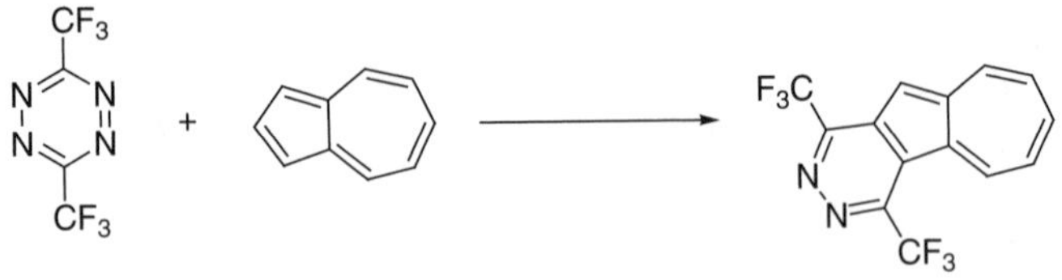

(221)   (222a)   (9 %)

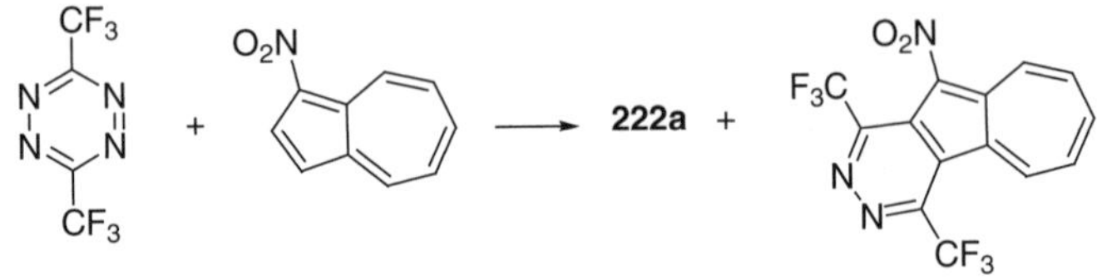

(221)   (21%)   (222b)   (7 %)

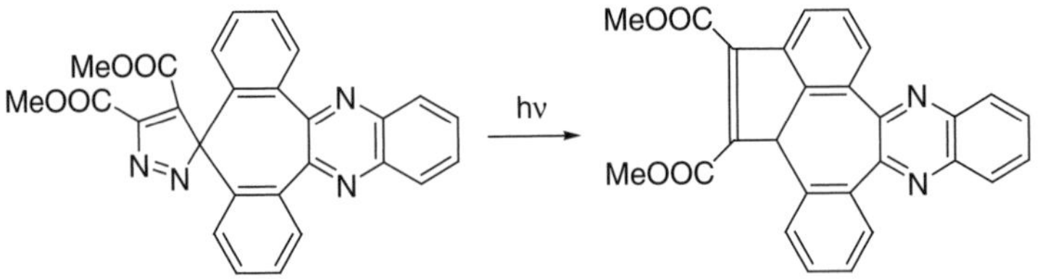

(223)   (224)

Scheme 51

partially undergoes an additional isomerization to yield azulenoquinoxa-
line **224** (77CC843).

2.4.3 Metal π-complexes

[8+2] Cycloaddition of 8-oxoheptafulvene and tricarbonyl (cycloheptatriene)iron yields, among other products, tricyclic ketone **225** (Scheme 52),
which is transformed by acetylation and dehydrogenation to cyclohept-
[*a*]azulene iron complex **226** (02JOM(642)80).

2.5 Summary

Table 2 lists compounds that were synthesized by methods given in
Sections 2.3 and 2.4.

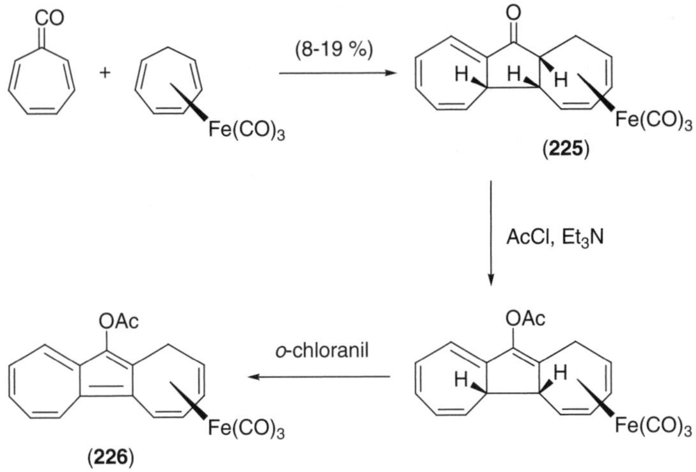

Scheme 52

3. STRUCTURE

3.1 Theoretical methods

Quantum-chemical calculations (cf. 59MI1, p. 177) on azulenohetero-cycles serve in predicting structural data, assisting studies, and corroborating results with the following fields:

- molecular geometry (cf. Section 3.2), for example, of thiophene **45a** (03OBC2572), pyridines **50** and **51** (89JF2,157), and pyridazine **220a** (78H(11)387),
- electronic spectra (cf. Section 3.3.3), for example, of *peri*-fused pyran, thiopyran, and pyridine parent systems **230a–230c** (see Scheme 55), calculated by the Pariser-Parr-Pople (PPP) method (68JPC3975, 68TCA247),
- aromatic stability (cf. Section 3.4.1), for example, of three isomers each of five-membered ring-fused azulenofurans, azulenothiophenes, and azulenopyrroles (87H(26)2025),
- tautomerism (cf. Section 3.6), for example, of *peri*-fused thiophenes **133a**, **133b** and **179c** (71JA2196),
- reactivity (cf. Section 4) relating to substitution (cf. 70PIC53), for example, on thiopyran **136** (75TL2077), and to cycloaddition, for example, onto unstable *o*-quinonoid furan **168** (83CC1025).

Moreover, the results of the detailed calculation of molecular geometry, electronic spectra, and stability have been reported regarding thiophenes **45a** and **69** (83JF2,1155) and azaazulenoazulene **88** (93TL835). According to theoretical consideration, among all of the *vic*-diazoazulenoquinones

Table 2 Azulenoheterocycle syntheses reviewed in Sections 2.3 and 2.4

Fused heterocycle	Sections or tables	Formula
Furan	2.3.1, 2.3.2, 2.3.3.3, 2.3.5.1, 2.3.7; Table 1	2, 8, 9, 47a, 59a, 60a, 149, 152
	2.4.1.1	164, 165, 168
Thiophene	2.3.1, 2.3.2, 2.3.3.3, 2.3.4, 2.3.5.2, 2.3.5.6, 2.3.6; Table 1	15, 16, 29, 43–46, 54, 55a–55c, 63, 64, 66, 68, 69, 121, 133, 134
	2.4.1.2	171, 173, 175, 176, 178, 179c
[1,3]Dithiole	2.4.1.3	180
Isoxazole	2.4.1.4	182, 184
Thiazole	2.3.2, 2.3.5.3, 2.3.5.4	23, 25–27, 78, 84, 85
	2.4.1.5	187–191, 193
Pyrrole	2.3.3.3, 2.3.4, 2.3.5.1, 2.3.5.4, 2.3.5.6, 2.3.7	47b, 55d, 55e, 59b, 60b, 82, 87, 88, 90, 122, 140
	2.4.1.6	194, 195, 198, 202, 203
Pyrazole	2.3.5.6; 2.4.1.7	119; 205, 206
Imidazole	2.3.5.3	72
Triazole	2.3.5.3	73
Pyran	2.3.3.2, 2.3.5.2, 2.3.7; 2.4.2.1	39, 70, 143; 208, 212
Thiopyran	2.3.6	136
[1,4]Thiazine	2.3.5.1	57
Pyridine	2.3.3.1, 2.3.3.3, 2.3.5.3, 2.3.5.5	31, 32, 50, 51, 81, 92–94, 96, 98–103, 105, 107, 108
	2.4.2.2	214, 216, 218
Pyridazine	2.3.5.7; 2.4.2.3	126, 127; 220, 222
Pyrimidine	2.3.3.1, 2.3.5.5	34, 36, 109, 111, 113, 115
Pyrazine	2.3.5.3, 2.3.5.5, 2.3.5.7; 2.4.2.4	74–76, 117, 124; 224
Thiepin	2.3.8	155c
Azepin	2.3.6	137
Phosphepin	2.3.8	155b
Silepin	2.3.8	155a
[1,4]Diazocine	2.3.5.7	128
Porphyrin	2.3.9	157, 159, 160, 161b
Cyclophane	2.3.9	162
Fe complex	2.4.3	226

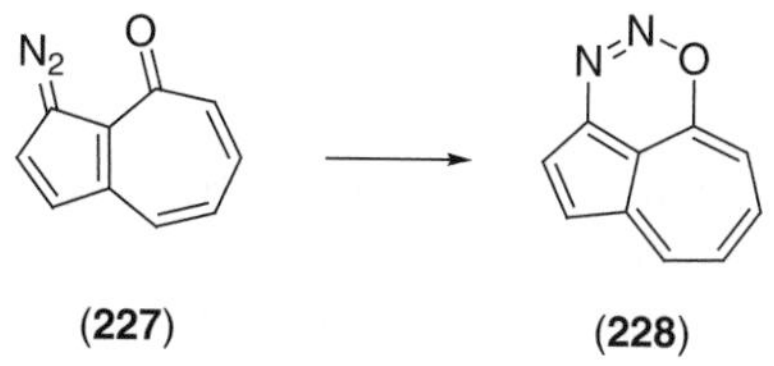

(227) (228)

Scheme 53

Table 3 X-ray diffraction of azulenoheterocycles

Ring system[a]	Formula	Reference[b]
Azuleno[1,2-*b*]thiophene (P)	45a	82AXB2729
	46, 256b	03OBC2572
	54	99TL6609
Azuleno[2,1-*b*]thiophene	55a	02T7653
Azuleno[1,2-*b*]benzothiophene	55b	04OBC1413
Azuleno[2,1-*b*]benzothiophene	55c	
Azuleno[1,8-*bc*]thiophene	133b, 134	71JA2196
Azuleno[2,1-*b*]indole	55d	06TL8535
Azuleno[1,8-*bc*]pyran	208	97MI1
Azuleno[8,1-*cd*]pyridazine	220a	78H(11)387
Azuliporphyrin[c], Pd(II) complex		07AXEm1351

[a]The substitution can be gathered from the formulas; P indicates parent substance (in following tables, too).
[b]Immediately successive identical references are not repeated (in following tables, too).
[c]See Section 2.3.9.

only the 1,8-compound (**227**; Scheme 53) may favor a cycloisomerization to produce oxadiazine **228** (00CJC224).

3.2 X-ray Diffraction

X-ray diffraction of azulenes has been widely used (cf. 59MI1, p. 205; 70PIC53; 85HOU(5/2c)127). Relevant work with azulenes fused to heterocycles, beginning in the 1970s, may be found in Table 3, arranged according to the ring systems.

As disclosed in several studies, the compounds concerned display planarity but significant bond-length alternation in the seven-membered ring (78H(11)387, 82AXB2729, 03OBC2572). Conjugation with an enone group, for instance, in thiophene **55c**, however, seems to result in the absence of a distinct alternation (04OBC1413). Thiophene **54** exists as a centrosymmetric-associated dimer formed by CH···OC hydrogen bonds (99TL6609).

3.3 Molecular spectra

3.3.1 ¹H-NMR spectra

Chemical shifts of various heterocyclicly fused azulenes are listed in Table 4. To get consistency throughout the NMR (Tables 4 and 5), by way of an exception a formal numbering system is used (examples **229a–229d**; Scheme 54), similar to that formerly proposed for the same purpose (67TL3443, 70BCJ509). It consequently bases on the numbering of parent azulene (**1**), different from the numbering given in formulas **2–5** and similar structures.

In Table 4, on proceeding from parent azulene position 1′ up to 8′, generally the same variation of the chemical shifts is observed as it is known with azulene itself, apart from some cases of heavily polarizing substituents (**133a** and **222a**). Distinct down-field shifts due to electron-attracting substituents of the five-membered azulene ring are observed with nearly all nuclear protons present in the tricyclic system. Substituents involved are, for instance, ethoxycarbonyl (in **188a** compared to **188b**, Table 4), formyl or methoxycarbonyl (in **7a** and **7b**, respectively, compared to **2**, Table 4 (05JNP248)), and nitro (in **222b** compared to **222a** (93AP29)). In the spectra of 2-pyridone **105b** and 4-pyridone **108**, the H-5 protons resonate at $\delta = 8.88$ and 10.31, respectively, in the latter case down-field shifted due to the neighboring carbonyl group (91BCJ2393).

General shifting of all the proton resonances of a molecule or moiety can be characterized in terms of the *average chemical shift* (δ_{AV}). Thus, δ_{AV} (7.58) of the azulene moiety of azulenopyridine **92a** is down-field compared to those of azulene (**1**) and benz[*a*]azulene, due to the electron-withdrawing pyridine ring (94JOC1309). Furthermore, δ_{AV} of the seven-membered ring of (acetylvinyl)azulenothiophene **54** is larger than that of the parent compound (**45a**), indicating a contribution of a dipole resonance structure (see Section 3.4.2) (99TL6609).

Shift assignment has also been reported with respect to tetracyclic azulenopyran **212** (82CL2027), azulenoazaazulene **88** (93TL831, 93TL835), pentacyclic diazulenopyridines **94a** and **94b** (01H(54)667), and azuleno-tropylidene iron complex **226** (02JOM(642)80).

¹H-NMR spectra allow conformational studies of azuleno-annulated pyridinophanes **93** (94JOC1309) (see Section 3.5). A large divergence of the vicinal spin–spin coupling constants in the seven-membered ring protons of azulenoheterocycles indicates bond alternation and double bond fixation (e.g., 81H(15)835, 91BCJ2393, 03OBC2383) (see Section 3.4.1).

The spectra of azulenes taken in a trifluoroacetic acid solution are those of the protonated species (conjugate acids, cf. Section 4.1.1), for instance, of that of thiophene **237** (67TL3443, 77MI1). They are characterized by down-field shifts of the methine protons (as known

Table 4 ¹H-NMR chemical shifts of azulenoheterocycles

Ring system[a]	Formula	δ (ppm)[b] in positions[c]												Solvent[d]	Reference
		1'	2'	3'	4'	5'	6'	7'	8'	α	β	γ	δ		
Azulene[e]	1	7.30	7.81	7.30	8.23	7.05	7.45	7.05	8.23	–	7.81	–	–	T	85HOU(5/2c) 127
[1,2]- or [2,1]-fusion (cf. 229a)															
Azuleno[1,2-b]furan (P)	47a	–	–	7.11	8.22		6.7–7.6[f]		8.36	–	7.81	6.81		C	86CL1021
Azuleno[2,1-b]furan	60a	–	–	7.39	8.33	7.16	7.56	7.21	8.42	7.19	–	–		C	81H(15)835
Azuleno[1,2-b]thiophene	171	–	–	7.29	(2.79)		7.07[f]		(2.84)	–	7.29	(2.54)		T	69BCJ1404
Azuleno[2,1-b]thiophene (P)	69	–	–	7.47	8.20		6.9–7.6[f]		8.39	7.32	7.69	–		C	81BCJ2537
Azuleno[1,2-c]thiophene	121a	–	–	(3.94)	8.85		6.87–6.99[f]		7.6	(2.82)	–	7.14		C	06IJB731
Azuleno[1,2-d]thiazole	23	–	–	7.33	8.22		7.0–7.7[f]		8.81	–	(2.90)	–		C	86BCJ3320
Azuleno[2,1-d]thiazole (P)	85b	–	–	7.65	8.33		7.0–7.7[f]		8.29	–	9.15	–		C	82CL707
Azuleno[1,2-b]pyrrole (P)	47b	–	–	7.26	8.23		6.7–7.4[f]		8.12	8.72	7.49	6.64		C	86CL1021
Azuleno[1,2-c]pyrrole	122a	–	–	(3.94)	8.01		7.34–7.43[f]		7.79	(2.44)	–	7.12		C	01H(54)647
Azuleno[1,2-d]pyrazole	119	–	–	7.03	7.98	7.37	7.25	7.37	8.01	(2.68)	–	–		C	02H(56)497
Azuleno[1,2-c]pyran	39	–	–	7.12	8.38	7.51	7.74	7.64	9.47	–	–	(2.42)	6.60	C	92H(34)429
Azuleno[1,2-b]thiazine	57	–	–	–	8.62	6.54	6.44	5.87		–	–	–	9.44	B	91H(32)213
Azuleno[1,2-b]pyridine	214	–	–	7.11	7.87	6.82	7.2	7.05	8.8	–	8.70	7.44	8.06	C	02EJO2547
Azuleno[1,2-d]pyridazine	222a	–	–	9.40	7.90	8.20	7.80	8.90	8.01	–	–	–	–	A	93AP29
[4,5]-fusion (cf. 229b)															
Azuleno[4,5-c]furan	149	6.55	6.93	7.02	–	–	–	6.69	7.11	8.16	–	8.14		C	03OBC2383
Azuleno[4,5-b]thiophene	16	7.29	7.57	(2.88)	–	–	7.68	6.94	(2.88)	–	7.07	(2.42)		C	77MI1
[5,6]- or [6,5]-fusion (cf. 229c)															
Azuleno[6,5-b]furan	2	6.95	7.11	(2.44)	8.14	–	–	7.03	(2.55)	(2.15)	7.37	–		A	98P1079
		7.04	7.25	(2.64)	8.10	–	–	7.28	(2.75)	(2.33)	7.13	–		T	81E442
	7a	7.32	8.09	–	9.99	–	–	7.80	(2.94)	(2.46)	7.56	–		C	05JNP248

Table 4 *(Continued)*

Ring system[a]	Formula	δ (ppm)[b] in positions[c]												Solvent[d]	Reference
		1'	2'	3'	4'	5'	6'	7'	8'	α	β	γ	δ		
Azuleno[5,6-c]furan (P)	152	6.80	6.92	6.70	7.72	–	–	6.61	6.97	8.03	–	7.81		C	03OBC2383
Azuleno[6,5-b]thiophene	15	7.19	7.45	(2.70)	8.41	–	–	7.42	(2.73)	(2.53)	7.00	–		C	77MI1
		7.02	7.25	(2.62)	8.20	–	–	7.25	(2.67)	(2.48)	6.87	–		T	67TL3443
	173b	(2.60)	7.61	(2.60)	8.40	–	–	7.52	(2.72)	(2.56)	7.05	–		T	70BCJ509
Azuleno[6,5-d]thiazole (P)	188b	7.35	7.81	7.42	8.80	–	–	7.90	8.27	–	9.10	–		C	96H(43)1049
	188a	–	8.79	–	10.46	–	–	8.50	9.78	–	9.33	–		C	
[1,8]- or [8,1]-fusion (cf. 229d)															
Azuleno[1,8-bc]thiophene	133a	(1.93)	(3.22)	–	–	6.72	5.60	(1.03)	6.07	–	–	–		C	71JA2196
Azuleno[8,1-bc]thiophene	178a	–	(3.45)	–	–	5.97	–	5.66	6.45	–	–	–		C	08JOC2256
Azuleno[8,1-bc]pyrrole	198c	–	–	–	–	7.72	8.37	7.90	9.08	–	11.48	–		S	88BCJ1225
Azuleno[1,8-bc]pyran	294	6.73	(2.35)	–	–	6.32	6.93	6.15	7.47	–	–	–		C	97MI1
Azuleno[1,8-bc]thiopyran	136	(2.3)	6.97	–	–	5.62	6.28	(1.1)	6.98	–	–	6.12		C	65TL1877
Azuleno[1,8-bc]pyridine	218a	(2.28)	5.81	–	–	5.24	5.71	(1.02)	6.40	(3.08)	(2.00)	4.99		C	64JA3137
Azuleno[8,1-cd]pyridazine	220b	6.67	6.88	–	–	5.66	6.37	5.46	7.18	(2.07)	–	(3.25)		C	78H(11)387
Azuleno[1,8-cd]azepin	137	(2.3)	7.08	–	–	5.97	6.20	(1.1)	6.81	–	–	–	5.83	C	65TL1877

[a] See Table 3, footnote a.
[b] Figures in parentheses refer to methyl or methylene groups.
[c] Formal positions according to Scheme 54.
[d] Solvents: A, acetone-d_6; B, C_6D_6; C, $CDCl_3$; S, DMSO-d_6; T, CCl_4.
[e] Azulene positions 1–8 for comparison.
[f] Positions 5', 6' and 7'.

Table 5 ^{13}C-NMR chemical shifts of azulenoheterocycles

Fused ring[a]	Formula		δ (ppm) in positions[b]													Solvent[c]	Reference
		1'	2'	3'	3a'	4'	5'	6'	7'	8'	8a'	α	β	γ		
—[d]	1	118.1	136.9	118.1	140.2	136.4	122.6	137.1	122.6	136.4	140.2	–	159.5	–	C	85HOU (5/2c)127
Thiazole	85b	122.0	166.5	107.8	142.9	138.1	123.3	136.7	123.0	133.8	130.6	–	–	–	C	82CL707
Furan	149	122.3	132.4	121.7	123.1	119.8	123.3	134.7	121.6	131.1	145.1	138.9	–	144.7	C	03OBC2383
	152	129.3	132.0	124.2	141.1	129.0	123.2	124.5	113.5	125.6	133.5	145.1	–	141.1	C	
	2	116.9	132.2	136.6[e]	127.1[e]	124.9	121.2	158.8	111.4	139.5	133.7	119.7	139.1	–	B	98P1079
	7a	117.5	141.3	135.1	117.4	131.5	126.6	159.4	117.4	141.5	141.7	120.4	141.8	–	C	05JNP248

[a]The structures can be gathered from the formulas (cf. Table 4).
[b]Formal positions according to Scheme 54 (Formulas 229a–229c).
[c]Solvents: B, C_6D_6; C, $CDCl_3$.
[d]Parent azulene positions 1–8a for comparison.
[e]Or interchanged.

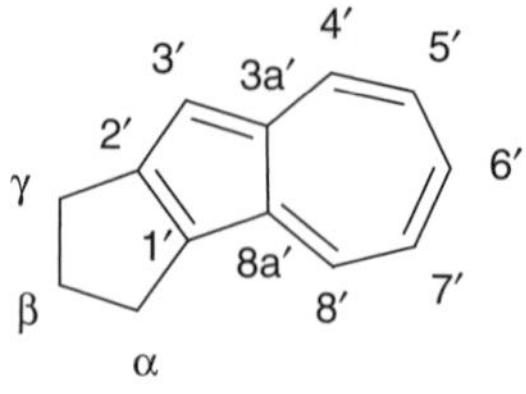

(229a) 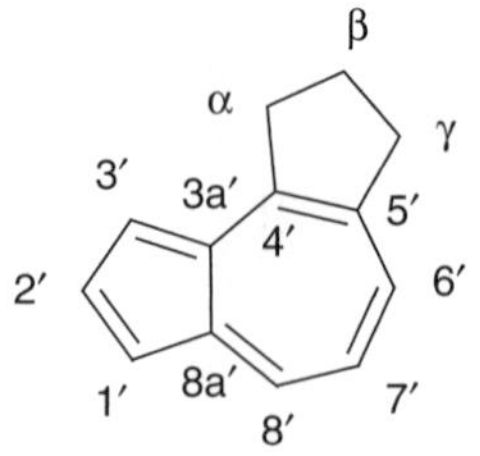   (229b)

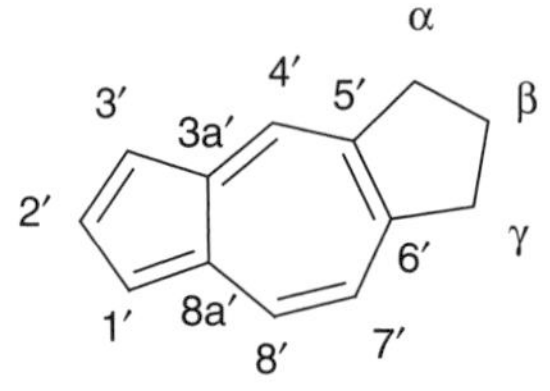

(229c) 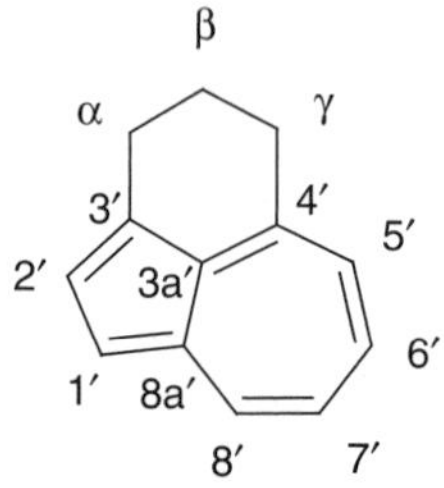   (229d)

Scheme 54

with carbonium ions) and by the presence of a five-membered ring methylene signal. In contrast, pyranone **70** when protonated gives upfield shifts (04BML63).

3.3.2 ^{13}C-NMR Spectra

^{13}C-NMR chemical shifts of relevant tricyclic compounds, as far as they have been assigned, are shown in Table 5. Assignments have also been published in the case of azulenoazaazulene **88** (93TL835), diazulenopyridines **94a** and **94b** (01H(54)667), and azulenotropylidene iron complex **226** (02JOM(642)80).

In the context of studies with methyl-substituted azulenes and *peri*-fused azulene derivatives, the vicinal coupling of ring hydrogen atoms to methyl carbon atoms in pyridazine **220a** was also determined (77T3127). As a result the coupling of sterically unperturbated methyl groups in seven-membered rings is not only dependent on the bond lengths but also on the CCH bond angles.

3.3.3 Electronic spectra

Visible spectra had been the first and for decades the most important means of characterizing azulenes and of elucidating their structure (cf. 50AG281; 59MI1, p. 218). Absorption maxima of numerous azulenes have been tabulated including those of functionalized and fused deriva-tives (52CRV127, p. 172; 55FCF(3)334, p. 375; 58AG419; 61FOR(19)32, p. 80). UV spectra are less characteristic of individual structure features.

Vis-spectral data of selected tricyclic compounds are compiled in Table 6. Similar data are also available in the case of tetra- and pentacyclic systems, for example, those of pyran **212** (82CL2027), pyridines **94a** and **94b** (01H(54)667), and iron complex **226** (02JOM(642)80).

Some typical complete electronic spectra are shown in Figure 2. It is the low-intensity, long-wave region (the visible region) of azulene spectra that provides the most valuable information. Thus, the azulenic character of pentacyclic compounds **155a–155c** (73CZ447, 76ZNB1116, 74CZ41, respectively) and even of the brown (in solution faint-yellow) colored pyrone **143** (74CB2956) was deduced from their long-wave maxima.

On the other hand, some *o*-quinonoid structures lack azulene properties. Yellow-brown azuleno[1,2-*c*]furan **44** (Table 6) is rather a heptafulvene than an azulene (83CL1721), similar to yellow 2-pyron **70**. Several brown-colored azuleno-fused imidazoles, triazoles, and pyrimi-dines (e.g., **72**, **73**, and **109**; Table 6), however, show typical azulene spectra (73BCJ3161), even when lacking distinct maxima in the visible region as in the case of triazoles **271**, **272a**, and **272b** (see Scheme 66) (85TL335) and, in addition, of pyridine **31a** (74BCJ1750). These substances predominantly absorb in the UV region.

The UV spectra of azulenoheterocycles were also extensively reviewed (50AG281; 52CRV127, p. 192; 55FCF(3)334, p. 379; 58AG419; 59MI1, p. 245). Annulation of aromatic rings onto azulene leads to strong *bathochromic shifts* (to a higher wave-length) of the UV maxima as a result of elongating the conjugate system, for instance, a shift from 291 to 325 nm with the couple vetivazulene (**170**)/azulenothiophene **171** (69BCJ1404).

The site of annulation determines both the shape of the electronic spectra and the position of the visible maxima (cf. Table 6). By analogy the decision on the structure of isomeric thiazoles **25** and **26** (λ_{max} 582 and 615, respectively) could be confirmed by comparison with the analogous couples of isomeric furans **2** and **8** (λ_{max} 562 and 607, respectively) and of thiophenes **15** (67MI1) and **16** (λ_{max} 576 and 622, respectively) (64UP1, 65MI1).

The influence of the ring substitution on the electronic spectra of azulenes is governed by *Plattner's displacement rules*, which are valuable tools of structure elucidation (55FCF(3)334, p. 375; 58AG419; 59MI1, p. 224; 66MI1, p. 202). The direction and extent of the maxima shift in the visible spectrum depends on both the position and nature of the new substituent. Roughly speaking, electron-donating groups (such as alkyl of any length) at odd ring positions cause bathochromic shifts, at even positions cause *hypsochromic shifts* (to a lower wave-length).

Table 6 Vis maxima of azulenoheterocycles

Ring system[a]	Formula	λ_{max}[b] in nm (log ε)	Color	Solvent[c]	Reference
Azulene[d]	1	557 (2.33) 580 (2.44) 604 (2.36) 633 (2.39) 660 (2.15) 698 (2.11)	Blue	CH	56LA(598)32
Azuleno[1,2-b]furan (P)	47a	619 (2.41) 652 (2.30) 750 (2.07)	Blue-green	MeOH	86CL1021
Azuleno[2,1-b]furan	60a	580 (2.63) 612 (2.65) 677 720 sh	Blue-green	MeOH	81H(15)835
Azuleno[1,2-b]thiophene (P)	45a	630 (2.71) 670 (2.65) 640 sh 690 (2.68) 768 (2.40)	Green	CH	83CL1721
	46	552 sh 607 (2.71) 650 (2.66) 736 sh	Green-blue	M	03OBC2572
	241a	532 sh 572 (2.58) 618 sh 687 sh	Blue	M	
Azuleno[2,1-b]thiophene (P)	69	570 (2.33) 612 (2.38) 662 (2.27) 718 (1.76)	Green-blue	MeOH	60CIL1302
Azuleno[1,2-c]thiophene	44	408 (3.77) 433 (3.92) 454 (3.82) 464 (3.85)			83CL1721
Azuleno[1,2-d]thiazole	23	512 (2.70) 555 (2.70) 608 (2.62)	Yellow-brown	CH	64UP1
Azuleno[2,1-d]thiazole (P)	85b	569 sh 608 (2.27) 670 (2.15) 750 (1.66)	Blue-black	CH	82CL707
Azuleno[1,2-b]pyrrole (P)	47b	615 (2.60) 650 (2.53) 671 (2.58) 774 (2.28)	Purple	CH	86CL1021
Azuleno[2,1-b]pyrrole	60b	590 (2.30) 642 (2.46) 708 (2.39) 794 (2.02)	Green	MeOH	81H(15)835
Azuleno[1,2-d]imidazole	72	581 (2.55) 635 (2.52) 720 sh	Green	MeOH	73BCJ3161
Azuleno[2,1-b]pyridine	92a	475 sh 515 (2.36) 550 sh	Brown-purple	EtOH	94JOC1309
Azuleno[1,2-d]pyridazine	222a	568 (2.93) 617 (2.89) 682 (2.53)	Green		93AP29
Azuleno[2,1-d]pyrimidine	109	418 (3.16) 514 (3.62) 550 (2.43) 610 (2.06)	Brown	M	73BCJ3161

Compound	No.					Color	Solvent	Ref.
Azuleno[1,2-*b*]pyrazine	**75**	505 (2.11)	550 (2.24)	<u>595</u> (2.28)	650 (2.16)	Violet		
Azuleno[4,5-*b*]furan	**74**	520 (2.38)	<u>554</u> (2.41)	600 sh	660 sh	Greenish		
Azuleno[4,5-*b*]furan	**8**	588 (2.85)	<u>607</u> (2.88)	659 (2.79)	727 (2.33)	Blue	L	63G395
Azuleno[4,5-*c*]furan	**149**	410 (3.93)	434 (3.72)	547 (2.31)		Yellow-green	H	03OBC2383
Azuleno[4,5-*b*]thiophene	**16**	<u>622</u> (2.72)	650 sh	683 (2.59)	725 sh	Blue	CH	79BCJ1549
Azuleno[5,4-*d*]thiazole	**26**	<u>615</u> (2.77)	678 sh	755 (2.14)		Violet-blue	CH	64UP1
Azuleno[6,5-*b*]furan	**165**	520 (2.50)	<u>545</u> (2.56)	567 (2.55)	595 (2.53)	Blue-violet	O	61CIL1715
	2	562 (2.70)	602 (2.60)	610 (2.60)	672 (2.25)	Purple	H	53CPB164
	9	<u>534</u> (2.71)	545 (2.69)	575 (2.64)	632 (2.34)	Red-purple	P	64JCS3577
Azuleno[5,6-*d*]thiazole	**25**	<u>582</u> (2.56)	608 sh	629 sh	706 (2.03)	Blue-violet	H	64UP1
Azuleno[1,8-*bc*]thiophene	**179c**	<u>647</u> (2.64)	692 (2.62)	716 (2.60)	773 (2.30)	Blue-green	CH	71JA2196
Azuleno[1,8-*bc*]pyran	**295**	<u>411</u> (2.94)	435 (2.87)			Green	MeOH	97MI1
Azuleno[1,8-*bc*]thiopyran	**136**	<u>420</u> (3.55)	450 (3.51)	480 (3.40)	925 (2.76)	Red-brown	CH	68JOC823
	250b	429 (3.92)	479 (4.03)	508 (4.13)	870 (2.69)		CH	75TL2077
Azuleno[1,8-*bc*]pyridine	**218a**	<u>422</u> (3.62)	449 (3.43)	763 (2.50)	862 (2.50)	Red-brown	CH	64JA3137
Azuleno[1,8-*cd*]pyridine	**216**	<u>593</u> (2.93)	640 (2.92)	705 (2.54)		Red-brown	MeOH	69ISJ435
Azuleno[8,1-*cd*]pyridazine	**220a**	<u>404</u> (2.87)	618 (2.81)	677 (2.81)	753 (2.51)	Deep blue	H	78H(11)387
Azuleno[1,8-*cd*]azepin	**137**	409 (3.42)	<u>434</u> (3.49)	462 (3.38)	823 (2.88)	Golden brown	CH	68JOC823

[a] See Table 3, footnote a.
[b] Partly including main maxima in the near IR region. Maximum absorptions in typical visible spectra are underlined.
[c] Solvents: CH, cyclohexane; H, hexane; L, ligroin; M, CH_2Cl_2; O, isooctane; P, heptane.
[d] Parent azulene for comparison.

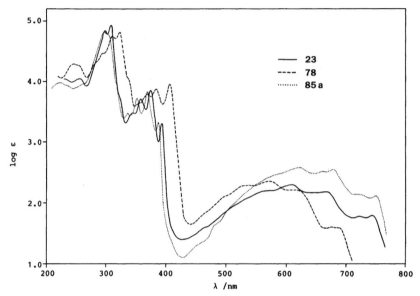

Figure 2   Electronic spectra of azulenothiazoles **23** and **78** (in cyclohexane) and **85a** (in hexane). (With permission from 86BCJ3320, Bulletin of the Chemical Society of Japan.)

Electron-withdrawing substituents behave contrarily. Apart from certain exceptions, the position-specific shifts (66CZ691) are additive and can be calculated and predicted. Plattner's rule generally can also be applied to azulenoheterocycles.

Most of the following examples are connected with data listed in Table 6, the principal (most intense) visible maximum being the point of reference (for formula numbers above **236** see Section 4). Thus, compared to parent azuleno[1,2-*b*]thiophene **45a** (λ_{max} 630 nm), hypsochromic shifts are demonstrated with 6'-isopropyl[2] compound **46** (607 nm), 3'-formyl compound **241a** (572 nm), and 4',8'-dimethyl compound **171** (593 nm) (69BCJ1404). Compared to parent azuleno[2,1-*b*]thiophene **69** (λ_{max} 612 nm), bathochromic shifts are observed with 5'- and 7'-isopropyl derivatives (**281b** and **281a**, respectively) (61NKZ1522) as well as 3'-dimethylaminomethyl and 3'-hydroxymethyl compounds (**247a** and **309**), whereas hypsochromic shifts occur with 3'-ethoxycarbonyl, 3'-cyano, and 3'-carbonamido substitution (compounds **282**, **310**, and **311**, respectively) (61NKZ1667). Hypsochromic shifts caused by 3'-ethoxycarbonyl groups are also observed with the couples **291** (599 nm)/**68** (574 nm)

---

[2] Positions 1' to 8' are those of the parent azulene as indicated in Scheme 54.

(81BCJ2537), **293/292** (61NKZ1517), and, in the azuleno[1,2-*b*]pyrazine series, **75** (595 nm)/**74** (554 nm).

The influence of methyl groups in different positions is exemplified by the series azuleno[6,5-*b*]furan **165** (λ_{max} 545 nm) – 8'-methyl compound **9** (ujacazulene, 534 nm) – 3',8'-dimetyl compound **2** (linderazulene, 562 nm). The differences (−11 and +28 nm, respectively) are exactly those given by Plattner's rules (66CZ691). Derivatives of azulenothiophene **15** (λ_{max} 576 nm, the sulfur analog of linderazulene) (67MI1), substituted by an acetyl in positions 1' (**173b**) or 2' (**173a**), have the maximum shifted to 554 and 624 nm, respectively (70BCJ509).

The absorption of azuleno[1,8-*bc*]pyran, -thiopyran, and -pyridine (**230a–230c**; Scheme 55) in the UV-vis region was calculated by means of the PPP procedure and the results were compared with the experimental spectra of thiopyran **136** and pyridine **218a** (68JPC3975, 68TCA247). Similar calculations with chromone **143** (74CB2956) and azaazulene **88** (93TL835) show good accordance with the measured values.

Azulenes are basic compounds (see Section 4.1.1). The formation of carbocyclic azulenium cations in the presence of strong acids is a transformation to tropylium salts that is accompanied by a large hypsochromic shift and the color changes to yellow (55FCF(3)334, p. 380; 66CZ691). Azulenes fused to heterocyclic nitrogen bases, which are primarily N-protonated, do not behave uniformly.

The principal visible maximum of azulenopyridine **92a** (568 nm) in the presence of trifluoroacetic acid (forming pyridinium salt **239**) undergoes a moderate hypsochromic shift to 532 nm (94JOC1309). On the other hand, after solving *peri*-fused azulenopyridine **218b** (λ_{max} 688 nm) in 10% sulfuric acid the highest-wave absorption band is found at 481 nm (64JA3137). Finally, azaazulene **88** combines two effects: monoprotonation (onto nitrogen to yield **240a**) only causes a small hypsochromic shift, whereas on diprotonation (forming ditropylium salt **240b**) the longest absorption band completely disappears (93TL835).

Quaternization of azulenopyridine **81** (to yield **270**) is accompanied by a general bathochromic shift and by an intensification of the visible bands (see Section 4.2).

Besides electronic spectra, Yamaguchi et al. measured and calculated photoelectron spectra of azulenopyridines **50** and **51** (89JF2,157),

(230a-c)   (X = O, S, NH)   (231)

Scheme 55

anomalous fluorescence spectra of furan **47a** (87SAA1377), pyrrole **47b** (90SAA1719), and pyridines **50** and **51** (87SAA1067), and anomalous fluorescence spectra together with magnetic circular dichroism of thiophenes **45a** and **69** (83JF2,1155).

3.3.4 Infrared spectra

The infrared spectrum of azulene is typical of an aromatic system (50AG281; 59MI1, p. 207). Among spectra of simple azulenoheterocycles, those of artemazulene (**8**) and azulenothiophene **15** have been published and discussed (63G395, 67MI1, respectively).

The NH bands of azulenoazole species have been assigned, for instance, in the case of pyrazoles **59b** and **60b** (3490 and 3410 cm^{-1}, respectively) (81H(15)835) as well as **194** (3220 cm^{-1}) (61CIL1715) and of pyrazoles **205** and **206** (3230 and 3135 cm^{-1}, respectively) (63BCJ1016). Substances **31a**, **31b**, and **32** have been confirmed to exist in the 4-hydroxy-2-pyridone forms by absorption of associated enolic OH (3300–2500 cm^{-1}) and CO (e.g., 1634 and 1626 cm^{-1} in **31a**) (74BCJ1750). In the pyrimidone series, compound **36** exhibits absorption at 3380 cm^{-1} (NH), 1670 and 1630 cm^{-1} (CO) (81H(15)839).

The low-frequency CO band (1637 cm^{-1}) of benzoylazulenothiophene **133a** is evidence for a strong thiophene–carbonyl mesomeric interaction (71JA2196), comparable with 2-benzoylthiophene (1636 cm^{-1}), but different from usual diarylketones (1670–1660 cm^{-1}). Compared with five-membered cyclic imides (for instance, N-methylphthalimide, 1780 and 1705 cm^{-1}), pyridinedione **216** (a six-membered imide) also exhibits low IR absorption (1695 and 1665 cm^{-1}) (69ISJ435), similar to that of naphthalene-1,8-dicarboxylic N-methylimide (1705 and 1665 cm^{-1}).

3.3.5 Mass spectra

In the mass spectra of methylazulenes M$^+$ and (M − CH$_3$)$^+$ ions predominate. In contrast, trimethylazulenothiophene **15** mainly produces M$^+$ and (M − 1)$^+$ ions. The latter process has been described as the formation of thiapyrylium ion **231** (67MI1) in accordance with the behavior of other methylthiophenes. Thiophene **173b** (an acetyl derivative of **15**) exhibits prominent fragment ions (M–CH$_3$)$^+$ and (M–CH$_3$CO)$^+$ (70BCJ509), whereas isopropylazulenothiophene **46** produces the (M–CH$_3$)$^+$ ion (03OBC2572).

3.4 Thermodynamic properties

3.4.1 Aromaticity and stability

Azulene is classified as a non-benzenoid aromatic compound that can be represented by the formulas **232a–232c** (Scheme 56) and other mesomeric structures (66CZ691, 66MI1). Its enthalpy of formation was found to be

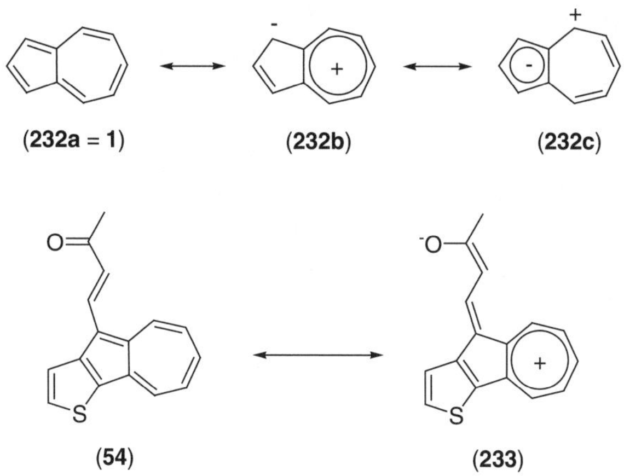

(232a = 1)   (232b)   (232c)

(54)   (233)

Scheme 56

about 32 kcal/mole higher than that of isomeric naphthalene (59MI1, 85HOU(5/2c)127).

Fusion of a benzene or heterocyclic ring onto the five-membered azulene ring tends to cause bond localization in the seven-membered ring (84MI1). Calculations suggest remarkable bond-length alternation (that is characteristic of polyenes) in the seven-membered rings of thiophenes **45a** and **69** (83JF2,1155) and pyridines **50** and **51**, which have been described as consisting of heptafulvene and pyridine (89JF2,157). Experimental support of a similar quantum chemical prediction in the case of thiophene **46** (03OBC2572) and *peri*-fused pyridazine **220a** (78H(11)387) was given by X-ray diffraction. Alternation in thiophenes **68** and **69** was deduced from the differences ΔJ (1.2 and 1.5 Hz, respectively) of the ^{1}H-NMR vicinal coupling constants J_{45} and J_{78} (81BCJ2537).

Bond-length alternation in the seven-membered ring reflects the large aromaticity of the heterocyclic rings in pyridine **286** (see Scheme 70; $\Delta J = 1.8$ Hz) (91BCJ2393) and in pyrrole **60b** ($\Delta J = 1.4$ Hz) in contrast with furan **60a** ($\Delta J = 0.1$ Hz) (81H(15)835). Alternation was also observed in both seven-membered rings of azulenoazaazulene **88** though in the azaazulene moiety to a smaller degree (93TL835).

The stability of azuleno[1,2-*b*]- and azuleno[2,1-*b*]thiophene (**45a** and **69**, respectively) is comparable to that of benz[*a*]azulene because the computed heats of formation of all three of them range from about 102 to 104 kcal/mole (83JF2,1155). In the groups of simple iso-π-electronic azuleno[1,2-*b*]-, azuleno[2,1-*b*]-, and azuleno[1,2-*c*]azoles, the stability has been predicted to decrease in the order of fused (benzene >) pyrrole > thiophene > furan, that is in the order of the stabilities of the heterocyclic moieties (87H(26)2025). Non-functionalized azuleno[6,5-*b*]pyrrole **195** is unstable, different from its ester **194** (61CIL1715); in the related indole series alkoxyazulene **140a** was proven to be a stable derivative of the easily oxidizable 4-hydroxyazulene (96CC937).

The azuleno[1,8-*bc*]pyridine system (e.g., in **218a**) appears to be more stable than its parent ring system 1-pyrindine or cyclohepta[*b*]pyridine (64JA3137). It seems to contain a delocalized π-electron system involving the "lone pair" on the nitrogen. Its stability suggests a close similarity to Hafner's hydrocarbon dimethylcyclopentadienoheptalene. On the other hand, the sulfur analogs (e.g., **136**) are even more stable; different from the pyridines, they can be chromatographed even on *activated* alumina without decomposition (68JOC823).

Most of the azulenoheterocycles having the azuleno-*o*-quinone dimethide structure (type **3**) are less stable than the other types (87H(26)2025). *o*-Quinonoid azulenothiophene **44** is rather unstable even at room temperature and could not be demethoxycarbonylated (83CL1721). *o*-Quinonoid furan **168** is not isolable at all (83CC1025). Furans **149** (rather unstable) and **152** show bond fixation, they are more stable than isobenzofuran, but behave similarly with respect to Diels–Alder cycloaddition (03OBC2383).

3.4.2 Polarity

The dipolar nature of azulene, symbolized by the participation of mesomeric forms **232b** and **232c**, is expressed by an experimental dipole moment μ of about 1.0 (59MI1, p. 208; 66MI1; 85HOU(5/2c)127).

The assignment of structures to isomers was assisted by the dipole moments of azulenothiophenes **218a** ($\mu = 2.01$ D) and **281b** ($\mu = 2.19$ D) (61NKZ1522). A calculation on the non-isolable azulenofuran **168** suggests a large dipole moment $\mu = 3.6$ D (83CC1025).

In acetovinyl derivative **54** (Scheme 56), the bond-length alternation (as it is known from parent azulenothiophene **45a**) almost disappears; this fact indicates a considerable contribution of the dipole resonance structure **233** (99TL6609).

3.5 Conformation

Spectra taken at various temperatures show dynamic behavior (bridge flipping) of the oligomethylene chains of pyridinophanes **93**. Strain increases as the chain becomes shorter, and the pyridine rings are deformed (94JOC1309).

Chromatographic R_F values suggest the presence of a boat-shaped conformation of the seven-membered ring in azulenophosphepin **155b** (76ZNB1116).

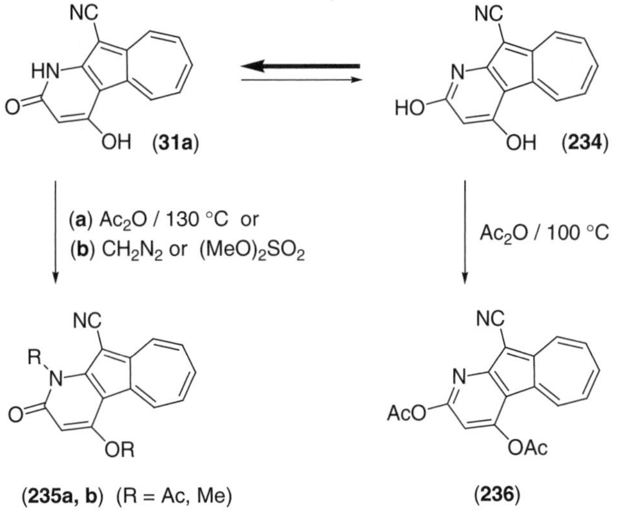

Scheme 57

3.6 Tautomerism

The question of prototropic tautomerism in azuleno[1,8-*bc*]thiophenes has been studied by calculation, X-ray crystallography, ^{1}H-NMR, UV-vis, and IR spectra (71JA2196). It turned out that the azulenodihydrothiophene system (as in **179c**) is normally more stable than the heptafulvenothiophene tautomer. Substitution of a strong conjugating group in the 2-position (as in **133**), however, reverses the order of stability.

Among examples of lactam–lactim tautomerism pyridine derivatives **98a**, **98b**, and **100** have been proven to be pyridones on the basis of their IR spectra, whereas UV and IR spectra suggest the aminopyridine structure **101** (63BCJ633). Spectral data similarly confirm the existence of substance **31a** (Scheme 57) in the 4-hydroxy-2-pyridone form rather than the 2,4-dihydroxypyridine form **234** (74BCJ1750). Consequently, its acetylation by acetic anhydride/pyridine at 130 °C or methylation yield N,O-disubstituted derivatives **235a** and **235b**, respectively. Acetylation at 100 °C, however, leads to O,O'-disubstitution (**236**).

4. REACTIVITY

4.1 Salts and complexes

4.1.1 Protonation
In the presence of strong acids (e.g., 60% sulfuric acid), azulene suffers reversible protonation at the 1- or 3-position to yield the azulenium ion

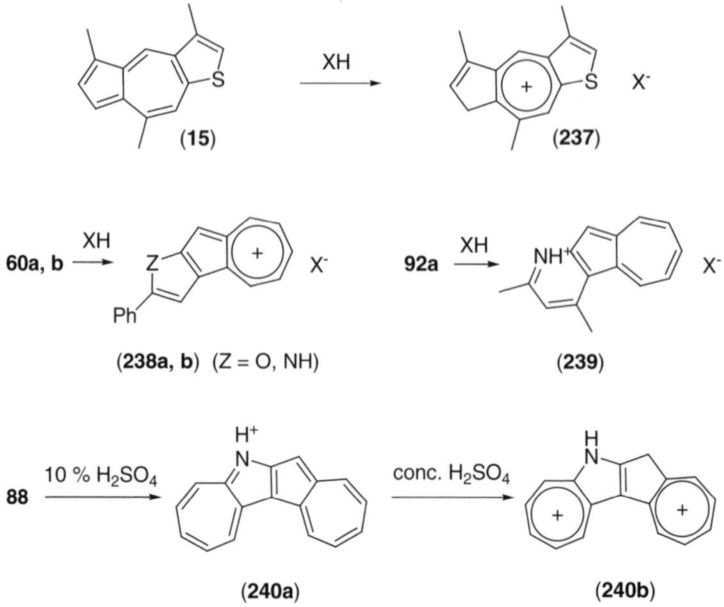

Scheme 58

(cyclopentadienotropylium ion) as the conjugate acid (55FCF(3)334, p. 366; 59MI1, p. 251; 66MI1).

In this way, azulenothiophenes (**15, 16, 171, 173a**) in trifluoroacetic acid exist also as azulenium salts, for instance, trifluoroacetate **237** (Scheme 58); the structures have been established by NMR spectra (67TL3443, 69BCJ1404, 70BCJ509, 77MI1, 79BCJ1549). Protonated azule-nofuran (**238a**) and -pyrrole (**238b**) are also azulenium salts rather than species protonated at the heteroatom (81H(15)835).

In contrast, azulenopyridine **92a** is protonated at the nitrogen atom to afford pyridinium salt **239** (94JOC1309). The pK_a values of the base (6.63) and pyridinophanes **93** (6.46–6.64) are similar to that of 2,4-dimethylpyr-idine; deformation of the pyridine ring does not affect the basicity. Azulenopyridines (e.g., **96a, 96b**, and **99**) form crystalline hydrochlorides (63BCJ633). *peri*-Fused azulenopyridines **218a** and **218b** are completely protonated in 5% hydrochloric acid; phenyl derivative **218c** is much less basic (64JA3137). Pyridazine **220a** is protonated in the 2-position (78H(11)387).

Thiazoloazulenes (**23, 25,** and **26**) are much less basic than the octahydrogenated precursors (e.g., **22**) (63ZC26). The latter are soluble in 10% sulfuric acid; thus, after dehydrogenation the non-reacted portions can easily be extracted from the organic solution. Then the targeted azulenes are extracted by means of 40% sulfuric acid and recovered by dilution with ice water.

Azulenoazaazulene **88** in 10% sulfuric acid forms N-protonated species **240a**, whereas concentrated sulfuric acid leads to ditropylium cation **240b** (93TL835). Diazulenopyridines **94a** and **94b** are protonated to yield pyridinium salts, which are stabilized in the form of monotropylium mesomers (01H(54)667).

Protonation of tetracyclic pyran **212** produces salt **211b** (82CL2027), and solving tricyclic pyrone **70** in trifluoroacetic acid seems to afford a 13C–12π electrons resonance system without the oxygen participating in the resonance (04BML63).

Properties of stable azulene-containing triarylmethyl cations will be dealt in Section 4.2.1.

4.1.2 Charge-transfer complexes

Crystalline complexes of azulenes and aromatic polynitro compounds (1,3,5-trinitrobenzene (TNB), 2,4,6-trinitrotoluene, picric acid, styphnic acid) have been frequently used for isolating, purifying, and characterizing azulenes (59MI2, 70PIC53); numerous examples have been tabulated in Refs. (50AG281; 52CRV127, p. 172; 55FCF(3)334, p. 388; 58AG419; 61FOR(19)32, p. 80).

In early pioneer work, TNB adducts, picrates, and styphnates of linder- and artemazulene (**2** and **8**, respectively) have been described (39YZ162, 54CCC792, 56JCS142), later on TNB derivatives and picrates of azulenothiophenes, for instance, **69** and **15** (60CIL1302, 67MI1, and respective further articles) and of azulenopyridines, for instance, **96a**, **96b**, and **98b** (63BCJ633).

4.2 Electrophilic reactions at ring atoms

4.2.1 Substitution at the azulene ring system

Besides protonation, electrophilic substitution at the 1- and 3-positions is another typical feature of azulene reactivity (55FCF(3)334, p. 367; 59MI2, p. 310; 61FOR(19)32; 66MI1; 84MI1; 85HOU(5/2c)127, p. 249). Assisted by their polarity (see structures **232**), azulenes have been even described as "pacemakers" (typical model compounds) of new electrophilic reactions (66CZ691); since they are colored, substitution can easily be observed and optimal conditions can be worked out. Examples mentioned earlier in the synthetic Section 2 are those of thiocyanation (Scheme 32) and alkylation (Scheme 38). In the case of occupied 1- and 3-positions reactions at C-5 and C-7 can be observed.

Vilsmeier reactions (cf. 97OR(49)1) of azulenothiophenes **45a/46** and **15** (Scheme 59) yield formyl derivatives **241a/241b** (03OBC2572) and acetyl compound **242** (70BCJ509), respectively. Thiophene **171**, which

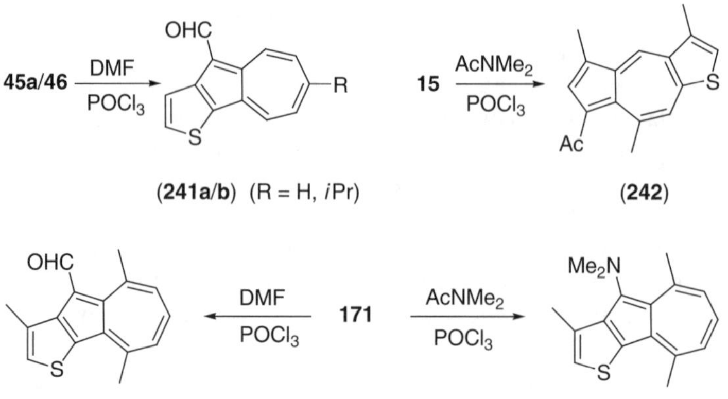

(241a/b) (R = H, iPr)   (242)

(243)   (244)

Scheme 59

with dimethylformamide yields an analogous formyl derivative (243), with dimethylacetamide in contrast forms dimethylaminoazulene 244 (69BCJ1404). This reaction was the first example at all of a dimethyl-amination proceeding in the course of a Vilsmeier acetylation.

Azulenothiophene 69 undergoes Vilsmeier reaction, azo coupling, and Mannich reactions to form compounds 245, 246, and 247a–247c, respectively (Scheme 60), but attempted bromination and nitration lead to non-separable mixtures (61NKZ1667). (For a successful method of brominating 69 see Section 4.2.3 and Scheme 68.) Methoxythiophene 248a yields diformyl product 248b, thiazole 85a (83H(20)1263) yields formyl derivative 249 and Mannich bases analogous to those of thiophene 69.

peri-Fused azulenothiopyran derivative 136 (Scheme 61) shows a different reactivity, which agrees with the results of the Hückel MO calculation (75TL2077). The most reactive site is C-9 followed by C-4. Vilsmeier and Friedel–Crafts reactions yield 9- and 4-formyl derivatives (250a and 250b) in about equal amounts and mono- and diacetyl compounds (252a and 252b), respectively. The Mannich reaction exclusively attacks the 9-position (to yield 251).

The condensation of azulenes with aromatic aldehydes leads to trisubstituted methane derivatives. Thus, the reaction of two molar amounts of azulenothiophene 45a with formyl compound 241a (Scheme 62) in acetic acid affords tris(azulenothienyl)methane 253 (03OBC2572). Subsequent hydride abstraction with DDQ yields the corresponding methyl cation salt 254. By analogy unsymmetrical salts 255, 256a, and 256b can be prepared starting from two moles of thiophene 45a and benzal-dehyde or from thiophenes 45a or 45b and benzhydrol, respectively.

The bonding situation and the thermodynamic stability of the cations were examined and calculated. The bond alternation in the fused azulene

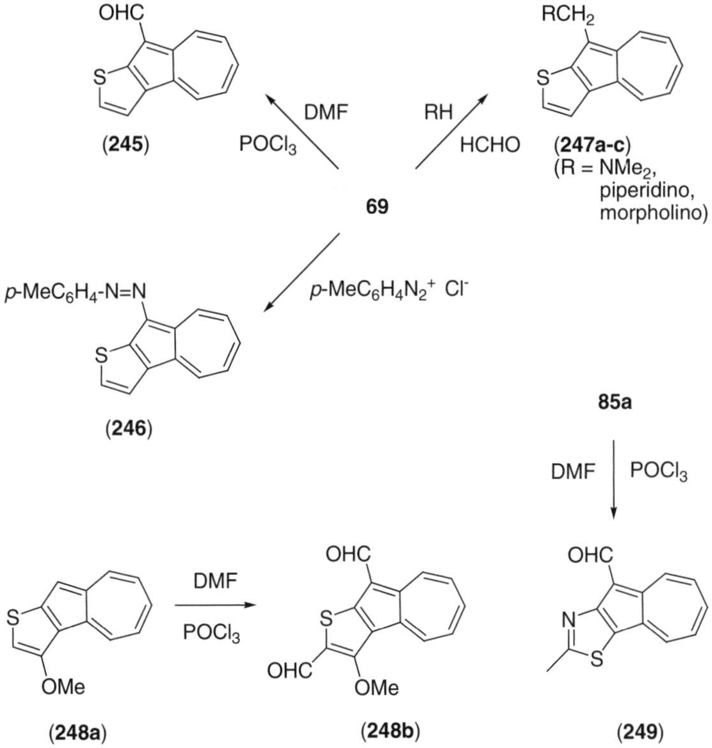

Scheme 60

systems decreases continuously with the decrease in the number of azulene rings (in the order of **254–255–256**). In cations **256a** and **256b**, a delocalized tropylium substructure has developed. The stability of the cations determined by pK_{R^+} values (cf. 96AHC(66)285, p. 324) and redox potentials is not significantly affected by the fused thiophene ring.

A number of patents describe the condensation of azulenes with squaric acid (**257**) or its equivalents and with other dicarbonyl compounds (or equivalents). Thus, azulenothiophene **45a** (Scheme 63) yields symmetrical squarate betain **258** (86EAA187015) or (with **259**) intermediate **260** and (after hydrolysis and reaction with 1,3,3-trimethyl-2-methyleneindole) unsymmetrical squarate **261** (89DEA3914151). Other examples are those of the two-step reactions of azulenothiophene **16** (Scheme 64) with glutaconedialdehyde or its cyclic equivalent **264** and another azulene to yield azulenium salts **263** and **266**, respectively, presumably through intermediates **262** and **265** (90USA4965178).

The cross-coupling of azulenodithiolethione **180** (Scheme 65) and dithiolones (e.g., **267**) by an ylide mechanism yields tetrathiafulvalenes

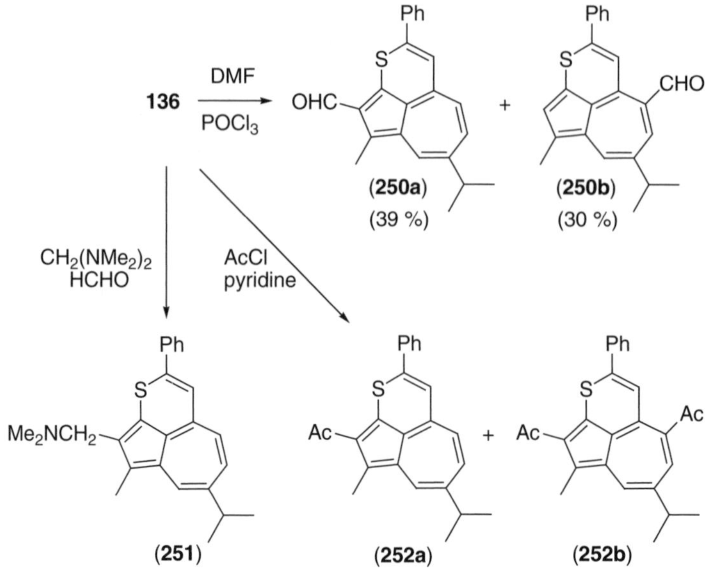

Scheme 61

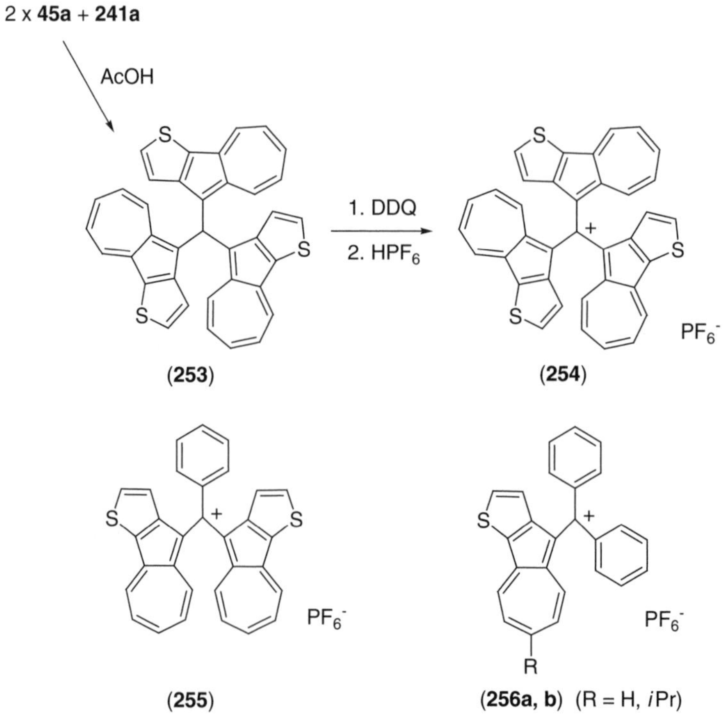

Scheme 62

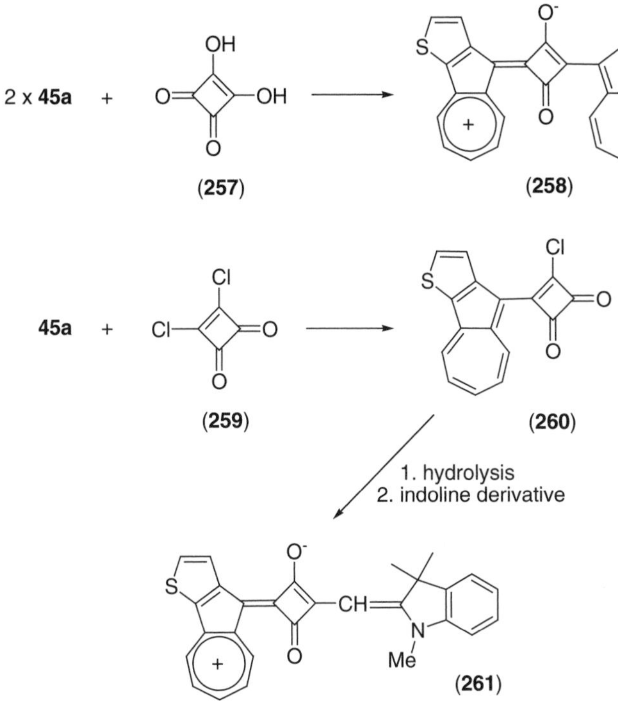

Scheme 63

(TTF), for instance, dithiane **268** (98JMC289). This dithiolethione–dithiolone strategy seems to be the optimum method to get dissymmetric TTF, due to the more marked capacity of dithiolethiones to form an ylide and to the greater facility of dithiolones to undergo nucleophilic attack on the carbon of the C=O bond (04CRV5133). Calculations, X-ray crystallography, spectra, and electrochemistry of the TTF have been described.

4.2.2 Reactions at nuclear nitrogen atoms

The reaction of pyrimidone **34** with dimethyl sulfate yields N-methyl compound **269a** (Scheme 66) (81H(15)839). N-Methylindole **55d** forms by the action of methyl iodide on the sodium salt of azulenoindole **269b** that is the product of N-debenzenesulfonylating derivative **55e** by means of tetrabutylammonium fluoride (TBAF) (07H(74)951).

Quaternization transforms pyridine **81** to pyridinium salt **270** (74CB2383), whereas thiazoles **23** and **25** with disacidified dimethyl sulfate yield methosulfates (e.g., **304**, see later) (65MI1).

Acetylation of pyrazoles **205** and **206** (63BCJ1016) and pyridone **32** (74BCJ1750) leads to the respective 1-acetyl compounds, whereas the

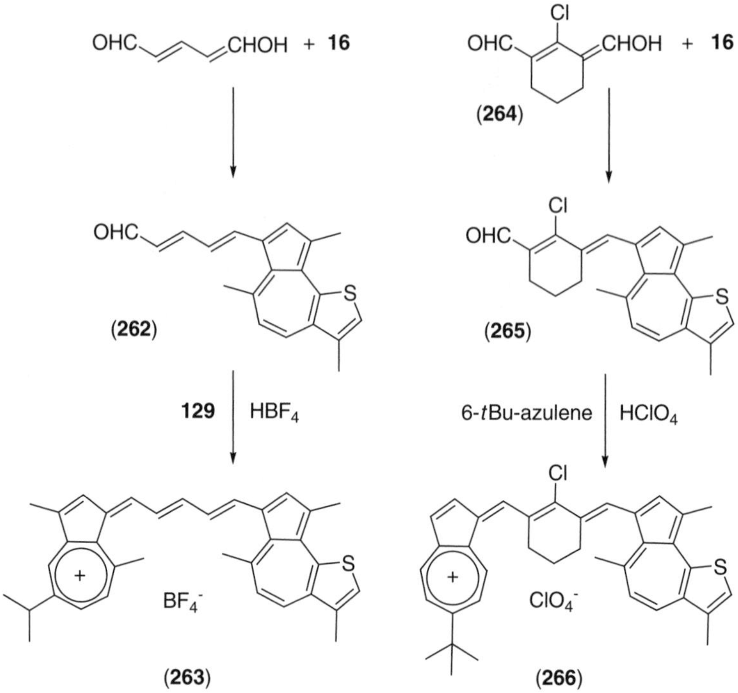

Scheme 64

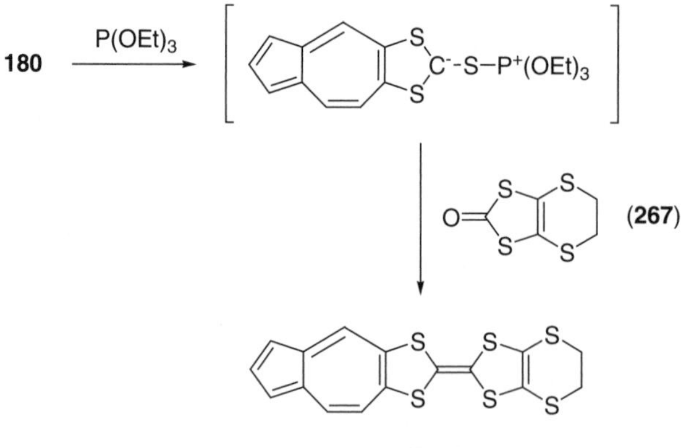

Scheme 65

10-cyano derivative (**31a**) of pyridone **32** shows a more differentiated behavior (see Section 3.6; Scheme 57).

N-Amination of triazole **271** (cf. Section 2.3.5.3) or the second NH-tautomer with O-(2,4-dinitrophenyl)hydroxylamine (DNPHA) affords a

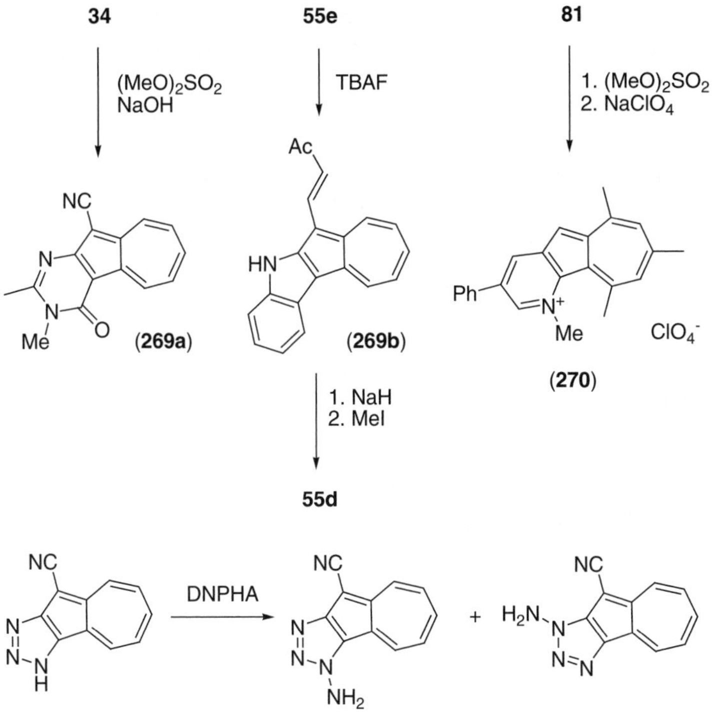

(271) (272a) (3 : 2) (272b)

Scheme 66

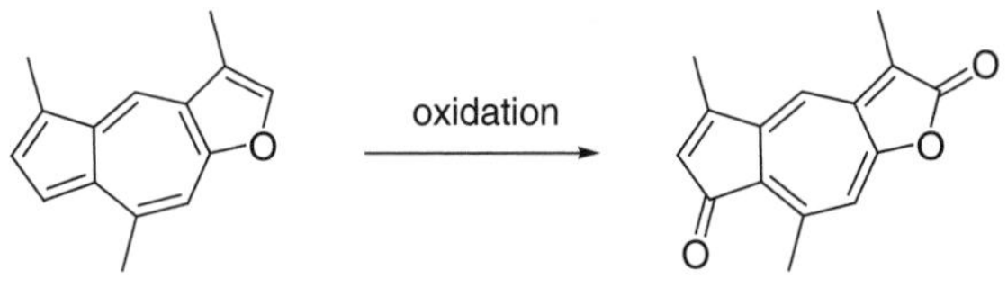

(2) (273)

Scheme 67

3:2 mixture of amino derivatives **272a** and **272b**, the assignation of which being tentative (85TL335).

4.2.3 Oxidation

Photooxidation (83TL4461) or oxidation by *m*-chloroperoxybenzoic acid (84TL2109) of linderazulene (**2**) leads to ketolactone **273** (Scheme 67).

peri-Fused azulenothiophene **133a** suffers oxidative dimerization (71JA2196); dimer **134** can also been obtained from the mutual precursor

132a (see Scheme 32). The reactive site is C-8; this situation is comparable to that with *peri*-fused azulenothiopyran **136** (see Section 4.2.1; Scheme 61).

The oxidative formation of azulenoquinones from linearly fused azulenoheterocycles, on the other hand, exactly follows the rules of electrophilic substitution on azulenes (cf. Section 4.2.1). Thiophene **45a** or furan **47a** (Scheme 68) can be oxidized by bromine/aqueous acetic acid, phenyltrimethylammonium perbromide (PTAB) or pyridinium perbromide to yield mixtures of quinones **274a** and **275a** or **274b** and **275b**, respectively (96BCJ1149, 03H(61)271).

Isomeric thiophene **69** by PTAB in excess is not only oxidized but also simultaneously brominated to yield quinones **276a** and **276b**. A possible mechanism involves the formation of a geminate-dibromo azulenium intermediate, its hydrolysis and dehydrobromination. The use of smaller amounts of PTAB results in the formation of mono- and dibromoazulenothiophenes **277a** and **277b**. The reaction of azulenothiazoles **188b** or **190** in aqueous THF with bromine/acetic acid forms bromoazulenoquinones **278a** and **278b**, respectively (96H(43)1049).

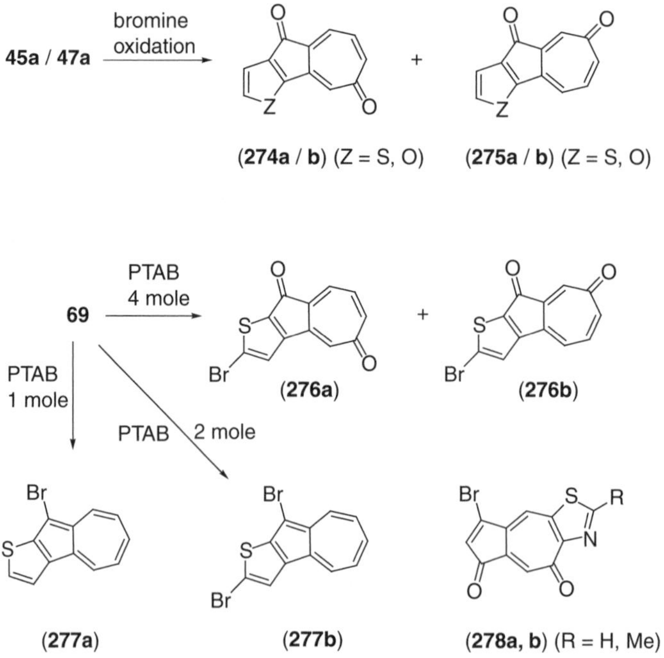

45a / 47a $\xrightarrow{\text{bromine oxidation}}$

(**274a / b**) (Z = S, O)   (**275a / b**) (Z = S, O)

69 $\xrightarrow{\text{PTAB 4 mole}}$

Br (**276a**)   Br (**276b**)

PTAB 1 mole   PTAB 2 mole

(**277a**)   (**277b**)   (**278a, b**) (R = H, Me)

Scheme 68

4.3 Nucleophilic substitution of functional nuclear substituents

Relevant reactions as far as they have been published are restricted to substituents at the heterocyclic rings. Matsui has described the transformation of 3-methoxyazuleno[2,1-*b*]thiophene (**248a**; Scheme 69) by ether cleavage, reduction of oxo compound **279**, and dehydration to yield parent compound **69** or, through Grignard reactions, to yield homologs **280a** and **280b** (60CIL1302, 61NKZ1520). Analogous reaction sequences lead from the respective isopropyl homologs of methoxy compound **279** to 5- and 7-isopropyl compounds **281a** and **281b** (61NKZ1522) or to the 6-isopropyl isomer (61NKZ1665) and from derivative **64** to ester **282** (61NKZ1667).

The reactions of azulenopyridones **98b** and **100** (Scheme 70) with phosphoryl chloride produce chloropyridines (e.g., **283**), which are precursors of several O- and N-functionalized derivatives (e.g., **284a** and **284b**; **285a** and **285b**, respectively) (63BCJ633). In a similar way, pyridones **105b** and **32** yield chloropyridine **286** (91BCJ2393) and dichloropyridine **287** (74BCJ1750), respectively, whereas pyrimidone **115** forms an analogous chloropyrimidine (81H(15)839).

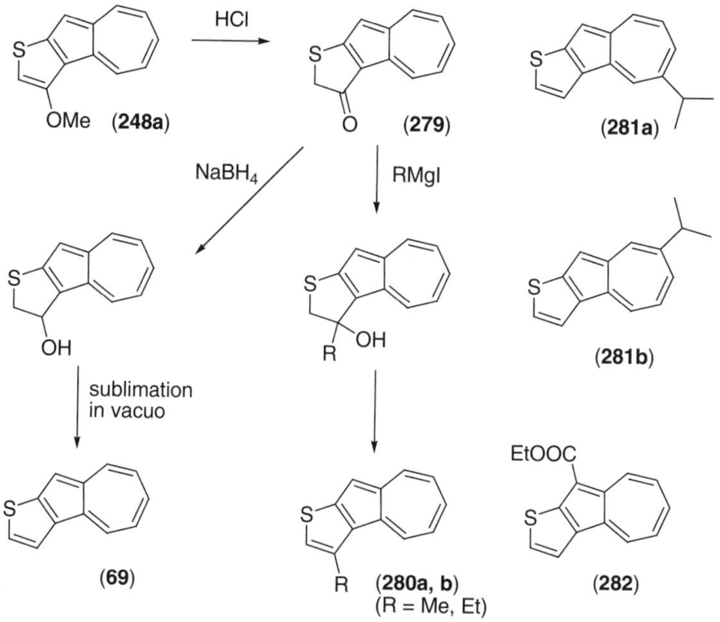

Scheme 69

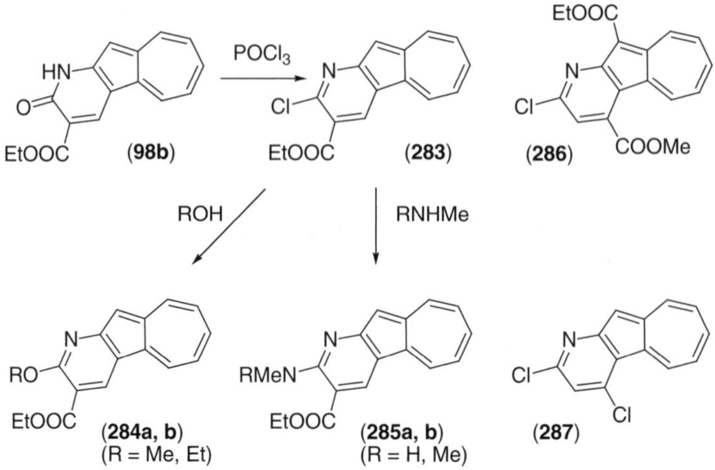

Scheme 70

4.4 Transformation of individual substituents

4.4.1 Hydroxy compounds

3-Hydroxyazuleno[2,1-*b*]thiophene **66** (Scheme 71) (61NKZ1517) and other derivatives (60CIL1302, 61NKZ1522) as well as thiazoles **191** and **193a** that bear hydroxyl groups at the five-membered azulene ring (96MI1) undergo O-methylation with diazomethane to yield, for example, methyl ether **288**. Acylation, for instance, benzoylation has also been described in Ref. (61NKZ1665).

4.4.2 Amines

The deamination of 2-aminoazulenothiazoles, such as compounds **84b** (82CL707) and **289** (96H(43)1049), proceeds by the action of inorganic or organic nitrites to yield, for instance, parent compound **188b**. Diamine **187** with isoamylnitrite yields a mixture of product **188a** (deaminated on both the five-membered rings) and monoamine **290** (96H(43)1049). Acetamide **189a** can be deacetylated by acid and subsequently deaminated by isoamylnitrite to form substance **189b**; acetamide **193b** (96MI1) reacts in the same way.

4.4.3 Carboxylic acid derivatives

Besides reduction/dehydration (Section 4.3) and deamination (Section 4.4.2), hydrolysis/decarboxylation of carboxylic esters or nitriles is the most widely used method to degrade primary synthetic azulene derivatives to form simpler species or parent ring systems. The reason is the easy accessibility and usefulness of mono- and di(alkoxycarbonylated)

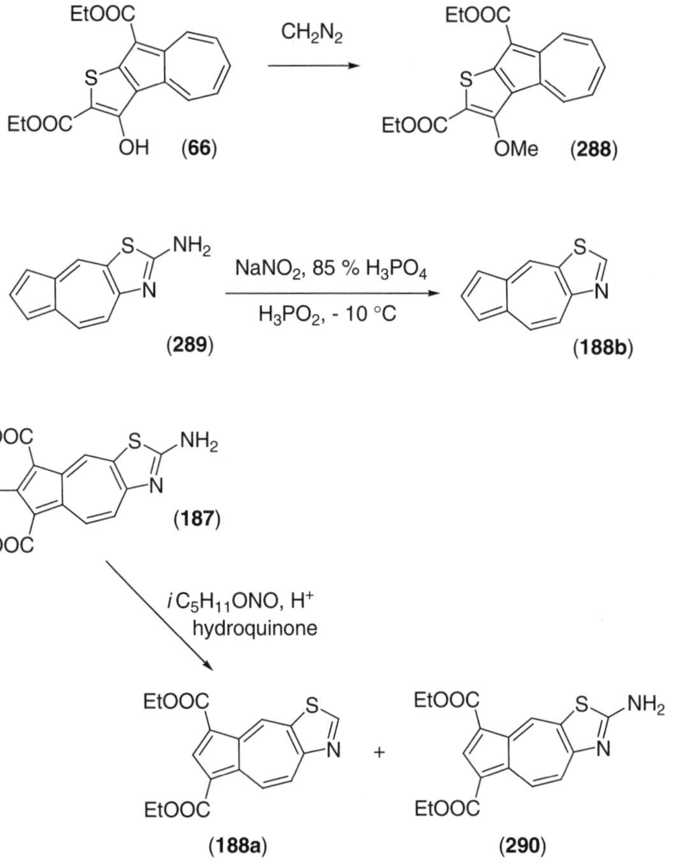

Scheme 71

azulenes and similar compounds as precursors (cf. Schemes 13, 16, 19, 28, 42–44).

Examples of both step-by-step transformation and, depending on the conditions, partial or total degradation can be found in Schemes 72 and 73, which show the reactions of thiophene **68** (81BCJ2537), furan **164** (61CIL1715), pyrazole **205** (63BCJ1016), and thiophene **288** (61NKZ1517). Azulenopyrrole **47b** forms by a two-step route analogous to that of thiophene **69** (86CL1021).

Table 7 lists further examples of one- or two-step degradation accomplished by hydrolysis, if necessary acidification, and decarboxylation. Azulenopyridine carboxylic ester **98b** can be hydrolyzed by base to yield the stable carboxylic acid (63BCJ633). Imidazole **72** and triazole **73** could not be successfully degraded (73BCJ3161). Degradation of nitriles **31a** (74BCJ1750) and **208** (97MI1) is depicted in Scheme 74.

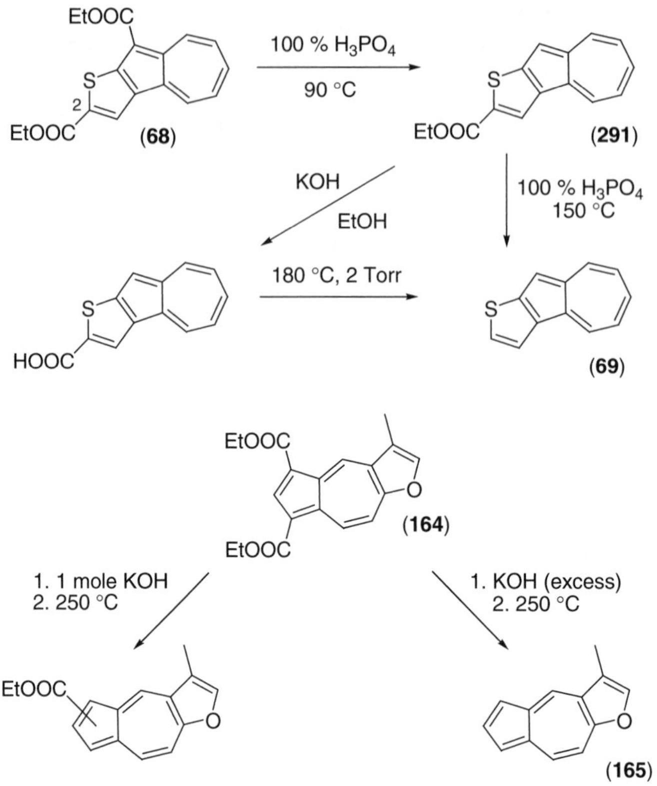

Scheme 72

4.5 Ring cleavage and transformation

4.5.1 Ring cleavage

The exhaustive oxidation of the azulene ring system to leave the heterocyclic moiety (Scheme 75) serves in the elucidation of structures. Thus, dicarboxylic acid **296** is obtained by oxidation of linderazulene (**2**) with hydrogen peroxide, against which the furan ring is fairly stable (53CPB164). Cleavage products **297** (63BCJ1016) and **298** (73BCJ3161) are the results of similar oxidative reactions.

Thiophene-ring opening by reductive desulfurization and subsequent dehydrogenation of partially hydrogenated intermediates to yield guaiazulene (**129**) or vetivazulene (**170**; Scheme 76) is the reversal of the reactions forming azulenothiophenes **15** (67MI1, cf. 70BCJ509) and **171** (69BCJ1404). Azulenopyrone **39** can be hydrolyzed by sodium methylate to yield azetonylazulene **38**, which had been isolated as an intermediate in the synthesis of pyrone **39** (92H(34)429).

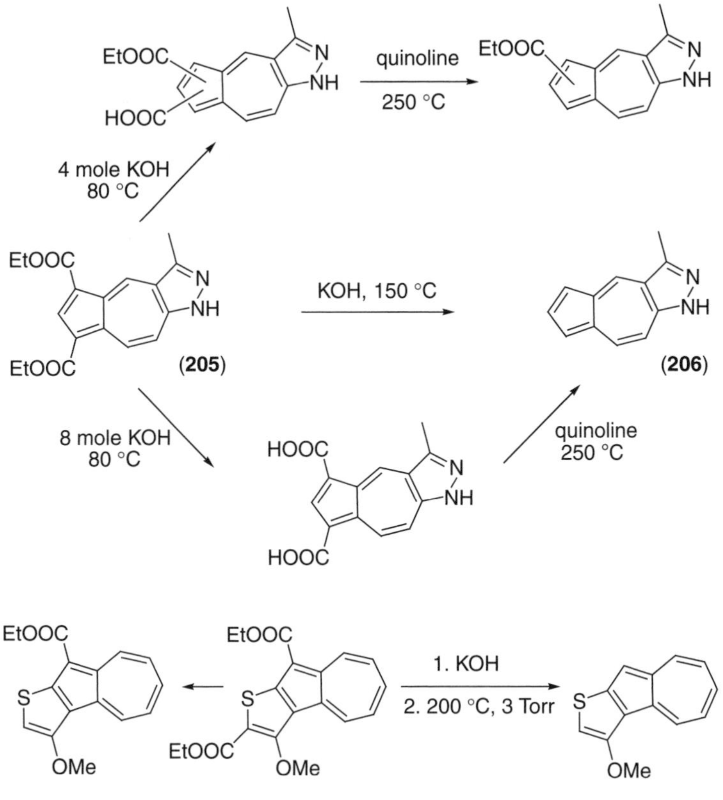

Scheme 73

The reaction of aminotriazoles **272a** and **272b** with lead tetraacetate unexpectedly led to heptafulvene **299** (85TL335). Any product of trapping the desired dehydroazulene derivative by cyclohexadiene (that was present in the reaction) could not be detected.

4.5.2 Ring transformation

Whereas bicyclic azulenes suffer thermal isomerization to yield naphthalene derivatives (52CRV127, p. 169; 85HOU(5/2c)127, p. 413), there are a few examples of transforming an azulene-fused heterocyclus to another heterocyclus or of rearranging it to benzene. The formation of azulenopyrrole **194** from azulenofuran **164** (61CIL1715) was mentioned previously (Section 2.4.1.6).

o-Quinonoid azuleno[1,2-*c*]pyrroles (such as **122a**; Scheme 77) having a diene moiety in the pyrrole ring undergo cycloaddition with

Table 7 Hydrolysis and decarboxylation of carboxylic esters

Ring system	Formula[a] Educt	Product	One-step method	Two-step method Step 1	Step 2	Reference
Azuleno[2,1-b]furan	59a	60a		KOH	100% H_3PO_4	81H(15)835
Azuleno[1,2-b]thiophene	43a, 43b	45a, 45b	PPA			83CL1721
	–	46	H_3PO_4			03OBC2572
Azuleno[2,1-b]thiophene	–	248a[b]		KOH	200°C/3 Torr	60CIL1302
Azuleno[1,2-d]thiazole	78	23	100% H_3PO_4			86BCJ3320
Azuleno[2,1-d]thiazole	–	85b	100% H_3PO_4			82CL707
	84b	–	100% H_3PO_4			
Azuleno[2,1-b]pyrrole	59b	60b	100% H_3PO_4			81H(15)835
Azuleno[1,2-b]pyrazine	74	75[c]		NaOH	Heating in vacuo	73BCJ3161
Azuleno[2,1-b]-1-azaazulene	–	88	100% H_3PO_4, 100°C			93TL831
Azuleno[6,5-d]thiazole	188a	188b[d]	100% PPA			96H(43)1049
	189b	190[e]	100% PPA			
	290	289	100% PPA			
Azuleno[6,5-b]pyrrole	194	195		KOH	250°C	61CIL1715

[a] As far as the respective formulas exist in this review.
[b] Likewise 5- and 7-isopropyl derivatives (61NKZ1522) and 6-isopropyl derivative (61NKZ1665) of **248a**.
[c] Likewise 2,3-diphenyl and 2,3-diphenyl-7-isopropyl analogs of **75**.
[d] Likewise 2-benzoylamino and 2-benzoylamino-7-methoxy derivatives of **188b** (96MI1).
[e] Likewise 7-methoxy derivative of **190** (96MI1).

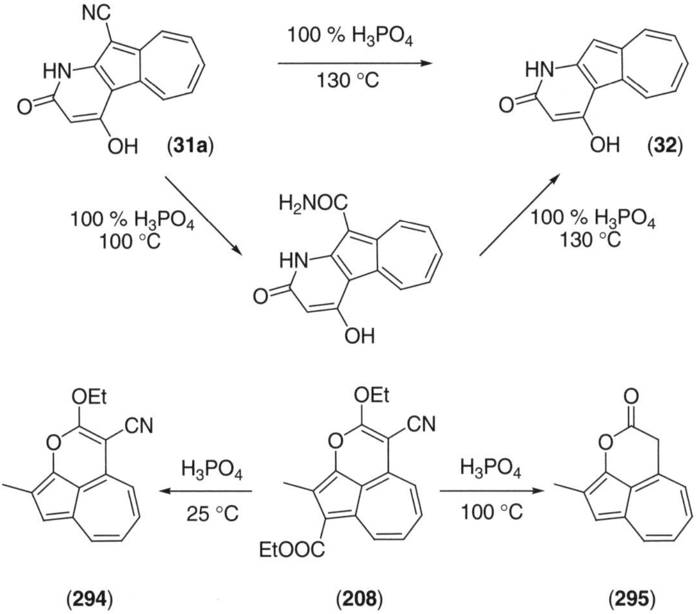

Scheme 74

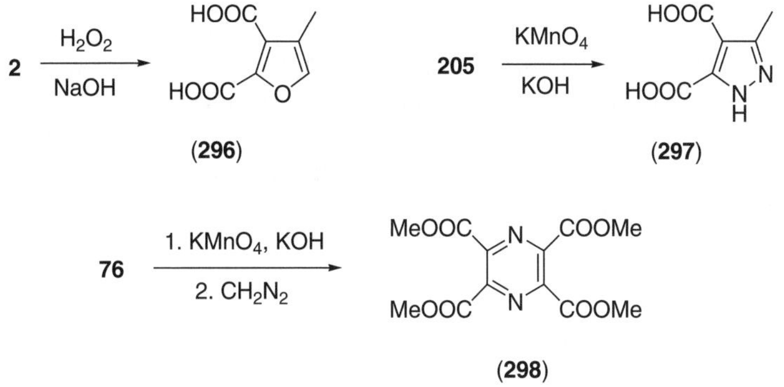

Scheme 75

acetylenedicarboxylic esters to yield, for instance, anilinobenzazulene **300** and by-product **301**, formally the result of a C-1 addition (01H(54)647). Both products, however, are presumed to arise from an intermediate cycloadduct by fission of *different* bonds and subsequent aromatization. Decomplexation of tricyclic azulene **226** is accompanied by oxidative ring contraction to yield formylbenzazulene **302** (02JOM(642)80).

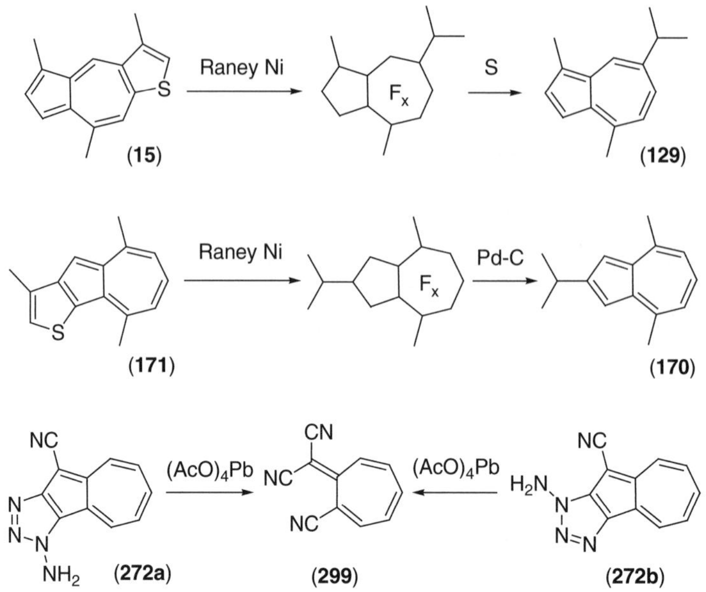

Scheme 76

4.5.3 Cycloaddition

o-Quinonoid azulenofurans undergo cycloaddition without any following aromatization. Non-isolable azuleno[4,5-c]furan **168** (Section 2.4.1.1) with maleic anhydride yields adduct **169** in a 5:2 ratio of *exo*- and *endo*-isomers (83CC1025). Chloro derivative **149** (which is a little more stable than parent substance **168**) with N-methylmaleimide forms analogous *exo*- and *endo*-isomers in a 1:1 ratio, whereas stable azuleno[5,6-c]furan **152** (Scheme 78) forms adduct **303** in an *exo:endo* ratio of 1:3 (03OBC2383).

4.6 Reactivity of side chains

4.6.1 Reactivity of methyl groups

Methyl groups in the 4- and 6-positions of azulene easily undergo deprotonation and subsequent reactions (77RCR530), as exemplified by syntheses depicted in the Schemes 32 and 49. Their reactivity can also be concluded from the occurrence of metabolites **7a** and **7b** accompanying linderazulene (**2**) in natural sources (05JNP248).

In fused azole or azine rings, methyl groups at carbon atoms adjacent to a nitrogen atom often have sufficient acidity to effect condensation with cyanine-dye forming synthons, especially after undergoing a reactivity-enhancing quaternization. Thus, for instance, methosulfates

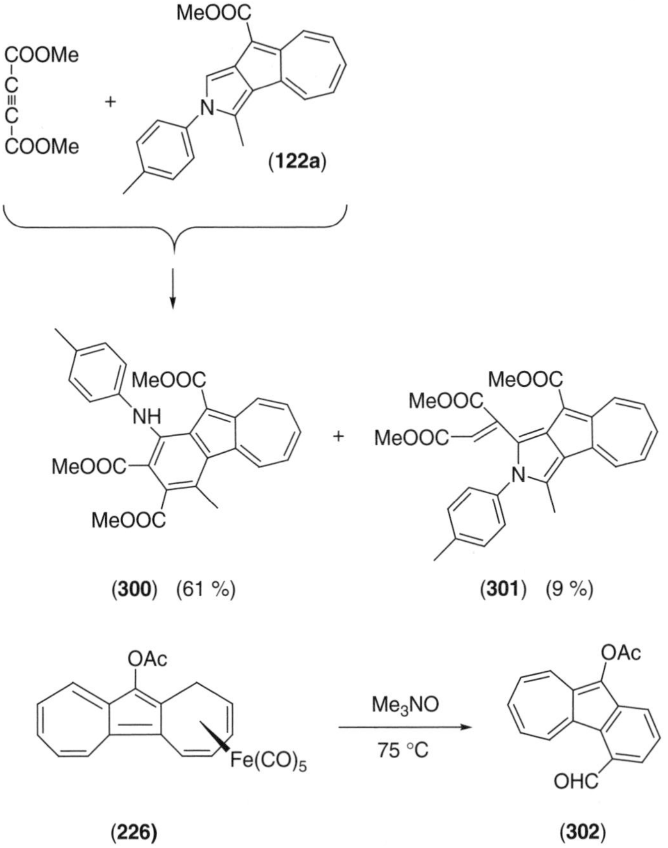

Scheme 77

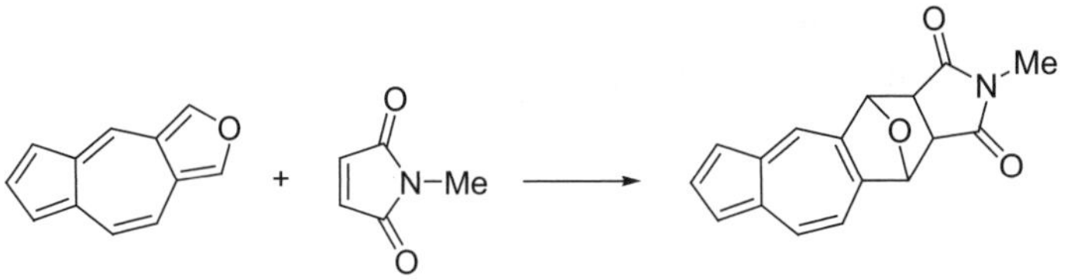

Scheme 78

304 of azulenothiazole **25** (Scheme 79) condenses with triethyl orthoformiate to afford symmetrical trimethine cyanine **305** and with reactive benzothiazolium salt **307** it forms unsymmetrical trimethine cyanine **306** (65MI1). The visible spectra of such dyes, compared with those of the corresponding benzothiazole dyes, reveal a bathochromic shift of the

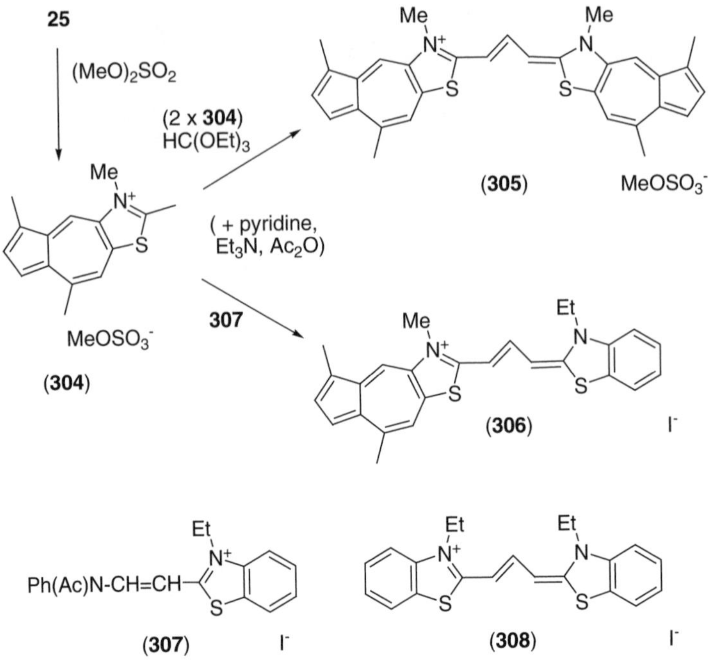

Scheme 79

principal maximum of 30 ± 5 nm per azulenothiazole moiety (e.g., comparison dye **308**, 560 nm; **306**, 590 nm; **305**, 630 nm).

4.6.2 Reactivity of functionalized side chains

Reduction and derivatization of the formyl group of azulenothiophene **245** are shown in Scheme 80 (61NKZ1667). Similar reactions are known with formylazulenothiophene **243** (69BCJ1404) and formylazulenothiazole **249** (83H(20)1263).

Azulenoindole **140a** with acetic anhydride or acyl chlorides yields esters, for instance, acetate **312** (01BML1981). Tertiary amine **247a** (Scheme 81) is quaternized by methyl iodide and the resulting salt **313** can be transformed by mercapto compounds to yield thioethers, for example, **314** (61NKZ1667).

It is possible to selectively reduce an unsaturated side-chain catalytically without reducing the azulene rings (66MI1). Recently, an effective metal-free method has been published for the selective reduction of the $C=C$ double bond even in azulenic enones, for instance, thiophenes **54** (Scheme 81) or **55a** and **55b**, using the hydride donor cycloheptatriene and protic acid (06OL3137).

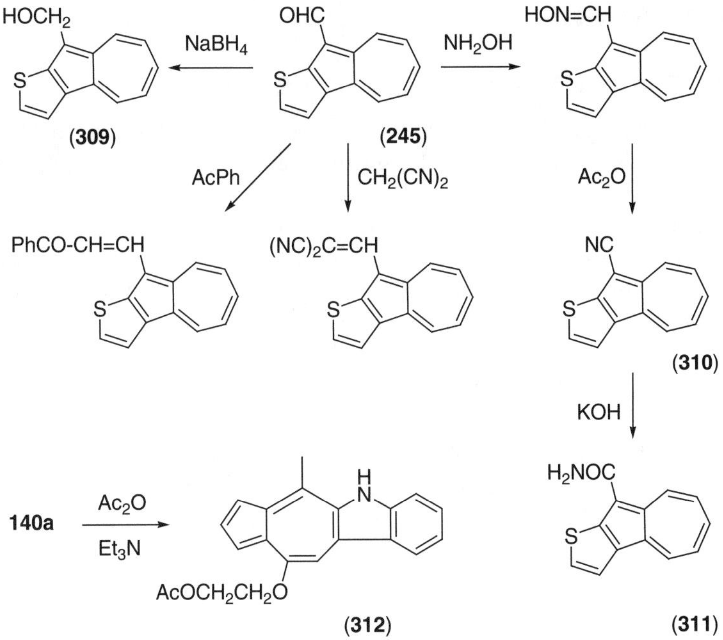

Scheme 80

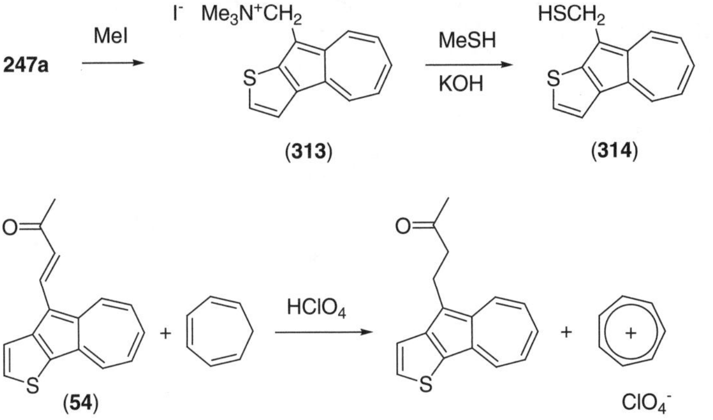

Scheme 81

5. APPLICATIONS

5.1 Pharmaceutical uses

The pharmacological properties of azulenes, especially the antiphlogistic action, have been studied for a long time (52CRV127, p. 184; 55FCF(3)334, p. 385).

Linderazulene (2) and its dihydro compound 6 have been found to exhibit antitumor, antifungal, and immunostimulating activities (87E624, 88WO191); they inhibit the cell division of fertilized ascidian eggs (89MI1, cf. 93CL2003). Two metabolites (oxidation products 7a and 7b) of linderazulene (05JNP248, 05USP6852754) and a number of azulenoindoles (e.g., 140a and 140b) (01BML1981) are antitumor agents as well. Nozoe et al. have claimed azulenopyrazoles (e.g., 205 and 206) as antiphlogistic (63JP25678), sedative, and analgesic agents (64JP3439) and azulenopyridines (e.g., 96a, 96b, 98a, and 98b) as anti-inflammatory agents (64JP18039).

5.2 Uses in information recording

Azulenothiazole-type cyanine dyes (e.g., 305 and 306) scarcely improve the spectral sensitivity of photographic silver halide emulsion (65MI1). peri-Fused azulenopyridines (e.g., 218a), azulenothiopyrans (136-type), and azulenoazepines (137-type) have been claimed for uses in high-sensitivity laser beam recording media having good storage stability (87JAK74692). The stabilization of the ionic state of azulenylmethyl

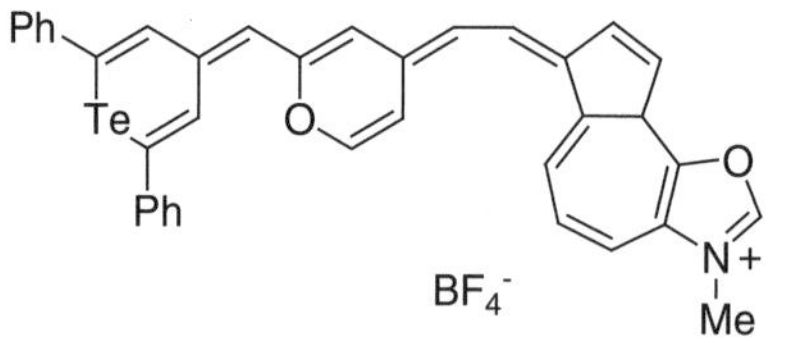

(315)

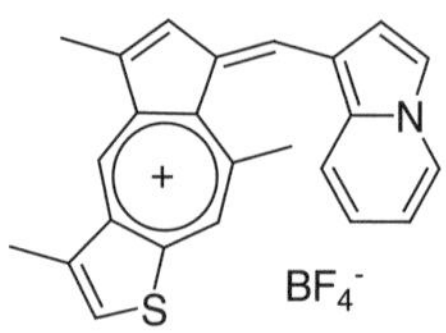

(317)

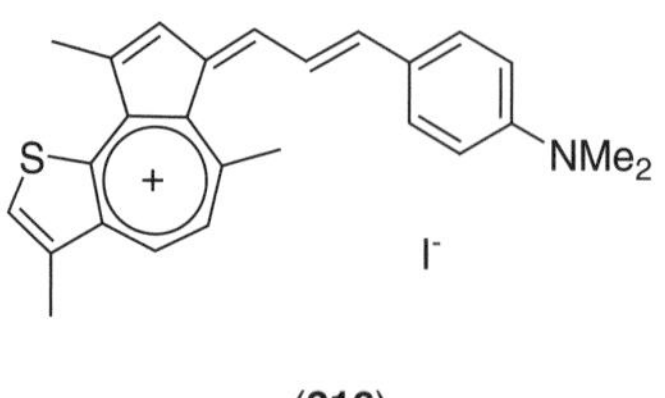

(316)

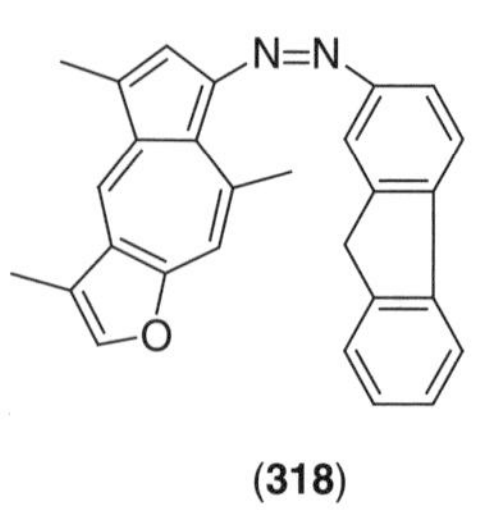

(318)

Scheme 82

compounds (Scheme 62) was studied in the context of creating novel advanced materials (03OBC2572, cf. 04MI1).

Polymethine dyes containing the azulenium moiety are suitable as charge generators for electrophotography, especially for semiconductor laser-recording media showing both high sensitivity in recording and stability in reading. Numerous patents claim dyes, such as mono-, bis-, and tris-azulenyl polymethine dyes including squaryl and croconyl compounds (e.g., **258**, **261**, **263**, and **266**; Schemes 63 and 64) (86EAA187015, 88EAA295144, 88JAK247086, 89DEA3914151, 90USA4965178), pyran derivatives (e.g., **315**; Scheme 82) (88JAK136054, 88JAK141068), and similar compounds, for instance, aniline **316** (85JAK262163) and indolizine **317** (86JAK15148). Symmetrical and unsymmetrical azulenylazo dyes (e.g., **318**) have also been claimed (86JAK223845).

REFERENCES

39YZ162	H. Kondo and K. Takeda, *Yakugaku Zasshi* **59**, 162 and 504 (1939).
44YZ(is.9A)32	K. Takeda and T. Shimada, *Yakugaku Zasshi*, **64**(9A), 32 (1944). [*CA*, **46**, 102 (1952)].
48FOR(5)40	A. J. Haagen-Smit, *Fortschr. Chem. Org. Naturst.*, **5**, 40 (1948).
48MI1	P. A. Plattner, in "*Newer Methods in Preparative Organic Chemistry*", p. 21, Interscience, New York (1948).
50AG281	H. Pommer, *Angew. Chem.*, **62**, 281 (1950).
52CRV127	M. Gordon, *Chem. Rev.*, **50**, 127 (1952).
52N571	E. Stahl, *Naturwissenschaften*, **39**, 571 (1952).
53CPB164	K. Takeda and W. Nagata, *Chem. Pharm. Bull.*, **1**, 164 (1953).
54CCC792	V. Herout and F. Šorm, *Collect. Czech. Chem. Commun.*, **19**, 792 (1954).
55FCF(3)334	W. Treibs, W. Kirchhof, and W. Ziegenbein, *Fortschr. Chem. Forsch.*, **3**, 334 (1955).
56CIL492	Y. Mazur and A. Meisels, *Chem. Ind. (London)*, 492 (1956).
56CIL1234	Z. Čekan, V. Herout, and F. Šorm, *Chem. Ind. (London)*, 1234 (1956).
56JCS142	D. H. R. Barton and P. de Mayo, *J. Chem. Soc.*, 142 (1956).
56LA(598)32	W. Treibs, M. Quarg, and E.-J. Poppe, *Liebigs Ann. Chem.*, **598**, 32 (1956).
57CCC1921	Z. Čekan, V. Herout, and F. Šorm, *Collect. Chem. Chem. Commun.*, **22**, 1921 (1957).
58AG419	K. Hafner, *Angew. Chem.*, **70**, 419 (1958).
58CCC2195	L. Dolejš, M. Souček, M. Horák, V. Herout, and F. Šorm, *Collect. Czech. Chem. Commun.*, **23**, 2195 (1958).
59CCC1548	F. Šorm, M. Suchý, and V. Herout, *Collect. Czech. Chem. Commun.*, **24**, 1548 (1959).
59CCC1554	Z. Čekan, V. Procházka, V. Herout, and F. Šorm, *Collect. Czech. Chem. Commun.*, **24**, 1554 (1959).
59CIL517	M. Suchý, V. Herout, and F. Šorm, *Chem. Ind. (London)*, 517 (1959).
59MI1	E. Heilbronner, in "*Non-Benzenoid Aromatic Compounds*" (D. Ginsburg, ed.), p. 171, Wiley (Interscience), New York and London (1959).

59MI2	W. Keller-Schierlein and E. Heilbronner, in "Non-Benzenoid Aromatic Compounds" (D. Ginsburg, ed.), p. 277, Wiley (Interscience), New York and London (1959).
60CCC507	M. Suchý, V. Herout, and F. Šorm, Collect. Czech. Chem. Commun., 25, 507 (1960).
60CCC1492	L. Novotny, V. Herout, and F. Šorm, Collect. Czech. Chem. Commun., 25, 1492 (1960).
60CCC1702	V. Sýkora and K. Vokáč, Collect. Czech. Chem. Commun., 25, 1702 (1960).
60CIL1302	K. Matsui and T. Nozoe, Chem. Ind. (London), 1302 (1960).
60G322	L. Mangoni and M. Belardini, Gazz. Chim. Ital., 50, 322 (1960).
61CIL1715	T. Nozoe, S. Seto, and S. Matsumura, Chem. Ind. (London), 1715 (1961).
61FOR(19)32	T. Nozoe and Sh. Itô, Fortschr. Chem. Org. Naturst., 19, 32 (1961).
61JA2326	A. Romo de Vivar and J. Romo, J. Am. Chem. Soc., 83, 2326 (1961).
61NKZ1517	K. Matsui, Nippon Kagaku Zasshi, 82, 1517 (1961). [CA, 59, 3862 (1963)].
61NKZ1520	K. Matsui, Nippon Kagaku Zasshi, 82, 1520 (1961). [CA, 59, 3863 (1963)].
61NKZ1522	K. Matsui, Nippon Kagaku Zasshi, 82, 1522 (1961). [CA, 59, 3863 (1963)].
61NKZ1665	K. Matsui, Nippon Kagaku Zasshi, 82, 1665 (1961). [CA, 59, 540 (1963)].
61NKZ1667	K. Matsui, Nippon Kagaku Zasshi, 82, 1667 (1961). [CA, 59, 2752 (1963)].
62JA2601	W. Herz, H. Watanabe, M. Miyazaki, and Y. Kishida, J. Am. Chem. Soc., 84, 2601 (1962).
62MI1	S. Dev, in "Recent Progress in the Chemistry of Natural and Synthetic Colouring Matters and Related Fields" (T. S. Gore, B. S. Joshi, S. V. Sunthankar, and B. D. Tihak, eds.), p. 59, Academic Press, New York (1962).
63BCJ633	T. Nozoe and K. Kikuchi, Bull. Chem. Soc. Jpn., 36, 633 (1963).
63BCJ1016	T. Nozoe, K. Takase, and M. Tada, Bull. Chem. Soc. Jpn., 36, 1016 (1963).
63CCC1202	V. Prochazka, Z. Čekan, and R. B. Bates, Collect. Czech. Chem. Commun., 28, 1202 (1963).
63G395	L. Mangoni and G. Bandiera, Gazz. Chim. Ital., 93, 395 (1963).
63JCS2591	K. Takeda, H. Minato, and M. Ishikawa, J. Chem. Soc., 2591 (1964).
63JP25678	T. Nozoe and K. Takase, Jpn. Pat. 38/25678 (1963). [CA, 60, 5510 (1964)].
63T1285	F. Sánchez-Viesca and J. Romo, Tetrahedron, 19, 1285 (1963).
63T2317	J. Romo, A. Romo de Vivar, and W. Herz, Tetrahedron, 22, 1499 (1963).
63ZC26	G. Fischer and E.-J. Poppe, Z. Chem., 3, 26 (1963).
63ZOB2734	K. S. Rybalko, Zh. Obshch. Khim., 33, 2734 (1963). [CA, 60, 5561 (1964)].
64BCJ859	Y. Kitahara and T. Kato, Bull. Chem. Soc. Jpn., 37, 859 (1964).
64CCC938	M. Holub, D. P. Popa, V. Herout, and F. Šorm, Collect. Czech. Chem. Commun., 29, 938 (1964).
64JA3137	L. L. Replogle, J. Am. Chem. Soc., 86, 3137 (1964).
64JCS3577	K. Takeda, H. Minato, K. Hamamoto, I. Horibe, T. Nagasaki, and M. Ikuta, J. Chem. Soc., 3577 (1964).

64JCS4578	K. Takeda, H. Minato, and M. Ishikawa, *J. Chem. Soc.*, 4578 (1964).
64JP3439	T. Nozoe and K. Takase, *Jpn. Pat.* 39/3439 (1964). [*CA*, **61**, 3116 (1964)].
64JP18039	T. Nozoe and K. Kikuchi, *Jpn. Pat.* 39/18039 (1964). [*CA*, **62**, 5258 (1965)].
64UP1	G. Fischer, *Unpublished results* (1964).
65CPB717	H. Minato, M. Ishikawa, and T. Nagasaki, *Chem. Pharm. Bull.*, **13**, 717 (1965).
65JOC118	W. Herz, A. Romo de Vivar, and M. V. Lakshimikantham, *J. Org. Chem.*, **30**, 118 (1965).
65MI1	G. Fischer and E.-J. Poppe, *Veröff. Wiss. Photo-Lab. Wolfen*, **10**, 123 (1965).
65TL1877	L. L. Replogle, K. Katsumoto, and T. C. Morrill, *Tetrahedron Lett.*, 1877 (1965).
65ZOB580	K. S. Rybalko and V. I. Sheichenko, *Zh. Obshch. Khim.*, **35**, 580 (1965). [*CA*, **63**, 3452 (1965)].
66CPB550	H. Hikino, H. Takahashi, Y. Sakurai, T. Takemoto, and N. S. Bhacca, *Chem. Pharm. Bull.*, **14**, 550 (1966).
66CZ691	W. Treibs, *Chem. Ztg.*, **90**, 691 (1966).
66MI1	D. Lloyd, "Carbocyclic Non-Benzenoid Aromatic Compounds", p. 182, Elsevier, Amsterdam, London, and New York (1966).
66T1499	J. Romo, P. Joseph-Nathan, and G. Siade, *Tetrahedron*, **22**, 1499 (1966).
67JOC507	W. Herz and P. S. Santhanam, *J. Org. Chem.*, **32**, 507 (1967).
67MI1	Sh. Hayashi, M. Okano, Sh. Kurokawa, and T. Matsuura, *J. Sci. Hiroshima Univ., Ser. A-II*, **31**(2), 79 (1967).
67TL3443	Sh. Hayashi, Sh. Kurokawa, M. Okano, and T. Matsuura, *Tetrahedron Lett.*, **35**, 3443 (1967).
67YZ70	H. Hikino, K. Meguro, G. Kusano, and T. Takemoto, *Yakugaku Zasshi*, **87**, 70 (1967). [*CA*, **66**, 95219 (1967)].
68BCJ2095	T. Nozoe, P. W. Yang, H. Ogawa, and T. Toda, *Bull. Chem. Soc. Jpn.*, **1**, 2095 (1968).
68BCJ2182	Sh. Hayashi, Sh. Kurokawa, and T. Matsuura, *Bull. Chem. Soc. Jpn.*, **1**, 2182 (1968).
68JOC823	L. L. Replogle, K. Katsumoto, T. C. Morrill, and C. A. Minor, *J. Org. Chem.*, **33**, 823 (1968).
68JPC3975	J. Fabian, A. Mehlhorn, and R. Zahradnik, *J. Phys. Chem.*, **72**, 3975 (1968).
68MI1	A. Romo de Vivar, *Rev. Soc. Quim. Mex.*, **12**, 212A (1968). [*CA*, **71**, 3493 (1969)].
68T5625	J. Romo, A. Romo de Vivar, and E. Díaz, *Tetrahedron*, **24**, 5625 (1968).
68TCA247	J. Fabian, A. Mehlhorn, and R. Zahradnik, *Theor. Chim. Acta*, **12**, 247 (1968).
69AQ285	J. Bermejo Barrera, C. Betancor, J. L. Breton Funes, and A. Gonzalez Gonzalez, *An. Quim.*, **65**, 285 (1969).
69BCJ1404	Sh. Hayashi, Sh. Kurokawa, and T. Matsuura, *Bull. Chem. Soc. Jpn.*, **42**, 1404 (1969).
69ISJ435	J. Altman, E. Babad, D. Ginsburg, and M. B. Rubin, *Isr. J. Chem.*, **7**, 435 (1969).
69TL2073	M. Yanagita, S. Inayama, T. Kawamata, and T. Okura, *Tetrahedron Lett.*, 2073 (1969).

70BCJ509	Sh. Kurokawa, *Bull. Chem. Soc. Jpn.*, **43**, 509 (1970).
70PAC181	K. Takeda, *Pure Appl. Chem.*, **21**, 181 (1970).
70PIC53	M. R. Churchill, *Prog. Inorg. Chem.*, **11**, 53 (1970).
71CPB676	K. Takeda, Y. Mori, and T. Sato, *Chem. Pharm. Bull.*, **19**, 676 (1971).
71JA2196	H. L. Ammon, L. L. Replogle, P. H. Watts, K. Katsumoto, and J. M. Stewart, *J. Am. Chem. Soc.*, **93**, 2196 (1971).
71KPS727	P. V. Chugunov, K. S. Rybalko, and A. I. Shreter, *Khim. Prir. Soedin.*, **7**, 727 (1971). [*CA*, **76**, 127168 (1972)].
71MI1	T. Asao and Y. Kitahara, *Kagaku No Kyoiki, Zokan*, (94), 87 (1971). [*CA*, **75**, 20053 (1971)].
73BCJ3161	T. Nozoe, T. Asao, and M. Kobayashi, *Bull. Chem. Soc. Jpn.*, **46**, 3161 (1973).
73CZ447	E. Müller and G. Zountsas, *Chem. Ztg.*, **97**, 447 (1973).
73MI1	T. Nozoe and I. Murata, *Int. Rev. Sci., Org. Chem, Ser. One*, **3**, 201 (1973).
74AHC(16)181	J. Ashby and C. C. Cook, *Adv. Heterocycl. Chem.*, **16**, 181 (1974).
74BCJ1750	T. Nozoe, K. Takase, T. Nakazawa, S. Sugita, and M. Saito, *Bull. Chem. Soc. Jpn.*, **47**, 1750 (1974).
74CB2383	Ch. Jutz and E. Schweiger, *Chem. Ber.*, **107**, 2383 (1974).
74CB2956	Ch. Jutz, E. Schweiger, H. Löbering, A. Kraatz, and W. Kosbahn, *Chem. Ber.*, **107**, 2956 (1974).
74CZ41	E. Müller and G. Zountsas, *Chem. Ztg.*, **98**, 41 (1974).
74JOU1320	Yu. N. Porshnev and E. M. Tereshchenko, *J. Org. Chem. USSR (Engl. Transl.)*, **10**, 1320 (1974).
75TL2077	T. C. Morrill, R. Opitz, L. L. Replogle, K. Katsumoto, W. Schroeder, and B. A. Hess, *Tetrahedron Lett.*, 2077 (1975).
76MI1	T. Nozoe and I. Murata, *Int. Rev. Sci., Org. Chem., Ser. Two*, **3**, 197 (1976).
76ZNB1116	W. Winter, *Z. Naturforsch. B*, **31**, 1116 (1976).
77CC843	H. Dürr, S. Fröhlich, B. Schley, and H. Weisgerber, *J. Chem. Soc. Chem. Commun.*, 843 (1977).
77JOC348	S. M. Kupchan, Y. Shizuri, R. L. Baxter, and H. R. Haynes, *J. Org. Chem.*, **42**, 348 (1977).
77MI1	H. Tachibana, K. Kohara, Y. Masuyama, and Y. Otani, *7th Int. Congr. Essent. Oils*, 345 (1977). [*CA*, **92**, 129114 (1980)].
77RCR530	V. B. Mochalin and Yu. N. Porshnev, *Russ. Chem. Rev. (Engl. Transl.)*, **46**, 530 (1977).
77T3127	S. Braun and J. Kinkeldei, *Tetrahedron*, **33**, 3127 (1977).
78H(11)387	K. Hafner, H. J. Lindner, and W. Wassem, *Heterocycles*, **11**, 387 (1978).
79BCJ1549	K. Kohara, H. Tachibana, Y. Masuyama, and Y. Otani, *Bull. Chem. Soc. Jpn.*, **52**, 1549 (1979).
79FOR(38)47	N. H. Fischer, E. J. Olivier, and H. D. Fischer, *Prog. Chem. Org. Nat. Prod.*, **38**, 47 (1979).
79MI1	T. Nozoe and Y. Yoshida, *Koen Yoshishu – Hibenzenkei Hokozoku Kagaku Toronkai Kozo Yuki Kagaku Toronkai, 12th*, 53 (1979). [*CA*, **92**, 215177 (1980)].
79MI2	Y. Nesumi, T. Nakazawa, and I. Murata, *Koen Yoshishu – Hibenzenkei Hokozoku Kagaku Toronkai Kozo Yuki Kagaku Toronkai 12th*, 145 (1979). [*CA*, **92**, 198157 (1980)].
80CL197	T. Morita, M. Karasawa, and K. Takase, *Chem. Lett.*, 197 (1980).

81BCJ2537	K. Yamane, K. Fujimori, and T. Takeuchi, *Bull. Chem. Soc. Jpn.*, **54**, 2537 (1981).
81E442	S. Imre, R. H. Thomson, and B. Yalhi, *Experientia*, **37**, 442 (1981).
81H(15)547	T. Nishiwaki and N. Abe, *Heterocycles*, **15**, 547 (1981).
81H(15)835	T. Morita, T. Nakadate, and K. Takase, *Heterocycles*, **15**, 835 (1981).
81H(15)839	K. Takase, T. Nakazawa, and T. Nozoe, *Heterocycles*, **15**, 839 (1981).
82AXB2729	S. Kashino, M. Haisa, K. Fujimori, and K. Yamane, *Acta Crystallogr. Sect. B*, **38**, 2729 (1982).
82CL707	K. Yamane, K. Fujimori, and Sh. Ichikawa, *Chem. Lett.*, 707 (1982).
82CL2027	M. Yasunami, A. Takagi, and K. Takase, *Chem. Lett.*, 2027 (1982).
82JOU132	Yu. N. Porshnev, T. N. Ivanova, L. V. Efimova, E. M. Tereshchenko, M. I. Cherkashin, and K. M. Dyumaev, *J. Org. Chem. USSR (Engl. Transl.)*, **18**, 132 (1982).
82PAC1907	R. K. Okuda, D. Klein, R. B. Kinnel, M. Li, and P. J. Scheuer, *Pure Appl. Chem.*, **54**, 1907 (1982).
82RCR1089	Yu. N. Porshnev, V. B. Mochalin, and M. I. Cherkashin, *Russ. Chem. Rev. (Engl. Transl.)*, **51**, 1089 (1982).
83CC1025	M. Sato, K. Tanaka, S. Ebine, and K. Takahashi, *J. Chem. Soc. Chem. Commun.*, 1025 (1983).
83CL1721	K. Fujimori, T. Fujita, K. Yamane, M. Yasunami, and K. Takase, *Chem. Lett.*, 1721 (1983).
83H(20)1263	K. Yamane, K. Fujimori, Sh. Ichikawa, Sh. Miyoshi, and K. Hashizume, *Heterocycles*, **20**, 1263 (1983).
83JF2,1155	H. Yamaguchi, H. Shimoishi, M. Ata, H. Baumann, Ch. Morini, and K. Yamane, *J. Chem. Soc. Faraday Trans. 2*, **79**, 1155 (1983).
83MI1	Y. Chen, J. Yu, and H. Fang, *Zhongcaoyao*, **14**, 534 (1983). [*CA*, **100**, 153828 (1984)].
83TL4461	B. Alpertunga, S. Imre, H. J. Cowe, Ph. J. Cox, and R. H. Thomson, *Tetrahedron Lett.*, **24**, 4461 (1983).
84MI1	D. Lloyd, *"Non-Benzenoid Conjugated Carbocyclic Compounds"*, p. 350, Elsevier, Amsterdam, Oxford, New York and Tokyo (1984).
84TL2109	M. K. W. Li and P. J. Scheuer, *Tetrahedron Lett.*, **25**, 2109 (1984).
85HOU(5/2c)127	K. Zeller, *Houben-Weyl Methoden Org. Chem.*, **5/2c**, 127 (1985).
85JAK262163	K. Katagiri, Y. Oguchi, T. Otake, K. Arao, M. Kitahara, and Y. Takasu, *Jpn. Kokai* 60,262,163 (1985). [*CA*, **105**, 32931 (1986)].
85TL335	T. Nakazawa, M. Kodama, S. Kinoshita, and I. Murata, *Tetrahedron Lett.*, **26**, 335 (1985).
86BCJ3320	K. Fujimori, H. Kitahashi, S. Koyama, and K. Yamane, *Bull. Chem. Soc. Jpn.*, **59**, 3320 (1986).
86CB2956	E. V. Dehmlow, D. Balschukat, P. P. Schmidt, and C. Gröning, *Chem. Ber.*, **119**, 2956 (1986).
86CL1021	K. Fujimori, H. Fukazawa, Y. Nezu, K. Yamane, M. Yasunami, and K. Takase, *Chem. Lett.*, 1021 (1986).
86EAA187015	K. Miura, J. Iwanami, and T. Ozawa, *Eur. Pat. Appl.*, 187,015 (1986). [*CA*, **106**, 129460 (1987)].
86JAK15148	K. Katagiri, Y. Oguchi, T. Otake, K. Arao, and Y. Takasu, *Jpn. Kokai*, 61,15,148 (1986). [*CA*, **105**, 52179 (1986)].
86JAK223845	Y. Oguchi, *Jpn. Kokai*, 61,223,845 (1986). [*CA*, **106**, 205142 (1987)].
87E624	S. Sakemi and T. Higa, *Experentia*, **43**, 624 (1987).
87H(26)2025	S. Nicolić, A. Jurić, and N. Trinajstić, *Heterocycles*, **26**, 2025 (1987).

87JAK74692   T. Ozawa, Sh. Maeda, and Y. Kurose, *Jpn. Kokai*, 62,74,692 (1987). [*CA*, **107**, 187621 (1987)].

87SAA1067   M. Higashi, H. Yamaguchi, and K. Fujimori, *Spectrochim. Acta, Part A*, **43**, 1067 (1987).

87SAA1377   H. Yamaguchi, M. Higashi, and K. Fujimori, *Spectrochim. Acta, Part A*, **43**, 1377 (1987).

88BCJ1225   N. Abe and T. Takehiro, *Bull. Chem. Soc. Jpn.*, **61**, 1225 (1988).

88EAA295144   Y. Oguchi and T. Santoh, *Eur. Pat. Appl.*, 295,144 (1988). [*CA*, **111**, 48215 (1989)].

88JAK136054   E. Kato, K. Ishii, T. Ukai, and H. Okada, *Jpn. Kokai*, 63,136,054 (1988). [*CA*, **110**, 85366 (1989)].

88JAK141068   E. Kato and K. Ishii, *Jpn. Kokai*, 63,141,068 (1988). [*CA*, **110**, 125301 (1989)].

88JAK247086   T. Santo and Y. Oguchi, *Jpn. Kokai*, 63,247,086 (1988). [*CA*, **111**, 87517 (1989)].

88WO191   T. Higa and Sh. Sakemi, *WO Pat.* 88 00,191. [*CA*, **109**, 104792 (1988)].

89BCJ1567   G. Tian, A. Mori, H. Takeshita, M. Higashi, and H. Yamaguchi, *Bull. Chem. Soc. Jpn.*, **62**, 1567 (1989).

89DEA3914151   T. Satoh, I. Shimizu, and Y. Ito, *Ger. Pat. Appl.* 3,914,151 (1989). [*CA*, **113**, 32042 (1990)].

89JF2,157   H. Yamaguchi, M. Higashi, K. Fujimori, and T. Kobayashi, *J. Chem. Soc. Faraday Trans. 2*, **85**, 157 (1989).

89MI1   K. Kawamura, H. Fujita, and M. Nakauchi, *Develop. Growth Differ.*, **29**, 627 (1989).

90PAC507   T. Asao, *Pure Appl. Chem.*, **62**, 507 (1990).

90SAA1719   H. Yamaguchi, M. Higachi, and K. Fujimori, *Spectrochim. Acta, Part A*, **46**, 1719 (1990).

90USA4965178   T. Santoh and Ch. Hioki, *US Pat. Appl.* 4,965,178 (1990). [cf. *CA*, **111**, 123960 (1989)].

91BCJ2393   N. Abe, *Bull. Chem. Soc. Jpn.*, **64**, 2393 (1991).

91H(32)213   T. Nozoe, H. Wakabayashi, K. Shindo, S. Ishikawa, Ch. Wu, and P. Yang, *Heterocycles*, **32**, 213 (1991).

92H(34)429   H. Wakabayashi, P. Yang, Ch. Wu, K. Shindo, S. Ishikawa, and T. Nozoe, *Heterocycles*, **34**, 429 (1992).

93AP29   R. Hoferichter, U. Reimers, and G. Seitz, *Arch. Pharm. (Weinheim, Ger.)*, **326**, 29 (1993).

93CL2003   M. Ochi, K. Kataoka, A. Tatsukawa, H. Kotsuki, and K. Shibata, *Chem. Lett.*, 2003 (1993).

93MI1   M. Nitta and Y. Iino, *Rikogaku Kenkyusho Hokoku, Waseda Daigaku*, **140**, 24 (1993).

93TL831   M. Nitta, Y. Iino, T. Sugiyama, and A. Akaogi, *Tetrahedron Lett.*, **34**, 831 (1993).

93TL835   M. Nitta, Y. Iino, T. Sugiyama, and A. Toyota, *Tetrahedron Lett.*, **34**, 835 (1993).

94HOU(E6a)186   W. Rudorf, *Houben-Weyl Methoden Org. Chem.*, **E6a**, 186 (1994).

94JOC1309   M. Nitta, T. Akie, and Y. Iino, *J. Org. Chem.*, **59**, 1309 (1994).

95AHC(64)81   G. Fischer, *Adv. Heterocycl. Chem.*, **64**, 81 (1995).

96AHC(66)285   G. Fischer, *Adv. Heterocycl. Chem.*, **66**, 285 (1996).

96BCJ1149   T. Nozoe and H. Takeshita, *Bull. Chem. Soc. Jpn.*, **69**, 1149 (1996).

96CC937   B. Hong and S. Sun, *J. Chem. Soc. Chem. Commun.*, 937 (1996).

96CHEC2(1–10)	A. R. Katritzky, E. F. V. Scriven, and C. W. Rees (eds.), *"Comprehensive Heterocyclic Chemistry"*, 2nd edn., Vol. 1–10, Pergamon Press, Oxford etc. (1996).
96H(43)1049	T. Huang, Y. Lin, and T. Nozoe, *Heterocycles*, **43**, 1049 (1996).
96MI1	T. Huang and Y. Lin, *J. Chin. Chem. Soc. (Taipei)*, **43**, 365 (1996).
96MI2	T. Asao and Sh. Ito, *Yuki Gosei Kagaku Kyokaishi*, **54**(1), 2 (1996).
97AGE839	T. D. Lash and S. T. Chaney, *Angew. Chem. Int. Ed. Engl.*, **36**, 839 (1997).
97JRM2434	N. Abe, H. Chijimatsu, S. Kondo, H. Watanabe, and K. Saito, *J. Chem. Res. (M)*, 2434 (1997); *J. Chem. Res. (S)*, 434 (1997).
97MI1	Ch. Wu, L. Cheng, Y. Wen, and Ch. Hsiao, *J. Chin. Chem. Soc. (Taipei)*, **44**, 265 (1997).
97OR(49)1	G. Jones and S. P. Stanforth, *Org. React.*, **49**, 1 (1997).
98JMC289	H. M. Yamamoto, J. Yamaura, and R. Kato, *J. Mater. Chem.*, **8**, 289 (1998).
98MI1	Chemical Abstracts Service, *Ring Systems Handbook*, 1998 edn., American Chemical Society, Columbus, OH, p. 1498RSF.
98P1079	U. Mölleken, V. Sinnwell, and K. Kubeczka, *Phytochemistry*, **47**, 1079 (1998).
98SL950	V. Nair and G. Anilkumar, *Synlett*, 950 (1998).
99TL6609	K. Yamamura, N. Kusuhara, Y. Houda, M. Sasabe, H. Takagi, and M. Hashimoto, *Tetrahedron Lett.*, **40**, 6609 (1999).
00CJC224	B. Wang, Y. Lin, J. Chang, and P. Wang, *Can. J. Chem.*, **78**, 224 (2000).
00MI1	H. Hopf, *"Classics in Hydrocarbon Chemistry"*, p. 281, Wiley-VCH, Weinheim, New York, Chichester, Brisbane, Singapore and Toronto (2000).
01BML1981	B. Hong, Y. Jiang, and E. S. Kumar, *Bioorg. Med. Chem. Lett.*, **11**, 1981 (2001).
01H(54)647	D. Wang and K. Imafuku, *Heterocycles*, **54**, 647 (2001).
01H(54)667	T. Okujima, T. Terazono, Sh. Ito, N. Morita, and T. Asao, *Heterocycles*, **54**, 667 (2001).
02AGE1371	Sh. R. Graham, D. A. Colby, and T. D. Lash, *Angew. Chem. Int. Ed. Engl.*, **41**, 1371 (2002).
02EJO2547	I. Dix, Ch. Doll, H. Hopf, and P. G. Jones, *Eur. J. Org. Chem.*, 2547 (2002).
02H(56)497	D. Wang and K. Imafuku, *Heterocycles*, **56**, 497 (2002).
02H(58)405	K. Imafuku and D. Wang, *Heterocycles*, **58**, 405 (2002).
02JOM(642)80	N. Morita, R. Yokoyama, T. Asao, M. Kurita, Sh. Kikuchi, and Sh. Ito, *J. Organomet. Chem.*, **642**, 80 (2002).
02JRM162	N. Abe, M. Mori, and H. Fujii, *J. Chem. Res. (M)*, 0162 (2002).
02T7653	K. Yamamura, N. Kusuhara, A. Kondou, and M. Hashimoto, *Tetrahedron*, **58**, 7653 (2002).
03H(61)271	H. Matsuo, K. Fujimori, A. Ohta, A. Kakehi, M. Yasunami, and T. Nozoe, *Heterocycles*, **61**, 271 (2003).
03MI1	Th. Eicher and S. Hauptmann, *"The Chemistry of Heterocycles"*, 2nd edn., Wiley/VCH, Weinheim (2003).
03OBC2383	A. D. Payne and D. Wege, *Org. Biomol. Chem.*, **1**, 2383 (2003).
03OBC2572	Sh. Ito, T. Kubo, M. Kondo, Ch. Kabuto, N. Morita, T. Asao, K. Fujimori, M. Watanabe, N. Harada, and M. Yasunami, *Org. Biomol. Chem.*, **1**, 2572 (2003).

04BML63	B. B. Lin, T. Morita, Y. Lin, and H. Chen, *Bioorg. Med. Chem. Lett.*, **14**, 63 (2004).
04CRV5133	J. M. Fabre, *Chem. Rev.*, **104**, 5133 (2004).
04EJO899	T. Asao, Sh. Itô, and I. Murata, *Eur. J. Org. Chem.*, 899 (2004).
04MI1	Sh. Itô, N. Morita, and T. Kubo, *Yuki Gosei Kagaku Kyokaishi*, **62**, 766 (2004). [*CA*, **142**, 76175 (2005)].
04OBC1413	K. Yamamura, Y. Houda, M. Hashimoto, T. Kimura, M. Kamezawa, and T. Otani, *Org. Biomol. Chem.*, **2**, 1413 (2004).
05JA13108	N. Sprutta, M. Swiderska, and L. Latos-Grażyński, *J. Am. Chem. Soc.*, **127**, 13108 (2005).
05JNP248	N. S. Reddy, J. K. Reed, R. E. Longley, and A. E. Wright, *J. Nat. Prod.*, **68**, 248 (2005).
05USP6852754	A. E. Wright, R. E. Longley, N. S. Reddy, and J. K. Reed, *US Pat.* 6,852,754 (2005). [*CA*, **142**, 173471 (2005)].
06IJB731	J. R. Hou, D. L. Wang, and K. Imafuku, *Indian J. Chem. Sect. B*, **45**, 731 (2006).
06MI1	O. V. Grechana, O. V. Mazulin, S. V. Sur, O. G. Vinogradova, and O. V. Prokopenko, *Farm. Zh. (Kiev)*, (2), 82 (2006). [*CA*, **146**, 180805 (2007)].
06OL3137	T. Kimura, T. Takahashi, M. Nishiura, and K. Yamamura, *Org. Lett.*, **8**, 3137 (2006).
06TL8535	I. Ueda, M. Nishiura, T. Takahashi, K. Eda, M. Hashimoto, and K. Yamamura, *Tetrahedron Lett.*, **47**, 8535 (2006).
07AXEm1351	G. M. Ferrence and T. D. Lash, *Acta Crystallogr. Sect. E*, **63**, m1351 (2007).
07EJO3981	J. A. El-Beck and T. D. Lash, *Eur. J. Org. Chem.*, 3981 (2007).
07EJO5461	T. D. Lash, *Eur. J. Org. Chem.*, 5461 (2007).
07H(74)951	M. Nishiura, I. Ueda, and K. Yamamura, *Heterocycles*, **74**, 951 (2007).
07JA13800	T. D. Lash, D. A. Colby, A. S. Idate, and R. N. Davis, *J. Am. Chem. Soc.*, **129**, 13800 (2007).
07JOC8402	T. D. Lash, J. A. El-Beck, and G. M. Ferrence, *J. Org. Chem.*, **72**, 8402 (2007).
08CHEC3(1–15)	A. R. Katritzky, C. A. Ramsded, E. F. V. Scriven, and R. J. K. Taylor (eds.), "*Comprehensive Heterocyclic Chemistry*", 3rd edn., Vol. 1–15, Elsevier, Amsterdam etc. (2008).
08JOC2256	Sh. Itô, T. Okujima, S. Kikuchi, T. Shoji, N. Morita, T. Asao, T. Ikoma, Sh. Tero-Kubota, J. Kawakami, and A. Tajiri, *J. Org. Chem.*, **73**, 2256 (2008).
08MI1	Zh. Zhang and T. D. Lash, *Abstracts of Papers, 235th ACS National Meeting*, ORGN-376 (2008).

CHAPTER **4**

Conjugates of Calixarenes and Heterocycles in the Design of Newer Chemical Entities

Subodh Kumar, Vijay Luxami and **Harjit Singh**

Contents

Department of Chemistry, Guru Nanak Dev University, Amritsar-143005, India

Advances in Heterocyclic Chemistry, Volume 97
ISSN 0065-2725, DOI 10.1016/S0065-2725(08)00204-3

1. INTRODUCTION

The designs of newer chemical entities (NCEs) capable of inducing selective binding and transport of ionic or neutral species or mimicking biological processes are of growing interest due to their exploitability in emerging technological processes as well as due to the role played by them in the understanding of a variety of biological phenomena (95MI1). Among the major categories of synthetic receptors, namely crown ethers, cyclophanes, cyclodextrins, etc., calix[n]arenes having a cyclic array of phenolic rings bound at meta positions with methylene linkers occupy a pivotal position because they allow the control of the shape of the NCE molecule. Structurally, calixarenes offer a preorganized scaffold with certain degree of flexibility. As a result, in a receptor designed on the calixarene platform, the desired topology of the molecule for its optimal-targeted functioning can be attained. In their structures, calixarenes uniquely elaborate a π–electron rich lypophilic macrocyclic cavity, display low energy conformational changes to form cone, partial cone, 1,2- and 1,3-alternate, etc. conformers (Figure 1; 98MI1).

On the rims of calix[n]arenes, variedly sized and functionalized substituents and pendants which greatly influence their structures can be implanted. When equipped with entities of diverse binding nature, these

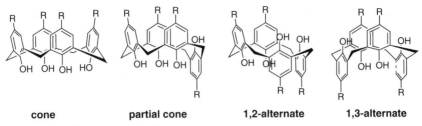

cone   partial cone   1,2-alternate   1,3-alternate

Figure 1   Conformations of calixarenes.

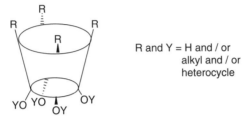

R and Y = H and / or
alkyl and / or
heterocycle

Figure 2   Conjugates of calixarenes and heterocycles.

modified calixarenes constitute promising hosts not only because of the binding ability of appendages but more so due to the role of the controlled mobility of calixarene platform generating directional pre-organization of functional binding groups and their capability to adjust the binding sites by low energy conformational changes such as those in enzymes. So, calix[*n*]arenes are ideally suited for catalytic role of the enzymes requiring geometrical changes during the dynamic catalytic processes as well as directional preorganization of active components. Thus, by equipping the upper and/or the lower rim of calixarene with chemical entities having appropriate structural and binding features, we can design calixarene scaffold-based NCEs as multifunctional enzyme models or as newer functional materials.

Heterocycles, as molecules of life, are ubiquitous in biological systems and are involved in many biological functions because of their ability to undergo varied noncovalent interactions with ions as well as neutral species. Heterocalixarenes (2005AHC67) constituted by replacement of phenolic unit(s) of calixarenes with various heterocyclic species and metallacalixarenes (2008AHC123) constituted by replacing methylene linkers in heterocalixarenes by metal species amply elaborate such versatile binding and structural features. It may be visualized that the conjugates of heterocycles and calixarenes could be formed by grafting varied types and number of heterocycles directly or through some linkers in place of R and/or H of OH, respectively, on the wider upper and narrow lower rim of vase-shaped calixarene molecules in a regioselective manner (Figure 2). These heterocyclic pendant moieties could be both spatially organized to constitute tailor-made ligands and also adjusted by rotation for effective noncovalent interactions of binding sites. The availability of a large variety of heterocycles of both purely synthetic and/or natural origin, point to unlimited possibilities of the designs of their conjugates with calixarenes which have the potential of creating NCEs of immense contemporary relevance. In addition to these features, the heterocycles in these conjugates in the form of their metal complexes could induct their inherent photochemical, electrochemical and catalytic properties in these NCEs.

In an overall attempt at a holistic presentation of the gamut of roles of heterocycles in designing calixarene-based NCEs, in continuation of our articles on heterocalixarenes (2005AHC67) and metallacalixarenes (2008AHC123), this third and last article addresses the various aspects of the design, synthesis and properties of conjugates of calixarenes and heterocycles. The calixarene component of these conjugates has been mainly confined to calix[4]arenes and only a few exploratory reports using calix[5]/[6]arenes are available. However, a large variety of heterocyclic components ranging from monocyclic-pyridine, imidazole, etc. to biheterocycles-bipyridyl, bithiazolyl, etc. to many complex systems such as sugars, porphyrins and tetrathiafulvenes, etc. have been used. Thus, in the following account of the conjugates of calixarenes and heterocycles, their classification is based on the nature of their heterocyclic components. Furthermore, in each of these categories, the calixarene–heterocycle conjugates could be classified into two subcategories based on the linkage of the heterocyclic moiety at the upper or lower rim of the calix[n]arene molecule and so have been discussed as two separate subsections based on these linkages.

2. CONJUGATES OF CALIXARENES WITH PYRIDINES

2.1 General

Pyridine has been conjugated at its nuclear nitrogen or a carbon center with a calixarene scaffold at both the lower and/or the upper rim mainly through various linkers. These conjugates deliver a variety of ionophoric materials with added chromogenic or fluorogenic properties. The complexes of some of these conjugate designs with appropriate metal ions constitute efficient chemical models of metalloenzymes. Some of the conjugates reveal unique noncovalent interactions with neutral organic molecules by invoking mainly hydrogen bonding.

2.2 Calixarene–pyridine upper rim systems

Conjugates of pyridine at the upper rim of calixarenes are relatively scarce. Except for one case of direct attachment of pyridine N, mostly calixarenes are attached at the 2-picolyl exocyclic carbon through $-N=$, $-CH=CH-$, $-CH_2-N-$ linkers or at the 2-picolinamide exocyclic N directly or through a CH_2 linker.

Chromogenic calix[4]arene–pyridinium cation conjugates **1–4** have been formed from an appropriate calix[4]arene having one or two NH_2 group(s) at the upper rim, by condensations with 2,4,6-triphenylpyrilium cation or a pyridinium-2/3/4-carbaldehyde. Compounds **1** and **2** acted as chromoionophores exhibiting optical responses in organic solvents upon complexation of Li^I, Na^I and K^I salts (96T639) and compounds **3** and **4** containing dissociable OH and pyridinium imino groups revealed Li^I, Na^I, Ca^{II} recognition in the presence of weak base (97T16867). The upper rim pyridyl calix[4]arene conjugates **5**, in which 4-pyridyl moieties are directly attached to calixarene, elaborated a rigid cone conformation and formed unique molecular capsules through metal-mediated self-assembly of two conjugate species and four *cis*-Pd^{II} complexes (2001JOC1002). The upper rim conjugates on 5,11,17,23-tetrakis (2-/3-/4-pyridylmethylamino)-25,26,27,28-tetrapropyloxycalix[4]arenes **6–8** were obtained by condensation of precursor tetraaminocalix[4]arene with a corresponding pyridine aldehyde and subsequent borohydride reduction. Conjugate **8** on interaction with sulphate anion in aqueous HCl formed a 1:1 complex with association constant $1770 \, M^{-1} \, dm^3$ (2007JMS48).

The picolinamide-based upper rim conjugates of calix[4]arene, **9** and **10**, were formed from its corresponding amino derivatives and picolinic acid pentafluorophenyl ester. Of these two ligands, **9** lacking the spacer showed high Am/Eu-binding selectivity of 13.8 (2005EJO2338).

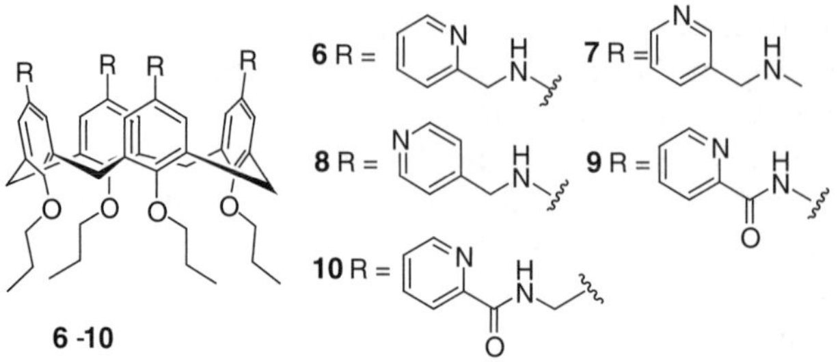

The mono-armed upper rim conjugate of pyridine and calix[4]arene **11**, in which one of the *p*-positions of calix[4]arene is linked with pyridine through an ethylene unit, was conveniently obtained through a Wittig reaction of monoformyl-tris-(bu^t)calix[4]arene and the phosphonium salt of 2-chloromethylpyridine (96TL6311). The tetra-armed conjugate **13** was obtained from tetra-formyl-*O*-propylcalix[4]arene and lithiated 2-picoline. As determined by vapor pressure osmometry molecular weight and fluorescence studies, the 4-pyridine groups of **13** and 4-carboxylic acid groups of another calix[4]arene derivative **12** formed a unique self-assembled dimeric molecular capsule **14** with an inner cage (Scheme 1; 94TL8255).

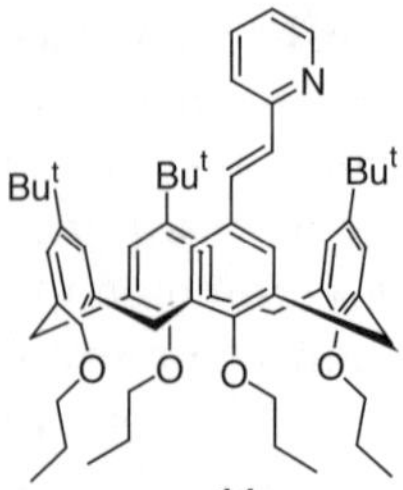

A series of upper rim pyridyl azo calix[*n*]arenes (*n* = 4, 6, 8) have been synthesized by coupling calix[*n*]arenes with diazonium salts derived from 3-, 4-aminopyridine. In these reactions, the dizonium salt of

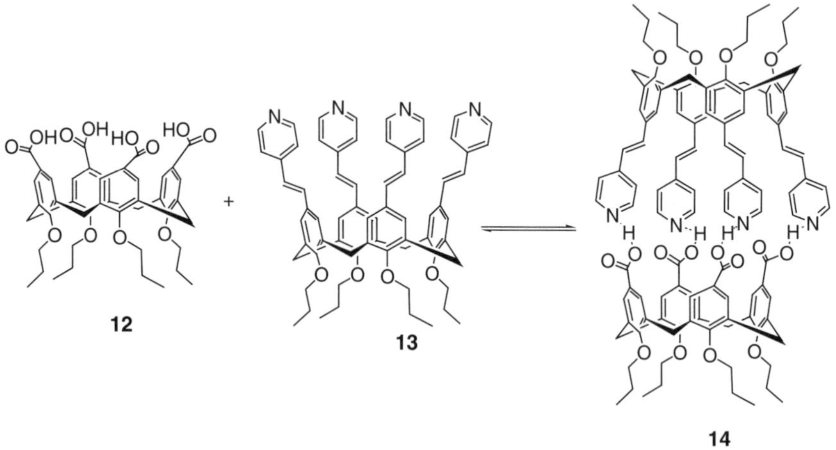

Scheme 1

3-aminopyridine gave tetra, hexa and octa substituted products. These conjugates showed good potential as selective ion filters for cesium ion. The pyridylazo derivatization has also been used as a photoactive monitoring handle for other calixarene-based receptors (2006T2901, 2006T9758).

On the upper rim of calix[4]arenes embroidered with –OCH$_2$CH$_2$ COOEt and/or a crown ether at the lower rim, up to three 2,6-bis(aminomethyl)pyridyl groups have been stitched at one aminomethyl group through methylene linker(s) to generate conjugates **15–18**. The zinc complexes of these conjugate systems have been investigated as models of metalloenzymes for catalyzing the cleavage of phosphate diesters. The trinuclear and dinuclear models **17**-[Zn]$_3$ and **16**-[Zn]$_2$, under neutral conditions, caused a rate acceleration of 32,000 and 23,000 in the transesterification of the RNA model substrate 2-hydroxypropyl-p-nitrophenyl phosphate **19**. The enhanced rate in the case of the trinuclear complex has been ascribed to the additional activation of β-hydroxyl group of **19** by the third Zn^{II}. The **18**-[Zn]$_2$, a rigid analog of **16**-[Zn]$_2$, exhibited both a lower substrate binding strength and a lower catalytic rate demonstrating the significance of a certain order of flexibility between the cooperating catalytic centers. The comparisons with the activities of mononuclear complex **15**-[Zn] and a reference complex **20** lacking the calixarene backbone showed that the catalysis was due to the cooperative action of the Zn^{II} ions and that the hydrophobic effects of calixarene contribute to the catalysis. The **16**-[Cu]$_2$ and **17**-[Cu]$_3$ complexes showed only a very low catalytic activity probably because of the different Cu^{II} coordination geometry disfavoring cooperative action in catalysis (99JOC3896, 97JA2948).

The **17**-[Zn]$_3$ efficiently catalyzed the cleavage of RNA dinucleotides **21**, by the cooperative action of the Zn^{II} centers, with high rate enhancement and significant nucleobase specificity. The catalytic rates for dinucleotides having different nucleobases was observed in the order of GpG$\gg$UpU$\gg$ApA. The heterotrinuclear complex **17**-[Zn$_2$Cu] has been found to be even more active (99AGE3189).

15 - 17   18 R = Pr^n

A head to head double calix[4]arene system linked at the 1,3-distal upper rim positions of tetrapropyloxy calix[4]arene with 2,6-diamino-pyridine groups was obtained by condensation of 1,3-diaminocalix[4]-arene with pyridine dialdehyde and showed binding ability toward viologen type guest molecules (2002T9019).

2.3 Calixarene–pyridine lower rim systems

The lower rim O-alkylation of a calixarene core with pyridine-based appendages would provide universal ligands for both hard and soft metal ions. Since minor changes in the regioselectivity of derivatization and conformation of modified calixarenes would influence their complexation, a variety of such pyridine conjugates varying in nature of linkers, site of attachment in pyridine, have been investigated.

A unique calix[4]arene–pyridinium cation conjugate 22 has been formed from 25-(4-bromobutoxy)-26,27,28-trihydroxycalix[4]arene and pyridine. As it combines both a host unit and a guest unit in one molecule, 22 has the potential of self-assembly to become an oligomer invoking cation–π interactions between a pyridinium unit of one molecule and a calixarene moiety of a neighboring molecule. The X-ray diffraction crystallographic study of 22 revealed that it is oriented in a one-dimensional structure in the solid state. ^{1}H NMR analysis indicated the formation of oligomer in solution and the restraint of segmental motion leads to stabilization of the cation–π interaction as compared with a bimolecular complex of the tetrahydroxycalix[4]arene and N-butylpyridinium bromide (2006TL181).

22   Oligomer of 22

The lower rim tetra O-2-picolyl derivatives of calix[4]arene **23** could attain cone, partial cone, 1,2-alternate and 1,3-alternate shapes of the calix[4] arene skeleton. Exhaustive alkylation of calix[4]arene with 2-(chloromethyl)pyridine hydrochloride in dimethyl formamide (DMF) was strongly affected by the nature and amount of the base. The use of an excess of NaH and Cs_2CO_3/K_2CO_3, respectively, formed cone and mixtures of partial cone and 1,3-alternate conformers (92JOC2611, 89JOC5407). The small amount of 1,2-alternate conformer was also formed on using Cs_2CO_3 (95JOC4576).

23 R =

24 R =

25 R =

All these conformers showed low ionophoric efficiency toward alkali metal cations. The highest phase transfer values were observed for cone conformer of tetra derivative **23** in the order of Na^I, K^I, Rb^I, Cs^I, Li^I, whereas the corresponding N-oxide had no activity. The tetra O-3-picolyl derivative of calix[4]arene **24**, because of the unfavorable geometry, was unable to form a complex even with Na^I, indirectly indicating its complexation in a 2-picolyl system through the cooperative effect of O and N donor atoms (95JOC4576).

Consequently upon the experimental findings that the complexation process involving **23** and alkali metal cations was fast in acetonitrile solution, thermodynamics of **23** and its geometrical isomers 3-picolyl **24** and 4-picolyl **25** derivatives in various solvents and their complexation properties with Ag^I in acetonitrile have been studied (98JCS(FT)3097). On using alcohols as solvent, as the aliphatic chain length increased, the solvent ligand interaction increased in the case of 2-picolyl derivatives, whereas for the 3- and 4-picolyl derivatives the opposite was observed. The results of protonation constants showed that the basic character of these ligands follow the sequence: 4-picolyl > 3-picolyl > 2-picolyl derivatives of calix[4]arene. Thermodynamic parameters for the complexation of these macrocycles and silver cation in acetonitrile reveal that as the distance between the ethereal oxygen and the pyridyl nitrogen increases, the strength of complexation decreases and the 2-picolyl derivative provides ethereal oxygens and pyridyl nitrogens as the active sites for complexation with silver. The 1:1 monoacetonitrile and Ag^I complex of **23** in its X-ray structure showed two conical parts, the hydrophobic calixarene part being filled with acetonitrile molecule and the hydrophilic pendant part encapsulating Ag^I through O and N. Thus, the ability of this calix[4]arene-derived macrocycle to complex Ag^I in its hydrophilic cavity is demonstrated (98JCS(FT)3097).

These results show that the free energies of complexation are free from any influence of side processes usually present in extraction experiments. Hence, molecular dynamic simulations were performed with tetra O-2-picolyl derivatives of calix[4]arene and its complexes with Na^I in the gas phase and in acetonitrile. These studies suggest that solvent molecule hosted by each monomer in the crystal does not play any role in complexation of the cation but is included only after the formation of the LNa^I complex (2000CP359).

The thermodynamic affects of complexation of 25, its monomeric component and 23 with metal ions monitored through ^{1}H NMR studies in nonaqueous solvents also show that in these conjugates, the positions of the pyridyl nitrogen and ethereal oxygen play a primary role in their hosting abilities for metal cations and as the distance of N and O increases, the ability of the ligand to coordinate decreases. Thus, 23 was able to interact with alkali metal cations, but this ability is lost for 25. The conductance measurements show that for all cations except Hg^{II}, the composition of complexes of 25 is 1:1. However, 25 hosts two Hg^{II} cations (2005JPC(B)14735).

In the bis-O-2-picolyl derivatives of calix[4]arene 26a–c, the two binding pendant functionalities can adopt syn-proximal (26a), syn-distal (26b) or anti-distal (26c) relationships. On using lower proportions of NaH mainly syn-proximal 26a and some syn-distal 26b products along with mono and tris derivatives were formed. 26a is a useful intermediate for further selective synthesis of tri and tetra alkylated cone conformers having the same or mixed pendants (92JOC2611).

| syn-proximal | syn-distal | anti-distal | |
| 26a | 26b | 26c | 27 |

The calixarene 26b experienced attractive π–π interactions between its two of four outside cavity faces and both faces of 1,4-diiodotetrafluoro-benzene (27) forming infinite one-dimensional noncovalent ribbons, where the two modules alternate. A two-dimensional supramolecular network was formed by cross-linking induced by donor–acceptor interactions between pyridine N and iodine atoms of 27 in two parallel chains (2000CEJ3495).

The syn-distal 4-picolyloxy calix[4]arene 28 synthesized by alkylating calix[4]arene with two equivalents of 4-chloromethylpyridine · HCl in acetonitrile in the presence of K_2CO_3 and NaI, formed complexes with resorcinol, phthalic acid and catechol by invoking hydrogen bonding and van der Waals interactions (2001JIPMC47).

The three derivatives of *syn*-distal di-*O*-2-picolyl calix[4]arene **29–31** have been successfully used as neutral carriers in Ag^I selective electrodes which exhibited good linear responses and good sensitivity and selectivity (2000MJ129). The ligand **31** on reaction with Cu^{II} perchlorate formed a bimetallic complex of stoichiometry [(**31**)Cu$_2$](ClO$_4$)$_2$], which is diamagnetic. However, **32** formed from **31** through a sequence of steps, gave only a mixture of complexes (92T9917) with Cu^{II} perchlorate.

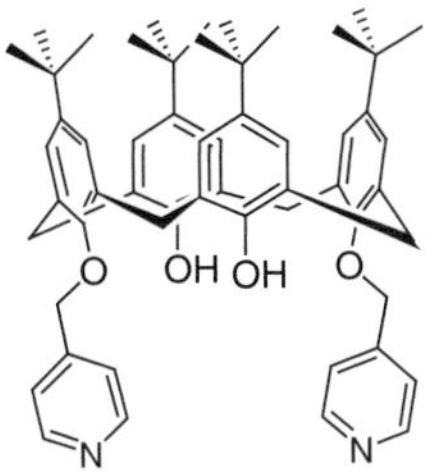

29 R = Bu^t, R$_1$ = CH$_2$PPh$_2$

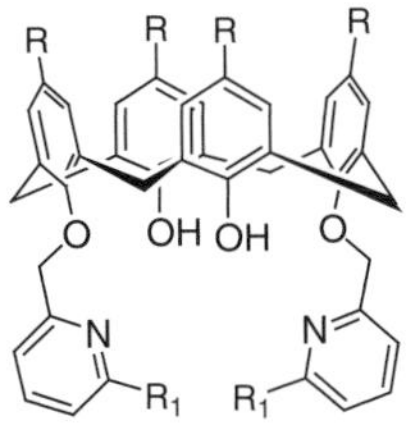

30 R = Bu^t, R$_1$ = CH$_2$OH

31 R = H, R$_1$ = CH$_2$OH

32 R = H, R$_1$, R$_1$ =

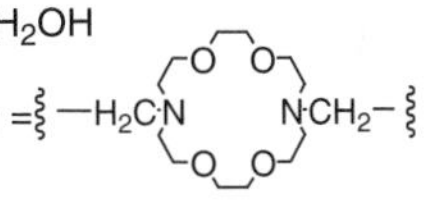

28   **29 - 32**

A unique assembly of a crystalline hybrid supramolecular salt was achieved by reaction of 2,4-bis{(3-pyridylmethyl)oxy}-(1,3)-*p*-bu^t-calix[4]-crown-6 (**33**), hexadecafluoro-1,8-diidooctane and CsI (2007T4951). In its solid state, the cesium ion of the supercation was encapsulated inside the cavity created by the crown ether loop, the picolyl and inverted phenyl moieties, whereas the two iodide ions of the superanion, formed a discrete five component aggregate held together by hydrogen and halogen bonds. Similar adduct existed in solution. A combination of this calix-crown **33** and octafluoro-1,4-diidobutane acted as an effective binary host for the selective extraction of CsI from aqueous to fluorous phase.

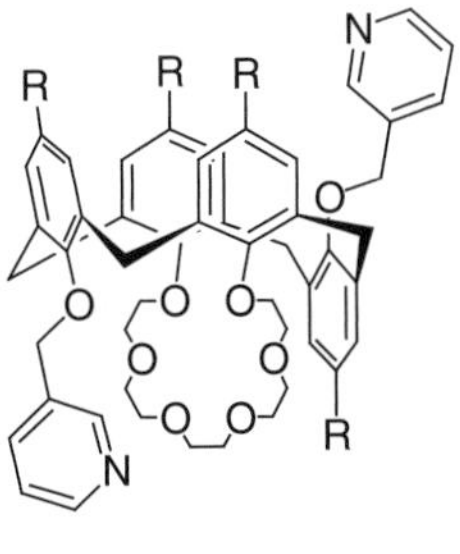

33 R = Bu^t 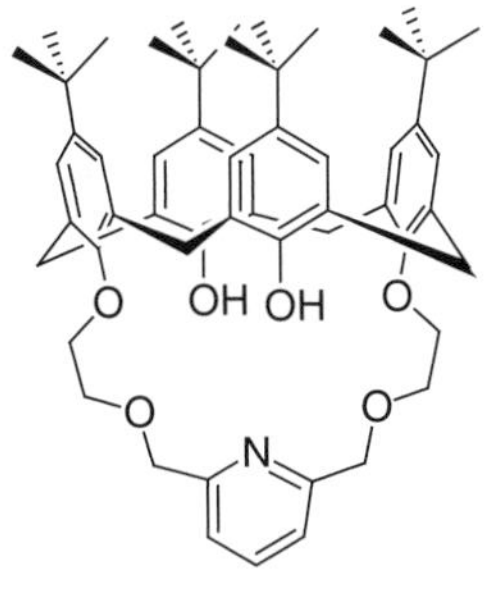   **34**

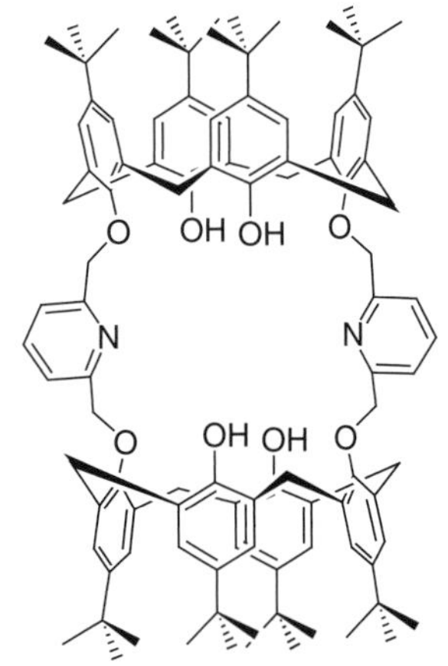

35

The calix[4]arene **34** and bis calix[4]arene **35** derivatives were generated by bridging their lower rim oxygen with one 2,6-lutidyl unit or two of these at 1,3-distal positions of calixarene scaffolds. The calixarenes **34** and **35** showed complexation toward alkali metal cations (2000T3121).

25,27-Bis(3-pyridylcarboxylate)-26,28-dihydroxy-calix[4]arene   **36** (2003A92) and its tetra-bu^t derivative **37** (2000MJ75) having two alternately placed nicotinoyl-binding species at the lower rim have been conveniently formed from the corresponding calix[4]arene and nicotinyl chloride hydrochloride in the presence of triethylamine. On crystallization, **36** revealed two types of crystals elaborating flattened partial cone (**36a**) and cone conformations (**36**; 2003A92). The liquid membrane transport and the silver selective electrode based on **37** showed excellent cation selectivities for Ag^I and Hg^{II} over other alkali, alkaline earth, transition metal and ammonium ions which might be attributed to the overriding binding influence of pyridine nitrogens (2000MJ75).

36 R = H
37 R = Bu^t

36a R = H

Picolinamide and thiopicolinamide have been introduced on the lower rims of calixarene scaffolds of various sizes in order to study whether the size of the macrocycle, the cooperation and stereochemical disposition of these binding groups affect the efficiency of extraction and selectivity in the actinide/lanthanide separation. Thus, picolina-mide conjugates of calix[4]arene **38–40** and those of calix[6]arene **42–44** and of calix[8]arene **45** and **46** have been formed from appropriate amino derivatives of the corresponding calixarenes and picolinic acid pentafluorophenyl ester. The thiopicolinamide analog **41** has been synthesized from **40** and Lawesson's reagent (2005EJO2338).

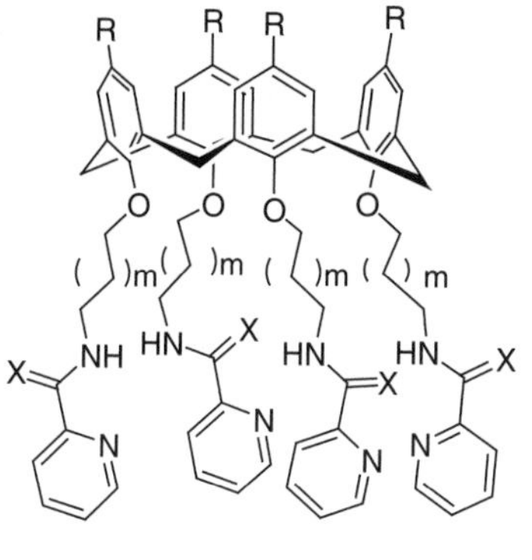

38 R = Bu^t, m = 2, X = O

39 R = H, m = 1, X = O

40 R = Bu^t, m = 1, X = O

41 R = Bu^t, m = 1, X = S

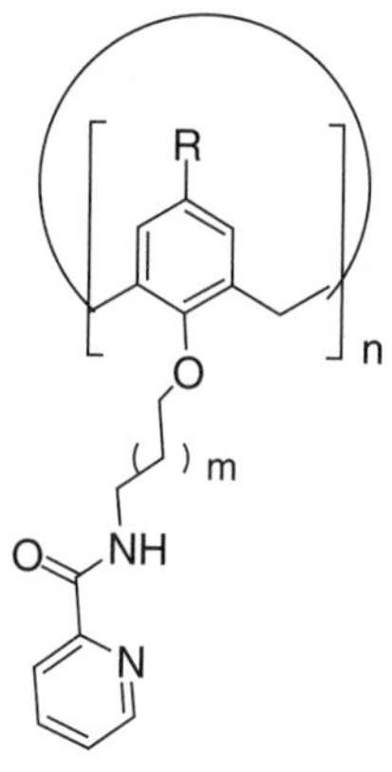

42 R = Bu^t, m = 1, n = 6

43 R = H, m = 1, n = 6

44 R = H, m = 2, n = 6

45 R = OBn, m = 1, n = 8

46 R = H, m = 1, n = 8

All calix[4]arene tetraamides **38–40** elaborated a cone conformation and calix[8]arene conjugates **45** and **46** were conformationally quite mobile in CDCl$_3$ solution. The conformational properties of calix[6]arene derivatives **42–44** depended strongly on the solvent, temperature or the substituents on the upper rim. Thus, the calix[6]arene picolinamides **43** and **44** lacking p-bu^t substituents were very mobile in solution on an ^{1}H NMR time scale and p-bu^t-calix[6]arene derivative **42** constituted a mixture of conformers with predominant 1,2,3-alternate conformation in dimethyl sulfoxide (DMSO)-d_6 solution. The X-ray studies on a single crystal of **42** also revealed a 1,2,3-alternate structure in which two crystal lattice water molecules were crucial in determining its molecular packing. With the exception of thio analog **41**, all these ligands displayed high extraction efficiency and Am/Eu selectivity but calix[8]arene systems were superior to the ones derived from calix[4]arene and calix[6]arene (2005EJO2338).

The lower rim tetraamido **47a, 47b** and 1,3-distal diamido **48** calix[4]arene derivatives having terminal 2-pyridyl groups in the pendant chains have been synthesized by aminolysis of the corresponding calixarene and chloride/ester with an appropriate pyridyl amine. The binding properties of **47a, 47b** and **48** toward alkali, alkaline earth metal ions and lanthanides have been determined (2006SC219, 2001TL3595).

47a m = 1
47b m = 2

48

The tetra-(2-(2-pyridylthio-N-oxide)ethoxy)calix[4]arene **49** synthesized from tetra-(2-bromoethoxy)calix[4]arene and 2-pyridylthiol-N-oxide constituted one of the better modified calix[4]arene extractants for Pt^{IV} (97ICA309).

The conjugates of pyridine and calix[5]arene **50a** have been of added interest because of their larger cavity than calix[4]arene while retaining the capability of assuming a cone conformation. The conjugates **50b–h** have been formed by base catalyzed alkylation of p-bu^t-calix[5]arene **50a** with 2-(chloromethyl)pyridine hydrochloride in DMF. The reactions of **50a** with varying amounts of alkylating agents (2–4 equiv.) in the presence of a base such as CsF, KHCO$_3$, BaO/Ba(OH)$_2$, K$_2$CO$_3$, NaH, etc. provided mixtures of products which were separated into pure components. The regioselective 1,2,4- or 1,2,3-tri-O-alkylation has been achieved by an appropriate choice of molar ratios, solvent and base. The pentaethers **50h–j**, endowed with 2-pyridyl, 3-pyridyl and 2-quinolyl pendant groups at the lower rim have also been prepared. The cone conformation in solution for all new compounds has been established by ^{1}H NMR spectroscopy and confirmed for 1,2,3-tri-O-alkylated **50e** by a single crystal X-ray analysis showing intramolecular hydrogen bonds between the phenolic oxygen and the proximal pyridyl nitrogen (96JOC2407).

49

50 a -j

50a	R$_1$ = R$_2$ = R$_3$ = R$_4$ = R$_5$ = H
50b	R$_1$ = 2-pic; R$_2$ = R$_3$ = R$_4$ = R$_5$ = H
50c	R$_1$ = R$_2$ = 2-pic; R$_3$ = R$_4$ = R$_5$ = H
50d	R$_1$ = R$_3$ = 2-Pic; R$_2$ = R$_4$ = R$_5$ = H
50e	R$_1$ = R$_2$ = R$_3$ = 2-pic; R$_4$ = R$_5$ = H
50f	R$_1$ = R$_2$ = R$_4$ = 2-pic; R$_3$ = R$_5$ = H
50g	R$_1$ = R$_2$ = R$_3$ = R$_4$ = 2-pic; R$_5$ = H
50h	R$_1$ = R$_2$ = R$_3$ = R$_4$ = R$_5$ = 2-pic
50i	R$_1$ = R$_2$ = R$_3$ = R$_4$ = R$_5$ = 3-pic
50j	R$_1$ = R$_2$ = R$_3$ = R$_4$ = R$_5$ = 2-CH$_2$quin

For achieving deep cavities in higher level supramolecular structures, lower rim conjugates **51–53** of calix[6]arene and pyridine having larger sized linkers have been obtained by 3-fold reactions of 2-[4-(bromomethyl)phenyl]pyridine, 5-(bromomethyl)-2-phenylpyridine and 5-(bromomethyl)-2-[9,9-di-n-hexylfluorenyl]pyridine with 1,3,5-trimethyl ether of the p-bu^t-calix[6]arene in the presence of NaH. These conjugates elaborated mainly cone conformations (2006JOC9589).

R = n-hexyl

51   **52**   **53**

The chemical models of metalloenzymes, known for transforming organic molecules with high efficiency and selectivity, operationally are a key for better understanding of the chemical processes involved in a biological catalytic cycle and could lead to the design of new tools for organic synthesis. In search for supramolecular biomimetic systems for metalloenzymes, the capacity of lower rim 2-picolyl calix[6]arene conjugates in forming mononuclear complexes with transition metals has been investigated.

In the reactions of 2-picolyl chloride hydrochloride and p-bu^t-calix[6]arene by tuning the amount of electrophile, nature of various bases and solvents, 10 of the 12 possible pyridine calix[6]arene conjugate homologues have been obtained. The selective 1,2,4,5-tetra- or 1,4-di-O-picolylation has been achieved in good yields by using NaH and Ba(OH)$_2$, respectively, whereas the reactions with limited amounts of electrophile and K$_2$CO$_3$ have afforded invariably 1,2,3-tri-O-picolyl derivatives as major product (93JOC1048).

A calix[6]arene-based model, 1,3,5-tripicolyl-2,4,6-trimethoxy-p-bu^t-calix[6]arene **54a** elaborating an N3 donor set that mimics a polyimidazole coordination site of mononuclear copper enzymes has been designed. It has an in-built facility of inward transposition of the coordination sites in the hydrophobic π-electron cone cavity funnel of calix[6]arene skeleton that would both protect the metal center of its metal complex from an undesirable interaction with another metal center and facilitate the approach or exit of an organic molecule. The tridendate ligand **54a** was synthesized by reaction of picolyl chloride with p-bu^t-calix[6]arene derivative, having alternate three OH groups which are protected as methoxy groups. On reaction with Cu(NCMe)$_4$PF$_6$, **54a** formed a cationic complex [Cu(p-Bu$^t_6$Me$_3$Pic$_3$)(NCMe)] PF$_6$ (**55**) in which cuprous ion is coordinated to three pyridyl residues and an acetonitrile molecule is located in the cavity. The fourth ligand was labile and exchanged easily with propionitrile and allyl nitrile molecules but not with benzonitrile or benzylnitrile. The complex was not oxidized under an atmosphere of O$_2$ in most solvents. The X-ray structure of [Cu(p-Bu$^t_6$Me$_3$Pic$_3$)(NCEt)]PF$_6$ showed four coordinate Cu^I with three pyridyl N projected toward the inside of the cone pushing out OCH$_3$ groups and the propionitrile molecule was buried deep inside the cavity of calix[6]arene (98AGE2732).

54a R = Bu^t
54b R = H

55

56

On keeping the complex **55** in chloroform solution, the acetonitrile was uniquely replaced by chloride ion to provide [Cu(p-Bu$^t_6$Me$_3$Pic$_3$)Cl], which was alternately obtained from **54a** and CuCl. This stable C$_3$ symmetrical complex existed as a pair of helical enantiomers and was insensitive to oxidation. Its low temperature ^{1}H NMR revealed that the chirality was transmitted to the calixarene skeleton, thereby providing a chiral cavity around the apical-binding site of the metal center (98NJC1143).

The reactions of stoichiometric amounts of Cu(CF$_3$SO$_3$)$_2$ or Cu(ClO$_4$)$_2$(H$_2$O)$_6$ with **54a** formed Cu^{II} funnel complexes of general formula [Cu(p-Bu$^t_6$Me$_3$N$_3$)(H$_2$O)L]$^{2+}$ in which the metal ion was coordinated to all three pyridine nitrogens of the ligand and to an exogenous labile molecule L which was a solvent molecule such as H$_2$O or CH$_3$CN. These complexes were relatively less stable than Cu^I complexes and displayed enhanced reactivity for oxidation of exogenous substrates such as aromatics and alcohols.

1,3,5-Trimethoxy-2,4,6-tris-O-2-picolylcalix[6]arene **54b**, a partially de-*tert*-butylated homolog of **54a**, was obtained from 2-picolyl chloride and partially de-*tert*-butylated 1,3,5-trimethoxycalix[6]arene. To evaluate the role of the cavity in these picoline-based mononuclear metaloenzyme models for binding organic molecules, cuprous complex **56** was obtained by the reaction of Cu(NCMe)PF$_6$ with **54b**. The complex **56** offered a larger pocket with a wider opening than homologous **55** due to the removal of the three bu^t groups from calix[6]arene. As a result, the recognition pattern for MeCN vs PhCN was inverted. The ^{1}H NMR-based study of acetonitrile exchange in **55** and **56** showed a dissociative pathway in both the cases. Evidently the difference between **55** and **56** stemmed from the presence of a door that entrapped the guest in **55** and the removal of bu^t groups in **56** led to a 100-fold acceleration of acetonitrile exchange rate. Hence, these supramolecular systems provide interesting models for the hydrophobic substrate channel giving access to a metalloenzyme active site (2002JA1334).

3. CONJUGATES OF CALIXARENES WITH PIPERIDINE, PYRROLIDINE AND MORPHOLINE

Lower rim calix[4]arene conjugates **57a–c** having perhydroheterocyclic pendant groups, piperidine, pyrrolidine and morpholine, have been obtained by reactions of p-bu^t-calix[4]arene with 1-(2-chloroethyl)piperidine, 1-(2-chloroethyl)pyrrolidine and 4-(2-chloroethyl)morpholine, respectively.

These conjugates have higher affinity for H$^+$ and in protonated form have a high potential to interact with anions which has been corroborated by the crystal structure of {tetrakis[1-(ethyl)piperidinium]oxyl}-p-bu^t-calix[4]arene-bis-[tetrachloroaurate(III)] dichloro dehydrate **58**. It showed that the protonated nitrogen atoms of the four piperidino pendant groups interact through hydrogen bond formation with two chloride ions and a water molecule, but the two AuCl$_4^-$ anions were located outside the hydrophilic cavity (99ICA142, 96JCS(FT)1731). Protonated alkyl ammonium forms of calix[4]arene diamide derivatives **59** effectively extracted

236   S. Kumar et al.

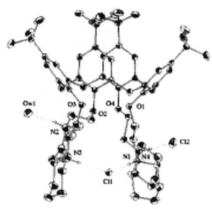

Figure 3 Ortep drawing of **58**. AuCl$_4^-$ and H atoms are omitted. [99ICA142] Reproduced by permission from Elsevier.

HCr$_2$O$_7^-$ and Cr$_2$O$_7^-$ from water (Figure 3; 2007SC159).

R = H, bu^t

59

4. CONJUGATES OF CALIXARENES WITH QUINOLINES, QUINAZOLINES AND PHENANTHROLINES

A lower rim calix[4]arene, and 8-hydroxyquinoline conjugate **60** formed from 5,11,17,23-tetra-bu^t-25,27-bis(3-bromopropoxyl)-26,28-(dihydroxy)-calix[4]arene and 8-hydroxyquinoline, on coordination with Eu^{III} and Tb^{III} ions exhibited luminescence and energy transfer properties. The spectral studies revealed that the attachment of 8-hydroxyquinoline groups to the calix[4]arene moiety provided new binding sites for coordination of metal ions facilitating their interaction with the calixarene lower rim oxygen atoms and allowing in this way, for efficient energy transfer from organic host to the metal center (2000NJC841). In the presence of Zn^{II} ions, coordination of the quinoline groups of **60** was observed to lead to characteristic changes in absorption, fluorescence and NMR spectra and it precluded the energy transfer from calix[4]arene moiety. In the resulting complex, a significant decrease in the emission quantum yields took place. However, by incorporating an additional batho-phenanthroline ligand, a great emission enhancement was detected revealing the occurrence of synergistic effects in the bis-(8-oxyquinoline)-calix[4]arene(batho-phenanthroline)zinc^{II} complex **61** (2003ICC288).

The 5-chloroquinoline luminophores have been appended at positions 2 and 8 through $-CH_2-$ and $-OCH_2CH_2-$linkers respectively onto the one lower rim oxygen of calix[4]arene also bearing three acetamide-binding groups on the remaining three lower rim oxygens in conjugates **62** and **63**. The binding properties of both these ligands investigated in chloroform and methanol solutions indicated that they are efficient fluoroionophores with selectivity for sodium and strontium between alkali and alkaline earth metal ions. Some differences in photophysical properties of **62** and **63** have been ascribed to participation of both N and O in **62** toward coordination, whereas in **63**, the N-atom was too far from a metal ion. Their lanthanide metal ion complexes formed in acetonitrile, showed good luminescence in the case of Nd^{III}, Yb^{III} and Er^{III} ions indicating efficient energy transfer (2003EJO1475).

Calix[5]arene conjugate **50j** having pentaether endowed with 2-quino-lylmethyl pendant groups at the lower rim of calix[5]arene and having a cone conformation in solution as established by ^{1}H NMR spectroscopy has also been reported (96JOC2407).

The reaction of p-bu^t-calix[4]arene with 4-chloro-2-phenylquinazoline gave a mixture of the *syn*- and *anti*-1,3-diether derivative **64**. In solution, the *syn* form adopted predominantly the cone conformation, whereas in its crystals an energetically unfavorable 1,3-alternate conformation was present (2003NJC236).

The calix[6]arene **65** bridged by a 1,10-phenanthroline was synthe-sized from p-bu^t-calix[6]arene and 2,9-bis(bromomethyl)-1,10-phenan-throline (97TL4539). Its complex with Cu^I has been beneficially used for Cu^I catalyzed cyclopropanation of styrene and indene. Several chiral derivatives of **65** have been synthesized but the stereoselectivity of cyclopropanation, catalyzed with their Cu^I complexes, could not be enhanced (2006EJO4717). A similar bridging of calix[5]arene and calix[8]arene has also been investigated (2005EJO2330).

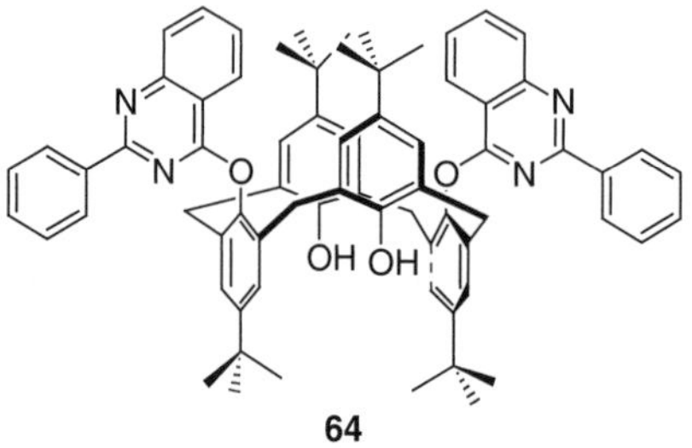

64

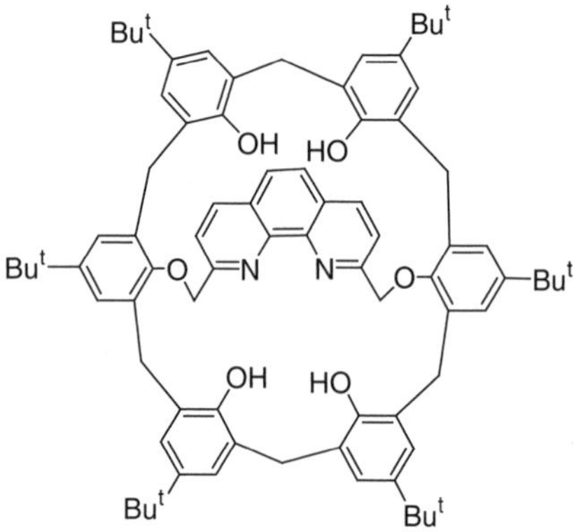

65

5. CONJUGATES OF CALIXARENES WITH IMIDAZOLES, BENZIMIDAZOLES AND BENZOTHIAZOLES

5.1 General

The imidazolyl group, a constituent of histidine, provides some crucial catalytic sites for enzymatic catalysis. The investigations on enzyme mimics constitute a major contemporary research area in supramolecular chemistry. The synthetic designs of enzyme mimics should necessarily possess a hydrophobic cavity and suitably preorganized functional catalytic sites to perform organic transformations. Calixarenes with their inbuilt hydrophobic cavity and due to directional preorganization of functional guest binding groups capable of adjusting rapidly by low energy conformational changes constitute promising host molecules for designing enzyme mimics. Consequently, conjugates of calixarenes and imidazoles have come into focus for use as enzyme mimics.

5.2 Calixarene–imidazole/benzimidazole/benzothiazole upper rim systems

On the basis of the catalytic roles of calix[4]arene derived model 18-[Zn^{II}]$_2$ (99JOC3896) and cis-diaqua Cu^{II} complexes for cleavage of phosphate diesters, Reinhoudt designed a calix[4]arene derivative 66-[Cu^{II}]$_2$ functionalized with two cis-diaqua Cu^{II} centers at the distal positions of the upper rim as a model for dinuclear metalloenzymes that catalyze chemical transformations of phosphate esters. It was synthesized from Cu(ClO$_4$)$_2$ and 5,17-bis(bis(1-methylimidazol-2-yl)hydroxymethyl)-25,26,27,28-tetrakis(2-ethoxyethoxy)calix[4]arene which was conveniently obtained from the precursor diester and lithiated 1-methylimidazole. In this model, the two Cu^{II} centers are well organized on the calixarene scaffold for performing synergistic action.

R = CH$_2$CH$_2$OEt

66-[Cu^{II}]$_2$ R$_1$ = H
67-[Cu^{II}]$_2$ R$_1$ = CH$_2$OH
68-[Cu^{II}]$_2$ R$_1$ = CH$_2$NH$_2$

69

As compared with mononuclear 69-Cu^{II}, it has been found that 66-[Cu^{II}]$_2$ efficiently catalyses with significant rate enhancement, the transesterification of RNA model substrate 19 and hydrolysis of DNA model ethyl-p-nitrophenyl phosphate. The high catalytic efficiency of this enzyme model has been attributed to a dynamic binding of the substrate and (pre)-transition state, possible by rapid low energy conformational changes of the flexible calixarene backbone (98JA6726). Reinhoudt also synthesized calix[4]arenes functionalized with two catalytic bis-imidazolyl-Cu^{II} centers and two additional hydroxymethyl 67-Cu$_2^{II}$ or aminomethyl 68-Cu$_2^{II}$ groups as models for enzymes that cleave phosphate diester bonds. These complexes also showed high rate enhancement in the catalysis of the intramolecular transesterification of

RNA model substrate **19**. The kinetics indicated bifunctional catalytic effects for **68**-Cu_2^{II} and that at the pH optimum of 7.4 at least one amine is protonated, which could assist as a general acid in the binding and activation of the substrate (99JOC6337).

Using well-defined multistep protocols bis-imidazole calix[4]arene conjugates, disubstituted distal **70** (cone) and **71** (partial cone) and proximal **72** (cone) and **73** (alternate) conformers were synthesized as metal-free enzyme mimics which exhibited transacyltransferase activity. Surprisingly, proximal and distal-substituted calix[4]arenes improved the hydrolysis of model substrate *p*-nitrophenyl esters in a similar manner and the acylated species exhibited higher catalytic activity than calixarenes bearing only hydroxyl groups at the lower rim (2001TL7837).

70 R = H
71 R = Bz

72 R = H
73 R = Bz

The conjugates having imidazolium pendants on the upper rim of calix[4]arene have been evolved as macrocyclic precursors for chelating N-heterocyclic ligands. The carbenes derived from 5,17-(bis-imidazolium)-substituted calix[4]arene **74** on reaction with Pd(OAC)$_2$ formed *cis*-palladium chelate complexes **75**. Using an *in situ* catalytic system consisting of the calixarene–imidazolium salt, Cs$_2$CO$_3$ and palladium salt, the Suzuki cross-coupling of 4-chlorotoluene and phenylboronic acid to yield 4-methylbiphenyl was performed (2004EJO607).

74 (a) R = H, (b) R = Bu^t
R$_1$ = Me, iso-Pr, mesityl

75
R$_1$ = Me, iso-Pr, mesityl

76

The phenanthroimidazole subunit based upon calix[4]arene diamide **76** exhibited Mg^{II} selective fluoroionophoric properties over other physiological relevant metal ions. A significant red shift in fluorescence emission allowed the ratiometric determination as well as naked eye detection of Mg^{II} ions (2007TL5397).

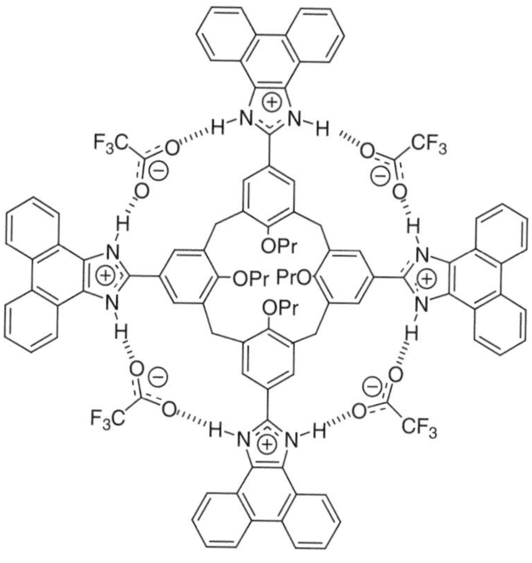

77

The reaction of tetra-*p*-formyltetra-*O*-propylcalix[4]arene with phenanthroquinone in the presence of NH$_4$OAc gave calix[4]arene derivative **77** with a deep cone-shaped macrocycle stabilized by linking the four coplanarly oriented 1H-phenanthro[9,10-*d*]imidazol-2-yl groups through H-bonded ion pairs with trifluoroacetic acid or DMSO–H$_2$O (1:1) (2004OL1091). The X-ray crystal structure of a sample obtained from CH$_2$Cl$_2$–MeOH–TFA mixture showed that two partially protonated calixarenes interdigitated in the solid state to give rise to a self-assembled face-to-face dimer, stabilized by π–π stacking interactions. In contrast to **77**, the calix[4]arenes **78–80** possessing two carbazol-9-yl or two 1,8-naphthalene imide moieties at 1,3-distal positions of the upper rim or four 1,8-naphthalene imide moieties (99JCS(P2)1749, 2002OL2901) at the upper rim of calix[4]arene gave deep cavity structures which collapsed into a pinched cone conformation with no permanent cavities for guest encapsulation. The X-ray structure of **80** revealed that the outer aromatic imides are almost perpendicular to the respective phenol rings, whereas the inner ones are twisted 60° to reach an optimal stacking and to minimize the repulsions between adjacent carbonyl groups.

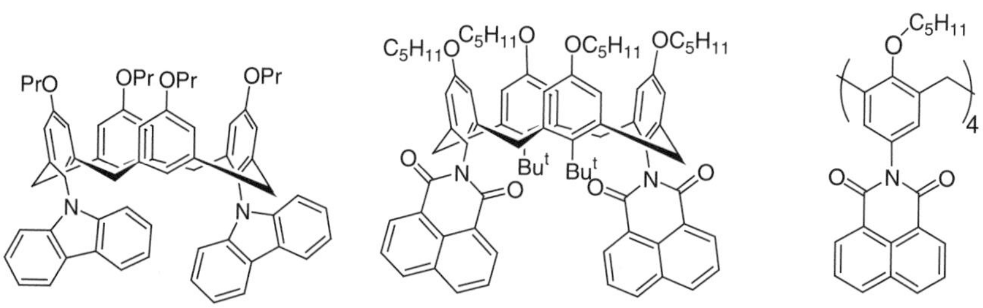

78

79

80

Calix[4]arene-crown-5-ether **81** possessing benzothiazolyl functionalities at the upper rim of 1,3-distal phenolic groups of calixarene exhibited a pronounced selective fluoroionophoric behavior toward Ca^{II} ions among physiologically important Na^{I}, K^{I} and Mg^{II} (2002TL3883). The calix[4]arene diamide **82** possessing two benzothiazolium units at the upper rim of 1,3-distal phenolic groups showed significant color change from light orange to purple selectively with Ca^{II} only. The other metal ions Na^{I}, K^{I} and Mg^{II} did not affect any color change or change in the spectrum (2003TL5299).

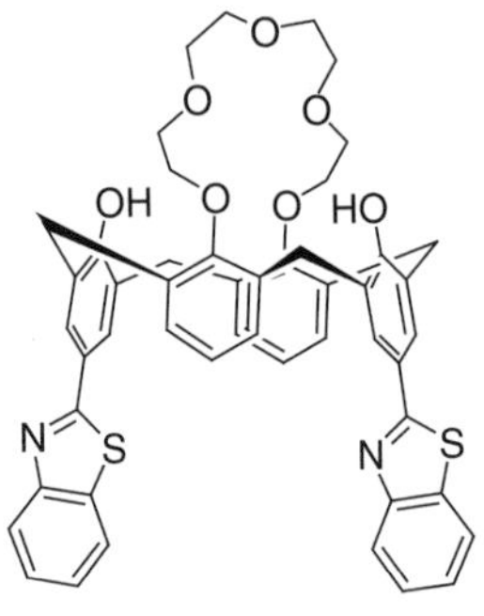

81 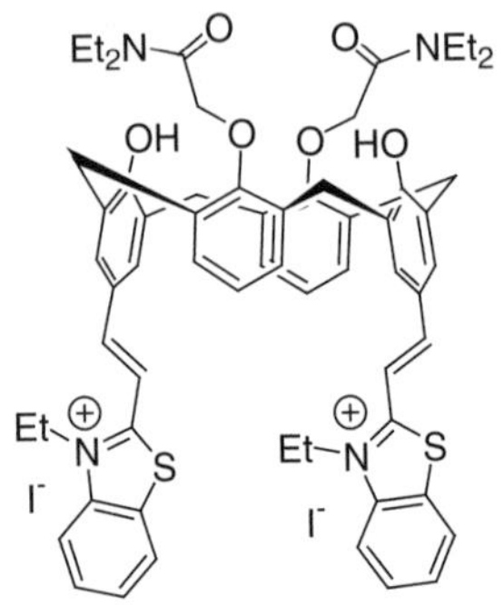   **82**

5.3 Calixarene–imidazole/benzimidazole/benzothiazole lower rim systems

The lower rim conjugates of calix[4]arene with these heterocycles are relatively scarce and mostly the larger cavity calix[6]arene scaffold has been used in designing the imidazole-based lower rim hybrids as biomimetic receptors of metalloenzymes.

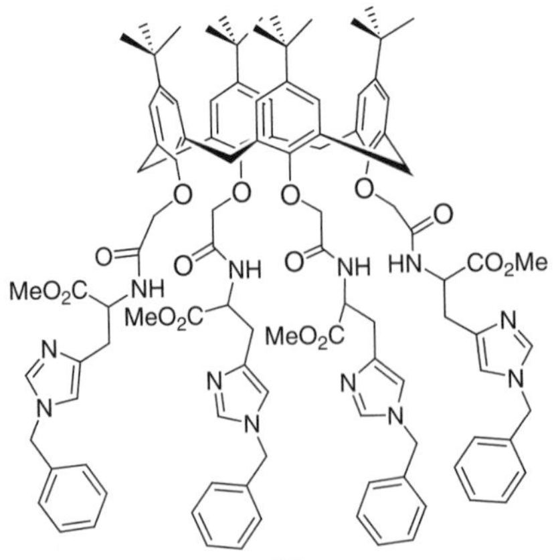

83 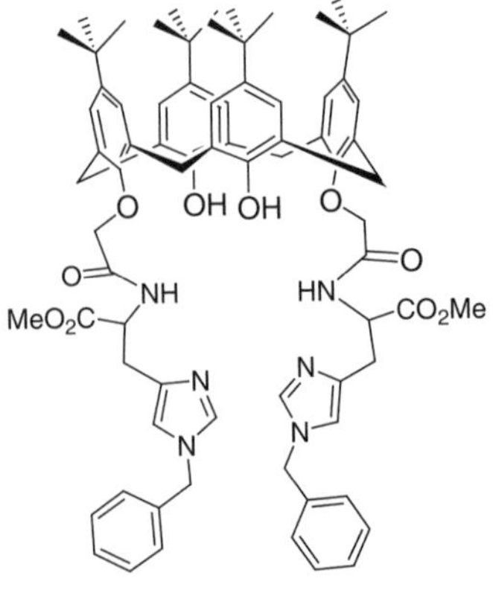   **84**

The tetrahistidinylcalix[4]arene **83** and dihistidinylcalix[4]arene **84** having histidyl amino acid residues on the lower rim have been synthesized and their Co^{II} complexes have been formed (99TL6383). The cationic species elaborating di(alkyl- and 4,4,4-trifluorobutyl imidazolium) substituents on the lower rims of p-bu^t-calix[4]arene **85** and p-bu^t-calix[4]arene-crown-5 **86** were prepared. These quaternary salts have high thermal stabilities and good solubilities in many solvents and based on their melting points, some of them could be classified as ionic liquids (2004IC7532).

85

R = Me, Bu^n, C$_3$H$_6$CF$_3$

X = PF$_6$, BF$_4$, N(SO$_2$CF$_3$)$_2$

86

X = PF$_6$, N(SO$_2$CF$_3$)$_2$

In mononuclear enzymes, tetrahedral Zn^{II} coordination with three histidine residues and a water ligand that is easily displaced by the substrate constitutes a commonly occurring active site. But in synthetic biomimetic models, it was difficult to stabilize tetrahedral dicationic Zn^{II}–aqua species. Reinaud's group extended their strategy of the use of tris-pyridinocalix[6]arene conjugates in creating mimics of the active site of mononuclear copper enzyme and replaced its pyridine moiety with imidazole. Thus, the conjugate 5,11,17,23,29,35-hexa-bu^t-37,39,41-tri-methoxy-38,40,42-tris[(1-methyl-2-imidazolyl)methoxy]calix[6]arene **87** on reaction with $Zn(H_2O)_6(ClO_4)_2$ was found to provide an air stable dicationic zinc–aqua complex **88**. This model has the advantage of the cone shape of calix[6]arene functionalized with three imidazole groups that could mimic both the hydrophobic pocket and the role of histidines of the enzyme active site. Furthermore, the geometry of the system would constrain the metal center in a tetrahedral environment orienting the fourth coordination site toward the inside of the hydrophobic cavity of calix[6]arene, disfavoring the coordination of a fifth molecule to the metal center but favoring the substitution of its labile aqua ligand. The ^1H NMR studies showed an easy exchange of the aqua ligand for amines, alcohols, amides and nitriles (2000JA6183).

The aqua complex **88**, in its X-ray structure showed tetrahedral Zn^{II} within the tris(imidazolyl) environment provided by calixarene-based ligand which is in a cone conformation (2001JA8442). The two water molecules are buried in the calixarene cavity, with one of them coordinated to Zn and revealing an extensive hydrogen-bond network with the host. Both water molecules are expulsed upon exchange with an organic ligand. The X-ray structures of **89** (L = C_2H_5CN and $C_7H_{15}NH_2$) depicted hydrogen bonds between acidic protons of the coordinated guests and the calixarene phenoxy group suggesting their important role in the stabilization of these dicationic complexes (2000JA6183). In addition to hydrogen bonding, the role of CH/π interactions between calixarene host and guest has been revealed in the X-ray structures of **89** (L = C_2H_5OH and NH_2CHO; 2001JCS(CC)984).

87

88 L = H_2O
89 L = C_2H_5CN, $C_7H_{15}NH_2$,
          C_2H_5OH, NH_2CHO

For a comparison of calix[6]arene-based N3 heterocyclic ligands in stabilizing a tetrahedral Zn^{II} in a biomimetic environment, in addition to **54** and **87**, conjugates of calix[6]arene and pendant pyrrolidine, pyrazole and benzimidazole moieties **90** and their stable Zn^{II} tetrahedral complexes have been formed under stoichiometric conditions in acetonitrile. Here again X-ray studies showed that Zn^{II} was coordinated to three nitrogen atoms of the heterocycles with CH_3CN as a fourth ligand included in calixarene conic pocket. It may be pointed out that pyridine-based conjugates are not good ligands for Zn^{II} under stoichiometric conditions, but have remarkable ability to stabilize Cu^I which may be related to its relative softness (2001EJI2597).

90 a-d   **a**   **b**   **c**   **d**

Since copper enzymes are involved in redox processes, it was pertinent to explore Cu^{II} complexes of **87** as the same with **54** were less stable. The first such stable complex **91** formed from **87** and Cu^{II} perchlorate constituted a typical mononuclear five coordinate species with three imidazole groups and two water molecules in the coordination sphere. The small sized alkylnitriles displaced water molecules and **92**, formed in acetonitrile, in its X-ray structure revealed a distorted tetragonal N_4O environment provided by the N3 ligand, an acetonitrile buried inside the cavity and a water molecule placed outside (Figure 4; 2000IC3436).

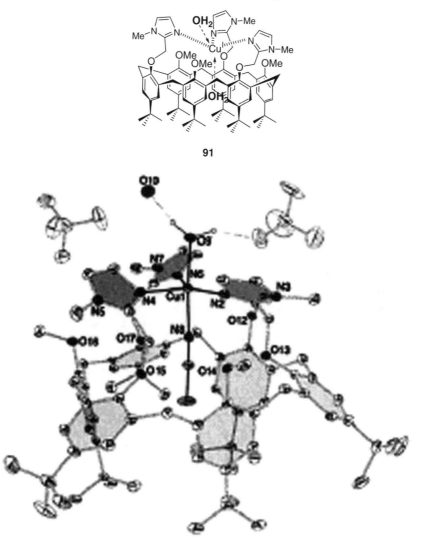

91

Figure 4 X-ray crystal structure of **92**. [2000IC3436] Reproduced by permission from American Chemical Society.

The air stable trisimidazolylcalix[6]arene-based Cu^I complexes having various substituents at phenylether O and imidazole N, obtained from the corresponding analogs of **87** and $[Cu(NCMe)_4]PF_6$, in contrast to their trispicolyl analogs **55** coordinated with CO to form stable mononuclear tetrahedral complexes **93** in chloroform (2000CEJ4218). In an extension of this work, using a trisimidazole calix[6]arene scaffold, **94** was obtained as the first stable water-soluble biomimetic cuprous complex demonstrating that through careful functionalization successful transposition of an organic solvent soluble biomimetic system into aqueous solvent could be implemented (2002AGE1044). These novel supramolecular systems reproduced not only the first coordination sphere encountered in many enzymes and the hydrophobic microenvironment of the active site but also the aqueous macroenvironment of a physiological medium (2000CEJ4218, 2000MI811).

93 L = CO   **94**

With the aim of substitution of the native metals, zinc and copper frequented in trishistidine-based metalloenzymes by other metals with different coordinating requirements, Co^{II} and Ni^{II} complexes of calix[6]arene-based trisimidazolyl systems have been studied. On reaction of **87** and $[Co(H_2O)_6(ClO_4)_2]$ in noncoordinating solvents, a tetrahedral mononuclear Co^{II} complex with the fourth water ligand in the hydrophobic calixarene cavity was formed. A water ligand could be exchanged with ligands such as EtOH, CH_3CN, EtCN and DMF. In the specific case of MeCN, due to coordination of a second nitrile ligand, a five coordinate species was evidenced. The N3 ligand **87**, on reaction with $[Ni(H_2O)_6(ClO_4)_2]$ in THF containing acetonitrile, formed a five coordinate Ni^{II} complex with a guest nitrile inside and a water molecule outside the cavity. Thus, it was evident that the trisimidazole calix[6]arene conjugate was flexible enough to allow the metal ion to adopt the most favorable geometry (2004EJI1817).

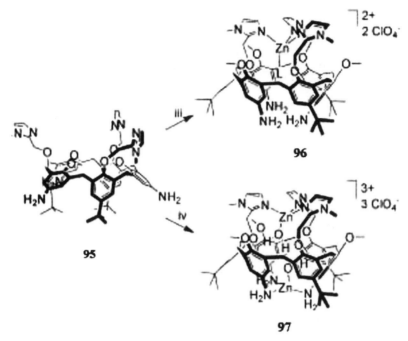

Calix[6]arene derivative **95** possessing a tris imidazole core at the lower rim and three NH_2 groups at the upper rim adopts a flattened cone conformation with aromatic units alternatively located at in and out positions relative to the cavity. With 1 equiv. of $ZnClO_4$, **95** formed a complex **96** with tetrahedral Zn^{II} bound to the three imidazolyl nitrogens and one N included in a guest ligand, which is acetonitrile solvent. The complexation of **95** with 2 equiv. of $Zn(ClO_4)_2$ $(H_2O)_6$ in THF formed complex **97** with two coordinative Zn^{II} ions, one at each rim. The first Zn^{II} stands in the tris imidazole core with four coordinate binding to an oxygen atom of a water molecule. The second Zn^{II} is also in a tetrahedral environment bound to three aniline NH_2 donors and another O of hydroxide anion. As a result, a $(HOH \cdots OH^-)$ unit is encapsulated in the aromatic cavity of the calixarene, closed at both rims by the Zn^{II} ions 5.806 Å away from each other. An additional H-bond between water and one oxygen of the calixarene skeleton further stabilizes the system. This complex represents the first example of a tetrahedral Zn^{II} ion coordinated to aniline donors (2006CC3924).

98   **99**   **100**

$R = Bu^t$—⟨benzene⟩—O

In another example of ditopic calix[6]arene **98** decorated with three N-methyl imidazoles at the lower rim and three 1,4-disubstituted-1,2,3-triazoles at the upper rim, the coordination of the first Zn^{II} cation selectively took place to form a trisimidazole complex with a fourth ligating site of Zn^{II} occupied by the solvent molecule. **98** formed mono and bimetallic complexes **99** and **100** with 1 and 2 equiv. of $Zn(Otf)_2$, respectively. The tris-triazole Zn^{II} ion shortened the size of calixarene cavity and oriented its labile site outside the cavity thus being easily accessible to exogenous binding although protected by the microenvironment defined by the triazole substituents. ^1H NMR studies showed that monometallic complex **99** encapsulated heptyl amine in the calixarene cavity. The higher shielding effect for positions 4–7 in the heptyl amine chain in the case of **99** than those for the Zn^{II} complex of **87** clearly indicated that the functionalization of the upper rim with a substituted triazole extended the cavity in **99**. The addition of acetonitrile and ethanol to **100** resulted in decoordination of the triazole–Zn^{II} complex and caused encapsulation of acetonitrile and ethanol molecules in the extended cavity of **99** (2007OL4987).

A series of calix[4]arene derivatives **101** decorated at 1,3-distal positions with benzothiazolylthiaalkoxy functions and having different spacers between the benzothiazolyl and calix[4]aryl groups have been found to possess a pinched cone conformation, based on their X-ray structures (2002T2647). Ion selective electrodes based on these calix[4]-arenes showed excellent Ag^I selectivity among the most interfering cations. The redox active calixarene **102** possessing (benzothiazolyl)thioalkoxy and quinone units has been coated on glassy carbon electrode and has been used for selective recognition of Hg^{II} through cyclic and square voltametric studies. Only, 500-, 50- and 100-fold molar excess of Pb^{II}, Ag^I and Cu^{II} ions, respectively, gave a voltametric response comparable with that of Hg^{II} and other cations did not pose any interference (2003Talanta553).

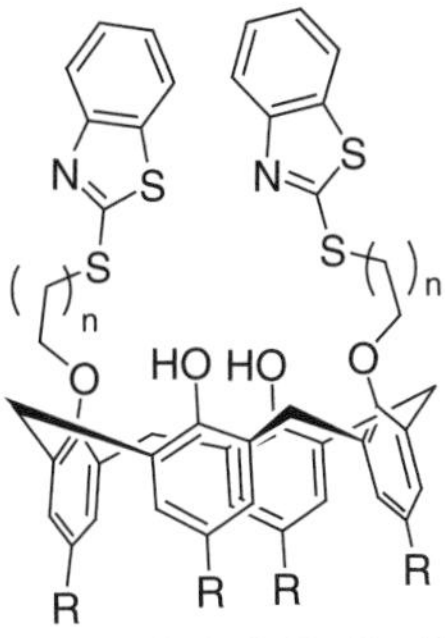

n = 1, 2, 3, 5; R = H, Bu^t

101

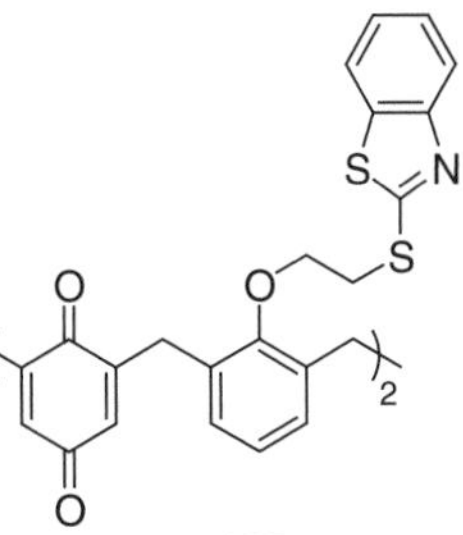

102

6. CONJUGATES OF CALIXARENES WITH NUCLEOBASES, NUCLEOSIDES AND NUCLEOTIDES

6.1 General

The nucleotides have the distinct structural ability to establish a variety of interactions such as complementary base pairing, multiple site hydrogen bonding, specific stacking and generalized electrostatic interactions. Such noncovalent interactions facilitate self assembly of nucleobase-derived NCEs which could provide chemical and biological insight into base-pairing processes and also offer a means of constructing well defined synthetic supramolecular motifs. The calixarene scaffold with inbuilt facility of controllable conformations when linked on either rim with a nucleobase, nucleoside or nucleotide pendant, could invoke complementary combinations to generate aggregated multiple cavity facility receptor systems with a potential for molecular recognition of guest molecules or ions of biological relevance.

6.2 Calixarene–nucleobase/nucleoside/nucleotide upper rim systems

The ureidopyrimidinyl calix[4]arene conjugates **103a** and **103b** generated by linking the 1,3-alternate calix[4]arene at the upper rim through a urea moiety attached at C_2 of uracil-derived motifs were synthesized from the 1,3-alternate calix[4]arene dibromide and appropriate 2-aminopyrimidinone derivatives through a sequence of steps. These conjugates elaborated dimeric species and each dimer existed in *syn–anti* isomeric forms. However, in a polar solvent or in the presence of an oxoanion, monomeric species prevailed (99AGE525).

103a $R_1 = n\text{-}C_9H_{19}$, $R_2 = H$
103b $R_1 = R_2 = (CH_2)_4$

The upper rim conjugates of calix[4]arene with thymine and uracil, **104–107** having –CH$_2$CONH-linkers were synthesized by reactions of para mono- or bis-amino (1,2-proximal and 1,3-distal) calix[4]arenes and thymin-1-ylacetic acid or uracil-1-yl acetic acid in the presence of dicyclohexyldicarbodiimide (DCC). The analogous calix[4]arene adenine conjugates **108–110** were obtained by an alternate mode involving the reactions of bromoacetylated aminocalix[4]arene derivatives with adenine in the presence of sodium hydride. The mono thymine and adenine calix[4]arene conjugates **104** and **108** revealed a distinct concentration dependence of the ^{1}H NMR signals of the nucleobase residues reflecting strong T–T and A–A self-association through intermolecular hydrogen bonding. Also Watson–Crick A–T base pairing recognition was evident in the ^{1}H NMR spectrum of a mixture of **104** and **108**. The bis-nucleoside base-derived systems showed a broadened, nondescript ^{1}H NMR spectrum due to conformational isomers and intermolecular aggregation (2001TL6179, 2003T2539).

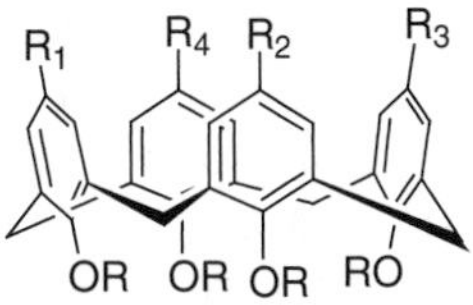

a $R = CH_2CH_2CH_3$
b $R = CH_2(CH_2)_8CH_3$

104 -112

104	$R_1, R_2, R_3 = H, R_4 = A$
105	$R_1, R_3 = A, R_4, R_2 = H$
106	$R_1, R_2 = A, R_4, R_3 = H$
107	$R, R_1, R_2, R_3 = H, R_4 = C$
108	$R_1, R_2, R_3 = H, R_4 = B$
109	$R_4, R_2 = H, R_1, R_3 = B$
110	$R_4, R_3 = H, R_1, R_2 = B$
111	$R_1, R_3 = D, R_4, R_2 = H$
112	$R_1, R_3 = E, R_4, R_2 = H$

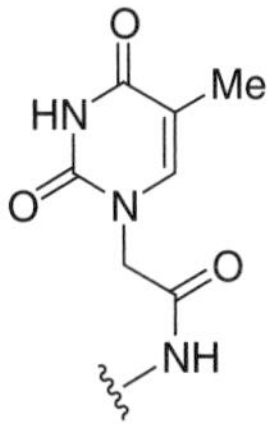

A 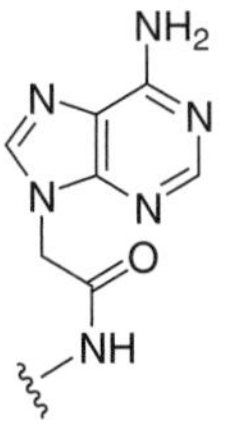　　　B 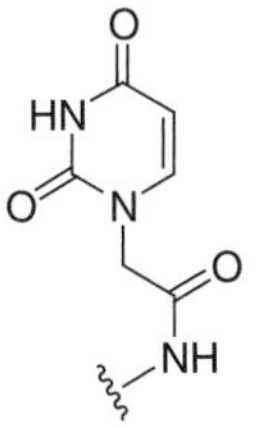　　　C 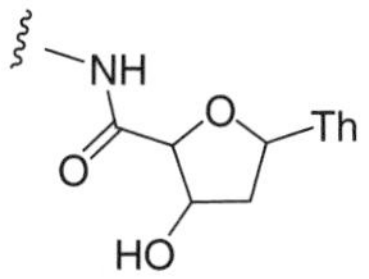　　　D 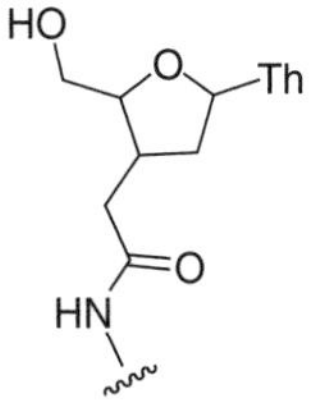　　　E

Significantly, ESI–MS investigations showed that **107** recognized adenine and adenosine from other nucleotides and nucleobases. The interaction of **107** with nucleotides and nucleobases was also indicated by surface pressure area isotherms at the air–water interface and the respective complexes were formed in the monolayer (2008CJC170).

The conjugates of calix[4]arene and nucleoside thymidine **111** and **112** have been synthesized by amide bond formation between amine functional groups of para-1,3-diaminocalix[4]arene and the carboxylic acid groups of the corresponding thymidine nucleosides. The X-ray crystal structure of homocoupled calixnucleoside **111** showed that the calix[4]arene moiety revealed a pinched cone conformation through four independent intermolecular hydrogen-bonding sets between thymine bases and amide linkages. Such an aggregation was also visible in the ^{1}H NMR spectra of these calixnucleosides in apolar solvents (2002TL6367).

6.3 Calixarene–nucleobase/nucleoside/nucleotide lower rim systems

The conjugates of various nucleosides and calix[4]arene **113** and **114**, in which one or two 3′-phosphorylated nucleoside moieties were anchored to the calixarene lower rim through phosphoester linkage, were synthesized by coupling the appropriate calixarene derivative with the correspondingly protected 2'-deoxynucleoside phosphoramidite followed by oxidation and deprotection. In analogy with similar systems bearing a nucleobase or a nucleoside residue, these compounds were stabilized as monomeric species in hydrogen bonding accepting solvents whereas they tended to undergo intermolecular self-association in apolar solvents. The conjugates **113a** and **113d** formed 1:1 complexes with *n*-tosyl-L-arginine methylester hydrochloride and sodium butyrate, respectively (2006TL3245). Among five conjugate systems in **114**, only **114a** was completely soluble in CDCl$_3$. Its ^{1}H NMR studies showed that its ammonium and sodium salts self-assembled through T–T hydrogen bonding in a triangular trimeric supramolecule, whereas on protonation of the nucleoside phosphate group, a dimeric species was formed (2007TL7974). The first water-soluble tetra nucleotide–calix[4]arene conjugates **115a** and **115b** were obtained by the above synthetic mode and tended to self-assemble in aqueous medium by stacking interactions. The inhibitory activity of **115a** and **115b** toward DNA replication during polymeric chain reaction (PCR) amplification was revealed (2007T10758).

Three amphophillic lower rim conjugates **116a**, **116b** and **116e** were synthesized by condensation of the corresponding calix[4]arenediamine with thymin-1-ylacetic acid, adenine-9-ylpropionic acid and uracil-1-ylacetic acid, respectively. The studies on interfacial interactions of these conjugates with complementary nucleosides on the subphase by L–B techniques revealed the formation of a stable monolayer at the air–water interface. The complementary nucleosides in the subphase were efficiently bound to monolayer and could be readily transferred onto the solid substrate along with the monolayers because of the strong intermolecular interactions by multiple hydrogen bonding (2003SC327).

113 R$_1$ = H, R$_2$, R$_3$ = C$_3$H$_7$, R = Y
114 R, R$_1$ = Y.; R$_2$, R$_3$ = C$_3$H$_7$
115 R$_1$, R$_2$, R$_3$, R = Y
116 R$_2$, R$_3$ = H.; R, R$_1$ = (CH$_2$)$_n$NHCOCH$_2$-**B** n = 1 or 2

B =   (a)   (b)   (c)   (d)   (e)   Y =

The calixarene–guanosine conjugate **117a** having an amidic linker and derived from 1,3-alternate calix[4]arene and guanosine in a multistep synthesis was structurally unique with respect to the possibilities of cation and hydrogen bonded self-assembly of four guanine units. Here, the calixarene platform of this G4-calix could orient orthogonal pairs of guanines in such a way that the four guanine units of a set of two consecutively disposed 1,3-alternate calix[4]arenes could form a hydrogen-bonded G_4 quartet which would be further stabilized in the presence of cations. Thus, **117a** reacted with $NaBPh_4$ in 1:1 CH_3CN/H_2O to form self-assembled nanotubes of composition $(117a \cdot Na^+)_n$ $(BPh_4^-)_n$ with diameters consistent with the G-quartet's dimensions (2000JCS(CC)2369.

To enhance the solubility of the above conjugate, a more lipophilic derivative **117b** was formed. In water-saturated $CDCl_3$, water induced the formation of its soluble discrete noncovalent dimer **118** of formula $(117b)_2 \cdot (H_2O)_n$ which exhibited the unique capability of extracting alkali halide ion pairs in organic solvent forming **119** having formula $(117b)_2 \cdot NaCl \cdot (H_2O)_n$. Here, the G-quartet formed by two molecules of **117b** provided a cation-binding site and the neighboring amide groups engaged the halide anion. The formation of **119** constituted a prime example of the cooperative interactions of host, solvent and guest where salt binding was enabled by water and the bound ion pair further stabilized the dimer. These noncovalent structures **118** and **119** were tunable. On changing the anion from a halide to the noncoordinating Ph_4B^- the assembly switched from a discrete dimer to a noncovalent polymer **120** of formula $(117b \cdot Na^+)_n$ $(BPh_4^-)_n$ revealing that the identity of the anion influenced the association process (2003JA15140).

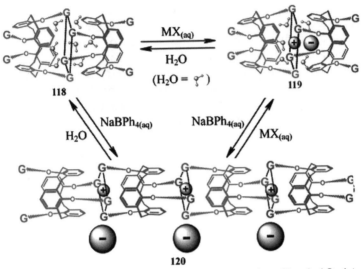

7. CONJUGATES OF CALIXARENES WITH TRIAZOLES, TETRAZOLES AND TRIAZINES

The upper rim of calix[4]arene has also been appended with 1,2,3-triazole and 1,2,3,4-tetrazole systems. The tetrakis-(1,2,3-triazole-5-aminosulfo-nyl) calix[4]arene **121** formed by the reaction of tetrakis-(azidosulfonyl)-calix[4]arene with 2-cyanoacetamides through a Dimroth rearrangement has been found to be a good receptor of chloride and nitrate anions (2004A31). C-glycoside clusters **122–124** having four 1,4-disubstituted 1,2,3-triazole rings linked to C-glucosyl fragments were constructed through multiple cycloadditions of azide/ethynyl functionalized calix[4]arene to ethynyl/azidomethyl C-glycosides (2006JOC7546).

R = CONHPh, CONHC$_6$H$_{11}$-cyclo

121

122 X, Y = A
123 X, Y = B
124 X = B, Y = H

A =

R = H, Bn, Ac,

B =

R = H, Bn, Ac,

The presence of two triazole rings at 1,3-distal *syn*-positions of the lower rim of calixarene has provided opportunities for metal ion receptors where binding could be studied by either color change or change in fluorescence. The chromogenic calix[4]arene **125** with triazoles as the metal-binding sites and the azophenol as both metal binding and coloration sites showed high sensitivity to Ca^{II} and Pb^{II} cations which could be detected by a naked eye as color changed from yellow to red (2007TL7274). Calixarene **126** possessing a 1,3-alternate conformation with two triazole rings on one side and a crown ring on the other side on addition of Hg^{II}, Cu^{II}, Cr^{III} and Pb^{II} underwent strong quenching of fluorescence. In the case of the complex of **126** with Pb^{II}, the addition of K^I, Ba^{II} or Zn^{II} caused revival of emission. ^1H NMR studies showed that Pb^{II} coordinated in a cavity made by two triazole rings caused quenching through activation of a PET phenomenon (2007OL3363). On addition of K^I, its encapsulation into a crown ring caused a conformational change in the **126**–Pb^{II} complex, which resulted in Pb^{II} out of cavity and revival of emission. The quenching of fluorescence could be attributed to complexation of triazole rings with Pb^{II}.

125   126

The upper rim conjugate tetrakistetrazolo calix[4]arene **127**, obtained from the corresponding tetraaminocalix[4]arene through a sequence of steps, on reaction with K_2PdCl_4 formed the complex $127 \cdot PdCl_2$ which revealed a dimeric structure. Such complexes are useful in evolving catalytic systems (2005T12282). Glycoclusters **128** and **129** possessing three or four ribosylmethyl-, galactosylmethyl- and gluco-sylmethyl-propoxytetrazole fragments on the upper rim of calix[4]arene platform have been prepared by coupling 1-ribosylmethyl/galactosylmethyl/glucosylmethyl-5-sulphonyltetrazole derivatives with tetrapropoxycalix[4]arene (2007T6339).

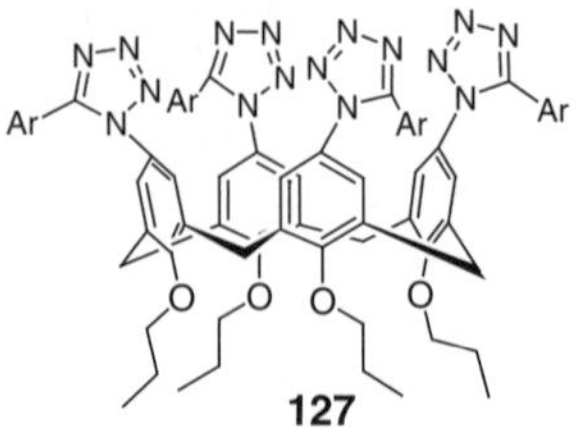

127

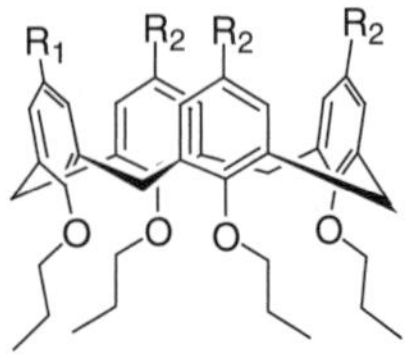

128 R$_1$, R$_2$ = A
129 R$_1$ = H, R$_2$ = A

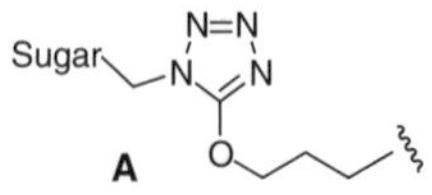

sugar =

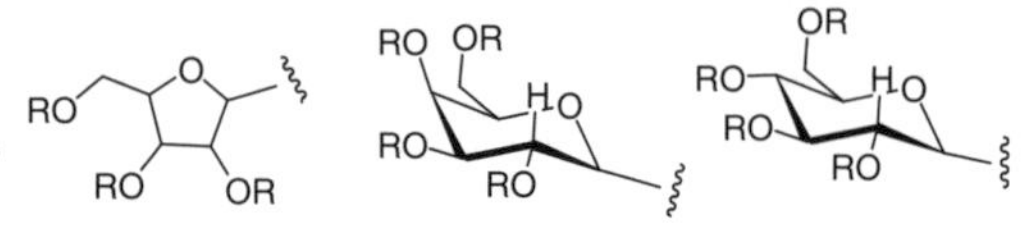

R = H, Bn

The upper rim calix[4]arene, diaminotriazine conjugates, 1,3-alternate 5,17-bis(2,6-diamino-1,3,5-triazine-4-yl)-25,26,27,28-tetra(ethoxyethoxy)-calix[4]arene **130** and 5,11,17,23-tetrakis(2,6-diamino-1,3,5-triazine-4-yl)-25, 26,27,28-tetra(ethoxyethoxy)-calix[4]arene **131** have been synthesized from calix[4]arene through a sequence of systematically performed and monitored steps. The crystal structure of disubstituted calixarene **130** was governed by the diaminotriazine moieties which formed infinite bonds through linking of each diaminotriazine moiety to two others by two hydrogen bonds and the calix[4]arene backbones were placed around the diaminotriazine bonds. In contrast, hydrogen bonding between the diaminotriazine moieties of the tetrasubstituted calixarene **131** favored face-to-face complexes resulting in infinite strands. The ^{1}H NMR of **130** in CDCl$_3$ resembled that of supramolecular polymers and could be explained due to hydrogen bonding between diaminotriazine moieties. But the spectra of both **130** and **131** in DMSO-d_6 showed the existence of monomeric species (2004NJC1335). The conjugate of calix[4]arene in a cone conformation with diaminotriazine showed intramolecular hydrogen bonding as well as formation of gels by the addition of suitable complex partners (97RTC363).

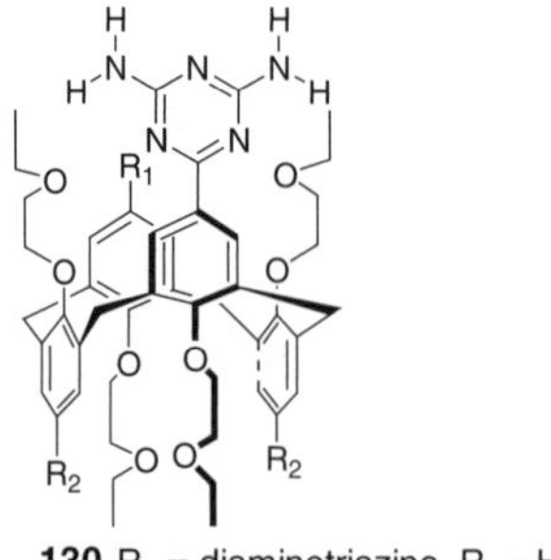

130 R$_1$ = diaminotriazine, R$_2$ = H
131 R$_1$ = R$_2$ = diaminotriaizine

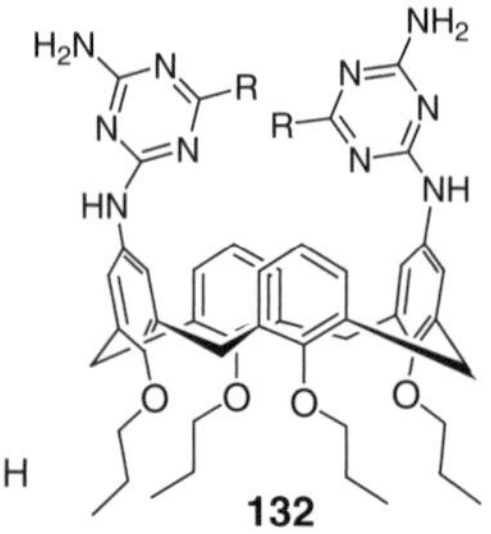

132

The 1,3-distal upper rim conjugates of tetrapropoxy calix[4]arene **132** bearing triamino 1,3,5-triazine derivatives linked at amino nitrogen when functionalized at another amino group with alkyl, aminoalkyl, ureido, pyridyl, carbohydrate, amino acid and peptide functionalities underwent self-assembly with barbituric acid or cyanuric acid derivatives to form well-defined hydrogen-bonded nanostructures (2005OBC3727). The thermodynamic stabilities of these self-assembled *exo*-receptors have been found to depend on several steric factors and the polarity of functional groups attached to assembly components. The multiple complementary recognition sites of these receptors have been used to complex multiple guests. Thus, ureido-functionalized receptors reacted with *p*-nitrophenol to form hydrogen bonds with all six recognition sites in a 1:6 binding mode. But phenols elaborating two OH groups formed two hydrogen bonds with two different ureido recognition sites in a 1:3 complexation mode. Aromatic carboxylic acids were complexed in the 1:6 ratio by a receptor having six amino groups. Uniquely, the receptor functionalized with small peptide unit Gly–Leu–Ser complexed *n*-octylgalactopyranose, a relatively complex guest (2005JOC8443).

8. CONJUGATES OF CALIXARENES WITH 2,2'-BIHETEROCYCLES

8.1 General

As compared with monoheterocycles coupled on the calixarene platform, the nitrogen and/or sulphur 2,2'-biheterocycles when put in the same dispositions could invoke the distinction of cooperative binding of their two hetero atoms with the guest species. Some such heterocycles are 2,2'-bipyridine, 2,2'-bithiophene, 2,2'-bithiazole (btz) and tetrathiafulvalene (TTF). The possibilities of attachment of such a heterocycle at any of its available nuclear carbons at either of the rims of the calixarene platform through various linkers could generate a rich structural variety of resultant receptor architectures. These heterocycles in their conjugates, would blend their inherent spectral and electronic character as well as the photochemical and electrochemical properties of their metal complexes with inbuilt binding and structural features of the macrocycle. These conjugates are presented with respect to the nature of the biheterocycle used in their designs.

8.2 Conjugates of calixarenes with 2,2'-bipyridines

8.2.1 General

In the following account, the upper rim conjugates of calix[4]arene and 2,2'-bipyridine having varied linkers, which have largely been transformed into their metal complexes for mainly anion binding have been described. Subsequently, the lower rim conjugates of calix[4]arene and 2,2′-bipyridine have been presented in the order of the attachment centers C6′, C5′ and C4′ of the 2,2′-bipyridyl moiety with a phenolic oxygen of calixarene through –CH₂-linkers. It is followed by an account of the Ru and Re complexes of this type of conjugates.

8.2.2 Calixarene–2,2′-bipyridine upper rim systems

The upper rim mono- and bis-2,2′-bipyridylcalix[4]arene conjugates **133** and **134** linked at the 6′-position of bipyridine through an ethylenic unit were synthesized by a Wittig reaction of the corresponding calix[4]arene aldehydes and 2-picolyl triphenyl phosphonium chloride. The hydrogenation of **133** and **134** smoothly provided the corresponding –CH₂-CH₂-linked systems. Upon reaction of **133** with $Cu(CH_3CN)_4PF_6$, the complex L₂M was formed. Both hydrogenated ligands formed Co^{II} complexes with $CoCl_2$ and in the mono-bipyridyl dichlorocobalt complex tetrahedral coordination involving the bipyridyl subunit and the two chlorine atoms was revealed (96TL6311, 2000EJI683).

133   134

The upper rim 2,2′-bipyridyl calix[4]arene conjugates **135–137** having NHCO-linkers, were synthesized from calix[4]arene 1,3- diamine or tetraamine derivatives and 4-chlorocarbonyl-2,2′-bipyridine in the presence of triethylamine. In the tris(bipyridyl)ruthenium complexes of these ligands formed by the above mentioned conjugates with (bipy)₂ $RuCl_2$, the topological positioning of the transition metal Lewis-acid centers and of the hydrogen bonding amide functionality invoked in them the anion sensitivity trend. Thus, the complex obtained from **135** allowed binding of anions such as halides, dihydrogenphosphate and bisulphate and the complexes of **136** and **137** selectively sensed through fluorescence emission and electrochemically the $H_2PO_4^-$ anion (94JCS(CC)1269, 96JCS(CC)689).

135 R_1, R_4 = H, R_2 = Me, R_3 = A
136 R_1, R_4 = H, R_2 =Ts, R_3 = A
137 R_1, R_2 = C_3H_7, R_3, R_4 = A

138a R = H
138b R = CH_2CO_2Et

An amide-linked upper rim 2,2'-bipyridyl biscalix[4]arene conjugate system was synthesized from the appropriate aminocalixarene derivative and 4,4-bis(chlorocarbonyl)-2,2'-bipyridine. The heteroditopic ligands 138a and 138b were formed by reactions of pentacarbonylchlororhenium[I] with the corresponding precursor conjugate system. These ligands have the capability to simultaneously complex anions at the upper rim and alkali metal cations at the lower rim. With both the ligands, co-bound lithium, sodium and potassium cations, significantly enhanced the strength of iodide binding in acetonitrile. The greatest positive cooperative iodide-binding effect was revealed with 138b and co-bound sodium cations which correlated with the known lower rim tetrasub-stituted ethyl ester showed a selectivity preference for the alkali metal cation (2000JCS(DT)2721).

8.2.3 Calixarene–2,2'-bipyridine lower rim systems

The lower rim attachment of two bipyridyl pendants at the 6'-position through a methylene linker at 1,3-phenolic oxygens of calix[4]arene fixed in its 1,3-alternate conformation provided a conjugate where only the donating bipyridyl groups were relatively free to rotate around C–C–O bonds. These conjugates formed both mononuclear and binuclear Cu[II] complexes which constituted good low molecular weight model systems for study of catalytic activity of copper enzymes in a nonaqueous environment. Conjugate 139a with copper perchlorate formed stable mononuclear and binuclear species [Cu · 139a]$^{2+}$ and [Cu$_2$ · 139a]$^{4+}$. The tetrahedral arrangement of nitrogen atoms around Cu[II] in the former accounted for the facile reversible reduction of this species (2003TL5415, 2004JCS(DT)3205). The mononuclear complex of appro-priately derivatized conjugate also formed self-assembled monolayers on Au[III] (2004JCS(CC)1812). Water-soluble tetrasulfonato derivative 139b formed primarily 1:1 complexes with Co[II] and Cu[II] in water and also with lower stability than those observed for **139a** in CH_3CN (2007TL8274).

The conjugate **140a** on reaction with Cu^I tetrakis (acetonitrile) hexafluorophosphate formed bimetallic $[(140a_2 \cdot Cu_2)PF_6]^+$ and mono-metallic $[140a \cdot Cu]^+$ species which displayed tetrahedral geometry. Since Ag^I adopted a pseudo-tetrahedral geometry with bipyridine and analogs, the conjugate **140b** equipped with two benzyl groups at distal positions of calix[4]arene lower rim quantitatively extracted Ag^I from a mixture of lead and silver nitrates (2001TL2681).

For introducing a hydrophilic behavior in the conjugates and their Cu^I complexes, a water-soluble 1,3-alternate calix[4]arene-based podand **140c** tethering two 4,4'-dicarboxy-2,2'-bipyridyl units at distal phenol rings has been synthesized. On reaction with $[Cu(MeCN)_4]PF_6$ in water, it formed a mononuclear Cu^I complex exhibiting considerable stability even in the presence of bovine serum albumin, promising interesting behavior in biological media (2001TL2799, 2004EJI2514). Some related calixarene–bis-(bipyridyl) conjugates incorporating four sodium carboxymethyl groups (**140d**) or two sodium(4-oxo)-butanoate units (**140e**) at the upper rim were synthesized (2006TL1895). The (4-oxo)-butanoate derivative bearing unsub-stituted bipyridyl arms was not soluble in water while its 4,4'-dicarboxylate analog was perfectly soluble as were also the carboxymethyl derivatives. The chelation of bipyridyl subunit coupled with water solubility resulted in their ability to complex the unstable Cu^I cation in water, even in the presence of bovine serum albumin. Another water-soluble calix[4]arenebi-pyridyl conjugate **140f** was prepared by incorporation of four sulphonate groups at the upper rim. Its hydrophilic and chelating properties were evaluated in the complexation of Cu^I in water (2002TL77).

139a R = H, R$_1$ = pr^n
139b R = SO$_3^-$ Na, R$_1$ = CH$_2$CH$_2$OEt

140a -140k

140a R$_1$ = R$_3$ = R$_4$ = R$_5$ = H, R$_2$ = Me
140b R$_1$ = CH$_2$Ph, R$_2$ = Me, R$_3$ = H, R$_4$, R$_5$ = Bu^t
140c R$_1$= R$_4$ = R$_5$ = H, R$_2$ = Me, R$_3$ = COONa
140d R$_1$ = H, R$_2$ = Me, R$_3$ = COONa, R$_4$, R$_5$ = CH$_2$COONa
140e R$_1$ = H, R$_2$ = Me, R$_3$ = COONa, R$_4$ = Bu^t,
　　　R$_5$ = COCH$_2$CH$_2$COONa
140f R$_1$ = Bu^n, R$_2$ = Me, R$_3$ = H, R$_4$ = R$_5$ = SO$_3$Na
140g R$_4$ = R$_5$ = -N=N-C$_6$H$_5$, R$_1$ = R$_3$ = H, R$_2$ = Me
140h R$_4$ = R$_5$ = -N=N-C$_6$H$_4$-NO$_2$ p, R$_1$ = R$_3$ = H, R$_2$ = Me
140i R1 = CH$_2$CONEt$_2$, R$_2$ = R$_3$ = R$_4$ = R$_5$ = H
140j R1 = CH$_2$CONEt$_2$, R$_2$ = R$_3$ = H, R$_4$ = R$_5$ = Bu^t
140k R$_1$ = R$_3$ = H, R$_4$, R$_5$ = P(=O)(OEt)$_2$, R$_2$ = Me

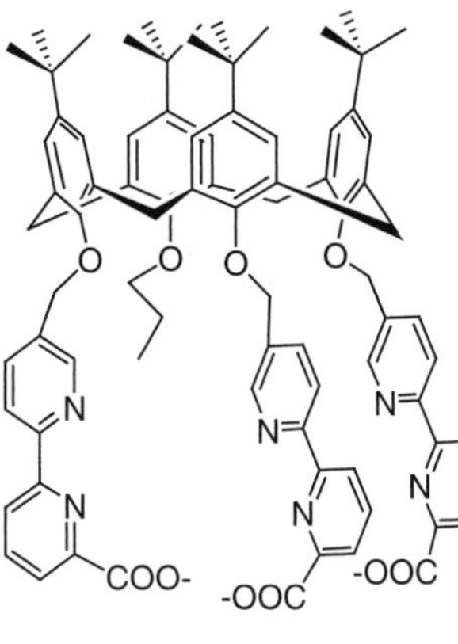

141

The lower rim conjugates of calix[4]arene and 2,2'-bipyridyl functionalized with azophenyl groups at the upper rim **140g, 140h, 142** and **143** have been synthesized. On complexation with Zn^{II}, the stoichiometry of 1:1 was found for the complexes Zn-**140g**, Zn-**140h** and Zn-**142** and 2:1 for the complex Zn-**143** as a function of the number of grafted bipyridyl groups (2003NJC644).

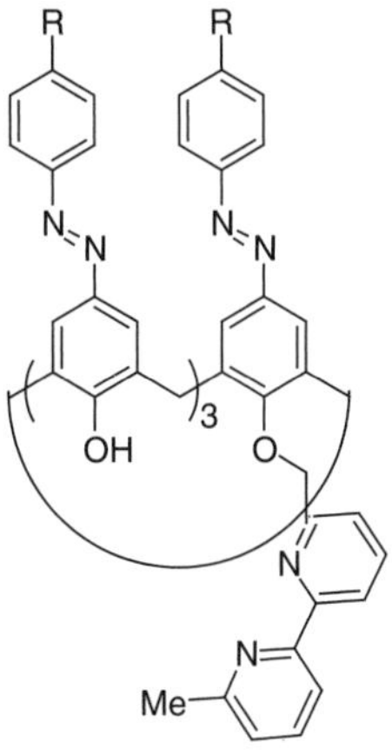

142a R = H
142b R = NO_2

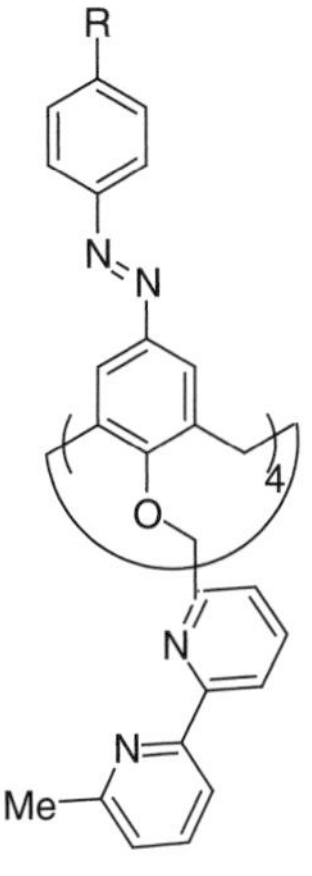

143a R = H
143b R = NO_2

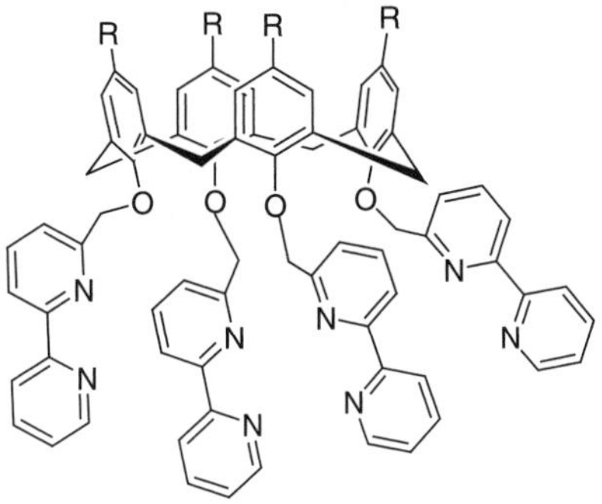

144 R = H, Bu^t

With the aim of sensitizing Eu^{III} and Tb^{III} luminescence on complexation, the conjugate **144** incorporating four bipyridyl chromophores and **140i** and **140j** bearing two amide and two bipyridyl groups have been synthesized. The Eu^{III} and Tb^{III} complexes of these ligands presented a more intense luminescence than the analogous tetra amide derivative of calix[4]arene (96JCS(P2)395). Water-soluble tetraphosphonato derivative **140k** showed complexation with Cu^I even in the presence of Cu^I proteins (2002TL7691). Tris(6-carboxylato-bipyridyl)calix[4]arene derivative **141** formed 1:1 and 2:2 stoichiometric complexes with Tb^{III}, the structures for which have been confirmed by X-ray crystallography (2003SC277).

The lower rim conjugates of calix[4]arene and 2,2'-bipyridine joined at its 5'-position through –CH$_2$-linker(s), **145a, 146–148**, were conveniently synthesized by appropriate base catalyzed reactions of p-bu^t-calix[4]arene with 5-methyl-5'-(bromomethyl)-2,2'-bipyridine and 5,5'-bis(bromomethyl)-2,2'-bipyridine. The complexation of **145a** and **146** with Eu^{III} showed a straightforward formation of new fluorescent complexes. The quinone derivatives formed by oxidation of **145a** and **147** have been shown to complex and recognize K^I electrochemically (94TL6299, 96MI247).

| 145a | R = H |
| 145b | R = CH$_2$COOMe |

146   **147**   **148**

The conjugate **149** has two sets of ion-binding groups, the hard hydroximate and 5'-linked soft 2,2'-bipyridyl ligating groups. When loaded with Fe^{III}, the hydroximate groups converged to embrace the metal ion while the soft groups diverged. On reduction, the ligand rearranged to engulf Fe^{II} with its soft bipyridyl groups and the hard groups diverged. However, the closely related analog of **149** in which the bipyridyl residues were attached at the 6'-position failed to form such complexes (96AGE2657). The bis(bipyridyl)calix[4]arenes **145a, 145b** modulated metal recognition and acted as a unique molecular gate at the guest recognition site. Both the conjugates bind Ag^I strongly. Due to the lack of ester groups, **145a** did not bind Na^I or K^I, but **145b** engaged Na^I strongly probably due to the size effect of the cavity. The addition of two equivalents of NaPF$_6$ to a solution of **145b** · Ag^I showed disappearance of Ag^I complex and appearance of Na^I complex with K$_a$(Na^I)/K$_a$(Ag^I) value 2.5. Hence, the silver complex released Ag^I and captured Na^I because of the lower affinity of the ligands to Ag^I than to Na^I. Interestingly even excess of K^I did not cause such a phenomenon.

Conseqently, **145b** constituted a multiregulation system of ion recognition regulating Ag^I and K^I binding by Na^I and Ag^I, respectively (2002OL3207).

149

150

With the aim of generating metallic complexes for use in homogenous catalysis, a lower rim conjugate **150** of 1,3-alternate p-bu^t-calix[4]arene and 4'-bonded bipyridyl group linked through a –CH$_2$ group has been synthesized. On reaction with 2 equiv. of Rh salts and PdCl$_2$(CH$_3$CN)$_2$, it formed bimetallic complexes. The nickel precursor NiBr$_2$(DME) reacted with **150** to form a bimetallic complex, which on crystallization from methanol gave a mononuclear complex. All these complexes displayed unique molecular packing arrangements based on hydrophobic/hydrophilic interactions (2003IC3160).

For designing luminescent species from calixarene–bipyridyl conjugates, the incorporation of a Ru(bipyridyl)$_3$ moiety, one of the most extensively used species for sensing purposes, has been employed. Thus, the lower as well as the upper rim calixarene–bipyridyl conjugates have been used to generate ruthenium bipyridyl complexes having this luminophore, as luminescent host molecules which could encapsulate luminescent lanthanide ions to lead to systems capable of exhibiting novel and unusual photophysical properties. The positive charge of these hosts could also induce in them anion-binding capability. The results of some such studies are depicted later (97CCR93).

The p-bu^t-calix[4]arene-linked mono and di-trisbipyridylruthenium^{II} complexes **151**, **152** formed by the reactions of cis-Ru^{II}(bipy)$_2$Cl$_2$ and lower rim conjugates of p-bu^t-calix[4]arene and 2,2'-bipyridyl unit(s) linked at 5'-position(s) through methylene linker(s), have free OH groups acting as acid-base sites. By operation of an intramolecular photoinduced electron transfer process, these complexes showed luminescent pH sensor action (94JCS(CC)185).

A lower rim 4′-linked 2,2′-bipyridyl calix[4]arene conjugate **153** having –CH$_2$CH$_2$NHCO-linkers was obtained from appropriately 1,3-distally substituted calix[4]arene diamine and 4,4′-bischlorocarbonyl-2,2′-bipyridine. The trisbipyridylruthenium^{II} complex of **153** formed by reaction with (bipy)$_2$RuCl$_2$ selectively electrochemically recognized H$_2$PO$_4^-$ in the presence of 10-fold excess amounts of HSO$_4^-$ and Cl$^-$ (94JCS(CC)1269).

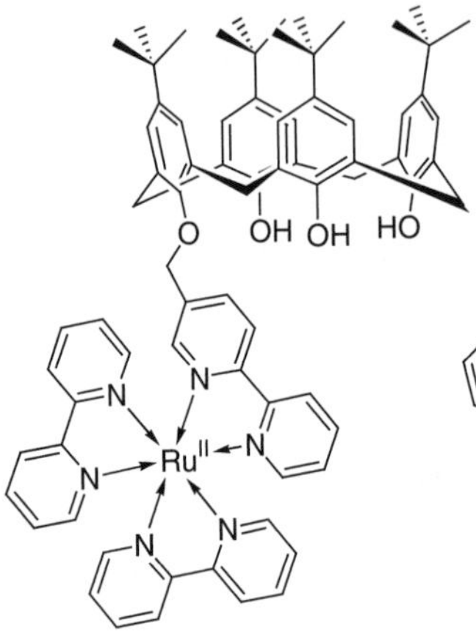

151

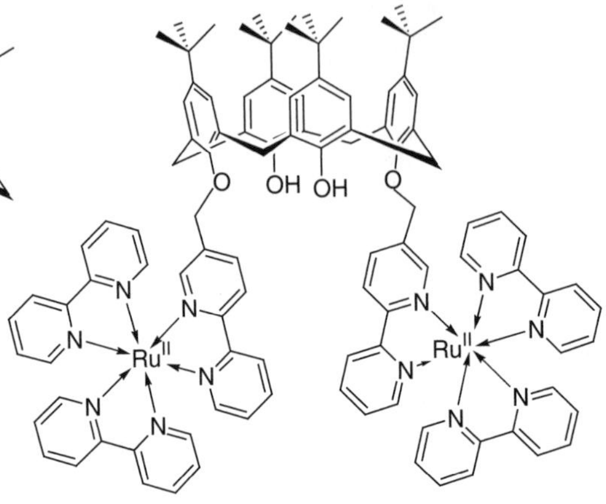

152

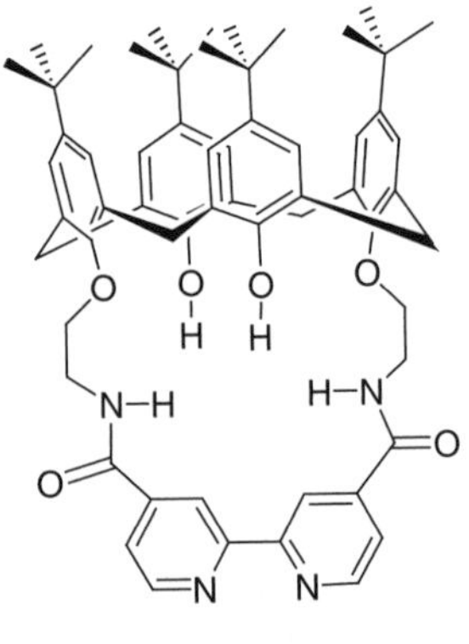

153

A series of heteroditopic Re^I and Ru^{II} bipyridyl calix[4]arene receptors **154a–h** having a Re or Ru bipyridyl moiety covalently linked at one lower rim oxygen of calix[4]arene through various diamidic linkers and three alkali metal cation-binding ester units have been synthesized in multistep procedures. The ^{1}H NMR titrations revealed that in these systems a lower rim ester co-bound alkali metal cation significantly enhanced the strength of bromide and iodide binding with the largest positive co-operative anion-binding effect of 60-fold being displayed by [Li · **154d**]$^{+}$ and bromide. The degree of halide binding enhancement for the neutral Re receptors is considerably larger than that for the charged Rh receptors. The solid/liquid extraction experiments showed that **154a** and **154h** solubilize NaCl and NaOAc in dichloromethane (2001JCS(DT)392).

154a M = Re(CO)$_3$Cl, R = (CH$_2$)$_2$
154b M = Re(CO)$_3$Cl, R = (CH$_2$)$_3$
154c M = Re(CO)$_3$Cl, R = (CH$_2$)$_4$

154d M = Re(CO)$_3$Cl, R =

154e M = (bipy)$_2$Ru^{II}(PF$_6^-$)$_2$, R = (CH$_2$)$_2$

154f M = (bipy)$_2$R^{II}(PF$_6^-$)$_2$, R = (CH$_2$)$_3$

154g M = (bipy)$_2$Ru^{II}(PF$_6^-$)$_2$, R = (CH$_2$)$_4$

154h M = (bipy)$_2$Ru^{II}(PF$_6^-$)$_2$, R =

154a-h

For coordinating and sensing luminescent lanthanide ions, a luminescent ruthenium bipyridyl complex has been covalently linked to one, two or six lower rim acid-amide modified calix[4]arene moieties in receptors **155–157**. All these complexes coordinated with lanthanide ions, Nd^{III}, Eu^{III} and Tb^{III}, with the formation of adducts of variable stoichiometries. The adduct formation affected the Ru luminescence, which was strongly quenched by Nd^{III} ion, increased by the Tb^{III} ion and moderately quenched or increased by Eu^{III} ion (2004IC3965).

155   **156**

157

8.3 Conjugates of calixarenes with 2,2′-oligothiophenes

The coupling of oligothiophenes, an important class of π-conjugated organic materials at either rim of calix[4]arene have provided conjugates in which thiophene oligomers are preorganized within a molecular framework for through-space interactions which could invoke novel functional and structural properties in the resulting materials.

The conjugates of calix[4]arene at all the p-positions of its upper rim and 2,2′-linked thiophenes **158** have been synthesized using palladium catalyzed Kumada coupling of thienylmagnesium bromide and bromo substituted calix[4]arene derivatives. The intramolecular interactions of proximal thiophenes in these systems lead to spectral widening, blue shift of absorption spectra, red shift of emission spectra and increased quenching in fluorescence quantum yield. Also the first oxidation potential of an oligothiophene assembly was lowered stabilizing the formation of a radical cation, which resulted in an increase in the subsequent voltametric oxidation and occurrence of the higher oxidation states as compared to the monomeric counterparts. These assemblies could serve as a model for the investigation of molecular interaction of π-conjugated systems (2006T7846).

$R = C_3H_7 / C_{10}H_{21}$
$n = 1\text{-}4$

158

The conjugates designed by functionalization of a polythiophene with a calixarene receptor scaffold would generate receptor-based conducting polymers which constitute a class of materials that could be utilized in the development of molecule-based sensory materials. Some cyclopenta[2,1-b;3,4-b']bithiophenes differently substituted at the 4-position with a calix[4]arene at its lower rim **159–161** have been synthesized and electrochemically polymerized by anodic coupling. These polymers showed strong affinity and selectivity for toluene and acetone from the gas phase. Their absorption capacity was higher by a magnitude of three orders than that of cyclopentabithiophene-based polymers devoid of a calixarene unit, indicating the role of calixarene units in the absorption process (2004JMC1804). Another category of polymers of bithiophene-functionalized lower rim calix[4]arene-based conjugates **162** having an inbuilt crown-like linker and variedly derivatized in the bithienyl unit have been synthesized. These polymers revealed ion selective voltammetric, chromic, fluorescent and resistive responses to Na^+ (95JA9842).

159   **160**   **161**

162

8.4 Conjugates of calixarenes with 2,2′-bithiazoles

Besides its double chelating property, btz has intrinsic blue fluorescence. The mono to tetrabithiazole derivatives of calix[4]arene **163a–e** have been prepared. Their comparative fluorescence studies showed a decrease in emission intensity of bithiazole from mono to tetra-substituted species for which intramolecular deactivation and calixarene effect have been proposed (96TL5889). Some chromogenic 2,2′-bithiazolylcalix[4]arene conjugates having one, two, three or four bithiazolyl pendants have been formed by varied base-induced condensations of azocalix[4]arene and 4-bromomethyl-4-methyl-2,2′-bithiazole (2001TL8177).

With the aim of creating a ligand able to complex two redox states of copper, the conjugate **163f** of p-bu^t-calix[4]arene bearing sets of two oppositely disposed picolyl and btz pendants at the lower rim, was synthesized. On reactions with [Cu(MeCN)$_4$PF$_6$] and cupric trifluoromethane sulphonate, **163f** formed the corresponding 1:1 complexes. In the Cu^I complex, the metal center revealed tetrahedral binding with 4N, one each of picolyl and btz moieties (99ICC44).

OR_3 $OR_4OR_2R_1O$
163a- 163f

163a R_1 = btz, R_2 = R_3 = R_4 = H
163b R_1 = R_2 = btz, R_3 = R_4 = H
163c R_1 = R_3 = btz, R_2 = R_4 = H
163d R_1 = R_2 = R_3 = btz, R_4 = H
163e R_1 = R_2 = R_3 = R_4 = btz
163f R_1 = R_3 = picolyl, R_2 = R_4 = btz

8.5 Conjugates of calixarenes with tetrathiafulvalenes

The conjugates of calix[4]arene and well known biheterocyclic electron-donor TTF species have been designed with the aim of creating redox-responsive ligands for exploring their electrochemical properties. The calix[4]arene amide-TTF assemblies **164–166** have been obtained from the corresponding 1,3-distal substituted calix[4]arene and succinimidyl-activated TTF ester. The additional amide unit could be introduced using *N,N*-diethyl α-chloroacetamide. The ^{1}H NMR and electrochemical studies revealed the binding ability of receptor **166** for $H_2PO_4^-$. In the upper rim conjugated *p*-(tetrathiafulvalenylmethylideneamino)calix[4]-arene where the coupling is through an imine bond, its good π –electron donating ability could be strengthened upon treatment with triethyla-mine (2005NJC1164, 2006CC2233, 97JCS(P2)2461).

The bis-calixarene systems bridged by a TTF framework **167** and **168** have been synthesized through triethyl phosphate-mediated self-cou-pling of the corresponding monomeric calix[4]arene-1,3-dithiol-2-thiones or their oxo analogs. The monomers were formed from 1,3-bis-(bromoethoxy/propoxy)calix[4]arene and bis(tetraethylammonium)-bis(1,3-dithiole-2-thione-4,5-dithiol)zincate. The X-ray structures of **168** and its precursor showed the cone conformation of calix[4]arene arranged around the tetrasubstituted TTF plane. The cyclic voltammo-grams (CV) of **167** and **168** showed a two-step redox behavior but displayed a CV deformation for the second redox process (2005JOC6254).

164 $R_1 = R_2 = H$
165 $R_1 = H, R_2 = CH_2-C(=O)NEt_2$
166 $R_1 = R_2 = CH_2-C(=O)NEt_2$

167 n = 1
168 n = 2

In the conjugate **169**, for coupling the anion–receptor ability of the calix[4]pyrrole system to the redox properties of TTF core, one TTF unit has been directly annulated to the upper rim of calix[4]pyrrole. This mono-TTF-calix[4]pyrrole system has been synthesized by condensation of a pyrrolo[3,4-*d*]TTF system with a bishydroxymethyltripyrrane derivative and has been found to be an efficient chemosensor for detection of anions by electrochemical means. It constitutes one of the receptors with the strongest binding affinities toward Br⁻, Cl⁻ and F⁻ ions yet recorded for a calix[4]pyrrole system (2003AGE187).

The first tetra-TTF calix[4]pyrrole conjugate **170** has been synthesized by the condensation of acetone and the corresponding monopyrrolo-TTF unit. This calix[4]pyrrole in its 1,3-alternate conformation acted as a host for neutral electron-deficient guests such as 1,3,5-trinitrobenzene, tetrafluoro-*p*-benzoquinone, tetrachloro-*p*-benzoquinone and *p*-benzoquinone. The addition of chloride ions served to effect release of these guests and thus neutral substrate recognition process could be blocked by the addition of chloride anion (2004JA16296).

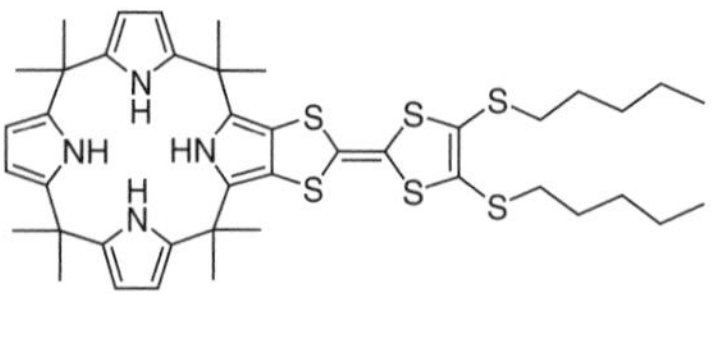

169

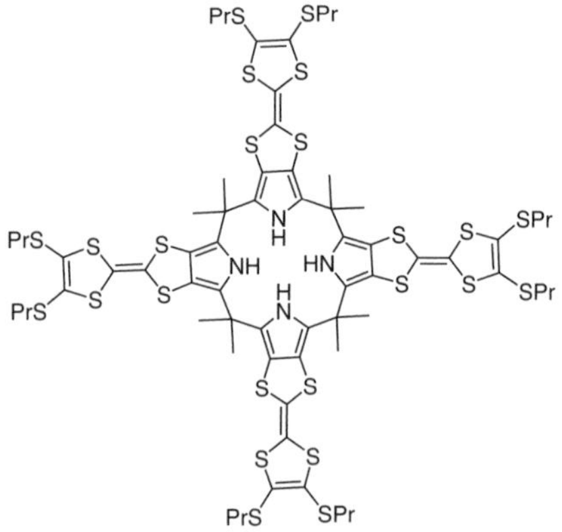

170

9. CONJUGATES OF CALIXARENES WITH SUGARS

9.1 General

Sugars, due to their chiral and polyhydroxylated nature, when coupled on either rim of a calixarene platform could constitute a unique class of water-soluble molecular receptors, which offer various opportunities for molecular recognition of chiral highly polar organic molecules in water and under physiological conditions. The synthesis of some such hybrids and their interactions are described.

9.2 Calixarene–sugar upper rim systems

By using ethyl 1-thio-*β*-D-galactopyranoside (94AGE2479) as glycosyl donor, 1,3-dihydroxymethyl/propyl and tetrahydroxymethyl/propyl tetrapropoxycalix[4]arene derivatives have been transformed into their corresponding upper rim pyranoside derivatives. These reactions were attended by the formation of ether-capped by-products. Alternately, the thiazolylgalactopyranasyl derivatives of calix[4]arenes were smoothly prepared by reactions of 1,3-dihydroxymethyl and tetrahydroxymethyl tetrapropoxycalix[4]arene derivatives with thiazolylgalactopyranose acetate (96JOC5155) and the thiazolyl moiety provided an inbuilt advantage of its convenient modification to a number of functional groups. The Suzuki reaction has also been used to prepare glycosidic conjugates by coupling *p*-bromophenyl glycosides to boronic acid derivatives of propoxycalix[4]arene. The phenoxy groups on the upper rim of calix[4]arene linking sugar moiety in these conjugates were responsible for creating deeper cavities (98TL9171).

In the above *O*-glucoconjugates, the acid sensitive nature of the acetal linkage holding the sugar to the macrocycle posed practical limitations on their use as receptors in aqueous medium. Consequently, the C-linked conjugates of carbohydrates and calixarene were synthesized by a Wittig olefination of calixarene aldehydes with sugar phosphoranes followed by reduction of the resulting alkenes to provide methylene isosteres of *O*-calixsugars (97TL7801, 2002JOC4722).

9.3 Calixarene–sugar lower rim systems

The lower rim tetravalent α-thiosialoside conjugate of *p*-bu^t-calix[4]arene **171a** was obtained by α-thiolated sialoside-induced nucleophilic substitution of cone-shaped halogenated calix[4]arene precursor formed in a multistep sequence from the parent calix[4]arene. In spite of its amphiphatic structure, **171c** was soluble in water. Its sodium salt as well as the methyl ester **171b** exhibited strong cross-linking ability with tetrameric wheat germ agglutinin, a plant lectin known to bind sialosides (96TL5469).

171a R = Me, R′ = Ac
171b R = Me, R′ = H
171c R = R′ = H

172

The dendrimeric lower rim conjugates of carbohydrates and calix[4]arene, loaded with multiamide functionalized linkers, have been designed for investigating their multivalent carbohydrate–protein interactions at the molecular level. The glycocalix[4] arenes **172–174** having suitable spacer substituted 2-acetamido-2-deoxy-α-D-galactopyranoside residues have been obtained through a multistep sequence involving appropriate protection, deprotection and condensation operations. These tetra **172**, octa **173** and hexadecavalent **174** glycocalix[4]arene dendrimers showed their lectin-binding properties toward *Vicia villosa* agglutinin (VVA) in aqueous environment (99AGE369).

173   **174**

The lower rim 1,3-distal D-gluconyl calix[4]arene derivative **175** having an amidic linker was smoothly formed by deacetylation of the product obtained from 1,3-aminoethoxy calix[4]arene preorganized in the cone conformation and 2,3,4,5,6- pentaacetyl-D-gluconyl chloride. Its ¹H NMR-monitored interaction with some monosaccharide derivatives revealed their 1:1 complexation (2000T1883).

Some rigid structures of calixsugars in which a cavity was formed by bridging two 1,3-distal lower rim sites of a calix[4]arene molecule with an oligosaccharide **176** or by assembling bis-calix[4]arene systems **177** through lower rim 3,3′ distal linking and 1,1′ of the oligosaccharide, have been synthesized as potential hosts for neutral molecules. The double calixsugar **177** showed selective receptive properties toward imidazole (2001TL3295).

175   **176**   **177**

10. CONJUGATES OF CALIXARENES WITH PORPHYRINS AND PHTHALOCYANINES

10.1 General

Similar to calix[n]arenes, porphyrins have been known to be one of the support pillars of supramolecular chemistry attributing suitable photo-active and electroactive properties to the molecular structures designed around them for building artificial molecular devices. Thus, various metallated and free base porphyrin–calixarene assemblies could afford attractive scaffolds for application in the areas of multipoint molecular recognition, receptors, host–guest chemistry, catalysis and photoinduced electron transfers.

10.2 Calixarene–porphyrin/phthalocyanine upper rim systems

The directly linked upper rim calix[4]arene–Zn metalloporphyrin conjugate **178** was synthesized by cross condensation of 5-formyl-17-nitro-25,27-dimethoxycalix[4]arene-26,28-diol, dipyrromethane, benzaldehyde and zinc acetate. ^{1}H NMR monitoring revealed its 1:1 noncovalent assembly with benzoquinone in which two phenolic hydroxyl groups of calixarene served as tweezers to capture the benzoquinone by two-point hydrogen bonding. The fluorescence of **178** in this assembly was partially quenched by an intra-ensemble electron transfer process involving the complexed quinone (99NJC977). The upper rim distal 1,3-bisporphyrin–calix[4]arene conjugate 5,17-bis [5-(2,8,13,17-tetraethyl-3,7,12,18-tetramethylporphyrinyl)]-25,26,27,28-tetrapropoxycalix[4]arene was synthesized through a multistep sequence from the corresponding bialdehyde 5,17-diformyl-25,26,27,28-tetrapropoxycalix[4]arene and its bis zinc^{II} complex was also obtained. The photophysical and ^{1}H NMR studies indicated fluxional process and a closed conformation was observed for the zinc complex (2005IC2836). The upper rim tetraporphorino–calix[4]arene conjugate **179** was conveniently obtained from a monobenzyl alcohol-substituted porphyrin · Ni formed from p-anisaldehyde and pyrrole in a multistep sequence. In its X-ray structure, the molecule adopted the cone ($\alpha,\alpha,\alpha,\alpha$) conformation. A predominant phenomenon of interaction between two neighboring molecules of **179** resulted in aggregative formation of a cogwheel arrangement leading to self-assembling supramolecular arrays (97AGE2497).

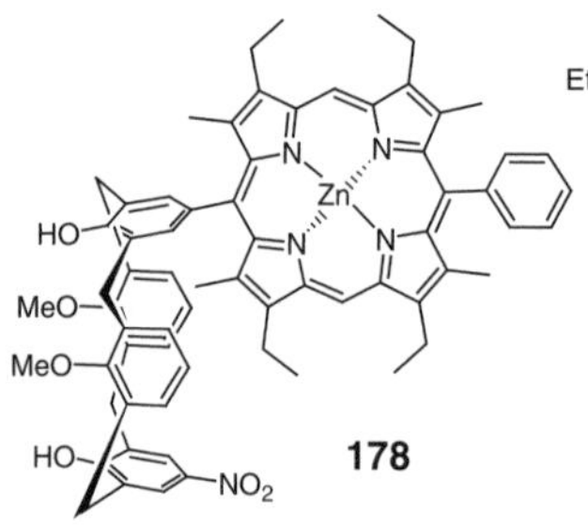

178

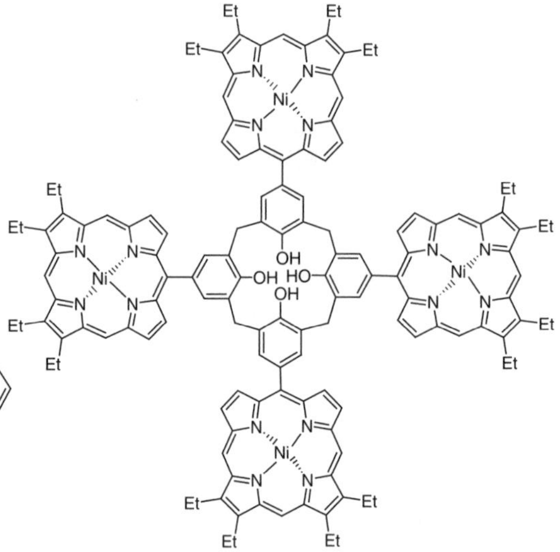

179

The upper rim cone conformed calix[4]arene–zinc porphyrin (ZP) conjugate **180** derivatized at ZP with a pyromellitimide (PI) electron acceptor at the distal position on calixarene with a carotenoid unit was designed and synthesized for studying the electric field effect of photogenerated ion pair on an adjacent chromophore. The photoinduced charge separation within a ZP–PI electron donor–acceptor pair having an 8.4 Å center–center distance in a linear orientation has been found to induce large electrochromic effects on the ground state absorption spectrum of the nearby carotenoid moiety (95JA2041).

180

For studying the long-range biological electron transfer events in a model system, the upper rim 1,3-distal ZP and PI derivatized calix[4]arene conjugate having cone **181** and 1,3-alternate **182** conformations were synthesized by elaboration of the corresponding 5-formyl-17-nitrocalix[4]arene into porphyrin and PI moieties. The fluorescence intensity of **181** and **182** was 95% and 3.5%, respectively, of the control upper rim conjugate of calixarene and Zn · porphyrin. Thus, in the 1,3-alternate conformer **182** due to intraensemble electron transfer through space, the fluorescence of Zn · porphyrin was quenched which was not feasible in **181** due to 15 Å separation between donor and acceptor. The electron transfer through bond mechanism was ruled out because the connecting bonds in both **181** and **182** were exactly same (2001JA10744).

181   182

For enhancement of noncovalent interactions between porphyrins and substrates complexed by the calixarenes, the directly linked upper rim bis-calix[4]arene–porphyrin conjugates **183** have an inbuilt scope of strengthening porphyrin–guest interaction by π–cation and π–π attractive contributions of calixarene moieties. It was synthesized by elaborating a porphyrin from two molecules of 5,11,17,23-tetrapropoxycalix[4]arene monoaldehyde with a dipyrrolyl derivative or by converting the monoaldehyde into a dipyrrolyl derivative followed by reactions with *p*-substituted benzaldehydes. The electron transfer character of the *p*-substituents in calix[4]arene phenyl units attached to the meso positions of the porphyrins influences the photophysical properties (99TL5949).

R = H, Me, Cl, CF$_3$, CN, NO$_2$

183

The conjugation of porphyrins with a calixarene scaffold at its upper rim has been used for evolving cofacial porphyrin pairs for enhancing interaction between porphyrins and substrates. The first cofacial porphyrin pair directly linked by two calix[4]arene moieties was synthesized by condensation of dipyrrylmethane and calix[4]arene 1,3-dialdehyde (93TL627). The upper rim cofacial bisporphyrin conjugates of preorganized 1,3-alternate **184, 185** and cone **186** calix[4]arene crown ethers linked with Zn · porphyrins through $-C \equiv C - C_6 H_4 -$ units have been synthesized by coupling of the corresponding diiodocalix[4]arene crown ethers with the coupled products of 5-(p-iodophenyl)-10,15,20-trismesitylporphinatozinc and trimethylsilyl acetylene. The cofacial preorganization of porphyrins in these conjugates was adequate for the formation of inclusion complexes with bidentate ligands such as 4,4′-diazabicyclo[2.2.2]octane (2002OL2129).

184 R = mesityl   **185** R = mesityl   **186** R = mesityl

Some novel anion receptors with inbuilt ability of UV-visible spectroscopic monitoring have been evolved by appending two tetraphenylporphyrin units at the upper rim 1,3-positions of a 1,3-alternate and both distal and proximal positions of cone conformers of a calix[4]arene system through the ureido functions. These receptors elaborate well organized cavities where anions could be held by synchronous hydrogen bonding interactions with NH groups. The binding constants of halides with these conjugates decreased with anion diameter reflecting their size recognition ability (2003OL149).

187

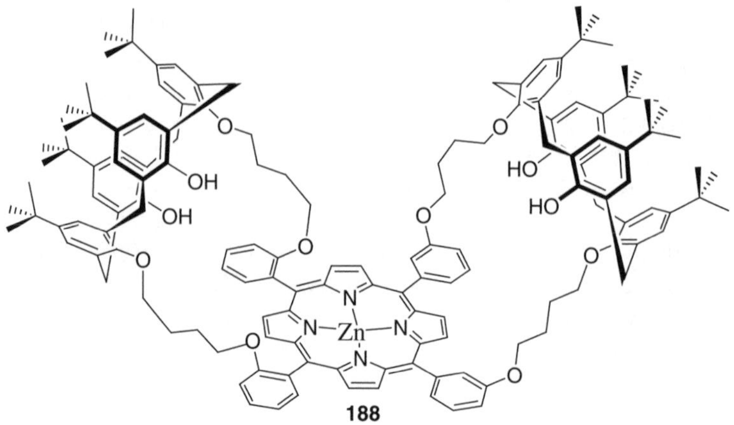

188

The large sized metalloreceptors in a preorganized molecular shape and ability of multifunctional complexation have been generated by covalent strapping of two structural modules, Zn·porphyrin and calixarene through suitable linkers. Both upper and lower rim U-shaped metalloreceptors biscalix[4]arene-Zn-tetraphenylporphyrins **187** and **188** have been synthesized by condensation of pyrrole with appropriate calix[4]arene-based dialdehyde followed by metallation with zinc acetate (94TL7131). A doubly calix[4]arene-capped Zn·porphyrin **189** having an egg shaped cavity has again been synthesized by stepwise condensation of pyrrole with upper rim calix[4]arene-derived appropriate dialdehyde followed by metallation. It is an excellent receptor for different aza-heterocycles and engaged pyridine derivatives, piperidine and N-methylimidzole by approximately 10–1000 times more strongly than the unsubstituted Zn·porphyrin (95JOC6585). A conjugate of calix[5]arene capped at its upper rim with Zn·porphyrin **190** was synthesized by amidation of syn-5,16,-bis(7-carboxy-1-naphthyl)-10,20-diphenylporphyrin with calix[5]arene having two amino functions and subsequent metallation. It binds 4-methyl pyridine much stronger than Zn·porphyrin (2002TL8191).

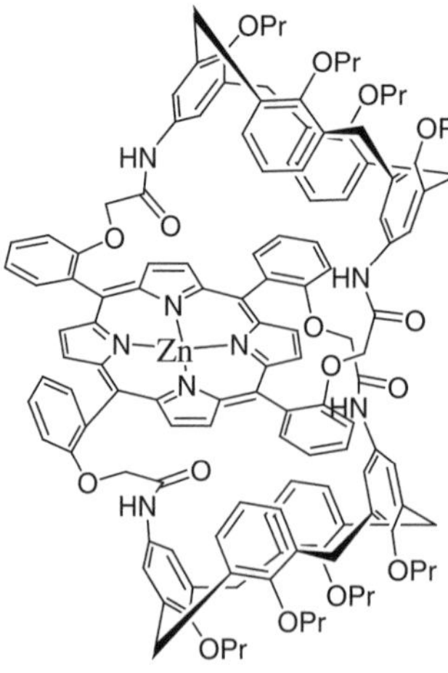

189

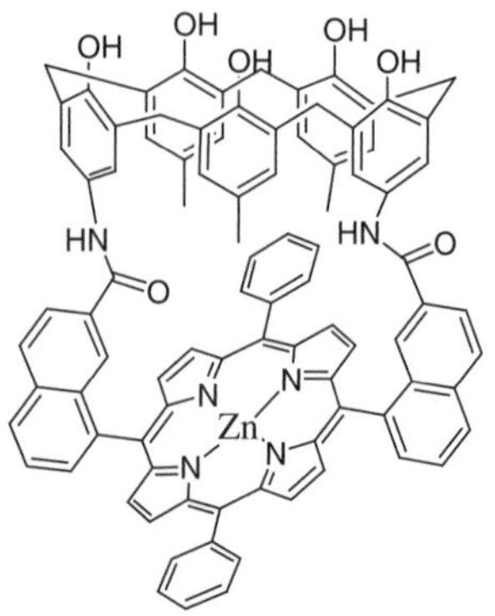

190

10.3 Calixarene–porphyrin/phthalocyanine lower rim systems

For devising an oxygen carrier model in these conjugates, in a unique structural set up, the calix[4]arene was capped through its lower rim oxygens at ortho positions of the phenyl substituents of the tetraphenylporphyrin through –CH$_2$-CH$_2$-O-linkers **191**. The fluorescence quenching in **191** with benzoquinone was 3.5 times more efficient than in tetraphenylporphyrin. The ferrous derivative of **191** was found to function as an oxygen carrier (94ICA1).

191   **192**

A lower rim conjugate in which the p-bu^t-calix[4]arene was covalently linked at 1,3-distal positions to tetraphenylporphyrin through –CH$_2$–CO–NH– at the para position of one of the phenyl rings was metallated using zinc acetate/triethyl amine to form **192**. It reacts with the bidentate ligand 1,4-diazabicyclo[2.2.2]octane (DABCO) to form a 1:2 complex in which both porphyrin units are ligated separately (2003T2409).

The conjugate **193** in which a porphyrin was linked through its two p-ethoxyphenyl units to the lower rims of two p-bu^t-calix[4]arene moieties was obtained by monoalkylations of two molecules of the latter with meso-di(p-bromoethoxy)phenyl porphyrin synthesized by condensation of dipyrrylmethane and p-bromoethoxybenzaldehyde followed by oxidation. It was also formed by selective monoalkylation of calix[4]arene with the correspondingly functionalized benzaldehyde followed by condensation with dipyrrylmethane. Here, the two calixarene units have been found to adopt cone shapes and accommodate neutral organic guests. Consequently, benzoquinone-induced fluorescence quenching of porphyrin was enhanced in **193** due to the presence of calixarene units (95TL2999).

193

The lower rim carbethoxy ethyl derivatives of *tert*-butylcalix[4]arene and *tert*- butylcalix[6]arene in which ester groups at 1,3 and 1,4 positions of calix[4]arene and calix[6]arene, respectively, were derivatized with the fluorophore Zn^{II} porphyrin and the quencher [60]fullerene groups have been synthesized. In the calix[6]arene-based system, a metal-induced 1,2,3-alternate to cone conformational change shortened the distance between the two pendants and the electron transfer efficiency was sharply increased. In the calix[4]arene-based system, the electron transfer efficiency was slightly increased in spite of a metal-induced rotation of the carbonyl group which enlarged the distance between these two groups. Thus, the electron transfer efficiency between $Zn^{II} \cdot$ porphyrin and [60]fullerene, a potential redox couple to mimic the photosynthetic system, could be controlled by a switch function (99TL8245).

The calixarene derivatives **194–195**, possessing one or two phthalocyanine units at the lower rim have been synthesized through substitution reactions of 1,3-dimethoxy calix[4]arene with 3-nitrophthalonitrile or 4-nitrophthalonitrile followed by condensation with phthalonitrile. These compounds characteristically lack the aggregation properties observed in phthalocyanines. The ^{1}H NMR spectrum of **194** revealed a partial cone conformation with two of the bu^t groups of the calix[4]arene lying above and below the phthalocyanine ring (2007TL681). The electrochemical properties of solid films obtained from **195** and its double-decker derivative **196** revealed the formation of suitable ion channels for counter ion movement in **196** (2007CP283). The 1,3-distal bis(dicyanophenoxy)calix[4]arene derivative **197** on self-condensation gave a ball type dimeric Zn^{II} phthalocyanine **198** which exhibited mixed valence behavior and nonArrhenius type dependence of conductivity. In the UV-Vis spectrum of **198** the presence of an additional week band centered at *ca.* 68 nm blue to the normal Q band showed exciton coupling between the two phthalocyanine units. The spin-coated films of **198** showed high sensitivity to toluene vapors at room temperature (2006CC320).

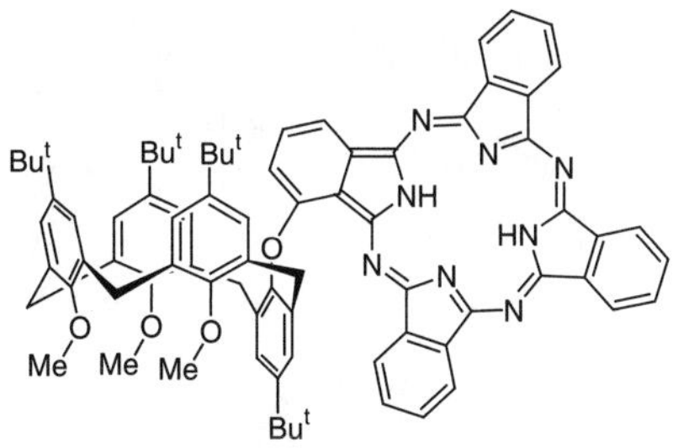

194

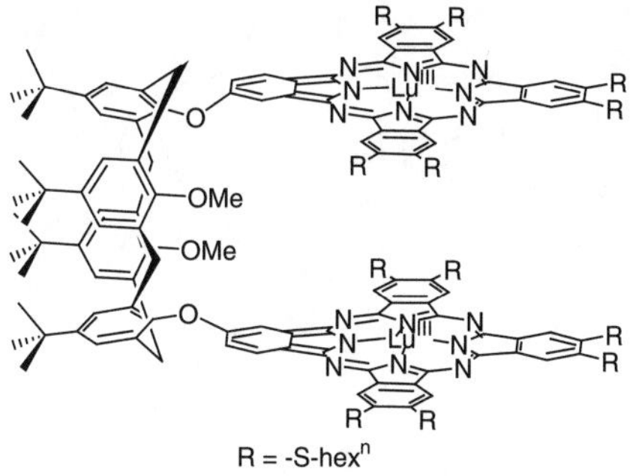

R = -S-hex^n

195

R = -S-hex^n

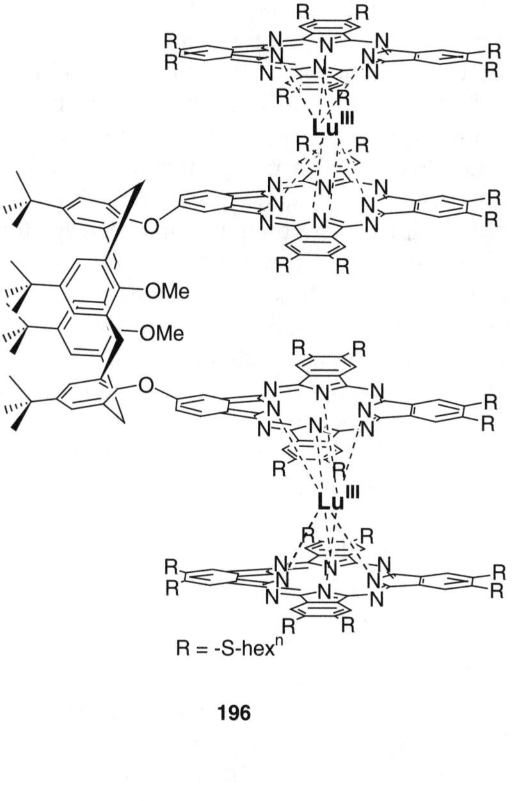

196

197   198

11. CONCLUSIONS

This account on the conjugates of heterocycles and calixarenes elaborates their design, synthesis and distinctive features as synthetic receptors. In these conjugates, the heterocyclic components, because of their placement outside the nuclear calixarene frame unlike that in hetero-calixarenes, in addition to their preorganization arising out of the conformational mobility of calixarene platforms, have the added facility of adjustability of the stereochemical and directional dispositions of their binding sites facilitated by their mobility. Except for a few highly innovative designs involving calix[6]arenes, in the majority of the conjugates calix[4]arenes have been used. But the utility of the larger size and flexibility of the calixarene hydrophobic cavity has been demonstrated in the designs of metalloenzymes. Their enhanced use in the designs of these conjugates would be of added interest.

Interestingly, a broad spectrum of heterocycles of varied nature and origin has been employed. As a natural consequence, the parent conjugates and their metal complexes display an equally wide range of structural and functional properties. These include recognition of anions, cations and neutral substrates; generation of photoactive and electro-active materials with even switch functions; self assembly triggered nanotubes; biological electron and energy transfer events, catalysis of organic reactions; models of metalloenzymes, etc. The designs of these conjugates using naturally occurring heterocyclic species could be uniquely versatile in the realm of synthetic receptors.

However, the investigations of the properties of these conjugates have been of limited nature and in the majority of cases the reports are confined to their procurement. Hence, the conjugates of calixarenes and heterocycles when designed with a deeper insight followed by detailed investigations on their targeted properties would provide NCEs not available in any other category of molecular architectures. This comprehensive attempt to highlight the immense potential of these conjugates in evolving NCEs of contemporary relevance hopefully will attract the exploratory attention of more supramolecular chemists.

LIST OF ABBREVIATIONS

A	Adenine
Ac	Acetyl
ApA	Adenylyl(3′→5′)adenosine
Bipy	Bipyridyl
Bn	Benzyl
Btz	Bithiazole
Bu^t	*tert*-butyl
Bz	Benzoyl
DABCO	1,4-diazabicyclo[2.2.2]octane
DCC	Dicyclohexyldicarbodiimide
DME	Dimethoxyethane
DMF	Dimethyl formamide
DMSO	Dimethyl sulfoxide
DNA	Deoxy ribonucleic acid
Et	Ethyl
GpG	Guanylyl(3′→5′)-guanosine
Hex-n	*n*-hexyl
K_a	Association constant
Me	Methyl
NCE	Newer chemical entities
TBS	*tert*-butyldimethylsilyl
PCR	Polymeric chain reaction
Pic	Picolyl
Pr	*n*-propyl
Py	Pyridyl
Quin	Quinoline
RNA	Ribonucleic acid
T	Thymine
THF	Tetrahydrofuran
TFA	Trifluoroacetic acid
TTF	Tetrathiafulvalene
UpU	Uridylyl(3′→5′)uridine

ACKNOWLEDGMENT

We thank INSA and CSIR, New Delhi, India for financial support.

REFERENCES

89JOC5407	F. Bottino, L. Giunta, and S. Pappalardo, *J. Org. Chem.*, **54**, 5407 (1989).
92JOC2611	S. Pappalardo, L. Giunta, M. Foti, G. Ferguson, J. F. Gallagher, and B. Kaitner, *J. Org. Chem.*, **57**, 2611 (1992).
92T9917	P. D. Beer, J. P. Martin, and G. B. Drew, *Tetrahedron*, **48**, 9917 (1992).
93JOC1048	P. Neri and S. Pappalardo, *J. Org. Chem.*, **58**, 1048 (1993).
93TL627	Z. Asfari, J. Vicens, and J. Weiss, *Tetrahedron Lett.*, **34**, 627 (1993).
94AGE2479	A. Marra, M. C. Scherrmann, A. Dandoni, A. Casnati, P. Minari, and R. Ungaro, *Angew. Chem. Int. Ed. Engl.*, **33**, 2479 (1994).
94JCS(CC)185	R. Grigg, J. M. Holmes, and S. K. Jones, W. D. J. A. Norbert, *J. Chem. Soc. Chem. Commun.*, 185 (1994).
94JCS(CC)1269	P. D. Beer, Z. Chen, A. J. Goulden, A. Grieve, D. Hesek, F. Szemes, and T. Wear, *J. Chem. Soc. Chem. Commun.*, 1269 (1994).
94ICA1	N. Kobayashi, K. Mizuno, and T. Osa, *Inorganica Chim. Acta*, **224**, 1 (1994).
94TL6299	G. Ulrich and R. Ziessel, *Tetrahedron Lett.*, **35**, 6299 (1994).
94TL7131	D. M. Rudkevich, W. Verboom, and D. N. Reinhoudt, *Tetrahedron Lett.*, **35**, 7131 (1994).
94TL8255	K. Koh, K. Araki, and S. Shinkai, *Tetrahedron Lett.*, **35**, 8255 (1994).
95MI1	J. M. Lehn, Supramolecular Chemistry, VCH, New York (1995).
95JA2041	D. Gosztola, H. Yamada, and M. R. Wasielewski, *J. Am. Chem. Soc.*, **117**, 2041 (1995).
95JA9842	M. J. Marsella, R. J. Newland, P. J Carroll, and T. M. Swager, *J. Am. Chem. Soc.*, **117**, 9842 (1995).
95JOC4576	S. Pappalardo, G. Ferguson, P. Neri, and C. Rocco, *J. Org. Chem.*, **60**, 4576 (1995).
95JOC6585	D. M. Rudkevich, W. Verboom, and D. N. Reinhoudt, *J. Org. Chem.*, **60**, 6585 (1995).
95TL2999	R. Milbradt and J. Weiss, *Tetrahedron Lett.*, **36**, 2999 (1995).
96MI247	H. C. Y. Bettega, J. C. Moutet, G. Ulrich, and R. Ziessel, *J. Electroanal. Chem.*, **406**, 247 (1996).
96AGE2657	C. Canevet, J. Libman, and A. Shanzer, *Angew. Chem. Int. Ed. Engl.*, **35**, 265 (1996).
96JCS(CC)689	P. D. Beer, *J. Chem. Soc. Chem. Commun.*, 689 (1996).
96JCS(FT)1731	A. F. Danil de Namor, F. F. Sueros Velarde, and M. C. Cabaleiro, *J. Chem. Soc. Faraday Trans.*, **92**, 1731 (1996).
96JCS(P2)395	A. Casnati, C. Fischer, M. Guardigli, A. Isernia, I. Manet, N. Sabbatini, and R. Ungaro, *J. Chem. Soc. Perkin Trans.*, **P2**, 395 (1996).
96JOC2407	S. Pappalardo and G. Ferguson, *J. Org. Chem.*, **61**, 2407 (1996).
96JOC5155	A. Marra, A. Dondoni, and F. Sansone, *J. Org. Chem.*, **61**, 5155 (1996).
96T639	I. Bitter, A. Grun, A. Szollosy, G. Horvath, B. Agai, and L. Toke, *Tetrahedron*, **52**, 639 (1996).

96TL5469	S. J. Meunier and R. Roy, *Tetrahedron Lett.*, **37**, 5469 (1996).
96TL5889	S. Pellet-Rostaing, J. B. Regnouf-de-Vains, and R. Lamartine, *Tetrahedron Lett.*, **37**, 5889 (1996).
96TL6311	J. B. Regnouf-de-Vains and R. Lamartine, *Tetrahedron Lett.*, **37**, 6311 (1996).
97AGE2497	R. G. Khoury, L. Jaquinod, K. Aoyagi, M. M. Olmstead, A. J. Fisher, and K. M. Smith, *Angew. Chem. Int. Ed. Engl.*, **36**, 2497 (1997).
97CCR93	C. Wieser, C. B. Dieleman, and D. Matt, *Coord. Chem. Rev.*, **165**, 93 (1997).
97ICA309	A. T. Yordanov and D. M. Roundhill, *Inorganica Chim. Acta*, **264**, 309 (1997).
97JA2948	P. Molenveld, S. Kapsabelis, J. F. C. Engbersen, and D. N. Reinhoudt, *J. Am. Chem. Soc.*, **119**, 2948 (1997).
97JCS(P2)2461	J.-B. Regnouf-de-Vains, M. Salle, and R. Lamartine, *J. Chem. Soc. Perkin Trans.*, **P2**, 2461 (1997).
97RTC363	R. H. Vreekamp, W. Verboom, and D. N. Reinhoudt, *Recl. Trav. Chim. Pays Bas*, **115**, 363 (1997).
97T16867	I. Bitter, A. Grun, L. Toke, G. Toth, B. Balazs, I. Mohammed-Ziegler, A. Grofcsik, and M. Kubinyi, *Tetrahedron*, **53**, 16867 (1997).
97TL4539	H. Ross and U. Luning, *Tetrahedron Lett.*, **38**, 4539 (1997).
97TL7801	A. Dondoni, M. Kleban, and A. Marra, *Tetrahedron Lett.*, **38**, 7801 (1997).
98MI1	C. D. Gutsche, Calixarenes Revisited, The Royal Society of Chemistry, Cambridge (1998).
98AGE2732	S. Blanchard, L. Le Clainche, M.-N. Rager, B. Chansou, J.-P. Tuchagues, A. F. Duprat, Y. Le Mest, and O. Reinaud, *Angew. Chem. Int. Ed.*, **37**, 2732 (1998).
98JA6726	P. Molenveld, J. F. J. Engbersen, H. Kooijman, A. L. Spek, and D. N. Reinhoudt, *J. Am. Chem. Soc.*, **120**, 6726 (1998).
98JCS(FT)3097	A. F. Danil de Namor, O. E. Piro, L. E. Pulcha Salazar, A. F. Anguilar-Cornejo, N. Al-Rawi, E. E. Castellano, and F. S. Sueros Valarde, *J. Chem. Soc. Faraday Trans.*, **94**, 3097 (1998).
98NJC1143	S. Blanchard, M. N. Rager, A. F. Duprat, and O. Reinaud, *New J. Chem.*, 1143 (1998).
98TL9171	C. Felix, H. Parrot-Lopez, V. Kalchenko, and A. W. Coleman, *Tertahedron Lett.*, **39**, 9171 (1998).
99AGE369	R. Roy and J. M. Kim, *Angew. Chem. Int. Ed. Engl.*, **38**, 369 (1999).
99AGE525	J. J. Gonzalez, P. Prados, and J. de Mendoza, *Angew. Chem. Int. Ed. Engl.*, **38**, 525 (1999).
99AGE3189	P. Molenveld, J. F. J. Engbersen, and D. N. Reinhoudt, *Angew. Chem. Int. Ed. Engl.*, **38**, 3189 (1999).
99ICC44	S. Pellet-Rostaing, J. B. Regnouf-de-Vains, R. Lamartine, and B. Fenet, *Inorganica Chem. Commun.*, **2**, 44 (1999).
99ICA142	A. F. Danil de Namor, O. E. Piro, E. E. Castellano, A. F. Aguilar-Cornejo, and F. J. Sueros Valarde, *Inorganica Chim. Acta*, **285**, 142 (1999).
99JCS(P2)1749	M. Larsen, F. C. Krebs, N. Harrit, and M. Jorgensen, *J. Chem. Soc. Perkin Trans.*, **P2**, 1749 (1999).
99JOC3896	P. Molenveld, W. M. G. Stikvoort, H. Kooijman, A. L. Spek, J. F. G. Engbersen, and D. N. Reinhoudt, *J. Org. Chem.*, **64**, 3896 (1999).

99JOC6337   P. Molenveld, J. F. J. Engbersen, and D. N. Reinhoudt, *J. Org. Chem.*, **64**, 6337 (1999).

99NJC977   T. Arimura, S. Ide, H. Sugihara, S. Murata, and J. L. Sessler, *New J. Chem.*, **23**, 977 (1999).

99TL5949   M. Dudic, P. Lhotak, V. Kral, K. Lang, and I. Stibor, *Tetrahedron Lett.*, **40**, 5949 (1999).

99TL6383   Y. Molard, C. Bureau, H. Parrot-Lopez, R. Lamartine, and J. B. Regnouf-de-Vains, *Tetrahedron Lett.*, **40**, 6383 (1999).

99TL8245   M. Kawaguchi, I. Ikeda, I. Hamachi, and S. Shinkai, *Tetrahedron Lett.*, **40**, 8245 (1999).

2000MI811   L. Le Clainche, Y. Rondelez, O. Seneque, S. Blanchard, M. Campion, M. Giorgi, A. F. Duprat, Y. Le Mest, and O. Reinaud, *C.R. Acad. Sci. Paris, Serie Iic, Chimie*, **3**, 811 (2000).

2000CEJ3495   M. T. Messina, P. Metrangolo, S. Pappalardo, M. F. Parisi, T. Pilati, and G. Resnati, *Chem. Eur. J.*, **6**, 3495 (2000).

2000CEJ4218   Y. Rondelez, O. Seneque, M. N. Rager, A. F. Duprat, and R. Reinaud, *Chem. Eur. J.*, **6**, 4218 (2000).

2000CP359   A. C. Fantoni and J. Maranon, *Chem. Phys.*, **262**, 359 (2000).

2000EJI683   J. O. Dalbavie, J. B. Regnouf-de-Vains, R. Lamartine, S. Lecocq, and M. Perrin, *Eur. J. Inorg. Chem.*, 683 (2000).

2000IC3436   L. Le Clainche, M. Giorgi, and O. Reinaud, *Inorganica Chem.*, **39**, 3436 (2000).

2000JCS(CC)2369 V. Sidorov, F. W. Kotch, M. El-Kouedi, and J. T. Davis, *J. Chem. Soc. Chem. Commun.*, 2369 (2000).

2000JCS(DT)2721 J. B. Cooper, M. G. B. Drew, and P. D. Beer, *J. Chem. Soc. Dalton Trans.*, 2721 (2000).

2000JA6183   O. Seneque, M. N. Rager, M. Giorgi, and O. Reinaud, *J. Am. Chem. Soc.*, **122**, 6183 (2000).

2000MJ75   Y. Liu, B. T. Zhao, L. X. Chen, and X. W. He, *Microchem. J.*, **65**, 75 (2000).

2000MJ129   L. Chen, X. Zeng, H. Zu, X. He, and Z. Zhang, *Microchem. J.*, **65**, 129 (2000).

2000NJC841   I. A. Bagatin and H. E. Toma, *New J. Chem.*, **24**, 841 (2000).

2000T1883   J. Budka, M. Tkadlecova, P. Lhotak, and I. Stibor, *Tetrahedron*, **56**, 1883 (2000).

2000T3121   A. P. Merchand, H. S. Chong, M. Takhi, and T. D. Power, *Tetrahedron*, **56**, 3121 (2000).

2001EJI2597   O. Seneque, Y. Rondelez, L. de Clainche, C. Inisan, M. N. Rager, M. Giorgi, and O. Reinaud, *Eur. J. Inorg. Chem.*, 2597 (2001).

2001JA8442   O. Seneque, M. N. Rager, M. Giorgi, and O. Reinaud, *J. Am. Chem. Soc.*, **123**, 8442 (2001).

2001JA10744   T. Arimura, S. Ide, Y. Suga, T. Nishioka, S. Murata, M. Tachiya, T. Nagamura, and H. Inoue, *J. Am. Chem. Soc.*, **123**, 10744 (2001).

2001JCS(CC)984   O. Seneque, M. Giorgi, and O. Reinaud, *J. Chem. Soc. Chem. Commun.*, 984 (2001).

2001JCS(DT)392   J. M. Cooper, M. G. B. Drew, and P. D. Beer, *J. Chem. Soc. Dalton Trans.*, 392 (2001).

2001JIPMC47   T. Tuntulani, G. Tumcharern, and V. Ruangpornvisuti, *J. Inclus. Phenom. Macrocl. Chem.*, **39**, 47 (2001).

2001JOC1002   Z. Zhong, A. Ikeda, M. Ayabe, S. Shinkai, S. Sakamoto, and K. Yamaguchi, *J. Org. Chem.*, **66**, 1002 (2001).

2001TL2681	J. B. Regnouf-de-Vains, J. O. Dalbavie, R. Lamartine, and B. Fenet, *Tetrahedron Lett.*, **42**, 2681 (2001).
2001TL2799	N. Psychogios and J. B. Regnouf-de-Vains, *Tetrahedron Lett.*, **42**, 2799 (2001).
2001TL3295	A. Dondoni, X. Hu, A. Marra, and D. Banks, *Tetrahedron Lett.*, **42**, 3295 (2001).
2001TL3595	A. Hamdi, R. Abidi, M. T. Ayadi, P. Thuery, M. Nierlich, Z. Asfari, and J. Vicens, *Tetrahedron Lett.*, **42**, 3595 (2001).
2001TL6179	C. Zeng, Y. Tang, Q. Zheng, L. Huang, B. Xin, and Z. Huang, *Tetrahedron Lett.*, **42**, 6179 (2001).
2001TL7837	G. Dopsil and J. Schatz, *Tetrahedron Lett.*, **42**, 7837 (2001).
2001TL8177	F. Oueslati, I. Dumazet-Bonnamour, and R. Lamartine, *Tetrahedron Lett.*, **42**, 8177 (2001).
2002AGE1044	Y. Rondelez, G. Bertho, and O. Reinaud, *Angew. Chem. Int. Eng.*, **41**, 1044 (2002).
2002JA1334	Y. Rondelez, M. N. Rager, A. Duprat, and O. Reinaud, *J. Am. Chem. Soc.*, **124**, 1334 (2002).
2002JOC4722	A. Dondoni, M. Kleban, X. Hu, A. Marra, and D. Banks, *J. Org. Chem.*, **67**, 4722 (2002).
2002OL2129	D. Jokiv, Z. Asfari, and J. Weiss, *Org. Lett.*, **4**, 2129 (2002).
2002OL2901	M. O. Vysotsky, V. Bohmer, F. Wurthner, C. C. You., and K. Rissanen, *Org. Lett.*, **4**, 2901 (2002).
2002OL3207	T. Nabeshima, T. Saiki, and K. Sumitomo, *Org. Lett.*, **4**, 3207 (2002).
2002T2647	X. Zeng, L. Weng, L. Chen, F. Xu, Q. Li, X. Leng, X. He, and Z. Z. Zhang, *Tetrahedron*, **58**, 2647 (2002).
2002T9019	G. T. Hwang and B. H. Kim, *Tetrahedron*, **58**, 9019 (2002).
2002TL77	N. Psychogios and J. B. Regnouf-de-Vains, *Tetrahedron Lett.*, **43**, 77 (2002).
2002TL3883	Y. H. Kim, N. R. Cha, and S.-K. Chang, *Tetrahedron Lett.*, **43**, 3883 (2002).
2002TL6367	S. J. Kim and B. Y. Kim, *Tetrahedron Lett.*, **43**, 6367 (2002).
2002TL7691	N. Psychogios and J. B. Regnouf-de-Vains, *Tetrahedron Lett.*, **43**, 7691 (2002).
2002TL8191	H. Iwamoto, Y. Yukimasa, and Y. Fukazawa, *Tetrahedron Lett.*, **43**, 8191 (2002).
2003A92	H.-Yi. Zhang, H. Wang, and Y. Liu, *Arkivoc*, 92 (2003).
2003AGE187	K. A. Nielsen, J. O. Jeppesen, E. Levillain, and J. Becher, *Angew. Chem. Int. Engl.*, **42**, 187 (2003).
2003EJO1475	A. Casnati, F. Sansone, A. Sartori, L. Prodi, M. Montalti, N. Zaccheroni, F. Ugozzoli, and R. Ungarro, *Eur. J. Org. Chem.*, 1475 (2003).
2003IC3160	R. Dorta, L. J. W. Shimon, H. Rozenberg, Y. Ben-David, and D. Milstein, *Inorganica Chem.*, **42**, 3160 (2003).
2003ICC288	I. A. Bagatin, E. S. de Souza, A. S. Ito, and H. E. Toma, *Inorganica Chem. Commun.*, **6**, 288 (2003).
2003JA15140	F. W. Kotch, V. Sidorov, Y.-F. Lam, K. J. Kayser, H. Li, M. S. Kaucher, and J. T. Davis, *J. Am. Chem. Soc.*, **125**, 15140 (2003).
2003NJC236	S. Simaan, K. Agbaria, I. Thondorf, and S. E. Biali, *New J. Chem.*, **27**, 236 (2003).
2003NJC644	F. Oueslati, I. Dumazet-Bonnamour, and R. Lamartine, *New J. Chem.*, **27**, 644 (2003).

2003OL149	M. Dudic, P. Lhotak, I. Stibor, K. Lang, and P. Proskova, *Org. Lett.*, **5**, 149 (2003).
2003SC277	R. F. Ziessel, L. J. Charbonniere, M. Cesario, T. Prange, M. Guarbigli, A. Roda, A. V. Dorsselaer, and H. Nierengarten, *Sup. Chem.*, **15**, 277 (2003).
2003SC327	Z.-S. Wang, G.-Y. Lu, X. Guo, and H.-M. Wu, *Sup. Chem.*, **15**, 327 (2003).
2003T2539	C.-C. Zeng, Qi.-Yu. Zheng, Ya.-L. Tang, and Z.-T. Huang, *Tetrahedron*, **59**, 2539 (2003).
2003T2409	M. Dudic, P. Lhotak, H. Petrickova, I. Stibor, K. Lang, and J. Sykora, *Tetrahedron*, **59**, 2409 (2003).
2003Talanta553	J. Lu, X. He, X. Zeng, Q. Wan, and Z. Zhang, *Talanta*, **59**, 553 (2003).
2003TL5299	E. J. Kim, J. I. Choe, and S. K. Chang, *Tetrahedron Lett.*, **44**, 5299 (2003).
2003TL5415	G. Arena, A. Contino, E. Longo, D. Sciotto, C. Sgarlata, and G. Spoto, *Tetrahedron Lett.*, **44**, 5415 (2003).
2004A31	Y. Yu. Morzerin, T. A. Pospelova, T. V. Gluhareva, and A. I. Matern, *Arkivoc*, 31 (2004).
2004EJI1817	O. Seneque, M. Campion, M. Giorgi, Y. L. Mest, and O. Reinaud, *Eur. J. Inorg. Chem.*, 1817 (2004).
2004EJI2514	N. Psychogios, J. B. Regnouf-de-Vains, and H. M. Stoeckli-Evans, *Eur. J. Inorg. Chem.*, 2514 (2004).
2004EJO607	M. Frank, G. Maas, and J. Schatz, *Eur. J. Org. Chem.*, 607 (2004).
2004IC7532	C. M. Jinn and J. M. Shreeve, *Inorganica Chem.*, **43**, 7532 (2004).
2004IC3965	P. D. Beer, F. Szemes, P. Passaniti, and M. Maestri, *Inorganica Chem.*, **43**, 3965 (2004).
2004JA16296	K. A. Nielsen, W. S. Cho, J. O. Jeppesen, V. M. Lynch, J. Becher, and J. L. Sessler, *J. Am. Chem. Soc.*, **126**, 16296 (2004).
2004JCS(DT)3205	G. Arena, R. P. Bonomo, A. Contino, C. Sgarlata, G. Spoto, and G. Tabbi, *J. Chem. Soc. Dalton Trans.*, 3205 (2004).
2004JCS(CC)1812	G. Arena, A. Contino, E. Longo, C. Sgarlata, G. Spoto, and V. Zito, *J. Chem. Soc. Chem. Commun.*, 1812 (2004).
2004JMC1804	S. Rizzo, F. Sannicolo, T. Benincori, G. Schiavon, S. Zecchin, and G. Zotti, *J. Mater. Chem.*, **14**, 1804 (2004).
2004OL1091	E. Botana, K. Nattinen, P. Prados, K. Rissanen, and J. D. Mendoza, *Org. Lett.*, **6**, 1091 (2004).
2004NJC1335	O. G. Barton, M. Schmidtmann, A. Muller, and J. Mattay, *New J. Chem.*, **24**, 1335 (2004).
2005AHC67	S. Kumar, D. Paul, and H. Singh, in *"Advances in Heterocyclic Chemistry"* (A. R. Katritzky, ed.), Vol. 89, p. 67 (2005).
2005EJO2330	S. Konrad, C. Nather, and U. Luning, *Eur. J. Org. Chem.*, 2330 (2005).
2005EJO2338	A. Casnati, N. D. Ca, M. Fontanella, F. Sansone, F. Ugozzoli, R. Ungaro, K. Liger, and J. F. Dozol, *Eur. J. Org. Chem.*, 2338 (2005).
2005IC2836	J. P. Tremblay-Morin, S. Faure, D. Samar, C. Stern, R. Guilard, and P. D. Harvey, *Inorganica Chem.*, **44**, 2836 (2005).
2005JOC6254	B. T. Zhao, M. J. Blesa, N. Mercier, F. Le Derf, and M. Salle, *J. Org. Chem.*, **70**, 6254 (2005).
2005JOC8443	M. G. J. ten Cate, D. N. Reinhoudt, and M. C. Calama, *J. Org. Chem.*, **70**, 8443 (2005).

2005JPC(B)14735	A. F. Danil de Namor, A. Aguilar-Cornejo, R. Soualhi, M. Shehab, K. B. Nolan, N. Ouazzani, and L. Mandi, *J. Phys. Chem. B*, **109**, 14735 (2005).
2005NJC1164	B. T. Zhao, M. J. Blesa, N. Mercier, F. Le Derf, and M. Salle, *New J. Chem.*, **29**, 1164 (2005).
2005OBC3727	M. G. J. ten Cate, M. Omerovic, G. V. Oshovsky, M. C. Calama, and D. N. Reinhoudt, *Org. Biomol. Chem.*, **3**, 3727 (2005).
2005T12282	V. Boyko, R. Rodik, O. Danylyuk, L. Tsymbal, Y. Lampeka, K. Suwinska, J. Lipkowski, and V. Kalchenko, *Tetrahedron*, **61**, 12282 (2005).
2006CC320	T. Ceyhan, A. Altindal, A. R. Ozkaya, M. K. Erbil, B. Salih, and O. Bekaroglu, *Chem. Commun.*, 320 (2006).
2006CC2233	J. Lyskawa, M. Salle, J. Y. Balandier, F. L. Derf, E. Levillain, M. Allain, P. Viel, and S. Palacin, *Chem. Commun.*, 2233 (2006).
2006CC3924	D. Coquiere, J. Marrot, and O. Reinaud, *Chem. Commun.*, 3924 (2006).
2006EJO4717	S. Konrad, M. Bolte, C. Nather, and U. Luning, *Eur. J. Org. Chem.*, 4717 (2006).
2006JOC7546	A. Dondoni and A. Marra, *J. Org. Chem.*, **71**, 7546 (2006).
2006JOC9589	X. Zeng, A. S. Batsanov, and M. R. Bryce, *J. Org. Chem.*, **71**, 9589 (2006).
2006SC219	I. M. Ziegler, A. Hamdi, R. Abidi, and J. Vincens, *Sup. Chem.*, **18**, 219 (2006).
2006T2901	H. M. Chawla, S. P. Singh, and S. Upreti, *Tetrahedron*, **62**, 2901 (2006).
2006T9758	H. M. Chawla, S. P. Singh, and S. Upreti, *Tetrahedron*, **62**, 9758 (2006).
2006T7846	X. H. Sun, C. S. Chan, M. S. Wong, and W. Y. Wong, *Tetrahedron*, **62**, 7846 (2006).
2006TL181	S. Ishahara and S. Shinkai, *Tetrahedron Lett.*, **47**, 181 (2006).
2006TL1895	M. Mourer and J. B. Regnouf-de-Vains, *Tetrahedron Lett.*, **47**, 1895 (2006).
2006TL3245	G. M. L. Consoli, G. G. E. Galante, F. Cunsolo, and C. Geraci, *Tetrahedron Lett.*, **47**, 3245 (2006).
2007CP283	A. Koca, T. Ceyhan, M. K. Erbin, A. R. Ozkaya, and O. Bekaroglu, *Chem. Phys.*, **340**, 283 (2007).
2007JMS48	J. Klimentova and P. Vojtisek, *J. Mol. Struct.*, **826**, 48 (2007).
2007OL3363	K. C. Chang, I. H. Su, A. Senthilvelan, and W. S. Chung, *Org. Lett.*, **9**, 3363 (2007).
2007OL4987	B. Colasson, M. Save, P. Milko, J. Roithova, D. Sehroder, and O. Reinard, *Org. Lett.*, **9**, 4987 (2007).
2007SC159	M. Durmaz, S. Bozkurt, A. Sirit, and M. Yilmez, *Sup. Chem.*, **19**, 159 (2007).
2007T4951	G. Gattuso, A. Pappalardo, M. F. Parisi, I. Pisagatti, F. Crea, R. Liantonio, P. Metrangolo, W. Navarrini, G. Resnati, T. Pilati, and S. Pappalardo, *Tetrahedron*, **63**, 4951 (2007).
2007T6339	A. Dondoni and A. Marra, *Tetrahedron*, **63**, 6339 (2007).
2007T10758	G. M. L. Consoli, G. Granata, E. Galante, I. D. Silvestro, L. Salafia, and C. Geraci, *Tetrahedron*, **63**, 10758 (2007).
2007TL681	S. O. Malley, N. Alhashimy, J. O. Mahomy, A. Kieran, M. Pryce, and K. Nolam, *Tetrahedron Lett.*, **48**, 681 (2007).

2007TL5397	K. C. Song, M. G. Choi, D. H. Ryu, K. N. Kim, and S. K. Chang, *Tetrahedron Lett.*, **48**, 5397 (2007).
2007TL7274	K. C. Chang, I. H Su, G. H. Lee, and W. S. Chung, *Tetrahedron Lett.*, **48**, 7274 (2007).
2007TL7974	G. M. L. Consoli, G. Geranata, D. Garozzo, T. Mecca, and C. Geraci, *Tetrahedron Lett.*, **48**, 7974 (2007).
2007TL8274	G. Arena, A. Contino, G. Maccarrone, D. Sciotto, and C. Sgarlata, *Tetrahedron Lett.*, **48**, 8274 (2007).
2008AHC123	S. Kumar, N. Kaur, and H. Singh, in *"Advances in Heterocyclic Chemistry"* (A. R. Katritzky, ed.), Vol. 96, p. 123 (2008).
2008CJC170	H. J. Shi, X. F. Shi, T. M. Yao, and L. N. Ji, *Chin. J. Chem.*, **26**, 170 (2008).

CHAPTER **5**

Hetarylazomethine Metal Complexes

**Alexander D. Garnovskii[*], Alexander P. Sadimenko[**],
Igor S. Vasilchenko[*], Dmitry A. Garnovskii[*],
Evgeniya V. Sennikova[*] and Vladimir I. Minkin[*]**

[*] Institute of Physical and Organic Chemistry of the Southern Federal University, Stachka Avenue, 194/2, Rostov on Don 344090, Russia
[**] Department of Chemistry, University of Fort Hare, Alice 5700, Republic of South Africa

Advances in Heterocyclic Chemistry, Volume 97
ISSN 0065-2725, DOI 10.1016/S0065-2725(08)00205-5

1. INTRODUCTION

Among various transformations of heterocyclic systems, reactions with metal-containing species (metals in the zero oxidation state, salts, and carbonyls) leading to the formation of the coordination compounds are of special interest (73RCR89, 87MI1, 93CCR237, 94RJCC74, 98AHC1, 98CCR31, 00AHC157, 03MI1, 03MI2, 05AHC126, 05CCR1857, 07AHC(93)185, 07AHC(94)107, 08AHC221). This reactivity pattern can be rather clearly illustrated by the interaction of the aforementioned metal sources with cyclic and acyclic azomethines. Complexes of acyclic azomethines involve chelates of Schiff bases (93CCR1, 97CCR191, 98CCR31, 02RCR943, 03MI3, 04CCR1717, 05RCR193, 08CCR1871), β-aminovinylketones (93CCR1, 03MI3, 07M175), and β-aminovinylimines (02CRV3031) containing heterocyclic moieties. Coordination compounds of cyclic azomethines with chelating six-membered metal rings and XM$\cdots$N constituents (X = NR, O, S, Se, Te) include complexes of 2-amino- (hydroxy-, hydrochalcogeno-) phenyl derivatives of azoles and azines (98AHC1, 00AHC157, 01CCR79, 07CHC1359). In the multinuclear coordination compounds under the present review, the heterocyclic frameworks play the role of the aldehyde, amine, and/or adduct-forming constituents. In the oligonuclear structures, heterocycles may also play the role of the intermetal bridges. The data collected in the present review are generalized and systematized with emphasis on the specified functions in the structure of azomethine ligands and heterocyclic frameworks.

2. HETARYLAZOMETHINE LIGANDS

2.1 Types of ligating azomethines

The most widespread acyclic azomethines include compounds **1–3** (81CJC1205, 00ARK382). Compounds **1** and **2** involve a five-membered intramolecular hydrogen bond (IHB). A rare case of a seven-membered IHB is given by **3** (00ARK382). Ligating compounds of the second type **4** are azomethines formed by heterocyclic aldehydes and aromatic amino-(hydroxy-, mercapto-) amines with the same size IHB. Schiff bases of the third type **5** include heterocyclic and aromatic constituents as well as heterocyclic and amine frameworks with a six-membered IHB. These may be exemplified by the azomethines of the monoheteroatomic rings (commonly, pyrrole, furan, and thiophene), azoles, and azines (00AHC157).

Chelates with a five-membered coordination unit are obtained from the ligating compounds of the first type. Hetarylazomethines of the second and third types lead to complexes with both five- and six-membered chelate units (05RCR193). Cyclic analogues of **5** are presented mainly by 2-amino- (hydroxy-, mercapto-) derivatives of azoles and azines (Equation (1); R = H, Alk, Ar) as well as 2-oxo- (thio-) indanone frameworks with six-membered rings (Equation (2); 07CHC1359). Such ligating systems are the basis for the metal chelates with six-membered coordination units. Compounds of type **1–5** are predominantly bidentate ligands and form mononuclear metal chelates. However, they can also serve as a source for di- and oligonuclear coordination structures.

$$(1)$$

$$(2)$$

2.2 Synthesis of hetarylazomethines

The basic synthetic method for the preparation of hetarylazomethines (79MI1) may be presented by Equations (3) (R = H, Alk, Ar, Het) and (4), where heterocyclic or aromatic aldehydes and amines are applied. The first method leads to the chelating **1** and **5**, whereas the second allows the preparation of ligands similar to **4**. The less common synthetic method for the preparation of hetarylazomethines, termed the Ehrlich–Sachs reaction, is expressed by Equation (5) and involves condensation of compounds containing a methyl-active group with nitroso-derivatives (63CRV489, 79MI1). Such a reaction is known for 9-methylacridine when Ar = $C_6H_4NMe_2$-p (63BCJ1477). Activation of the methyl group in 2-methylpyridine may be achieved by N-oxidation followed by the reaction expressed by Equation (6) leading to a hetarylazomethine (63JCS4600). Obviously, interaction of the quaternary salts of 2-methyl derivatives of azoles and azines should lead to Schiff bases followed by the thermal transformation described by Equation (7) (where X = NTs, O, S and R = H, Alk, Ar). However, to the best of our knowledge, there is no experimental proof.

$$\text{Het}-\overset{O}{\underset{H}{\text{C}}} + R-NH_2 \longrightarrow \text{Het}-\underset{H}{\text{C}}=N-R + H_2O \qquad (3)$$

$$Ar-\underset{H}{\overset{O}{\diagdown}} + Het-NH_2 \longrightarrow Ar-\underset{H}{C}=N-Het + H_2O \qquad (4)$$

$$Het-Me + O=NAr \longrightarrow Het-\underset{H}{C}=NAr + H_2O \qquad (5)$$

$$\underset{\underset{O^-}{N^+}}{\overset{}{\bigcirc}}-Me + O=NAr \longrightarrow \underset{\underset{O^-}{N^+}}{\overset{}{\bigcirc}}-\underset{H}{C}=N-Ar + H_2O \qquad (6)$$

$$(7)$$

2.3 Tautomerism and structure of hetarylazomethine derivatives

Three kinds of tautomeric equilibria are characteristic for the ligating hetarylazomethines; prototropic annular, prototropic photo- and thermal, and ring-chain. They are described in detail in a series of review publications (63AHC339, 76AHC(S)1, 95UK705, 96AHC1, 96MCJ209, 00AHC1, 00AHC85, 00AHC157, 06AHC1).

2.3.1 Tautomerism of the ligating amino- (hydroxy-, mercapto-) hetarylazomethines

For ligands of type 5, prototropic annular tautomerism is common (00AHC157). For example, for 2-hydroxy-3-aldimino derivatives of benzothiophene, the tautomeric equilibrium (Equation (8)) occurs where a ketoamino tautomer predominates (74JPR971, 00AHC85). A similar tautomerism (Equation (9)) is observed in a series of 3-hydroxy-(mercapto-, hydroseleno-)-2-aldimine derivatives of monoheteroatomic five-membered systems (A = NR, O, S, Se) (93CCR1). It is also observed for the derivatives of indazole (X = NR) (85KGS921), benzofuran (X = O), and benzothiophene (X = S) (92IZV917) as confirmed by X-ray structural data. An equilibrium of (E)- and (Z)-isomers has been studied (90ZOR2389, 96MCJ209). The rate of (E)–(Z) isomerization decreases in the following sequence of A: NMe > S > O. In the case of X = O and S, (E)- and (Z)-conformers were isolated in the crystalline state, which confirms their existence. In contrast to the situation expressed by

Equation (9), for the azomethine derivatives in Equation (10) (R = H, t-Bu, Br, NO$_2$) containing the aldimine fragment in an aromatic nucleus, the enolimine tautomer predominates (04RKZ30).

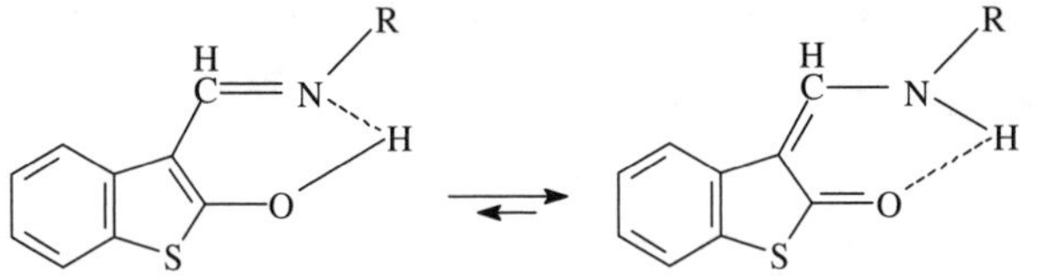

(8)

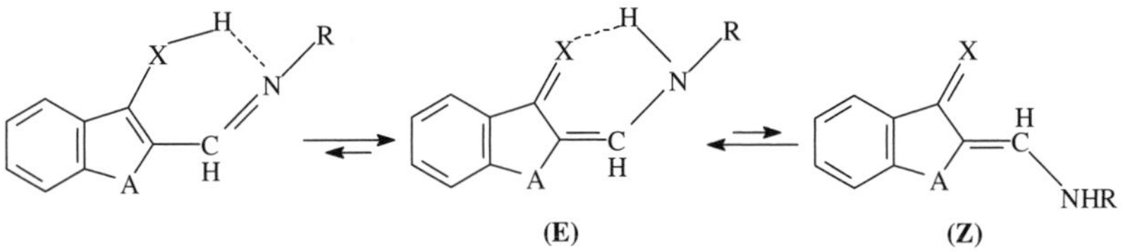

(E)   (Z)

(9)

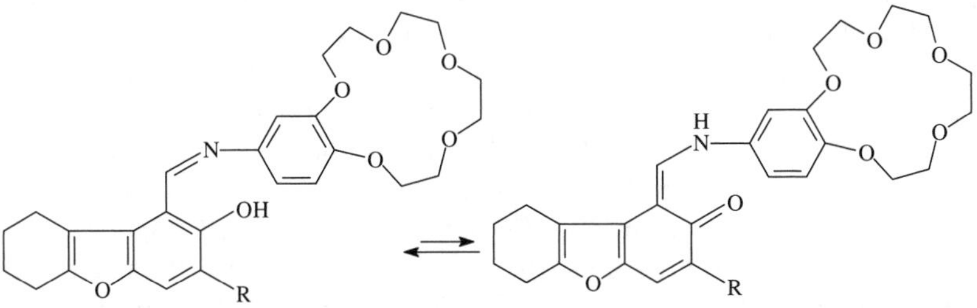

(10)

For azomethine derivatives of the azole series (X = NH, O, S, Se), aminomethylene oxo-, thio-, and seleno-tautomeric forms predominate in equilibria described by Equation (11) (96MCJ209). Illustrative examples are derivatives of pyrazole (A = NR^1, B = N, D = CR$_3$) (74ZOR2210, 94ZOB657), imidazole (A = NR^1, B = N, D = CR$_3$) (78KGS1677), and triazole (A = NR^1, B = D = N) (84JHC1603). Along with these forms, the zwitterionic structure is also discussed (74ZOR2210, 92SR321, 94ZOB657).

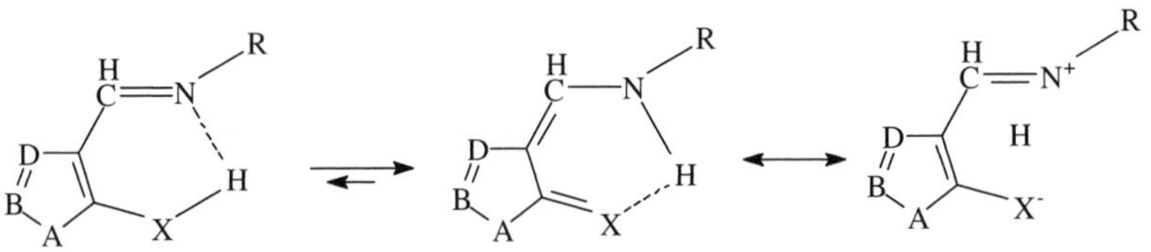

(11)

In a series of ligating pyrazole azomethines, the thio derivatives are widely represented, for which the enaminothione structures (Equation (11)) are observed both in solution (07JCC1493) and crystalline phase (00K850, 01ZK251, 02AX(E)1365, 03AX(E)430). The enamine structure was established by X-ray analysis for the salen-bis(thiopyrazolone) **6** (99EJI1393). Tautomerism and structure of the ligating azomethines with aromatic amino- (hydroxy-, mercapto-) aldehyde and heterocyclic imino fragments of type **7** (Het = azoles) have been studied in detail (92CCR67).

6   7

The general scheme of tautomerism is described by an equilibrium in Equation (12) (where $X = NR^3$, O, S; $R = H$, Alk, Ar, Het; $R^1 = H$, Alk, Ar; $R^2 = H$, Alk, Hal, NR_2, NO_2; $R^3 = H$, Alk, Ar, Ts). In contrast to the situation in Equation (11), an enolimine tautomer prevails both in solution and crystalline phase. This has been established for azomethines containing various heterocyclic O-donors at the nitrogen atom of the $C=N$ bond: antipyrine **8** ($R = H$, 3,4-*cyclo*-C_4H_4, 15-crown-5) (00ZSK1095, 01AX(C)117), carbohydrates **9** ($X = NTs$, O) (01CAR133, 02CAR79, 02CAR1477), and crown-ethers **10** (94CRV279, 00AS553). For structures of type **11** ($R = H$, Br; $X = NTs$, O), tosylaminoimine ($X = NTs$), and enolimine ($X = O$) forms are realized in the crystalline phase according to X-ray structural analysis (04DOK(395)626, 04DOK(398)62).

(12)

8   9   10

11

Azomethine ligands with a pyridine fragment are of interest and for them both enolimine and ketoamino tautomers have been structurally characterized (Equation (13)), where R = H, *cyclo*-C_4H_4 (00JMS(524)241). Among the ligating tetradentate salen-type compounds, the dihydroxy systems of type **12**, whose structure has been proven by X-ray analysis, are important (00STC361). Enolimine moieties enter the composition of tetrahydroxyl octadentate pyridine **13** (00ZN(B)1037).

$$(13)$$

12   **13**

Other hetarylazomethines containing several ligating tautomeric moieties may be illustrated by tautomeric equilibrium (14), in which *o*-tosylamino(hydroxy)aldimines containing 2-amino- (hydroxy-, mercapto-, hydroseleno-) substituted benzazoles (X = NTs, O; Y = NR; R = H, Alk, Ar) participate (07JCC1493). When Y = NR, the enolimine 2-amino-tautomer is realized. For X = O, S, and Se, the 2-keto, thio-, and selenoazole tautomers are more expressed.

(14)

2.3.2 Photo- and thermal prototropic tautomerism

This type of prototropic tautomeric transformations (Equation (15)) is typical for the ligating 2-(2′-amino- (hydroxy-, mercapto-) phenyl) derivatives (A = CH, N, NR^2, O, S; X = NR^3, O, S; R = H, Alk, Ar; R^1 = H, Alk, NO$_2$, Hal; R^2 = H, Alk, Ar; R^3 = H, Alk, Ts) heterocyclic analogues of Schiff bases (00AHC157, 07CHC1359). In equilibrium (15) for benzazole derivatives, the amino- (hydroxy-, mercapto-) tautomer prevails. The same tautomer is observed for 2-hydroxybenzazole derivatives (A = NR^2, O, S; X = O; R–R = cyclo-C$_4$H$_4$; R^1 = R^2 = H). According to X-ray structural data, this tautomer is realized in the crystalline state for 2-(2′-tosylaminophenyl)benzimidazole (07MC164). For 2-(2′-hydroxyphenyl)pyridine, the same tautomer is formed both in solution (97T11936) and crystalline phase (05ZSK388).

$$hv \qquad (15)$$

2.3.3 Ring-chain tautomerism

This type of tautomerism is principally important for the preparation of chelates with five-membered metallacycles of composition MN$_2$S$_2$ from benzothiazoline-ligating compounds (Equation (16)). Ring-chain tautomerism in a series of 2-aminophenylbenzimidazoles (Equation (17)) has been studied by dynamic NMR spectroscopy (02KGS1428, 04RKZ103).

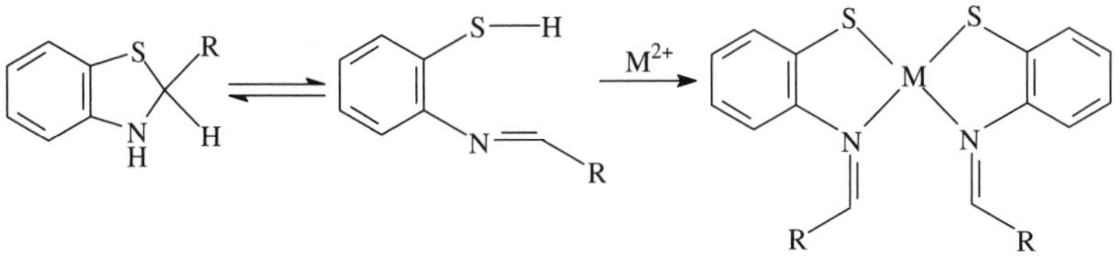

(16)

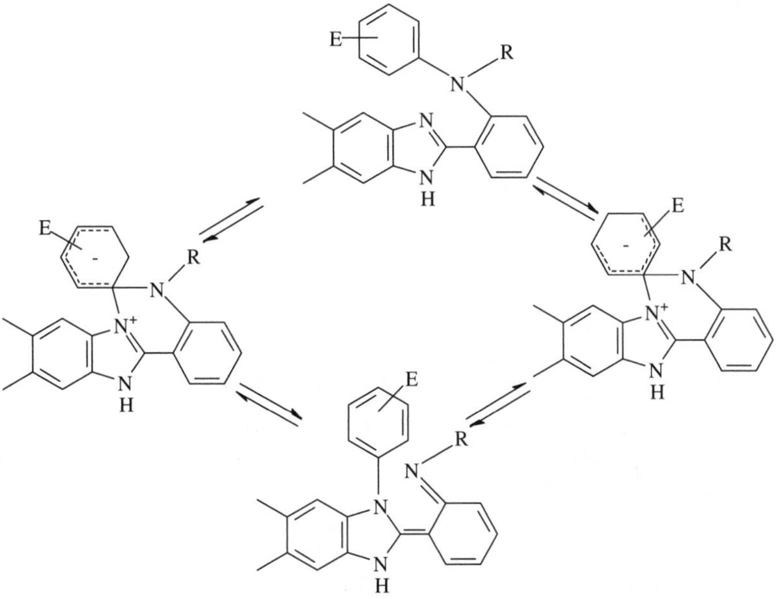

9

(17)

3. COORDINATION COMPOUNDS OF HETARYLAZOMETHINE LIGANDS

3.1 Types of metal complexes of hetarylazomethines

The basic types of complexes of ligating systems under consideration are presented by the mononuclear chelates **14–18**. Di-, oligo-, mono-, and heteronuclear complexes containing hetarylazomethine ligands are also known. They are considered as a function of heterocyclic azomethine fragments (fundamental five-membered heterocycles – pyrrole, furan, thiophene, selenophene; azoles and azines) (03MI2).

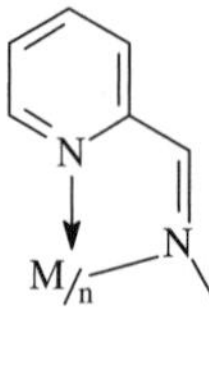

14 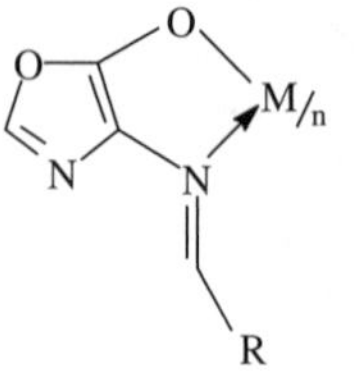   **15** 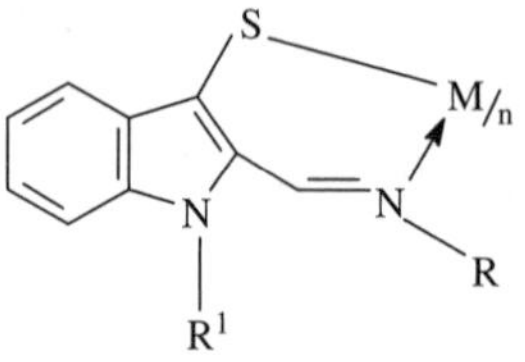   **16**

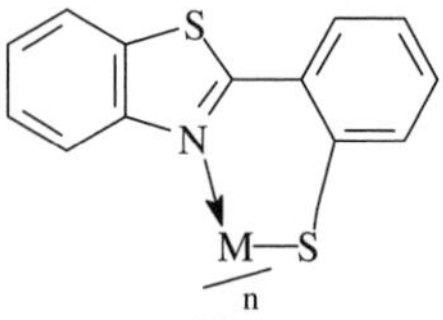

17 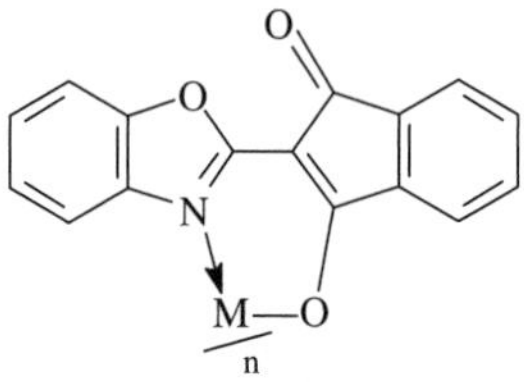   **18**

3.2 Synthesis of hetarylazomethine complexes

To prepare metal chelates **14–18**, methods of direct interaction of the ligating compounds and metal salts (Equations (18–20)), template synthesis (Equation (21)), ligand (Equation (22)) and metal (Equation (23)) exchange, as well as direct synthesis on the basis of metals having zero oxidation number (Equation (24)) are used (99MI1, 03MI2, 07JCC1435). Direct interaction (Equation (18)) is widely used for the preparation of chelates based on pyrrole-2-aldehydes (where L = OCOMe, Hal, NO_3). Similar synthetic transformations (Equations (19) and (20)) are applicable to prepare complexes of types **15** and **16**.

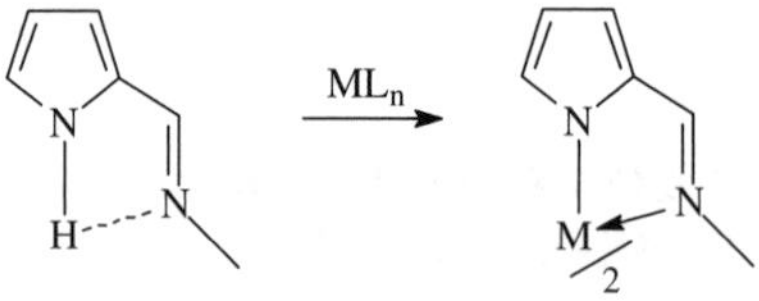

$$(18)$$

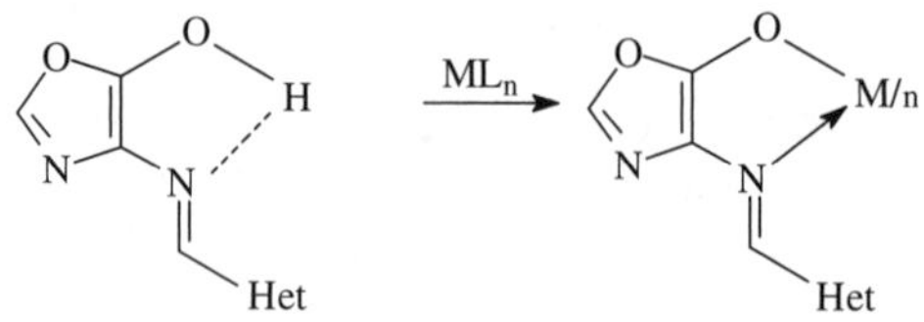

$$(19)$$

$$(20)$$

Template syntheses (Equation (21)) include two approaches (99MI2, 03MI2): synthesis with the use of a metal matrix (Equation (21a)) and modification of the ligands (Equation (21b). The principle of template synthesis is well illustrated by a [1+1]-cyclocondensation of the dialde-hyde derivatives of 1,10-phenanthroline with polyamines in the presence of $MnCl_2 \cdot 4H_2O$ to yield a series of macrocycles (08ICA1415). A similar approach was applied for the preparation of manganese(II) complexes with tripodal tetramines (00IC5878, 03ICA(355)286, 04ICA1283) and cadmium with linear triamines (07ICA2298).

$$(21)$$

Ligand-exchange reactions leading to chelates of hetaryl Schiff bases involve those with acetylacetonate (Equation (22a)) and nitrile (Equation (22b)) complexes (where $A = CH, N$; $R = Alk, Ar, cyclo\text{-}C_4H_4$; $R^1 = Alk, Ar, Het$; $R^2 = Me, Ph$).

$$(22)$$

Metal exchange is most often applied in the synthesis of complexes of hetarylazomethines from low-stable ligating systems, e.g., hydrochalco-geno azomethines and their cyclic analogues. Such syntheses are illus-trated by Equation (23) ($R = 4,5\text{-}cyclo\text{-}C_4H_4$; $X = S, Se, Te$; $M = Zn, Cd, Hg$) (76ZOB2706, 98IC2663, 99EJI1229, 99JOM(577)243).

$$(23)$$

Syntheses of chelates **18** are performed by chemical (Equation (19)) and electrochemical (Equation (24)) methods. Reactions expressed by Equations (18–21) often involve metal acetates in a weakly acidic medium (pH ~6) on refluxing in methanol. Application of nitrates and halides requires cautious alkalinization to prevent the formation of hydroxides. Reactions expressed by Equation (22a) occur in high-boiling solvents (xylene) followed by distillation of acetylacetone. Reactions illustrated by Equation (22b) occur in refluxing acetone (96MCJ19). Direct synthesis of hetarylazomethine complexes involves electrochemical transformation of a dissolving anode (99JCC219). These reactions usually occur at room temperature in acetonitrile using a platinum anode in the presence of conducting quaternary ammonium salts. In Equation (24), LH are ligands and ML_n are complexes.

$$\text{Cathode}: n\,LH + ne^- \rightarrow n\,L^- + \frac{n}{2}H_2$$
$$\text{Anode}: M^0 - ne^- \rightarrow M^{n+} \tag{24}$$
$$\text{Overall}: M^{n+} + n\,L^- \rightarrow [ML_n]$$

In the template electrochemical synthesis, several intermediate products inaccessible in chemical synthesis were isolated. Complexes in Equation (25) (M = Co, Cu, Ni, Zn, Cd) of the tridentate Schiff base (96ZOB1546, 99ZOB1533) were prepared electrochemically with pyridine 2-aldehyde. With the electrochemical approach, a metal-exchange method has been elaborated (99POL2651) that allows the transformation expressed by Equation (26) (R, R^1, R^2 = H, Alk, Ar; M = Co, Ni, Cu, Zn) under mild conditions (04POL1909). Application of electrosynthesis enhances the mobility of the protons of a weakly acidic benzimidazole moiety and allows the preparation of chelates of hetarylazomethines in Equation (27) (M = Co, Ni, Cu) (76ZOB675). Complex-formation by direct interaction of the ligating 3-hydroxyphenyl-1,2,4-oxadiazole is shown in Equation (28) (87KK869). Currently, activation methods for the direct synthesis are widely applied, i.e., Riecke metals, ultrasound and ultraviolet irradiation (00AA52, 00JOC2322, 04ICC1269, 05SR755).

$$\text{(26)}$$

$$\text{(27)}$$

$$\text{(28)}$$

3.3 Coordination compounds of the Schiff bases of pyrrole-2-aldehyde and its analogues

3.3.1 Mononuclear chelates

Trans-square planar complexes of type **19** (R = H, R^1 = *t*-Bu, M = Co, Cu) have been prepared long ago (72IC1100, 72IC2315, 80UK1234). The most widespread are the mononuclear complexes **19** and **20**. They include structurally characterized bis-chelates **19** with various metals: cobalt (M = Co, R = H, R^1 = *t*-Bu) (72IC1100) and (M = Co, R = H, Me, R^1 = Ph, 2,4-Me$_2$C$_6$H$_3$, 2,4-*i*-Pr$_2$C$_6$H$_3$) (07IC6880), nickel (M = Ni, R = H, R^1 = C$_6$H$_2$-2,4,6-Me$_3$) (06JCS(D)5362) and (M = Ni, R = H, R^1 = Me; R = R^1 = Me; R = Me, R^1 = Et; R = Me, R^1 = *n*-Pr) (94POL2055), palladium (M = Pd, R = H, R^1 = CH$_2$Ph) (04JUCR71) and (M = Pd, R = H, R^1 = C$_6$H$_3$R^{2}R^3; R^2 = H, Me, *i*-Pr; R^3 = H, Me, *i*-Pr, *t*-Bu) (04POL1619), and copper (M = Cu, R = R^1 = H) (71AX(B)1644), (M = Cu, R^1 = H, R^1 = *n*-Pr, *t*-Bu) (72IC2315), (M = Cu, R = H, R^1 = CH$_2$Ph, CHPh$_2$) (03JCC975), (M = Cu, R = Me, R^1 = H, CH$_2$Ph) (92ICA213), and (M = Cu, R = Me, R^1 = Et) (90AX(C)1434). Coordination compounds of cobalt, nickel, palladium, and copper have a *trans*-planar structure, while the zinc complex is characterized by a tetrahedral configuration (02EJI1060).

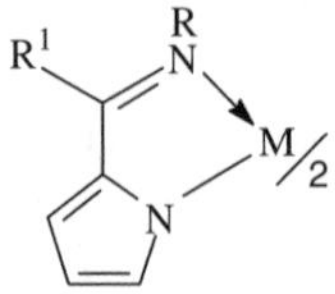

19 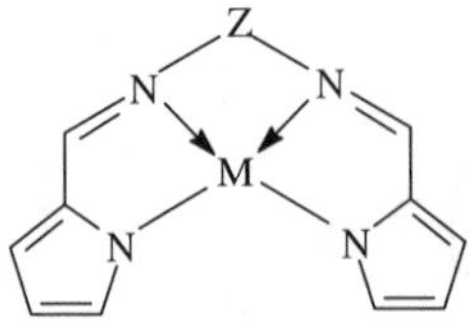   20

In contrast, for tricyclic coordination compounds **20** and other salen complexes (93CCR1, 01CRV37, 03CRV283, 03CR3071, 03MI3, 04CSR410), the *cis*-structures with various degrees of distortion of the metallacycles are observed. When $Z = (CR_2)_2$ (R = H, Me) complex compounds have practically planar structures irrespective of the electronic configuration of the metals iron (06AX(E)2668) or nickel (76CSC447, 84CL573). Such a structural situation is observed for **21** (04ICA1161) and **22** (00JOM86). In spite of the increase in the size of the internitrogen bridge Z, the tricyclic square-planar structure is retained for palladium **20** (M = Pd, Z = (CH$_2$)$_3$) (03ICA(342)229) and nearly planar for copper **20** (M = Cu, Z = (CH$_2$)$_3$, (CH$_2$)$_4$) (04EJI1478).

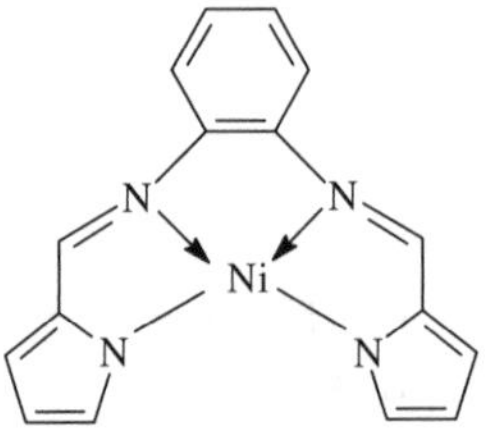

21 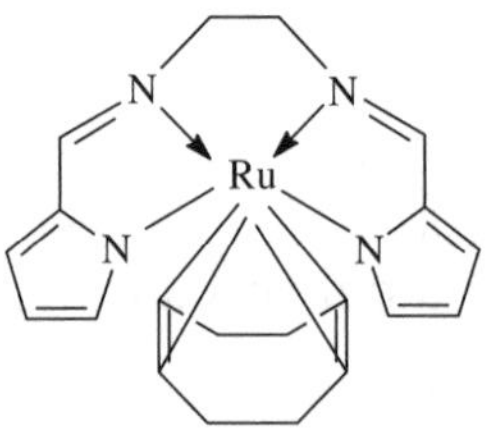   22

A series of salen-pyrrole ligands containing the coordinatively active N-donor sites in the internitrogen bridge has been synthesized and structurally characterized. These nitrogen atoms, according to the X-ray data, may or may not participate in the bonding to a metal center. Chelates of the first type are represented by **23** (M = Ni, R = H, A = ClO$_4$; M = Cu, R = H, A = NO$_3$; M = ReO$_2$Me, R = Me, A = PF$_6$) of nickel, copper (93JCSR473), and rhenium (90POL615). Other representatives are compounds containing diazo-salen ligands **24** (89POL491). Participation of the nitrogen atom of azo-salen ligands in metal bonding usually leads to an increase in the coordination number relative to that in chelates of the C-salen ligands **20–22** observed in **24** (coordination number six). In this respect, the fact that in chelate **23** only one pyrrole nitrogen atom is coordinated is unexpected. A similar effect is observed in other metal chelates where a salen-moiety does not take part in metal binding, as in coordinatively unsaturated tris-pyrrole manganese **25** (81JA241) and **26** (05PR(B)214408), and iron **27** complexes (78IC1288). Only one pyrrole fragment and one oxygen atom are coordinated in the chelate **28** (84JCS(D)2741). A limited range is presented by chelates with coordination-active R^1 substituents, increasing the metal coordination number, e.g., **29** (93JCC125). Another source of complexes with an enhanced coordination number are the adducts of type **30** (00JOM86).

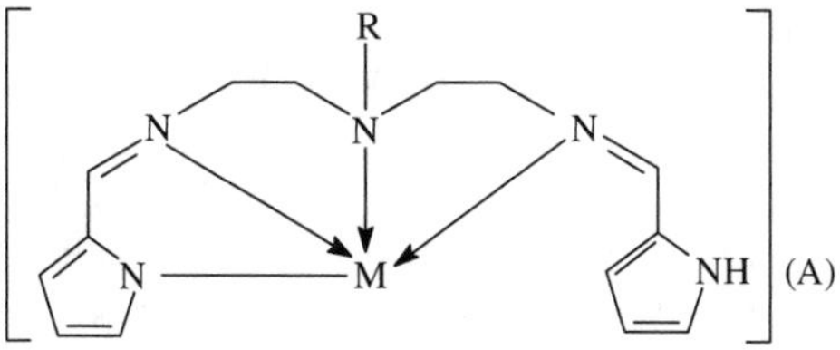

23

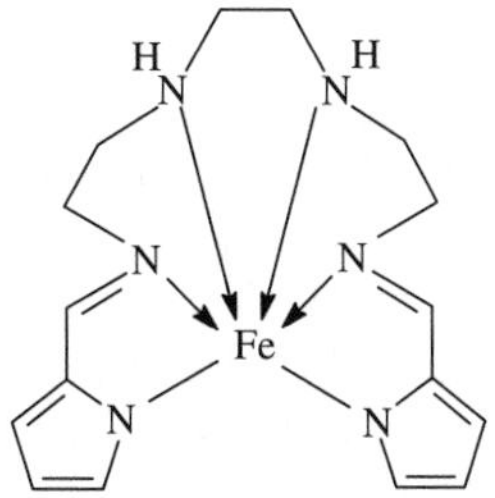

24

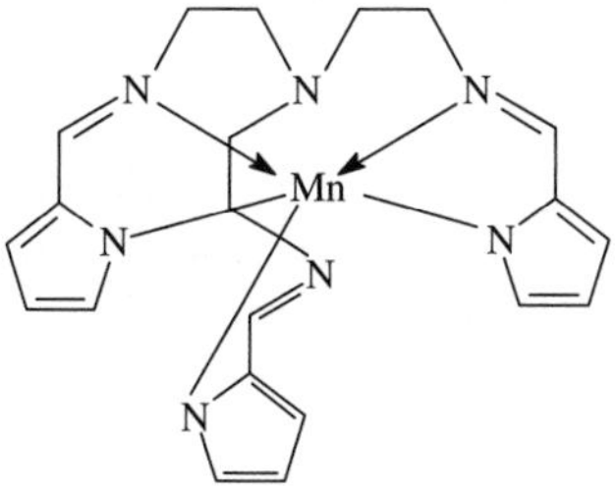

25

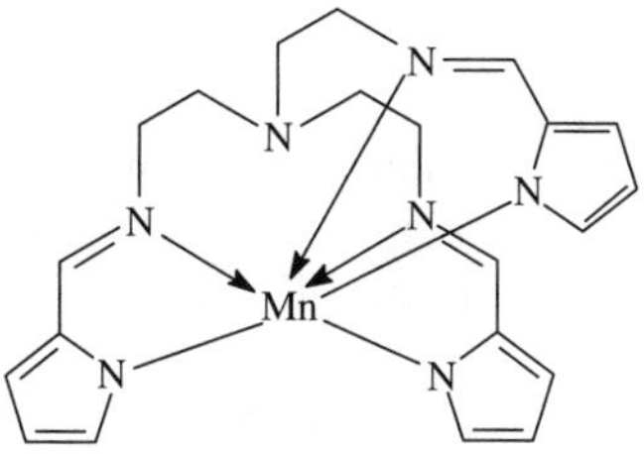

26

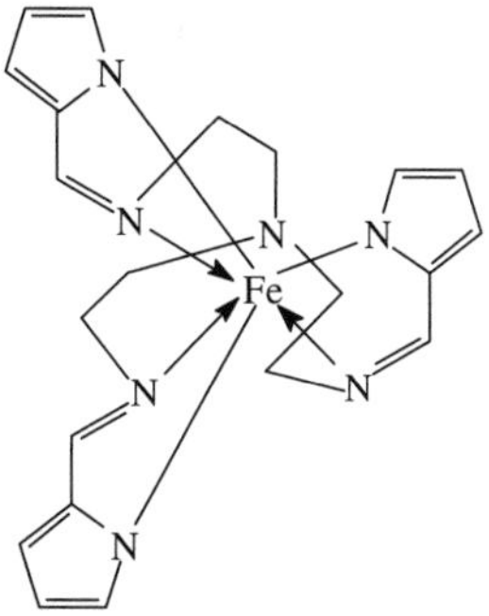

27

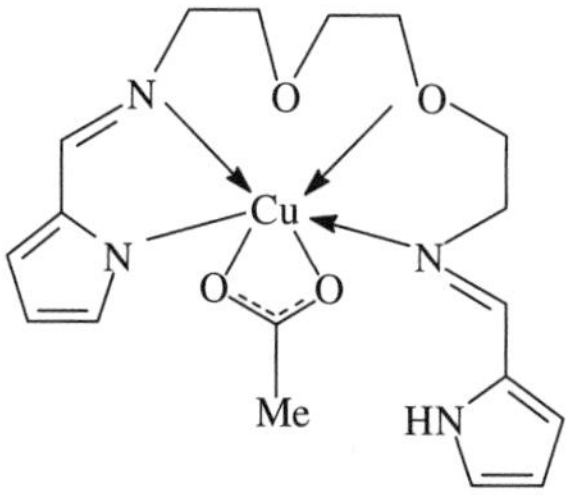

28

29   30

Examples of complexes of pyrrolyl azomethines with metals having oxidation number +3 are scarce, e.g., cobalt species **31** (92POL235). Little attention is paid to the monocyclic pyrrole azomethine chelates, although they are represented by a variety of illustrations. These include cobalt **32** (93OM3677), nickel **33** (R = H, R^1 = $C_6H_3Pr-i_2$-2,6) (03JOM(667)185), (R = Me, R^1 = Mes) (03JCS(D)4431), palladium **34** (M = Pd, R = Me, R^1 = py) (04POL1619), and aluminum **34** (M = Al, R = R^1 = Me) compounds (03JOM(667)185). Complexes of platinum **35** (03JA12674) and iridium **36** (00POL1519) are scarce, whereas those of ruthenium **37** (03ICC1140), **38** (04ZAAC91), and **39** (05OM5015), as well as rhodium **40** (Hal = Cl) (00POL1519), **40** (Hal = Br, J) (03POL2639), **41** (86ICA(112)65), **42** (98T(A)3763) are widely represented. Such a representation is due to the variation of the ligand environment, not only halide **40**, organic **37–41**, adduct-forming **35, 38, 39, 42**, but also heterocyclic **38** and **39**. Application of tris-pyrazolylborate **39** and diphosphine **42** ligands increasing the number of metallacycles is of special interest. Scarce representatives of monocyclic chelates of hafnium **43** (05OM3375), molybdenum **44** (82POL617), and zinc **45** (05CC4935) have been structurally characterized.

31   32   33

34   35   36

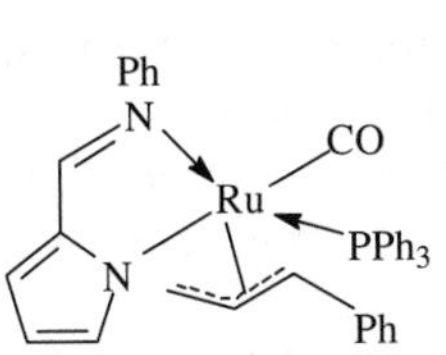

37 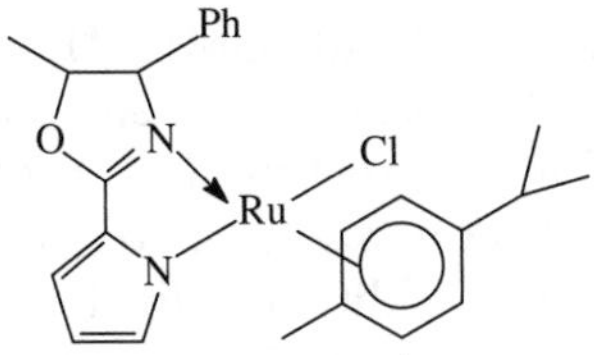   **38**

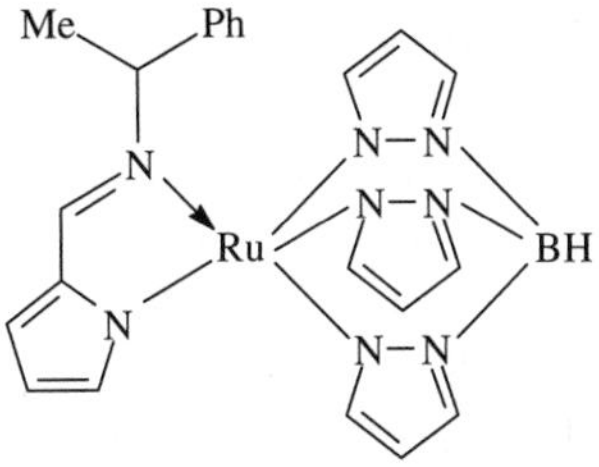

39 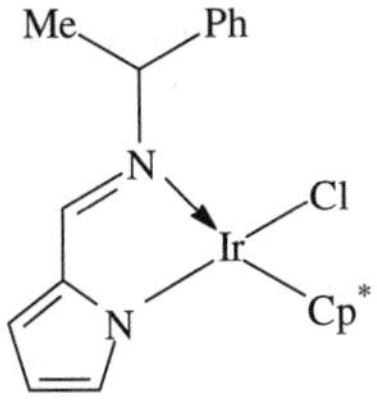   **40**

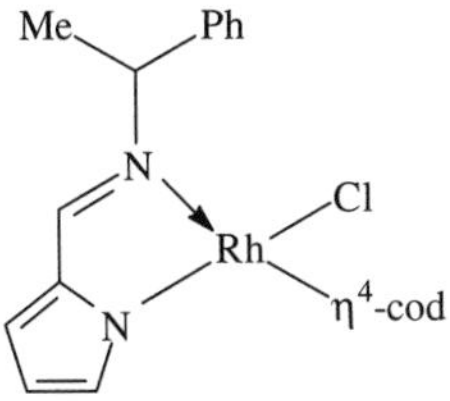

41 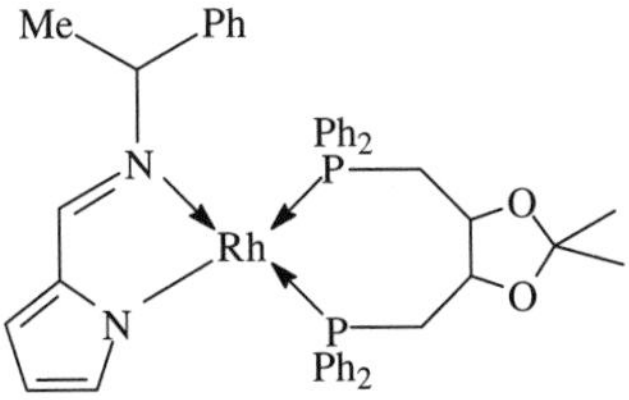   **42** 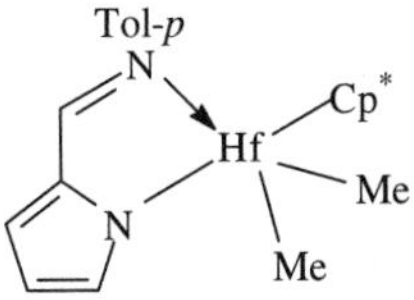   **43**

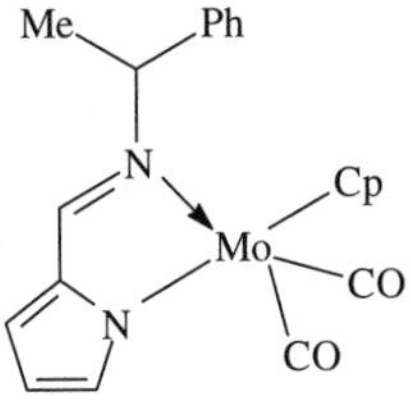

44 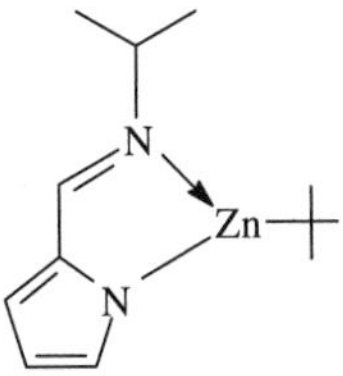   **45**

Along with monocyclic examples, a series of bicyclic chelates exists with a variable ligand environment, including carbon-, amino-, and chlorine-containing fragments in the vicinity of metal cations. Illustrations include complexes of titanium **46** (R = Et) (01OM4793), **46** (R = Ph, Cy) (00CL1270, 01OM4793) and zirconium **47** (A = Cl, C_6H_4OMe-p) (00CL1114), **47** (A = CH_2Ph, R = 2,6-C_6H_3Pr-i_2) (03CL756), **47** (A = $(NMe_2)_2$, R = 2,6-C_6H_3Pr-i_2) (00JCS(D)459). Mononuclear pyrrole azomethine coordination compounds also include mixed-ligand chelates with β-aminovinylketones **48** (85ICA7), Schiff bases with six- **49** (R = Ph, C_6F_5)

(05CC3150) and five- **50** (06JOM1321) membered metallacycles. The latter prepared by an electrochemical method in the presence of (Et$_3$NH)Cl has a rather unusual ionic structure. In complexes **51** (M = Ni, Cu) containing an additional N-donor site, two pyrrole fragments are coordinated (07IC3548). When additional ligating moieties are included in the composition of the azomethine ligands, one **52** (03MRC61) or two **53** (94JCR2469) pyrrolic constituents may not participate in coordination. Chelates of azomethines of 2-pyrrole aldehyde may contain in position 5 coordinatively active fragments. They may participate (**54** (00JCS(D)459), **55** (01OM3510), **56** (04OM2797), **57** (M = Y, Lu), and **58** (08EJI1475)), or may not participate (**59** (98KK164)) in the coordination interaction. The preference of coordination number four for zinc may be explained by the lack of coordination of the carbonyl groups in complex **60** (98KK164). Tetracoordination of nickel in tris-pyrrole azomethine **61** (96CB1195) is of particular interest. Autooxidation of the bis-azomethine dipyrrole **62** (M = Pd, Pt) by molecular oxygen proceeds differently to the tris-chelates **63** for M = Pt and **64** for M = Pd (07IC5465).

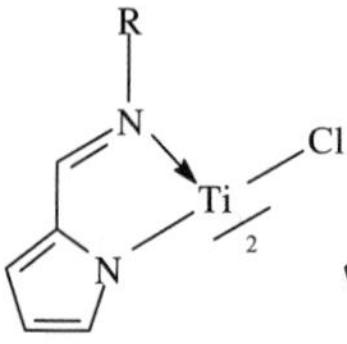

46 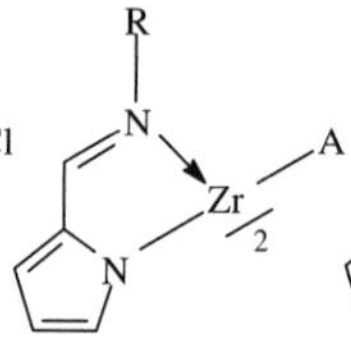   47 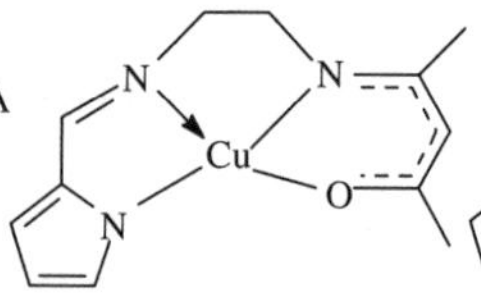   48 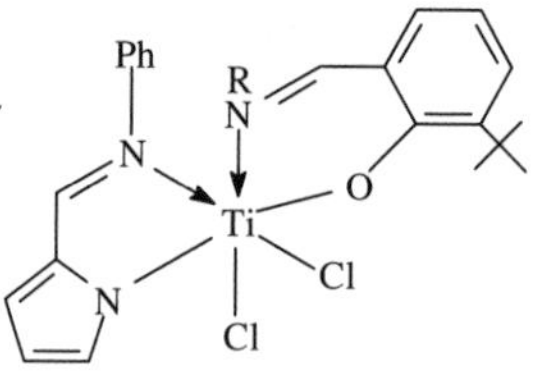   49

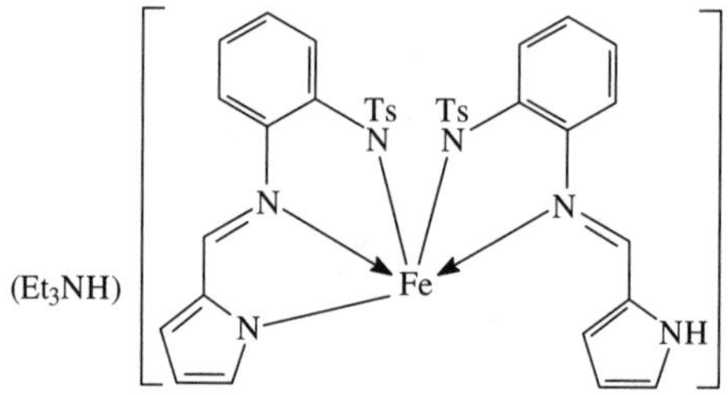

50 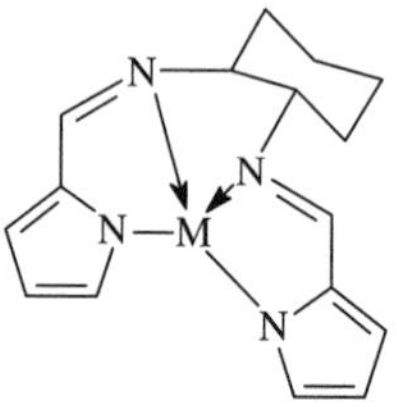   51

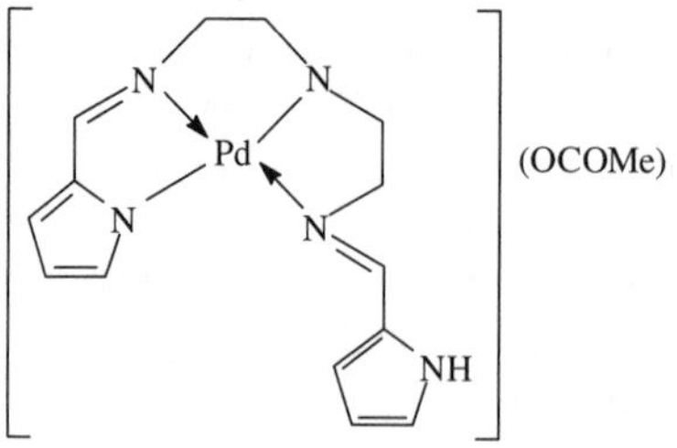

52 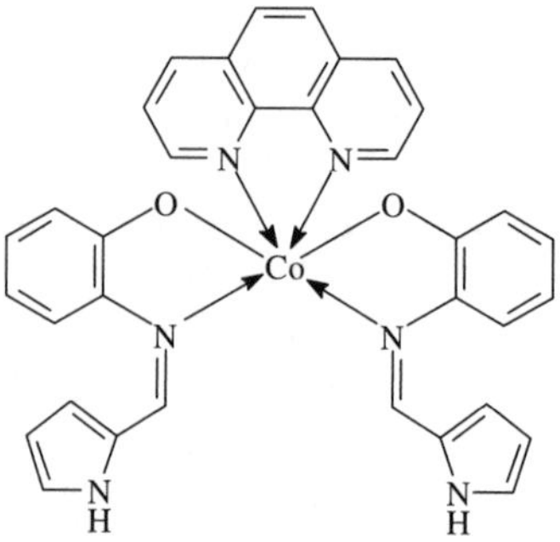   53

54   **55**   **56**

57   **58**

59   **60**   **61**

62   **63**   **64**

N-Pyrrolylphosphino-*N'*-aldimine ligands are sources for complexes **65** (where M = Ni, Pd, Rh; R = Ph, NPr-*n*; Ar = Ph, 3,5-Me$_2$C$_6$H$_3$, 2,4,6-Me$_3$C$_6$H$_2$, 2,6-*i*-Pr$_2$C$_6$H$_3$; X = Cl, Br) (06JCS(D)5362). 4,5-Benzannulated azomethine pyrroles (indoles) are represented by the complexes of Schiff bases of 7-formylindole **66** (06OM5800) and **67** (01CL566). An Ar-salen complex with azomethine indole has been obtained (85CC1172). Cyclopalladation of the diazomethine derivative of isoindole leads to a bis-chelate with a CN_2Cl-coordinated complex-forming agent **68** (05IC6476). 1,3-Bis(4,6-dimethylpyridyl-2-imino)isoindoline reacts with palladium(II) acetate in following two stages: formation of the classical *N,N,N*-chelate **69** and then a cyclometalated tetranuclear dimer **70** (07JCS(D)1101). Other examples of the chelates of pyrrole azomethine ligands can be found in reviews (71ZC81, 80UK1234, 98AHC1, 02RCR943) and monographs (99MI1, 03MI2).

3.3.2 Di- and oligonuclear metal complexes

In early publications on the coordination chemistry of Schiff bases (66PIC66, 71PIC214, 80UK1234, 87MI2, 93CCR1, 02RCR943, 03MI3, 05RCR193), including those on hetarylazomethines (98AHC1), di- and oligonuclear complexes of pyrrole aldimines have not been discussed. Exceptions are the oligonuclear chelates of macrocyclic azomethines with pyrrole ligating moieties (04CCR1717, 07CCR1311). However, numerous di- and oligo-nuclear metal complexes of acyclic pyrrole azomethine ligands are known. The first representative of such dinuclear copper chelates **71** was prepared with a pyrrole azomethine containing the *n*-propanol moiety at the nitrogen atom of the C$=$N bond (72ICA248). This complex may be regarded as the basis of binucleating azomethine ligands having variable amine fragments. The most typical representatives are salen-coordination compounds **72**. The latter include metal complexes of manganese **72** (M = Mn, Z = (CH$_2$)$_2$, R = R^1 = R^2 = H) (01CEJ1468), copper **72** (M = Cu, Z = (CH$_2$)$_2$, (CH$_2$)$_6$, R = R^1 = R^2 = H) (04EJI1478), and zinc **72** (M = Zn, Z = (CH$_2$)$_2$, R = COOEt, R^1 = Me, R^2 = Et) (03T10037, 04ICC249), **72** (M = Zn, Z = (CH$_2$)$_4$, R = R^1 = R^2 = H) (03T10037), **72** (M = Zn, Z = C$_6$H$_4$CH$_2$C$_6$H$_4$, R = COOEt, R^1 = Me, R^2 = Et) (03AGE3271). The salen-pyrrole dinuclear chelates are represented by complexes with bridging acetate **73** (99IC5633) and oxo-hydroxy **74** (01CEJ1468) groups. A methoxycarbonylmethyl group serves as an intermetal bridge in pyrrole aldimine **75** containing the Pd–C bond (02OM1462). Lithium 2-((2,6-di-*i*-propylphenyl)imino)pyrrolide with [RuCl$_2$(PPh$_3$)$_3$] affords the dinuclear **76** with a chloride intermetal bridge (Ar = 2,6-*i*-Pr$_2$C$_6$H$_3$) (06IC10293).

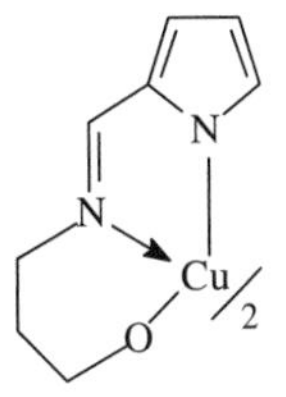

71

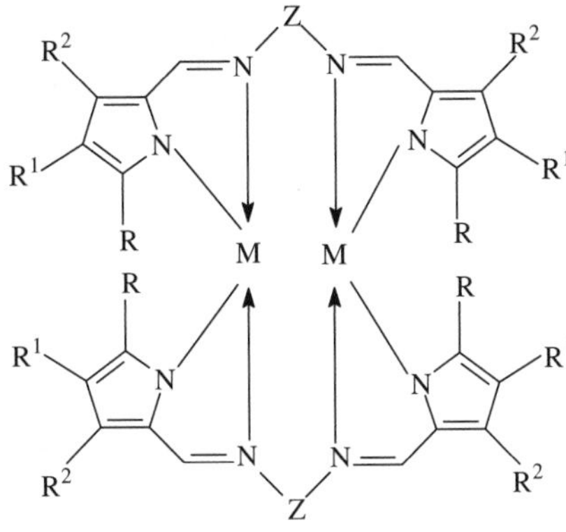

72

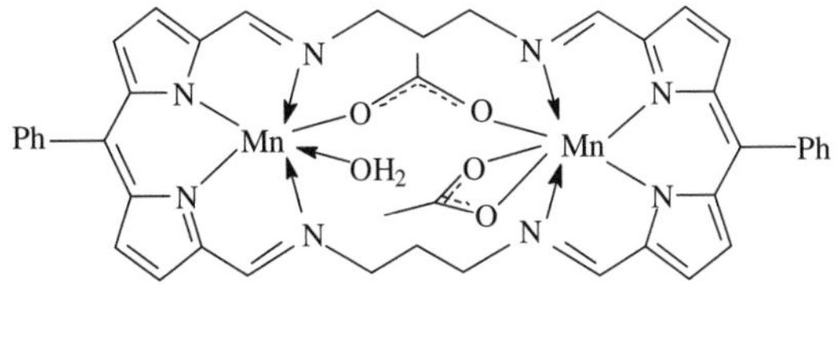

73

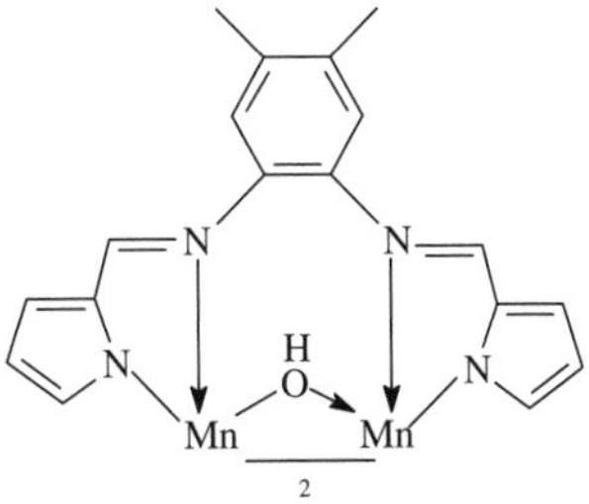

74

75   76

A series of complexes with tetradentate bis-pyrrolyl azomethine ligands does not contain intermetal groups 77 (M = Co, Zn) (06IC636), 77 (M = Fe) (03JCS(D)4387) or have intermetal oxygen atoms 78 (R = Me, R^1 = OMe), 78 (R = Et, R^1 = H) (04IC1220), 79 (07ICA273), and 80 (07AGE584). Structures 77–80 contain four pyrrole azomethine constituents. Formation of the oligonuclear pyrrole azomethine structures may also be realized at the expense of the C–M 81 (00JOM86) and MNCSM 82 (95AX(C)1522) units. In these complex compounds, pyridine nitrogen atoms participate in coordination. However, cases are known where the sterically remote pyridine substituent appears to be unavailable for such coordination, e.g., 83 (99JCS(D)2031). Ligands based on pyrrole-2,5-azomethine serve for the preparation of dinuclear chelates, e.g., 84 (01IZV2334). This structure contains two acetate intermetal bridges.

77   78

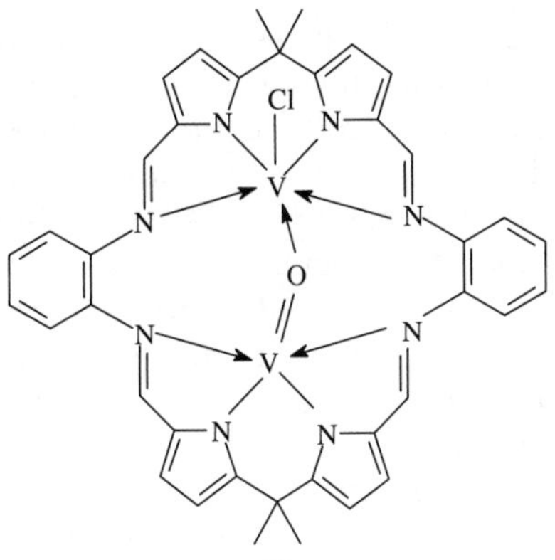

79

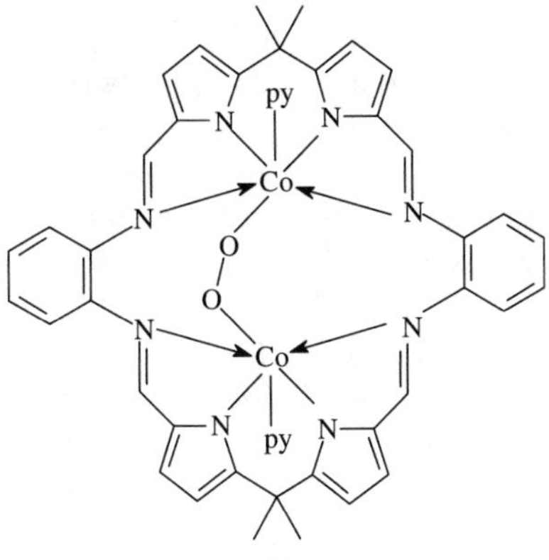

80

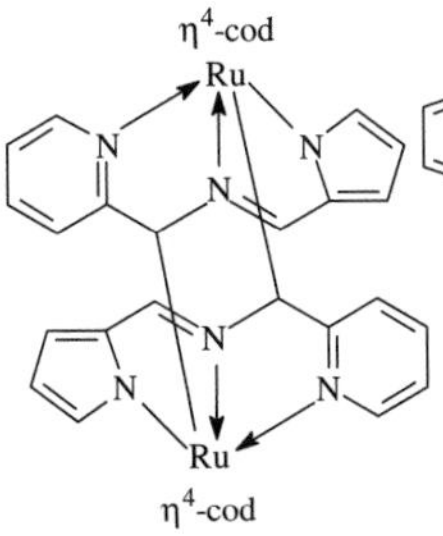

81

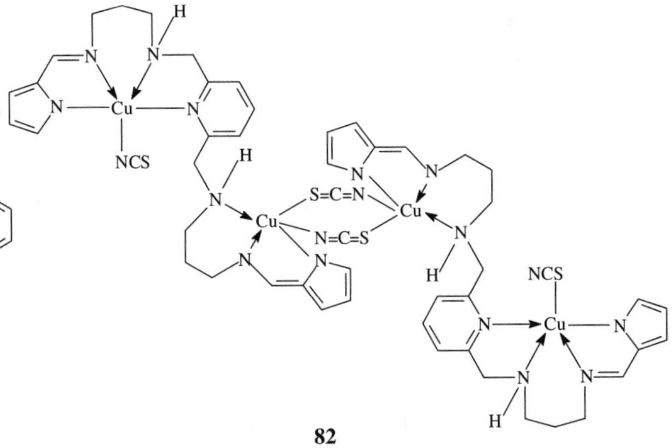

82

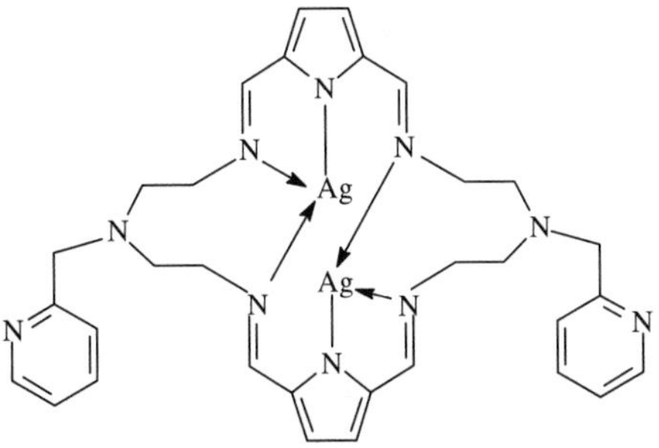

83

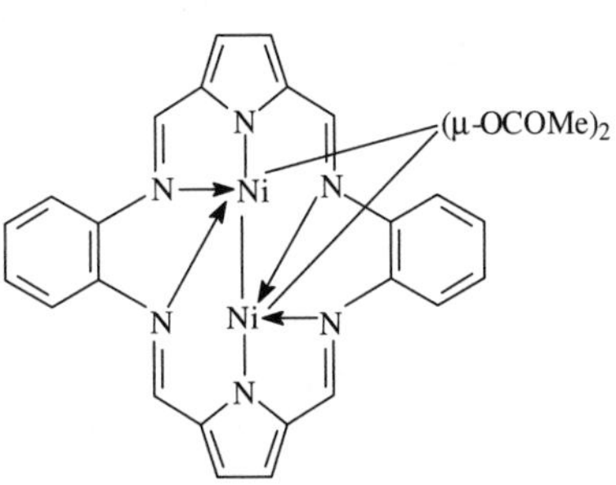

84

An unusual metal source [(Me$_3$SiN)$_2$Sm(THF)$_2$] gives rise to different types of samarium complexes of the pyrrole-salen ligands **85–87** prepared by ligand exchange (03OM434). The structural type of complexes **85–87** depends on the nature of the intermetal bridge of the salen ligands. Several examples of structurally characterized heteronuclear pyrrole azomethine complexes are known in which formation of dimetallic moieties occurs due to the M–C$_\pi$ **88** (06OM5210), M–O **89** (06JA9610), and M–Cl **90** (M = Zr, Hf) (03JOM(667)185) bond formation. The data in this section extend our knowledge on the reaction ability of pyrrole derivatives in their transformations through the NH fragment in which electrophilic agents are metal cations. The coordination compounds, products of these reactions, are represented in modern heterocyclic (96M11, 08M11) and coordination (98AHC1, 03MI1) chemistry basically by the chelates of the tetrapyrrole ligand systems, porphyrines and phthalocyanines (98MI1, 98MI2, 99MI3, 03MI4, 03MI5, 04RCR5).

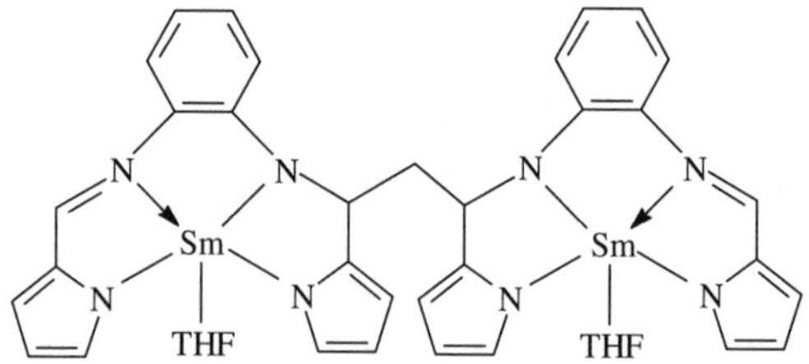

85

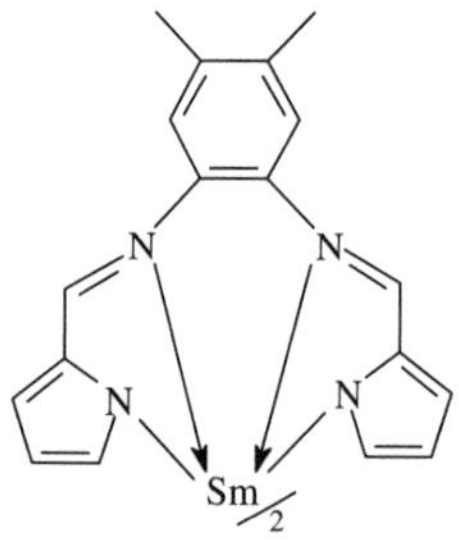

86 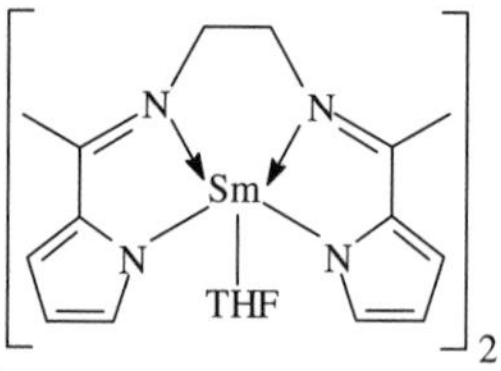   87 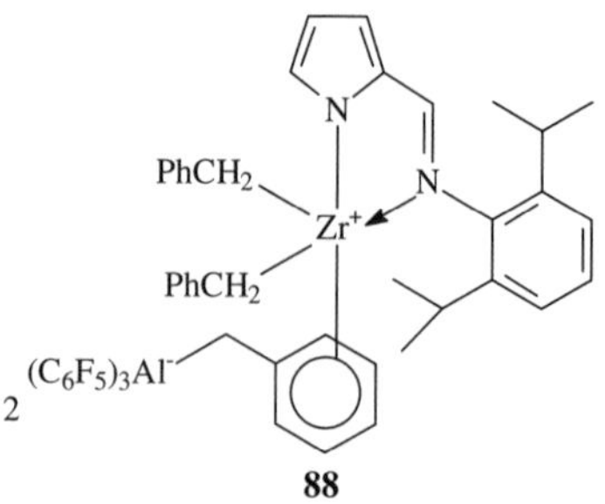   88

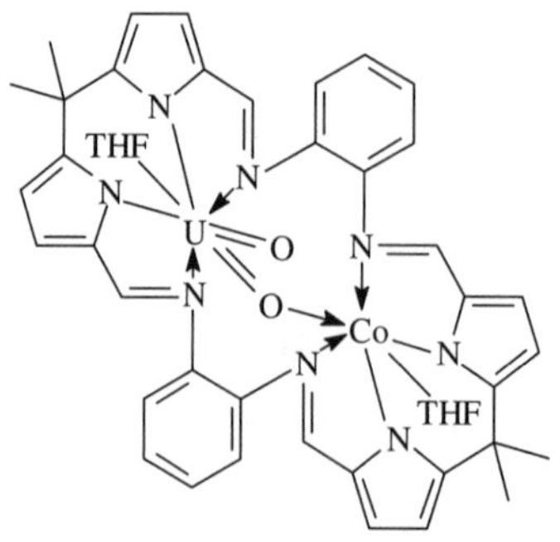

89 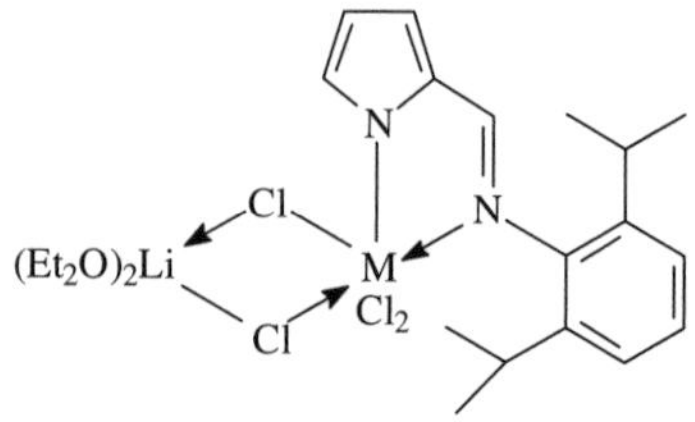   90

3.4 Complexes of hetarylazomethines with five-membered metal rings

Mono-chelate structures include complexes of azomethines of pyridine-2-aldehyde **91** of chromium ($ML_n = Cr(CO)_4$, R = *i*-Pr) (81JCS(D)1524), ($ML_n = Cr(CO)_4$, R = Ar) (07POL2433), ($ML_n = Cr(CO)(NO)$, R = C(H) (Me, Ph)) (97JOM255), molybdenum ($ML_n = Mo(CO)_3L$, L = CO, CNBu-*t*, PPh$_3$, R = 4-R^{1}OC$_6$H$_4$, R = H, Me$_3$Si) (01CEJ2922), ($ML_n = Mo(CO)_3$ (P(OEt)$_3$), R = (H)(Ph)(Me)) (96POL1723), Group VI metals (M = M(CO)$_4$, M = Cr, Mo, W, R = 4-HO-C$_6$H$_4$) (02JCS(D)2379), rhenium ($ML_n = Re$ (CO)$_3$Cl, R = Ph) (91IC4754), ruthenium ($ML_n = Ru(H)(Cl)(CO)(PPh_3)$, R = Ph, 4-ClC$_6H_4$, *p*-Tol) (98IC5968), ($ML_n = Ru(CO)_2Cl_2$, R = (CH$_2$)$_2$OMe) (93IC5528), rhodium and iridium ($ML_n = Rh(CO)I$, R = Ar) (03OM1047), ($ML_n = (\eta^5\text{-Cp*})MCl$, M = Rh, Ir, R = 3-XC$_6H_4$, X = H, Me, OMe, Cl, NO$_2$) (05POL1710), palladium ($ML_n = Pd(R^1)(X)$, R = *i*-Pr, R^1 = Me, COMe, X = Cl, BF$_4$) (96JOM109), and platinum ($ML_n = PtMe_2$, R = *n*-Pr) (96OM2108), ($ML_n = PtMe_2$, R = 3-SO$_3$Na-C$_6$H$_4$) (03JOM(673)110), ($ML_n = Pt(Me(Cl)(\eta^2\text{-CH}\equiv\text{CH}))$, R = Ph) (93JCS(D)1927), ($ML_n = Pt(\eta^2\text{-olefin})$, olefin = maleic anhydride, fumaronitrile, naphthoquinone, R = *t*-Bu) (02JCS(D)3696). Some representatives include substituted pyridine heterorings, namely cobalt **92** (X = Cl, Br; R = Alk, R^1 = Ar, Het) (04EJI1204) and **93** (03OM2545), as well as platinum **94** (01OM3635). Some of the neutral mono-chelates deserve special attention since they contain specific substituents at the azomethine nitrogen, which usually leads to various technical applications of the products. Thus, 2-aminomethylpyridine or 2-aminoethylpyridine with ferrocenecarbox-aldehyde form chelates **95** and **96**, respectively (08JOM619). Monoazo-methine complexes with ferrocene-containing amine counterparts **97–99** are also formed by platinum(II) (07OM5406). Complexes with aryl-ethynyl substituents **100** and **101** ($ML_n = MnBr(CO)_3$, ReBr(CO)$_3$, Mo(CO)$_4$, MoCl(η^3-C$_3$H$_4$Me)(CO)$_2$) embrace chromium and manganese groups (06JOM3434). Complexes **102** (R^1 = R^2 = H, R^1 = H, R^2 = OC$_{18}$H$_{37}$-*n*; R^1 = OC$_{18}$H$_{37}$-*n*, R^2 = H; R^1 = R^2 = OC$_{18}$H$_{37}$-*n*) as well as species with the same set of ligands (L) [Ru(bipy)$_2$L](PF$_6$)$_2$ are of special interest (07OM5423). Palladium(0) mono-chelate **103** contains a carbohydrate-derived substituent (00JCS(D)2545). Aminoalkyl platinum(IV) product **104** acts as the N,N,C-ligand (01OM408). Formation of tungsten mono-chelates in solution is often retarded by mono-coordinated precursors as illustrated by Equation (29) (R = *i*-Pr, *n*-Bu, *t*-Bu, Ph) (87JCS(D)513).

91   **92**   **93**

94   95   96

97   98   99

100   101   102

103   104

(29)

A wide range of cationic mono-chelate compounds includes **105**, those of molybdenum ($ML_n = (\eta^5\text{-}Cp)Mo(CO)_2$, R = i-Pr, A = PF_6) (80IC891), ruthenium ($ML_n = (\eta^5\text{-}Cp)Ru(CO)$, R = Ph, A = PF_6) (02OM4891), ($ML_n = (\eta^5\text{-indenyl})Ru(PPh_3)$, R = Ph, A = BF_4) (04JOM1249), palladium ($ML_n = Pd(\eta^3\text{-}C_3H_5)$, R = $C_6H_4OMe\text{-}4$, A = BF_4) (94JCS(D)1145), and platinum ($ML_n = (\eta^4\text{-cod})PtCl$, R = $C_6H_4OMe\text{-}4$, t-Bu, A = BF_4) (02JCS(D)212), ($ML_n = (\eta^2\text{-dimethylfumarate})Pt(I)(py)$, R = Ph, A = NO_3) (00EJI1717). Palladium **106** contains a substituted pyridine (01ICA(315)172) and ruthenium mono-chelate **107** has a specific azomethine N-substituent. Ferrocenyl-substituted pyridine azomethines form the cationic rhodium(I) **108** with non-coordinated N-azomethine atoms (06JOM4573).

105	**106**	**107**

108

Complexes of iminopyridines tend to enter into oxidative addition reactions, Equation (30) (R = H, Me; X = Cl, Br) (04EJI2053). An interesting application of the copper(II) complex of N-(2-pyridylmethyl)-pyridine-2-methylketimine is the synthesis of the 4'-(2-pyridyl)-2,2':6',2''-terpyridine ligand using the reaction sequence illustrated by

Equation (31) (06IC7994). A diazomethine based on a pyridine-2-aldehyde containing a phenyl spacer forms the mononuclear mono-chelate ruthenium(II) **109** (04JOM3612). Azomethines based on pyridyl amines as a rule have the N,N,C-chelate ring as in rhodium complex **110** ($n = 2$, 3) (01JA751). Formation of the N,N,C-chelated ruthenium complex **111**, however, proceeds through a four-membered mono-chelate cationic **112** and the product has a non-azomethine nature (03EJI1883). Complexes of pyridine azomethines with coordination-active substituents are scarce. Tetra-chelates **113** (M = Fe, Ni, Zn) and **114** contain the associated thiocyanato-ion (05IC8916), while a peculiar feature of tri-chelate **115** is the spectator ligand containing the hexahydropyrimidine ring. Dendrimers containing nickel and palladium complexes of N,N'-pyridylimine are useful materials (05CC5217, 06OM3045, 06OM3876). Pyridylimine ligands react with [PtCl$_2$(SMe$_2$)$_2$] in methylene chloride to yield **116** (R = H, Me) (08JOM278). The reaction of the carbosilane-containing oligomers of pyridylimines with [PtCl$_2$(SMe$_2$)$_2$] in methylene chloride or [Mo(CO)$_3$(1,3,5-C$_6$H$_3$Me$_3$)] in acetonitrile produces several species: **117** (R = Me, ML$_n$ = PtCl$_2$; R = H, M = Mo(CO)$_3$(AN)) and **118** (R = H, ML$_n$ = Mo(CO)$_3$(AN)).

(30)

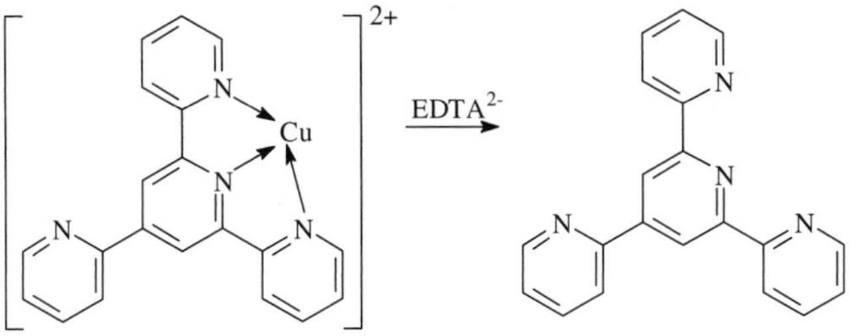

(31)

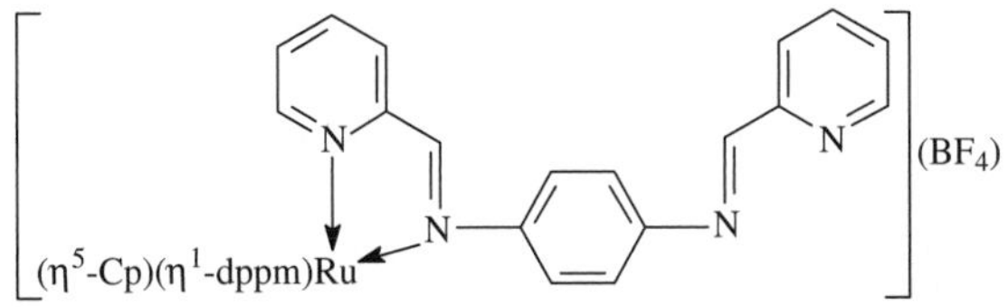

109

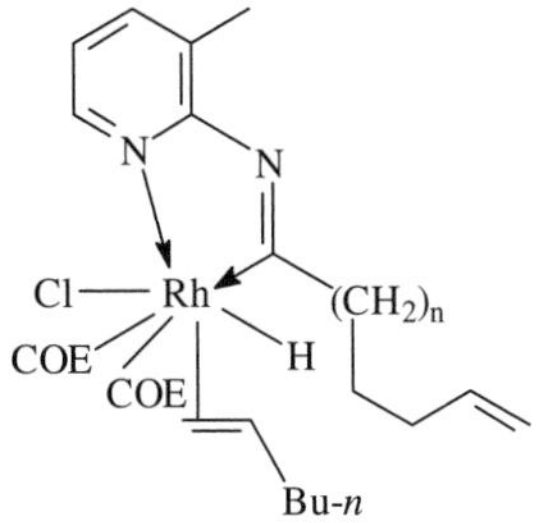

110 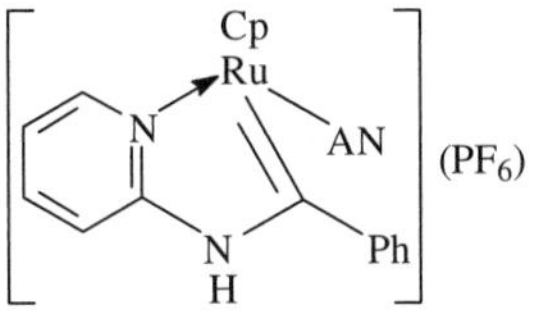   **111** 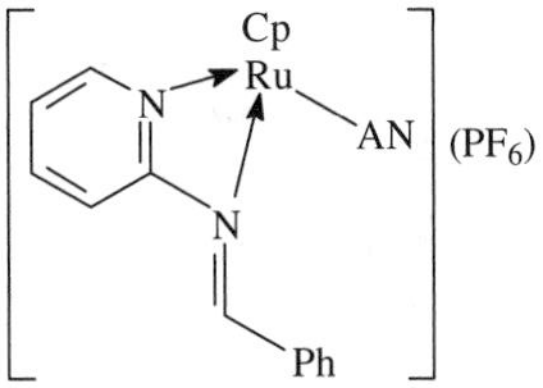   **112**

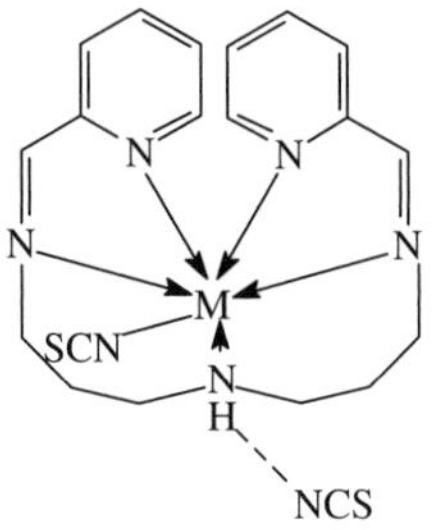

113 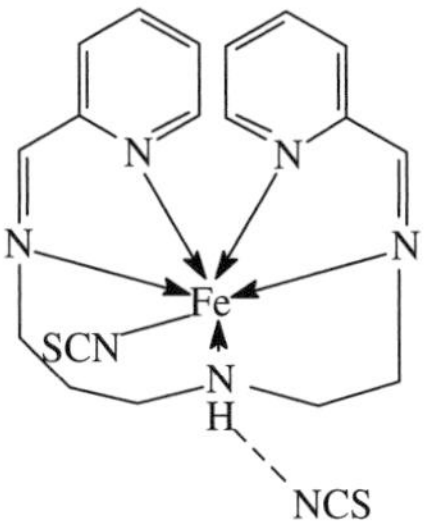   **114** 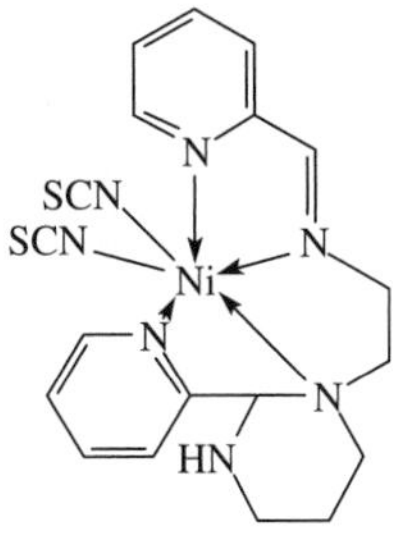   **115**

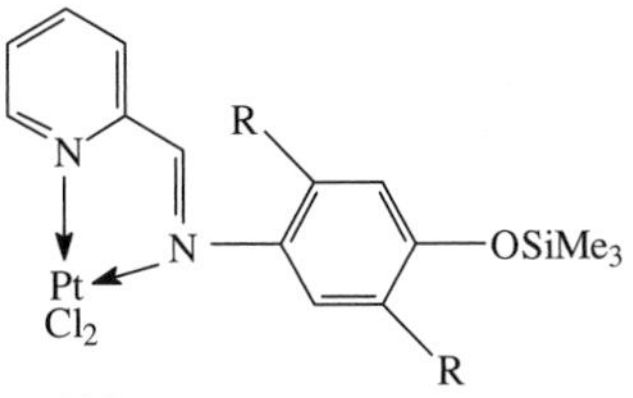

116 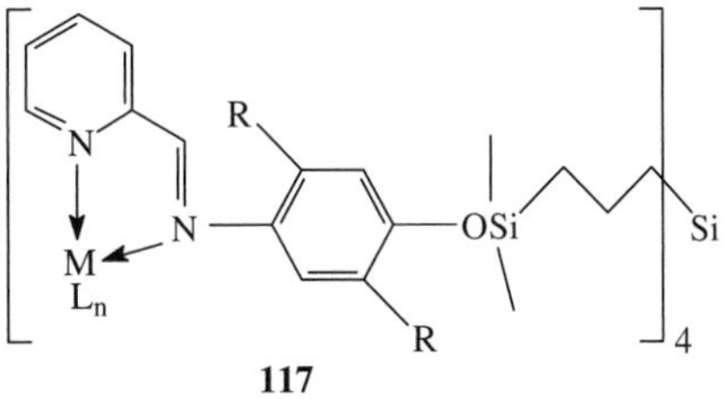   **117**

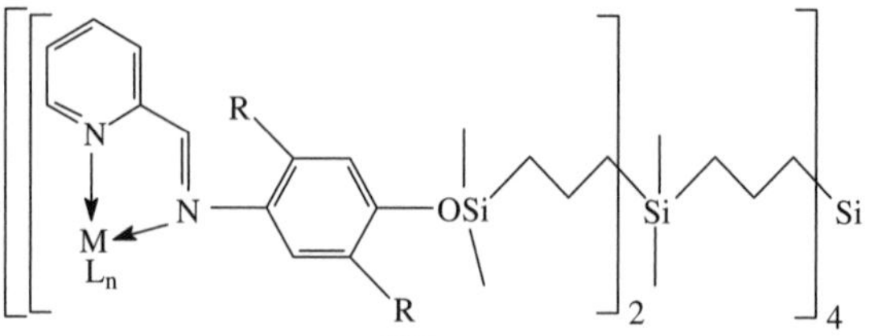

118

Complexes of iron(II), cobalt(II), and nickel(II) of 2,9-diimino-1,10-phenanthrolinyls **119** (M = Fe, Co, Ni; R = H, Me; Ar = 2,6-Me$_2$C$_6$H$_3$, 2,6-i-Pr$_2$C$_6$H$_3$, R^1 = Alk, X = Hal) (02JOM62, 06JOM4196, 06OM666, 07EJI3816, 07JMC(A)85, 08JOM483), N-((pyridine-2-yl)-methylene)-quinoline-8-imines **120** (M = Fe, Co, Ni, R = R^1 = R^2 = Alk, X = Hal) (07OM4781), 2-quinoxalinyl-6-iminopyridines **121** (M = Fe, Co, Ni; Ar = 2,6-Me$_2$C$_6$H$_3$, 2,6-i-Pr$_2$C$_6$H$_3$, X = Hal) (07JOM3532, 07JOM4506), and 2-(1-methyl-2-benzimidazolyl)-6-(1-aryliminoethyl)pyridines **122** (M = Fe, Co, Ni; Ar = 2,6-Me$_2$C$_6$H$_3$, 2,6-i-Pr$_2$C$_6$H$_3$, X = Hal) (07OM2439, 07OM2720) are of interest as polymerization catalysts. Also included are the complexes prepared from [CrCl$_3$(THF)$_3$] in methylene chloride, **122** (instead of MX$_2$–CrCl$_3$; R = i-Pr; R^1 = Me, Et, i-Pr, F, Cl, R^2 = H; R^1 = R^2 = Me) (08JOM750), as well as MCl$_2$ in ethanol, **122** (M = Fe, Co, Ni; R = i-Pr; R^1 = Me, Et, i-Pr, Cl, R^2 = H; R^1 = R^2 = Me; R^1 = Me, R^2 = Br) (08JOM1829).

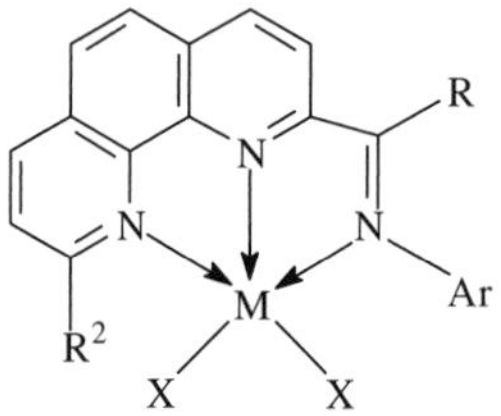

119

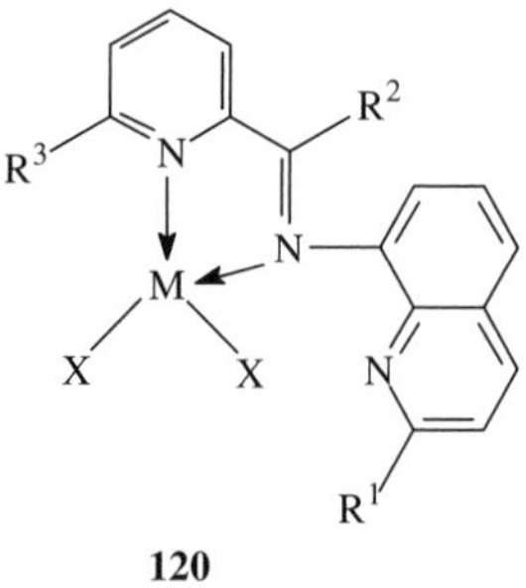

120

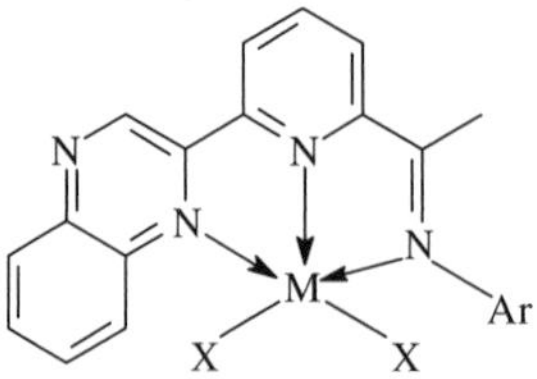

121

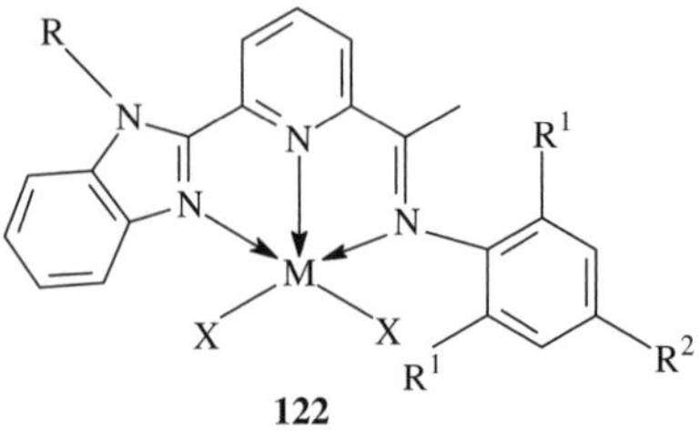

122

From the imines of pyridine-2-aldehyde, a series of dinuclear coordination compounds can be prepared. They include dinuclear Group VI metal complexes **123** (M = Cr, Mo, W; R = polystyrene-Si(i-Pr)$_2$) (04EJI3498), rhenium species **124** with a phenyl spacer (04IC622), dinuclear ruthenium polypyridyl complex with a bis(pyridylimine) ligand **125** that contains a diphenylmethane spacer (08JCS(D)667), rhodium complex with a phenyl spacer **126** (02JOM43), palladium species with two chloride bridges and uncoordinated but cyclometalated quinoline **127** (79JCS(D)1899), platinum **128** with a cyclohexyl spacer (99OM4373), and tetraazomethine coordination compounds **129** (M = Ag, Cu) (06IC1445). Dinuclear platinum(II) **130** are interesting because they readily oxidatively add methyl iodide to yield a mixture of the platinum(IV) products **131** and **132** (00OM2482). Among the polycarbonyl and cluster compounds, rhenium dinuclear **133** and osmium clusters **134–136** are illustrative (91IC42, 99JOM(572)271, 99JOM(573)121, 99OM4380, 00OM4310, 01EJI223).

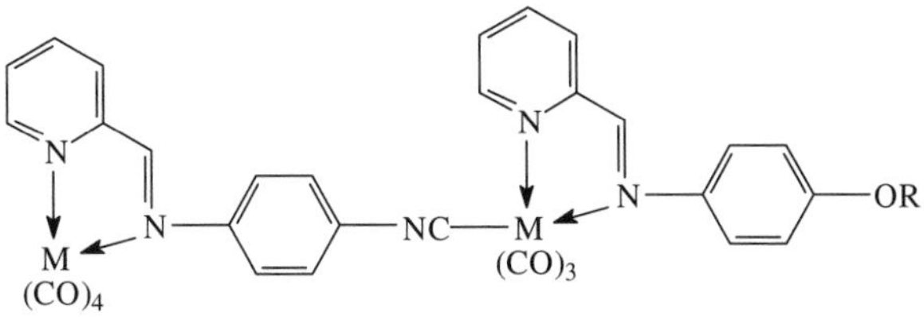

123

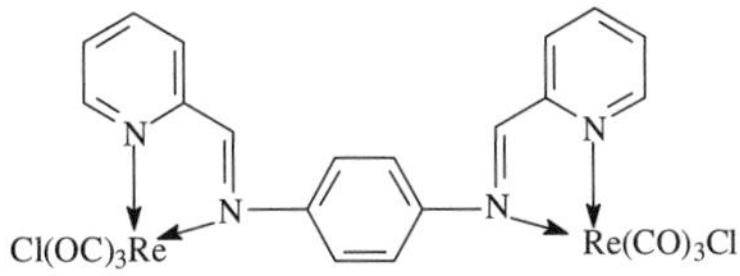

124 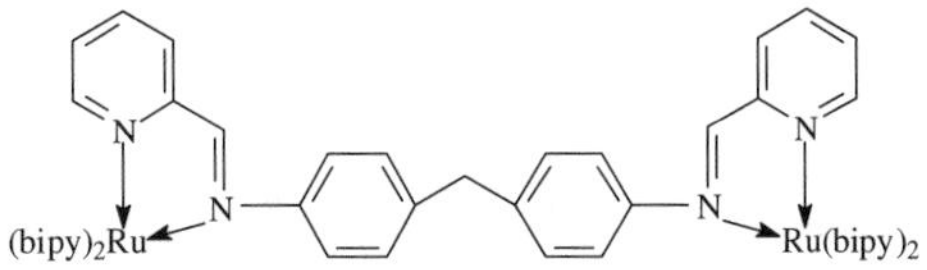   125

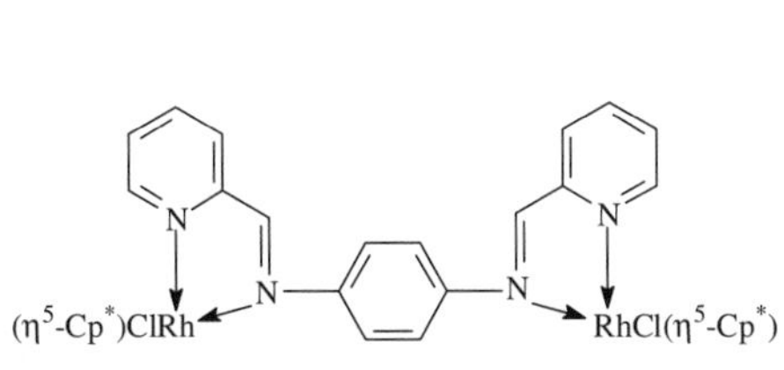

126 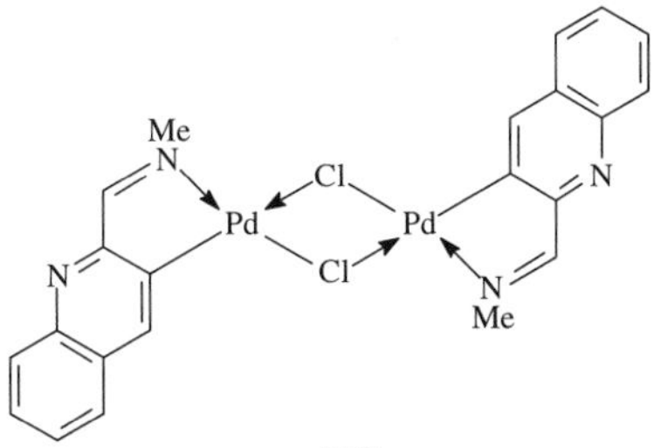   127

128   **129**   (PF$_6$)$_2$

130   **131**   **132**

133   **134**   **135**   **136**

Coordination compounds of the ligating azomethines containing an additional XH substituent (X = O, S, Se) in the aromatic or heteroaromatic counterpart have a limited range. The first complexes of this kind were prepared on the basis of ring-chain tautomerism of benzothiazolines (Equation (16), R = py) in the presence of metal cations at the end of the 1960s (67ICA365). Newer complexes of benzazole derivatives **137** (A = NMe, S; M = Pd, Zn, Cd) and **138** (M = Zn, Cd) were prepared and characterized (02RCR943). Tris((2-((imidazol-4-yl)methylidene)amino) ethyl)amine with *trans*-[Co^{III}Cl$_2$(py)$_4$]$^+$ yields **139** (07IC8170). Other chelates with pyridine fragments include **140** (R = H, M = ReO, L = L^1 = Cl; R = Me, M = Ru, L = Cl, L^1 = PPh$_3$) (02RCR943) and **141** (05RCR193). Molecular complexes **142** are similar in terms of their chelating centers and the nature of metallacycles (71JCM139, 06ICA1103). Chelates containing non-coordinated pyrrole **143** and benzimidazole **144** substituents R^1 are other illustrations of this structural type (02RCR943). A similar complex with a five-membered coordination unit is **145** containing the pyridine aldehyde group (04ICA3697). Other hetarylazomethine complexes containing five- and six-membered metallacycles are represented by 3-hydroxypyridin-2-yl-salicylideneimine **146** (X = NTs, O; R = H) and its 5-bromo-substituted derivative **145** (X = NTs, O; R = 5-Br) and dimethyltin(IV) **147** (R = H, 5-Br) (03ZK492). Complexes **147** have a trigonal-bipyramidal structure at their coordination site. Tris-chelate structures of pyridine-salen **148** exist for nickel (M = Ni, no L) (98AX(C)725), copper (M = Cu, no L) (97AS519), and zinc (M = Zn, L = py) (04JCS(D)1731, 05EJI4626), (M = Zn, no L) (07IC5829). Direct interaction of the ligating derivative **146** (R = H, X = O) with copper(II) acetate in ethanol yields the dimeric dinuclear **149** having two five-, two six-, and one four-membered metallacycles (08ZOB1002). Similar **149** (R = H, X = NTs, O) can be prepared by template synthesis from 2-amino-3-hydroxypyridine (98M455), N-tosylaminobenzaldehyde, and copper(II) acetate (08ZOB1002). Both coordination compounds are characterized by their antiferromagnetic exchange interactions. Additional illustrations are palladium **150** (X = O, S) (02EJI2179) and samarium **151** (08ICC349).

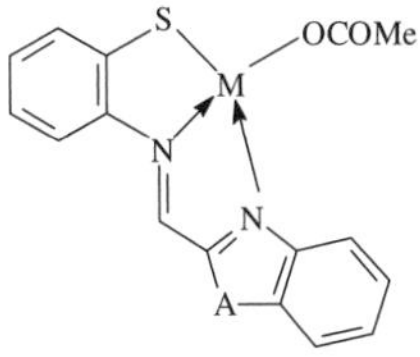

137 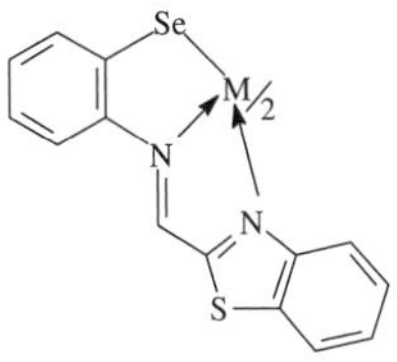   138 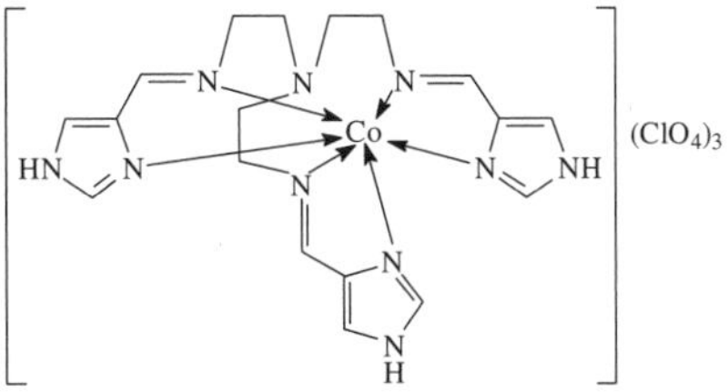   139

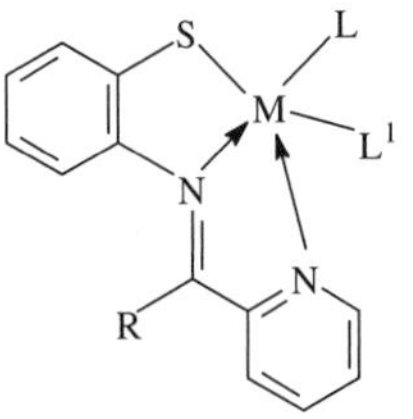

140 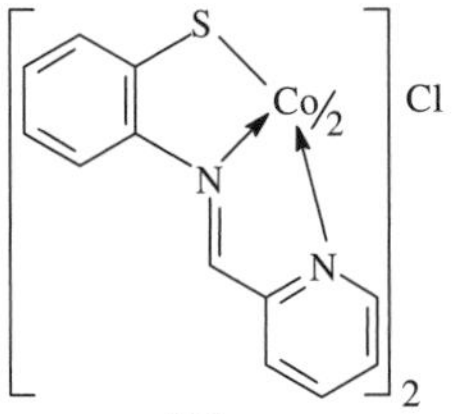   141 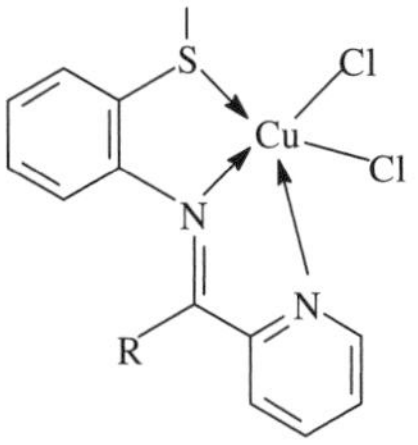   142 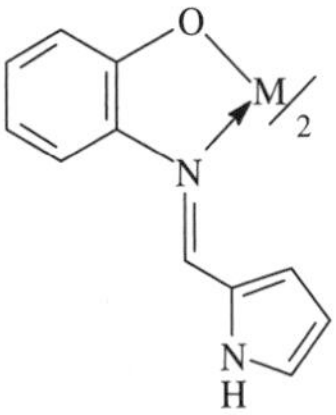   143

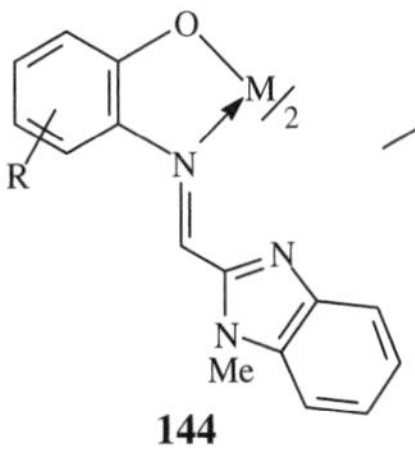

144 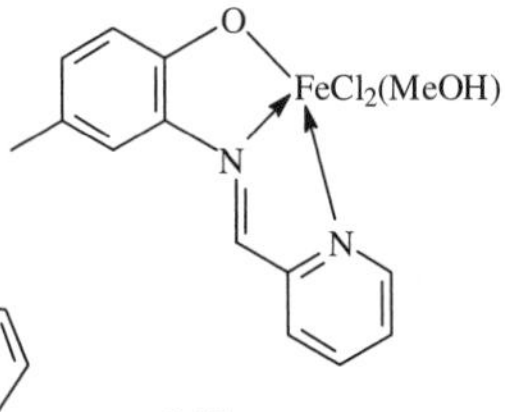   145 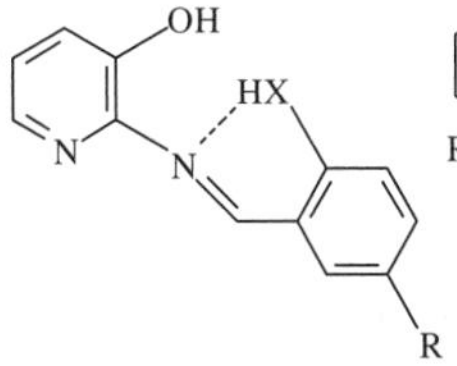   146 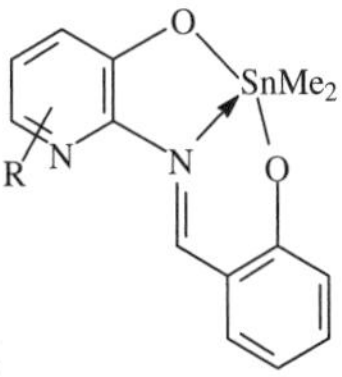   147

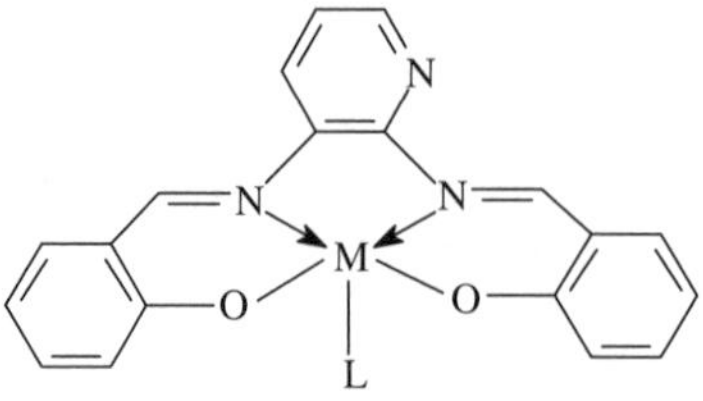

148

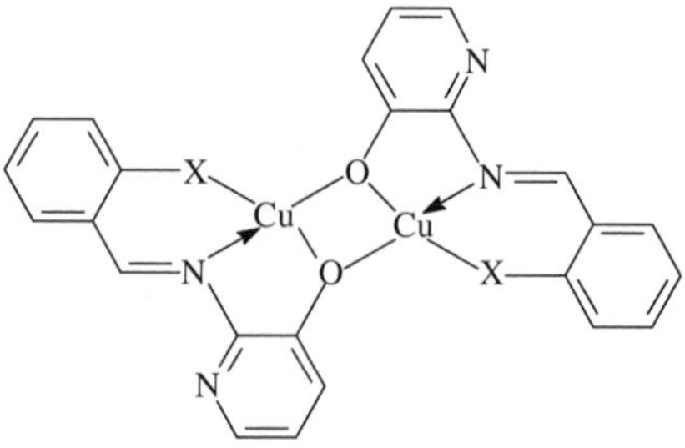

149

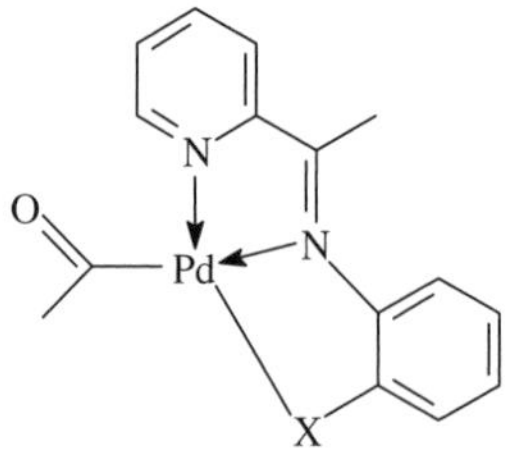

150

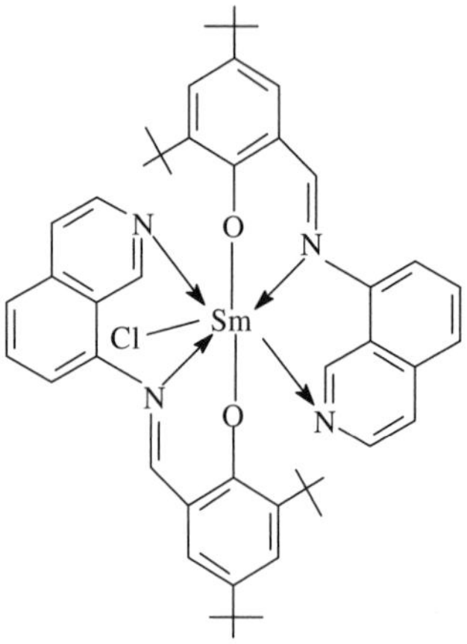

151

Complexes of hetaryldiazomethines include the bis-chelates of 2,5-dialdimines of the five-membered monoheteroaromatic rings, e.g., thiophene **152** (M = Ni, Pd; A = ClO_4, BF_4), where S-coordination has not been structurally proven (07ICC925). A majority of similar complexes can be prepared from 2,6-dialdimines of pyridine (05CCR2156). They include bis-chelate structures **153** (R = o-OH, p-OH; M = Co, Ni, Cu, X = Cl; M = Zn, X = Br; no L; M = Fe, R = o-OH, m-OH, L = H_2O; M = Co, Ni, Zn, X = Br; M = Cu, X = Cl, R = o-OH, m-OH, L = 2-$NH_2C_5H_4N$) (05SA(A)1059, 06SA(A)188). Other illustrations are bis-chelates **154** of manganese (M = Mn, Ar = 2,6-i-$Pr_2C_6H_3$, R = R^1 = CH_2CMe_2Ph, CH_2Ph, $CH_2CH=CH_2$) (07OM1104), iron (M = Fe, Ar = Ph, C_6Me_5, 2,4, 6-$Me_3C_6H_2$, R = $(CH_2SiMe_3)_2$, R^1 = H) (05OM4878), (M = Fe, Ar = 2,6-i-$Pr_2C_6H_3$, R = CH_2SiMe_3, R^1 = H) (08OM109), cobalt(I) (M = Co, Ar = 2,6-$Me_2C_6H_3$, R = Et, n-Pr, n-Bu, n-C_6H_{13}, R^1 = H) (04OM5503), rhodium (M = Rh, Ar = 2,6-$Me_2C_6H_3$, R = η^2-C_2H_4, R^1 = H) (01OM4345), and iridium (M = Ir, Ar = 2,6-$Me_2C_6H_3$, R = Ph, R^1 = H) (03AGE1632).

Cationic species include those of cobalt(II) **155** (Ar = 2,6-*i*-Pr$_2$C$_6$H$_3$) (05OM2039) and iron(II) **156** (Ar = 2,6-*i*-Pr$_2$C$_6$H$_3$) (05OM3664). Bis (imino)pyridine bis(dinitrogen) iron complex **157** (Ar = 2,6-*i*-Pr$_2$C$_6$H$_3$) is the source of the numerous neutral ligand-substitution products, such as *t*-butylisocyanide, 1,2-bis(diethylphosphino)ethane, *t*-butylamine, tetrahy-drothiophene, or triethylphosphine, e.g., **158** (07IC7055). Tetra-chelate complexes are represented by coordination compounds with N$_2$S$_2$ chelate environment **159** (72CC683, 73IC2316). Complexes with four five-membered metallacycles contain five nitrogen donor centers **160** (M = Zn, Cd; L = H$_2$O) (03EJI3193). Azomethines of 2,6-diformylpyridine are the sources for penta-chelate complexes **161** (74CC727), **162** (07IC4114), and **163** (01ICA(317)45). 2,6-Dialdimines of pyridine allow the preparation of hexa-chelate structures, e.g., **164** (79CC1033). Complex **165** based on 6-bis(1-(2-methylanisolylimino)ethyl)pyridine contains two five- and two six-membered chelate rings (07JCS(D)4644). Dimerization of a ligand in the process of complexation (03JCS(D)221) using methods based on condensation techniques leads to new systems as illustrated by **166** (M = Fe, Co; R = 2,6-*i*-Pr$_2$, 2,4,6-Me$_3$) and **167** (R = 2,4,6-Me$_3$) (07OM4639). Tris-azomethine complexes are interesting because in **166** only three out of five and in **167** only four out of five nitrogen atoms are coordinated.

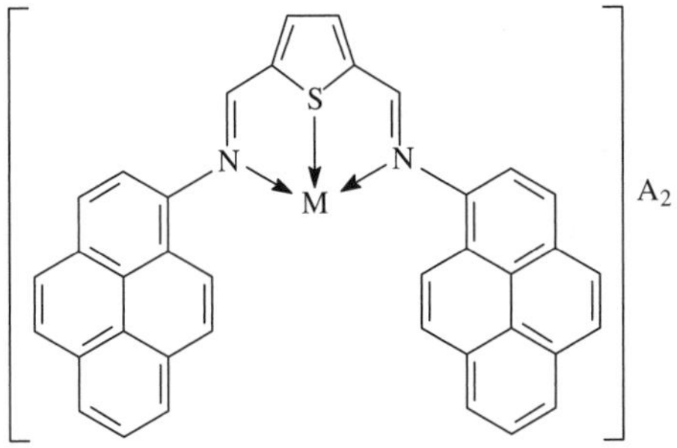

152 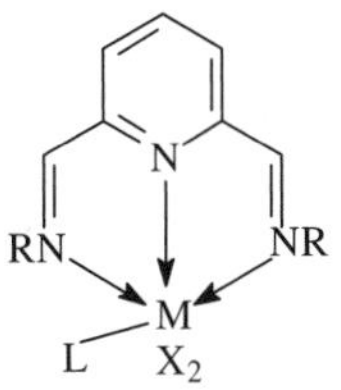   153 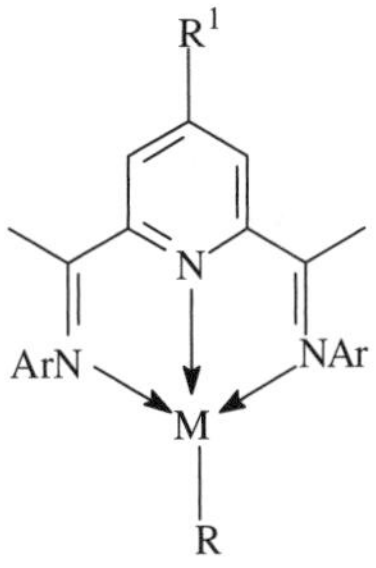   154

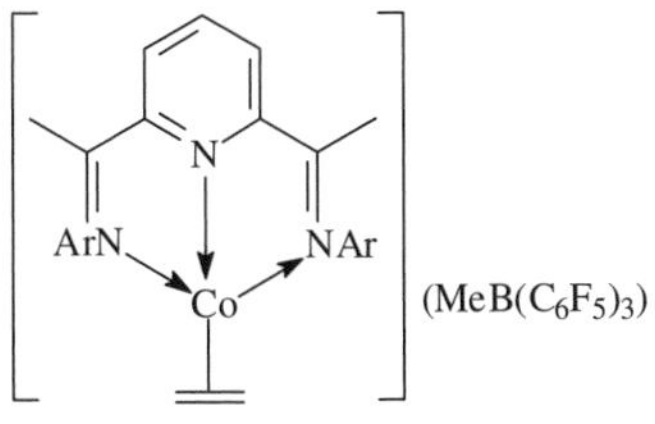

155 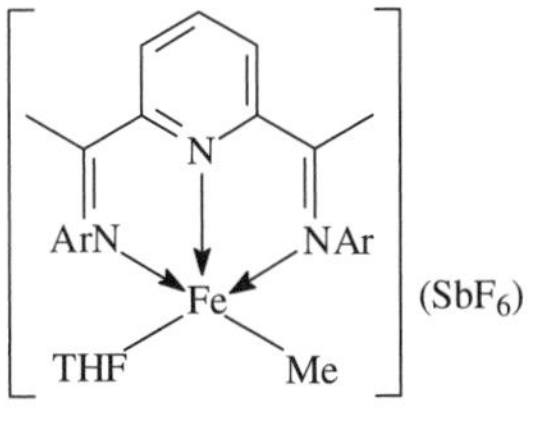   156 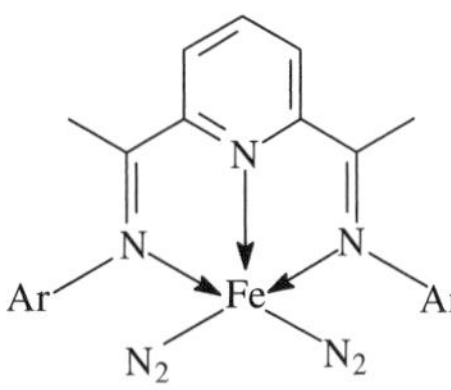   157

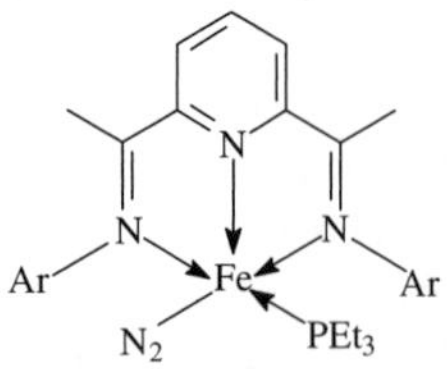

158

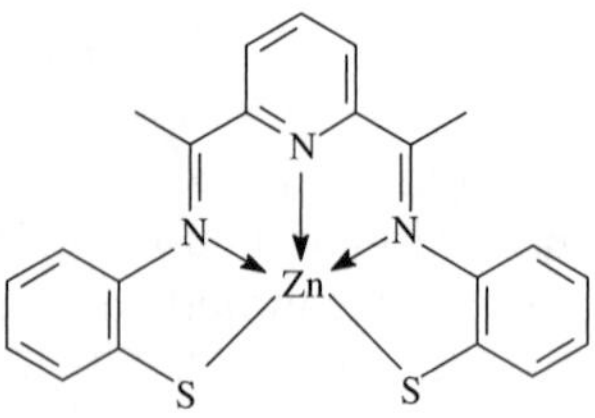

159

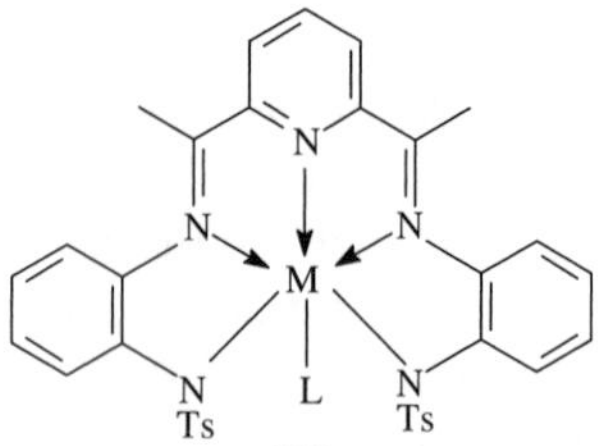

160

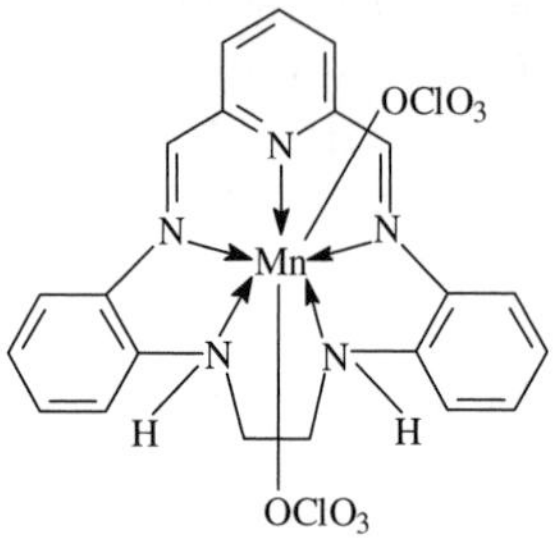

161

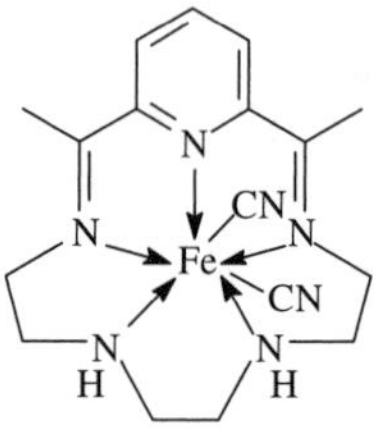

162

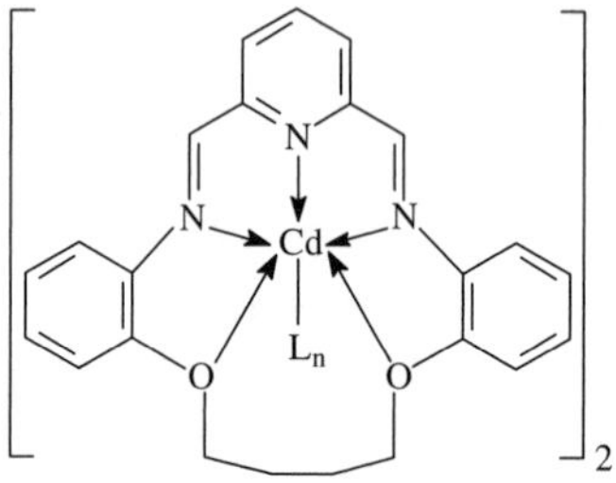

163

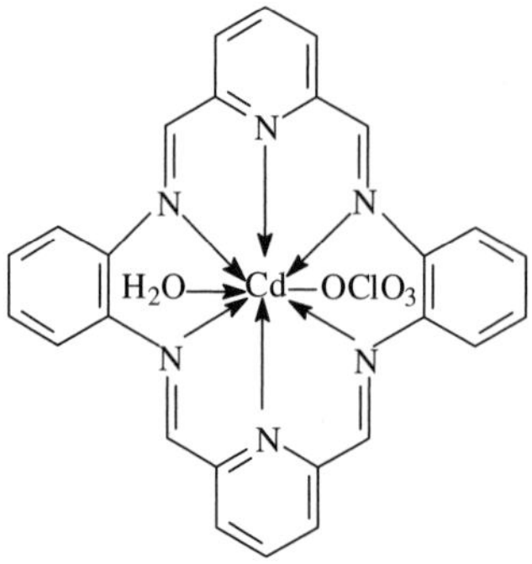

164

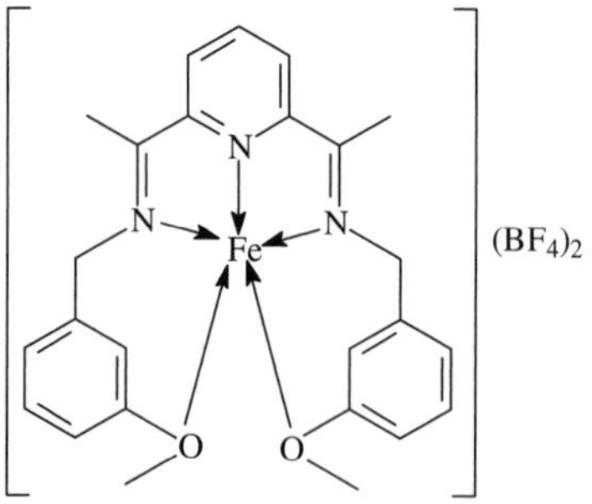

165

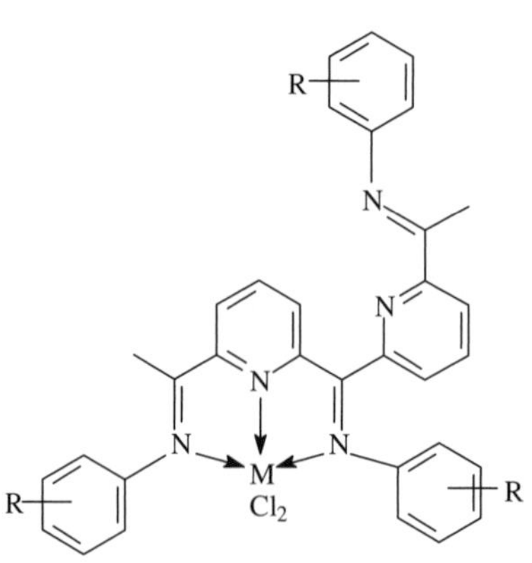

166

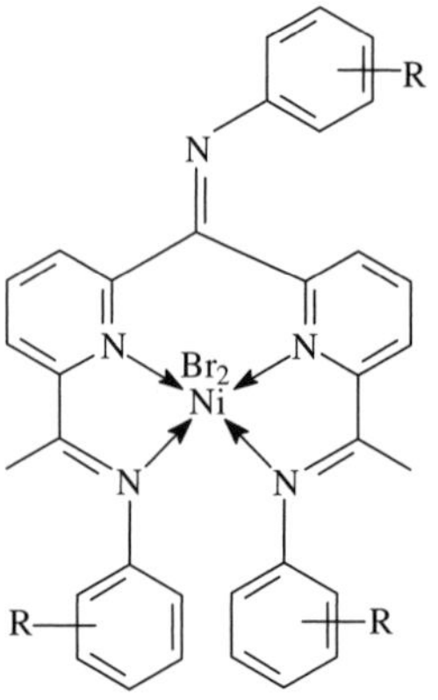

167

Metal complexes of the bis(imino)pyridine ligand are characterized by a variety of ligand-centered reactions. Those occurring with electron transfer yield the radical-anion **168**. Alkylations at the heteroring afford **169** and dimerization processes give rise to **170** employed as catalysts in polymerization processes (06JCS(D)5442). The latter aspect is related to a number of various, sometimes unique, dinuclear products. Sodium hydride reduction of **154** (M = Cr, Ar = 2,6-i-Pr$_2$C$_6$H$_3$, R = Cl, R^1 = H) proceeds through a number of stages dependent on the reaction time, among them rather unusual heteronuclear **171** and **172** capable of nitrogen fixation (07IC7040). Bis(imino)pyridinato chromium(III) complexes, e.g., **154** (M = Cr, Ar = 2,6-i-Pr$_2$C$_6$H$_3$, R = Cl$_3$, R^1 = H) tend to transform into chromium(II) **154** (M = Cr, Ar = 2,6-i-Pr$_2$C$_6$H$_3$, R = Cl$_2$, R^1 = H) under the action of even mild reducing agents, such as trimethyl aluminum (02JA12268). When being treated with methyl lithium, **154** form the dinuclear chromium(I) – lithium **173** (07OM3201). On treatment with sodium hydride, both chromium(III) and chromium(II) derivatives produce chromium(I) **154** (M = Cr, Ar = 2,6-i-Pr$_2$C$_6$H$_3$, R = Cl, R^1 = H) possessing extremely high polymerization catalytic activity. The reaction ability of the product is remarkable. Thus, with LiCH$_2$SiMe$_3$ in THF it yields the chromium(I) **154** (M = Cr, Ar = 2,6-i-Pr$_2$C$_6$H$_3$, R = THF, R^1 = H), while with methyl lithium in ether – the dinuclear chromium(I) – lithium **174**. Both of them contain the reduced methyl group and bis(imine) has transformed to an imino–amino ligand. Furthermore, with trimethyl aluminum a mixture of dinuclear chromium(I) – aluminum **175** and mononuclear chromium(I) – methyl derivatives result. The reaction with the catalyst activator i-butylaluminoxane in toluene yields the trinuclear product containing chromium in the zero oxidation state, **176**. This is a unique case where the heteroring at the azomethine group is η^5-coordinated.

168   **169**   **170**

171   **172**

173   **174**   **175**

176

On the basis of pyridine 2,6-dialdimines, homo- and heteropoly-nuclear structures of iron can be generated. Sodium hydride reduction of the iron(II) bis(iminepyridine) complex $(2,6\text{-}(2,6\text{-}i\text{-}Pr_2C_6H_3N=CMe)_2(C_5H_3N))FeCl_2$ in THF leads to a wide variety of products where N_2-activation takes place (08IC896). One is the remarkable heteropoly-nuclear **177** (Ar = $2,6\text{-}i\text{-}Pr_2C_6H_3$) with π-delocalized azomethine units. Activated 2,6-bis(imino)pyridyl iron dichloride **154** (M = Fe, Ar = $2,6\text{-}Me_2C_6H_3$, $2,6\text{-}i\text{-}Pr_2C_6H_3$, R = Cl_2, R^1 = H) (04OM6087) are the basis for the polynuclear complexes of pyridylazomethines, e.g., the macrocyclic trinuclear iron(II) bis(imino)pyridyl **178** (05MAC2559). Methylene-bridged bis(imino)pyridyl ligands with iron(II) chloride in THF yield **179** (R = R^1 = i-Pr; R = i-Pr, R^1 = Me; R = R^1 = Me) (08ICA1843). The iron–aluminum heterodinuclear **180** (Ar = $2,6\text{-}Me_2C_6H_3$) is an important catalytic material (04OM5375). Interesting developments are related to the dimerization of **154** (M = Fe, Ar = $2,6\text{-}i\text{-}Pr_2C_6H_3$, R = Cl_2, R^1 = H) in a reaction with $LiCH_2SiMe_3$ in THF leading to **181** (05JA13019). Other homo- and heterodimetallic complexes are **182** (M = M' = Fe, X = Cl; M = Cp, M' = Co, X = Cl; M = M' = Co, X = Cl, Br; M = Co, X = Cl, M' = Ni, X = Br) and **183** (M = Fe, Co) (07OM4639). Dinuclear **184** and

185 (M = Fe Co Zn, X = Cl; M = Ni, X = Br; Ar = 2,6-*i*-Pr$_2$C$_6$H$_3$) are also interesting in this respect (06IC9890). Sodium hydride reduction of the vanadium(III) pyridylimine complexes containing the VCl$_3$ fragment leads to the dinuclear nitrogen-bridged **186** and **187** (Ar = 2,6-*i*-Pr$_2$C$_6$H$_3$) (05IC1187).

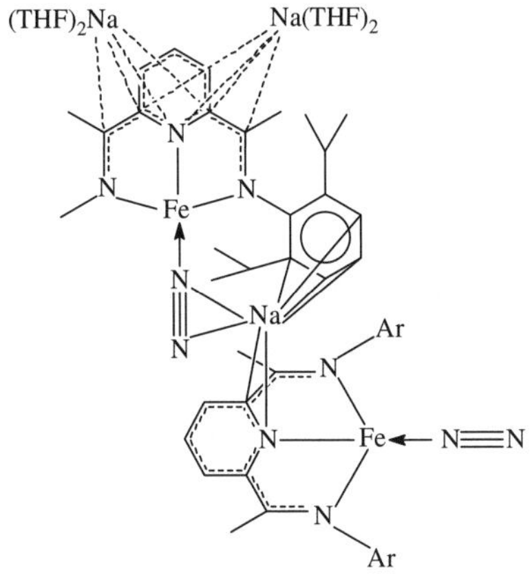

177

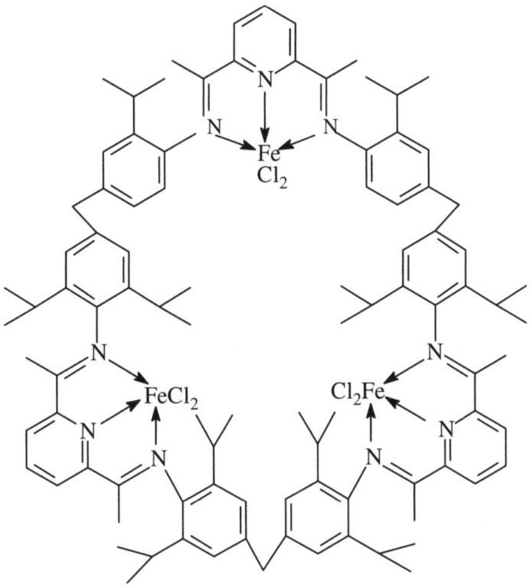

178

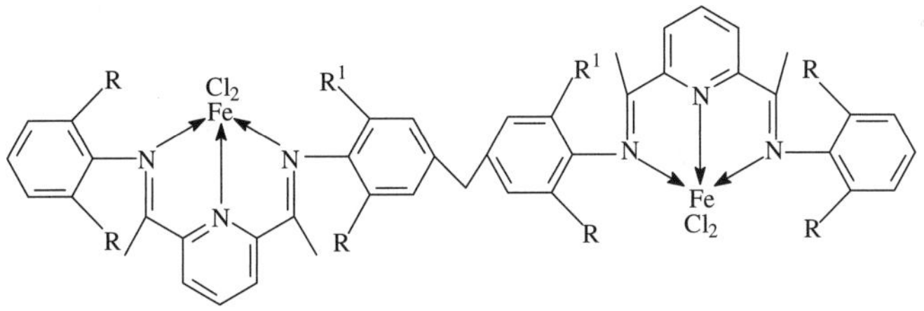

179

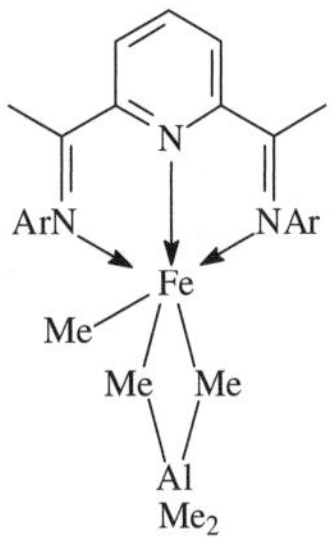

180

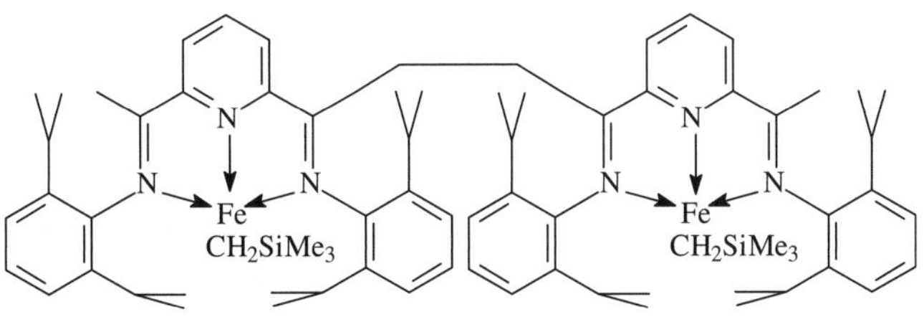

181

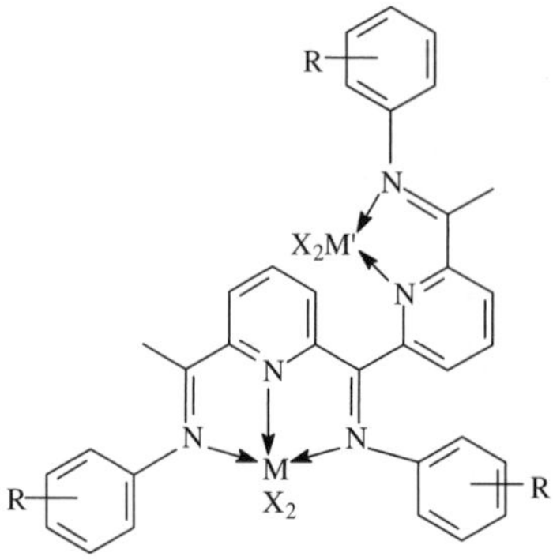

182

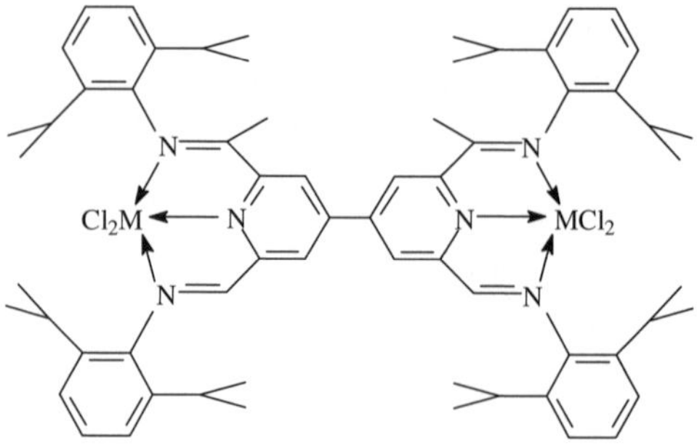

183

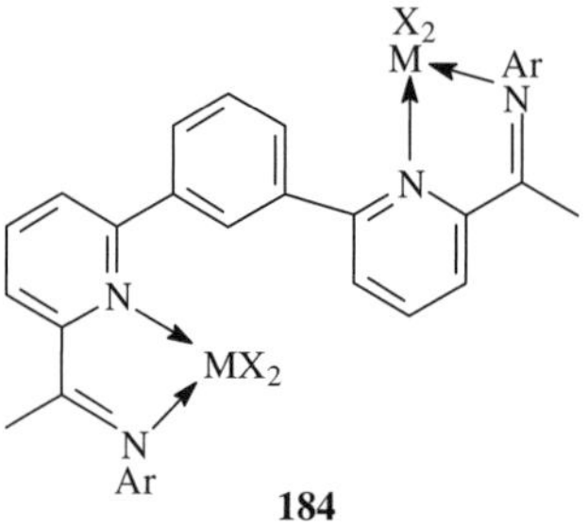

184

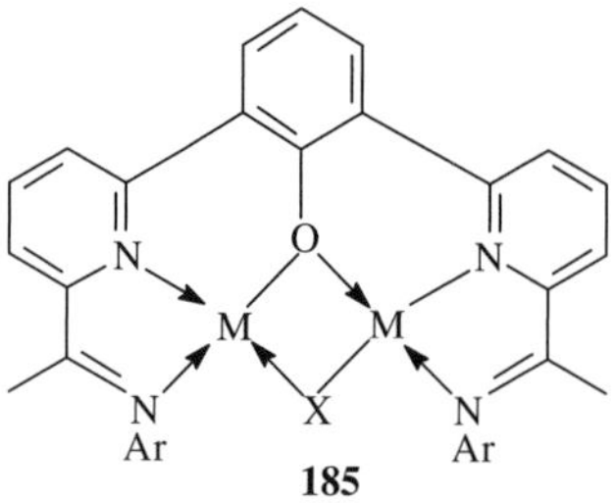

185

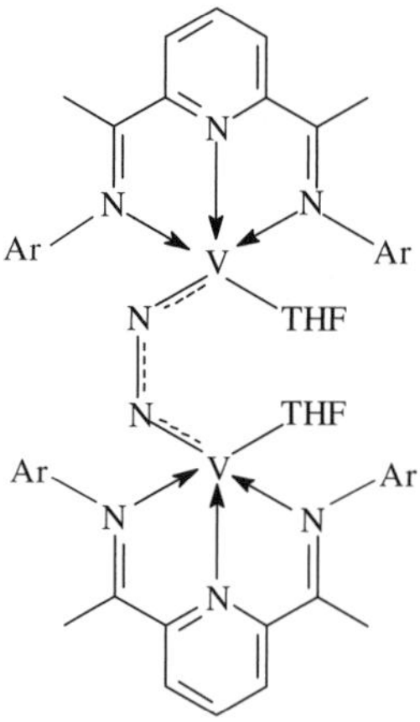

186

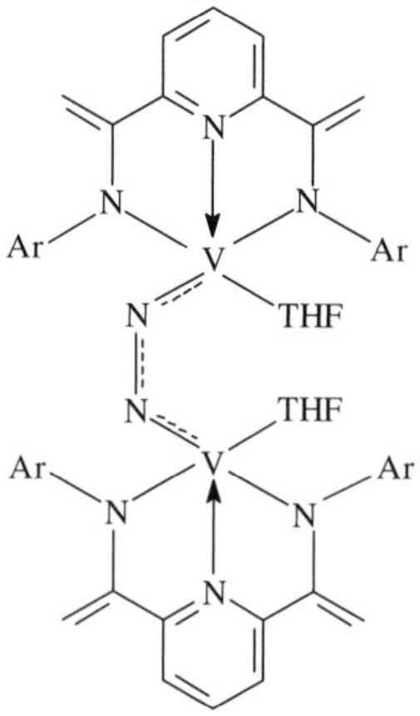

187

Cambridge crystallographic data bank lacks the data on the structurally characterized bis-chelates with heterocyclic aldehyde moieties. They may be obtained from precursor amines and their metal complexes. Such synthetic components **188** and **189** which may lead to complexes **190** were described for 2-amino-3-hydroxypyridines (M = Co,

Fe, Cu, Pd, Ru, Rh) (98M455). However, the synthesis of **190** (R = Ar, Het) requires additional experimental verification.

188   **189**   **190**

Complexes with imidazoline-imino moieties contain five-membered metallacycles and include mono- and dinuclear structures. Neutral lanthanide **191** (Ln = Y, Er, Lu) based on the bis(imidazoline-2-imino) pyridine are bis-chelate with tridentate coordination of the carbene-derivatized ligand (08ICA2236). Copper(I) complexes with an ethylene-bridged bis(imidazoline-2-imine) ligand **192** (R = i-Pr; X = PF$_6$, SbF$_6$) have potential for the catalytic polymerization of styrene (08JCS(D)887). Dizinc trichloride **193** (Ar = 2,6-i-Pr$_2$C$_6$H$_3$) reveals an interesting reactivity pattern with acetate to yield the acetate-bridged salt **194**, dibenzoylmethanate to produce **195**, ethoxide to afford the hydroxide-bridged **196**, and 2-pyridonate to yield a neutral trimetallic zinc **197** (07JCS(D)4565). Dinuclearity may be achieved by forming the internitrogen bridge, e.g., in the zinc(II) complexes of iso-acrydine azomethine **198** (07JCS(D)4788).

191   **192**

193   **194**   **195**

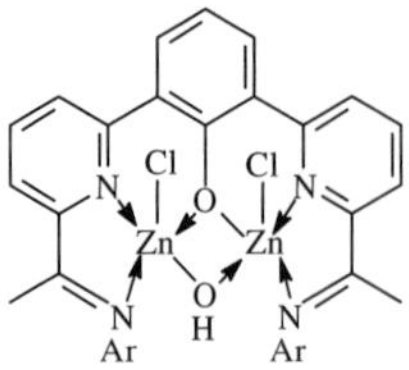

196 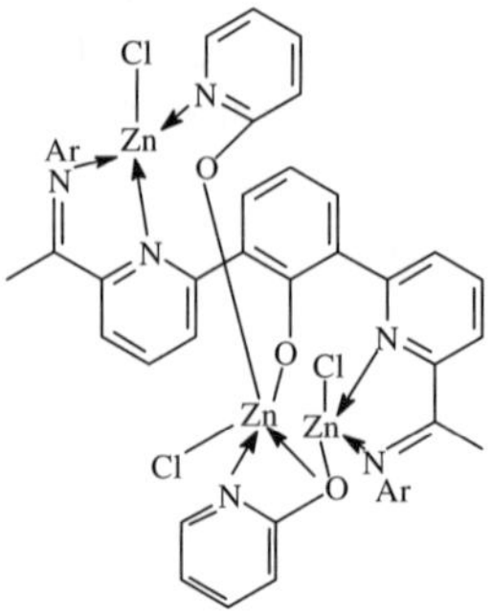   197 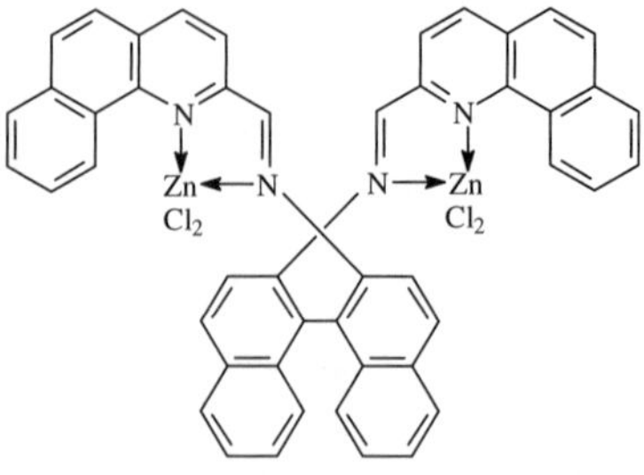   198

3.5 Complexes of hetarylazomethines with six-membered metal rings

Metal chelates are present in the complexes of Schiff bases of amino- $(X = NR_2)$, hydroxy- $(X = O)$, and mercapto- $(X = S)$ derivatives of monoheteroaromatic five-membered systems, azoles, and azines. A few publications on the complexes of azomethines of the monoheteroaromatic five-membered systems have appeared only recently. The X-ray structural study of copper(II) bis(2-N-n-octyliminomethyl)benzo[b]thiophene-3-olate) **199** (04ZNK1696) is interesting in the sense that its square-planar structure is complemented by an extended octahedral one due to the intermolecular contacts of the thiophene sulfur with the copper site. Among the azole complexes, azomethine derivatives of pyrazole (Equation (32)) prevail (05RCR193). There are several types of coordination units (N,N-, N,O-, N,S-, N,Se-) created by variation of the donor sites X in position 5 of the pyrazole ring. When X = O, S, Se, tautomer **b** is realized, whereas when $X = NR^4$, tautomer **a** predominates (Equation (32)). However, irrespective of the type of tautomer, in the chelates the coordination units have practically equalized bonds.

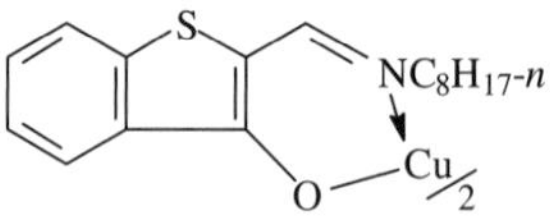

199

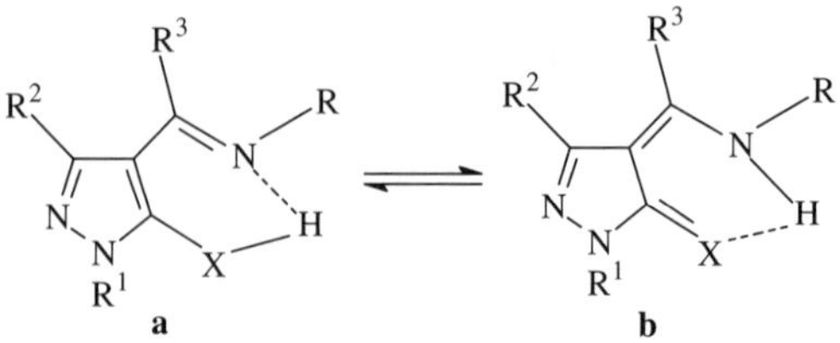

(32)

a   b

5-Amino-derivatives (X = NH) give rise to two types of structures in their metal complexes: classical bis-chelates **200** (97JCS(D)2045) and tricyclic salen complexes **201** (97JCS(D)4539). Chelates of aldimines of 5-oxopyrazoles (X = O) are represented by coordination compounds **202** and **203**. Complexes **202** include bis-chelates of cobalt (M = Co, R = C_6H_4-3-Cl, R^1 = Ph, R^2 = H, no L) (04CJAC49), (M = Co, R = $C_{10}H_7$, R^1 = R^2 = Ph, L = DMF) (04CJIC439), nickel (M = Ni, R = R^1 = R^2 = Ph, L = H_2O) (07AX(E)330), (M = Ni, R = R^1 = R^2 = Ph, L = EtOH) (06EPJ928), copper (M = Cu, R = Ph, R^1 = n-C_6H_{13}, R^2 = Et, no L) (05AX(C)318), and zinc (M = Zn, R = C_6H_4-4-t-Bu, R^1 = Ph, R^2 = H, no L) (05MI4). The tricyclic salen chelate contains the nitride group at the metal ion, **203** (05NJC283). The majority of pyrazole azomethine complexes have been prepared from 5-thioamino-derivatives (X = S), forming coordination compounds with two **204** or three **205** metallacycles. Metal chelates **204** include complexes of cobalt (M = Co, R = t-Bu, Ph, C_6H_4-4-t-Bu, R^1 = Ph, R^2 = Me) (05MI1), nickel (M = Ni, R = R^1 = Ph, R^2 = Me) (04MI1), (M = Ni, R = benzo-15-crown-5, R^1 = Ph, R^2 = Me) (04RKZ38), palladium (M = Pd, R = Ph, R^1 = i-Pr, R^2 = Me) (93DOK54), copper (M = Cu, R = CH_2Ph, R^1 = Ph, R^2 = Me) (00IZV1891), (M = Cu, R = 2'-C_5H_4N, R^1 = Ph, R^2 = Me) (92MC30), and zinc (M = Zn, R = C_6H_3-2,6-Me_2, R^1 = Me, R^2 = Ph) (97JCS(D)111), (M = Zn, R = R^1 = Ph, R^2 = Me) (05MI2), (M = Zn, R = C_6H_4-t-Bu, R^1 = Ph, R^2 = Me) (04MI1). For palladium bis-chelates, cis-(R = C_5H_4N, R^1 = i-Pr, R^2 = Me) or trans-(R = Cy, R^1 = i-Pr, R^2 = Me) isomers are realized depending on the nature of R (03MI3). Tricyclic salen-coordination compounds **205** (04CSR410) include a broad series of nickel tris-chelates (M = Ni, R = H, R^1 = R^2 = Me, Z = $(CH_2)_2$) (97JCS(D)2045), (M = Ni, R = H, R^1 = R^2 = Me, Z = $(CH_2)_4$) (97JCS(D)121), (M = Ni, R = R^1 = Ph, R^2 = Me) (03M16). Complexes with a diphenyl fragment (Z) include the seven-membered metallacycle **206** (95IC4467). Copper pyrazole-salen chelates, along with the complexes containing internitrogen alkyl bridges **205** (R = R^1 = Ph, R^2 = Me, Z = $(CH_2)_2$, $(CH_2)_3$) (99EJI1393), are represented by coordination compounds with aryl N,N-binding moieties **207** (R = H, 2,3-cyclo-C_4H_4) (99EJI1393). Zinc complexes **205** (M = Zn, R = H, R^1 = R^2 = Me, Z = 1,2-cyclo-C_8H_8) (05MI2) are scarce.

200   **201**

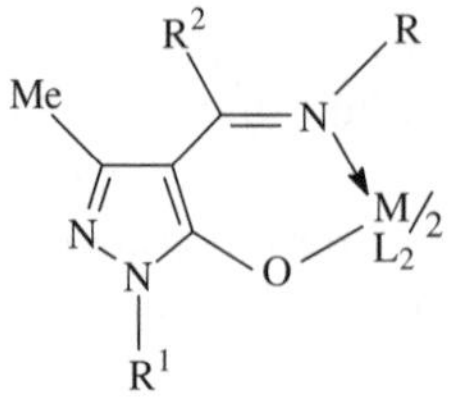

202

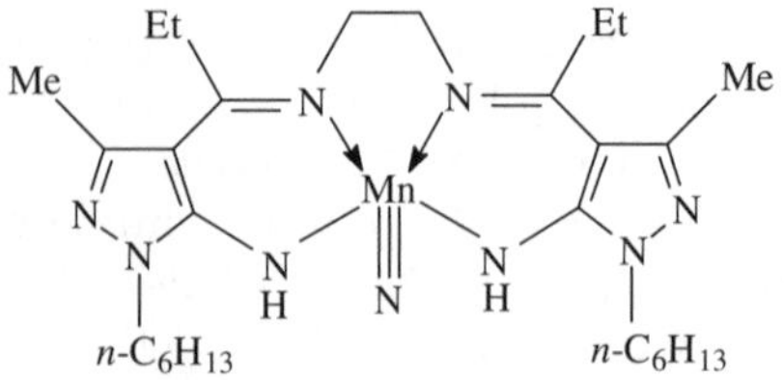

203

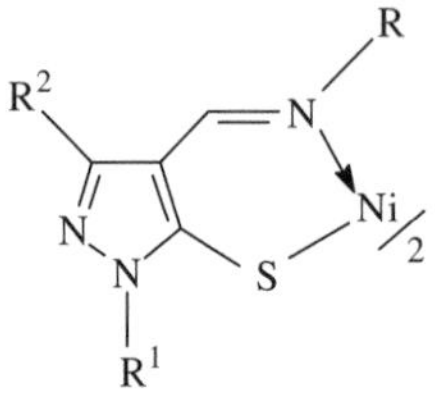

204

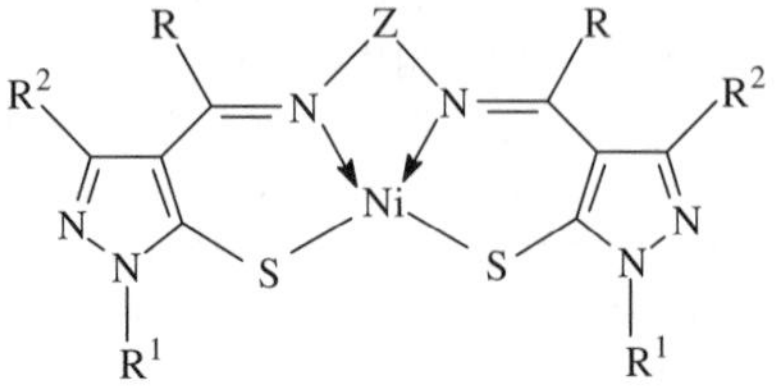

205

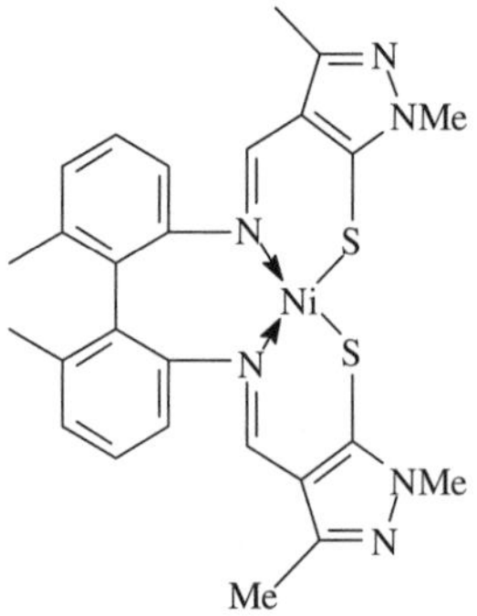

206

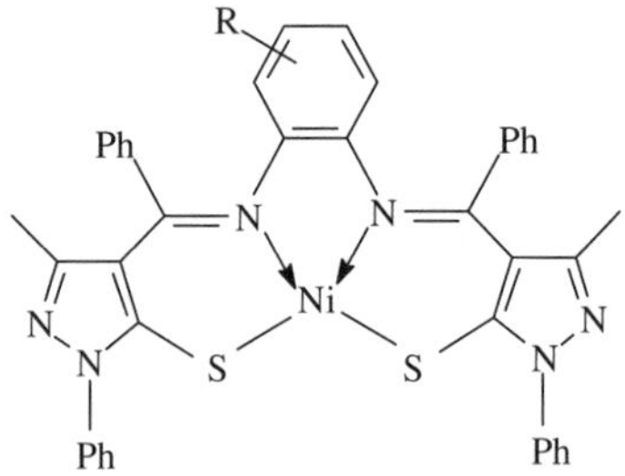

207

A limited number of oligonuclear structures are based on oxo- and thio-pyrazolone aminomethylene ligands. Copper dimers with oxygen phenol bridges **208** (05MI2) and **209** (04ACS2329) serve as illustrations. Dinuclear structures may be formed in cation **210** (97CC1711, 97DOK212). Trinuclear complexes can be prepared from the pyrazole azomethine metal ligand containing the tricarbonylchromium moiety in the N-phenyl substituent **211** (89JOM303).

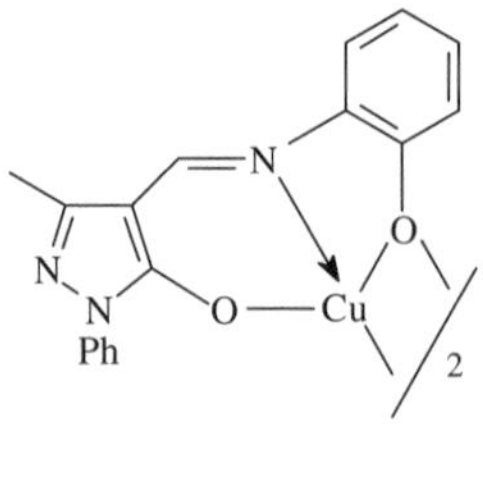

208

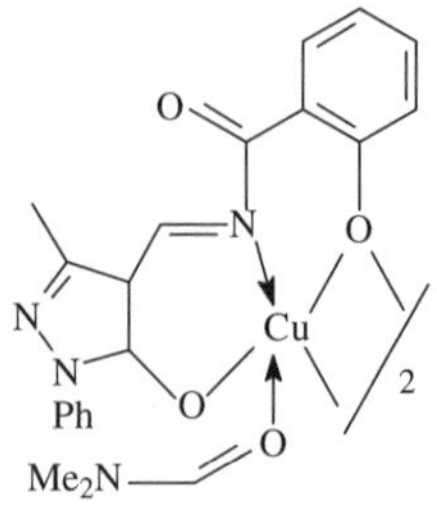

209

210   211

Coordination compounds of the azomethines of hydroxy- (mercapto-) derivatives of the azine series are poorly studied (93CCR1). They include neutral azomethine bis-chelates of pyridine **212** (M = Cu, Zn, R = *n*-Bu, *t*-Bu, CH$_2$Ph) (82ICA213) and quinoline **213** (M = Cu, Zn) (91MC78) as well as pyridine cation **214** (M = Zn, Cd, R = Ar, (CH$_2$)$_2$) (86ICA(124)121, 89JCS(D)1979). Hetarylazomethines containing the quinoline substituent in the aldehyde counterpart are represented by **215** (M = Ru, X = Cl, X^1 = NO; M = Cr, X = Cl, no X^1; M = Zn, no X and X^1) (03OM850). Another bis-chelate is based on the bis(2-pyridylimino) isoindolato ligand whose palladium complex has structure **216** (04EJI3424).

212   213   214

215   216

3.6 Metal chelates with heterocyclic N-substituents

N-Substituents in hetarylazomethine ligands are represented by three- and five-membered rings with one heteroatom as well as azole and azine heterocycles. They may not always participate in the formation of metallacycles and thus play the role of coordination-active or inactive constituents of the hetarylazomethine ligands. A great majority of coordination compounds includes mononuclear complexes, although di- and oligonuclear structures are also known.

3.6.1 Three- and five-membered monoheteroatomic rings

Three-membered n-propylaziridine substituents play the role of coordination-active fragments and form octahedral chelate units of iron in cationic **217** (93IC2670). In the nickel chelates (90KK395, 99T12045, 06JCCR61), e.g., **218** (99T12045), the pyrroline nitrogen is coordinatively active. A series of chelates contains pyrrole or indole N-substituents not bonded to metals. Examples are **219** (M = Co, Ni) (96ZN(B)757) and **220** (M = Ni, R = Me, Et) (89JCC273, 90ICA(176)261), **220** (M = Cu, R = Me) (89JCC273, 92ZK61). No coordination of the oxygen atom in the complexes of the O-containing five-membered heterocycles furan **221** (05CJSC909) and 1,3-dioxolane **222** (87IC349) was observed. When N-substituents in five-membered heterocycles include the coordinatively active oxygen-containing moieties, complementary O–M-bonding occurs. Such chelates are represented by indole **223** and thiophene **224** fragments. Coordination compounds **223** have been prepared and structurally characterized for nickel (M = Ni, L = H_2O, $n = 3$) (90AX(C)1414) and copper (M = Cu, L = H_2O, $n = 2$) (96POL4407), (M = Cu, L = $NH_2C_5H_4N$, $n = 2$) (01POL2877). The oxygen atom of the carboxyl group participates in coordination in **224** prepared from azomethines of the substituted salicylaldehydes and 2-amino-3-carbethoxy-4,5-diemthylthiophene (M = Co, Ni, Cu, Zn; R = R^1 = R^2 = R^3 = H, Me, Cl) (06RJCC879).

217

218

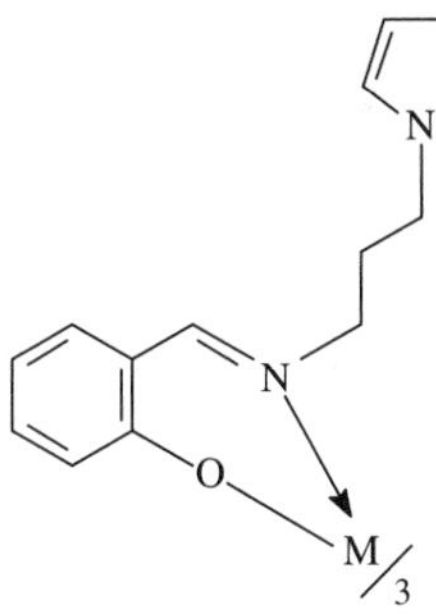

219 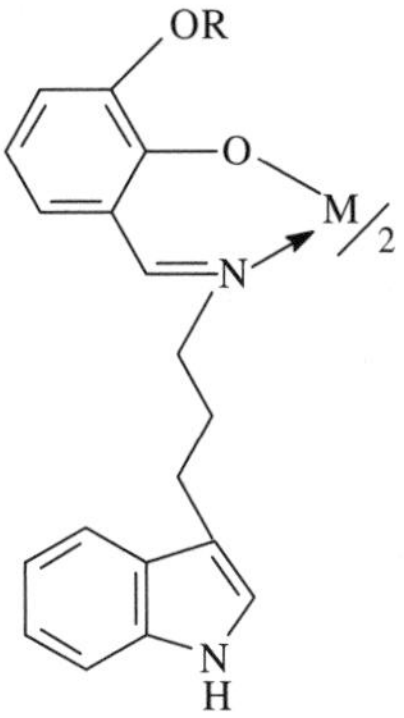   **220**

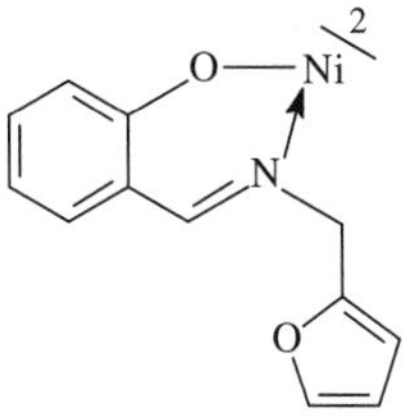

212 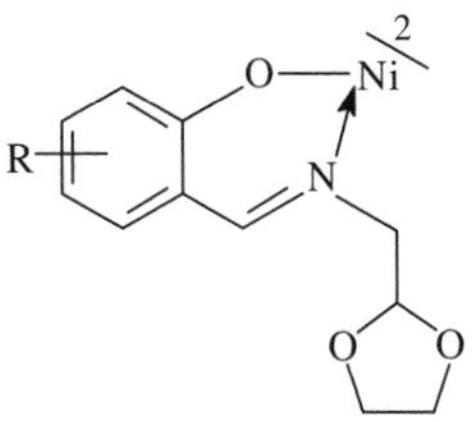   **222**

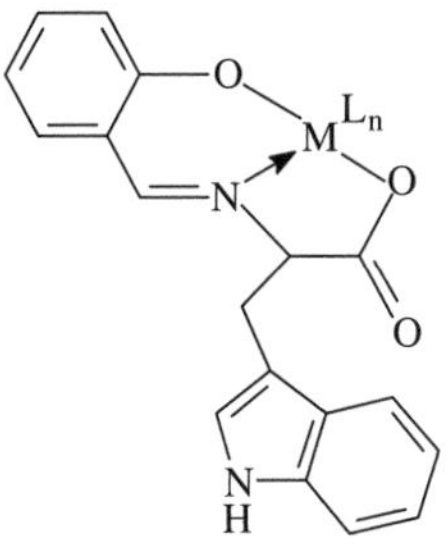

223 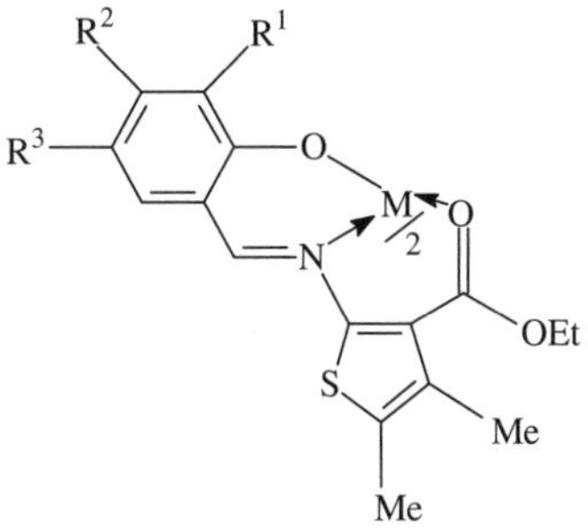   **224**

3.6.2 Azole N-substituents

Azole substituents at the nitrogen atom of an exocyclic C=N bond are predominantly coordination-active. The site for coordination is the nitrogen atom of the pyridine type. However, for the ring closure to stable five- or six-membered metallacycles, spatially favorable conditions are necessary. Illustrations include chelates of pyrazoles **225** (97JCS(D)2155) and **226** (X = O, S) (00ICA(310)183). Complexes of an imidazole azomethine involve chelates with different numbers of coordinated azole moieties. Nickel monoimidazole chelates are represented by **227** (99JA6956). The same number of imidazole rings is present in neutral **228** (95IC2108) and cationic **229** (Z = (CH$_2$)$_2$, R = R^1 = H, A = NO$_3$, L = H$_2$O) (00ICA(298)256), **229** (Z = (CH$_2$)$_3$, R = R^1 = H, A = BPh$_4$, no L) (88JCS(D)1943), **229** (Z = (CH$_2$)$_2$, R = H, R^1 = Me, A = ClO$_4$, no L) (97BCJ2461) copper complexes. Several cationic compounds contain two coordinatively active imidazole moieties, e.g., tetra-chelate octahedral manganese **230** (M = Mn, A = PF$_6$) (07ICA557), (M = Mn, A = ClO$_4$) (96BCJ1573) and iron (M = Fe, A = PF$_6$) (86IC394).

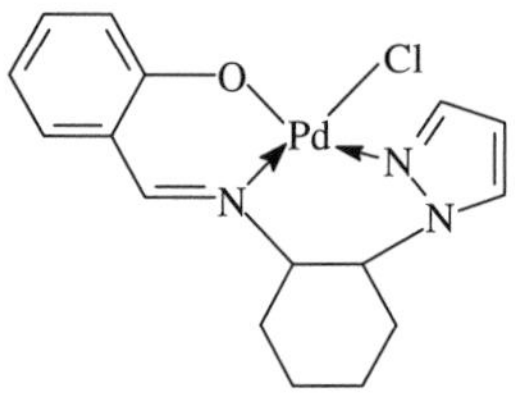

225

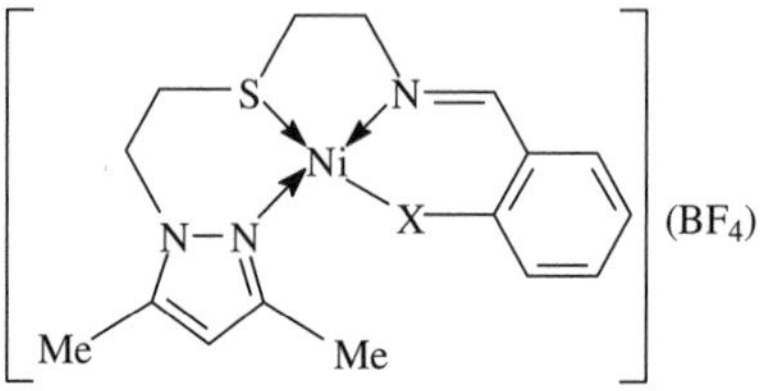

226

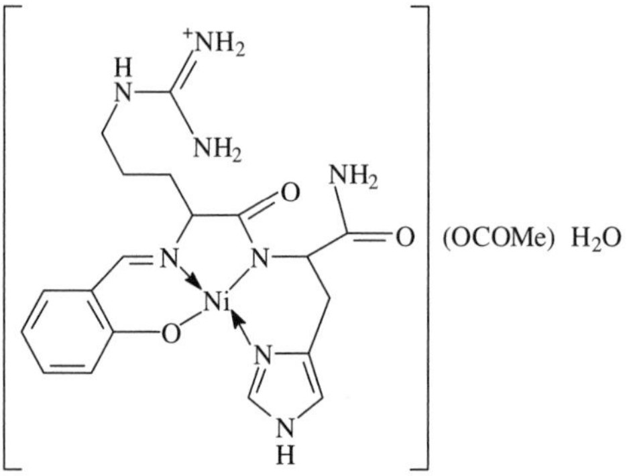

227

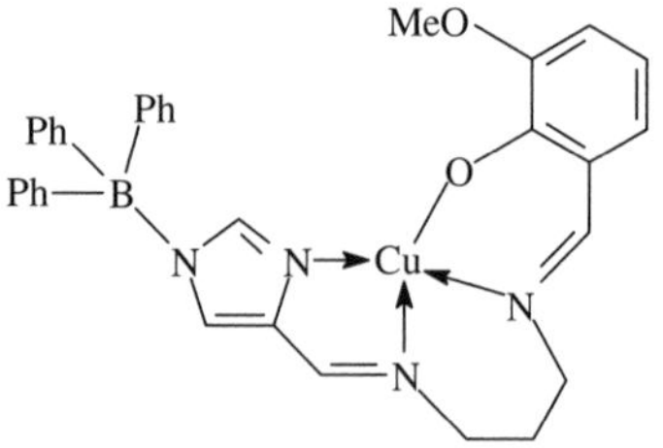

228

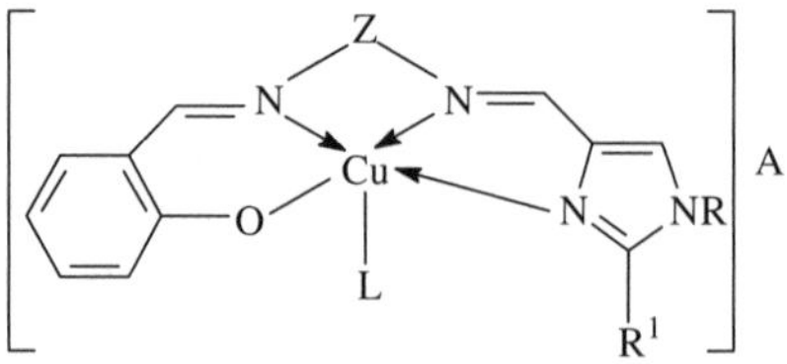

229

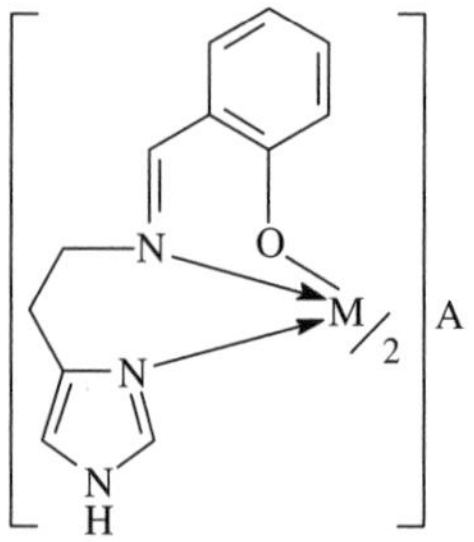

230

Coordinatively active *N*-benzimidazole substituents are present in the bis-chelate azomethine neutral complexes dioxovanadium **231** (M = VO$_2$, R = H, no L) (05MI3), copper (M = Cu, R = 4-Cl, L = NCS) (06AX(E)977), (M = Cu, R = 4-NO$_2$, L = Cl) (04AX(E)1855), and zinc (M = Zn, R = 4-Br, L = Cl) (05AX(E)2302). Cationic **232** contains coordinated benzimidazole and phenanthroline moieties (06CJSC1343). Imidazole N-substituents participate in coordination in the mixed-ligand complexes of vanadyl **233** (R = R^1 = H) (92IC1981, 92JA9925), (R = R^1 = –C≡C–Ph) (03IC1663) and **234** (92IC2035). *N*-Benzimidazole azomethine acetylacetonate tetrachelate **235** is similar (05MI3). A Schiff base prepared from salicylaldehyde and 2-aminomethyl- or 2-aminoethylbenzimidazole (HL) reacts with [VO(acac)$_2$] to yield the vanadium(IV) complexes [VO(acac)(η^3(N,N,O)-L)] (06IC5924). The products can be oxidized in air to a dioxovanadium(V) [VO$_2$(η^3(N,N,O)-L)]. A Schiff base derived from 2-aminoethylbenzimida-zole also reacts with VOSO$_4$ to yield a mixture of products, among them **236–238**.

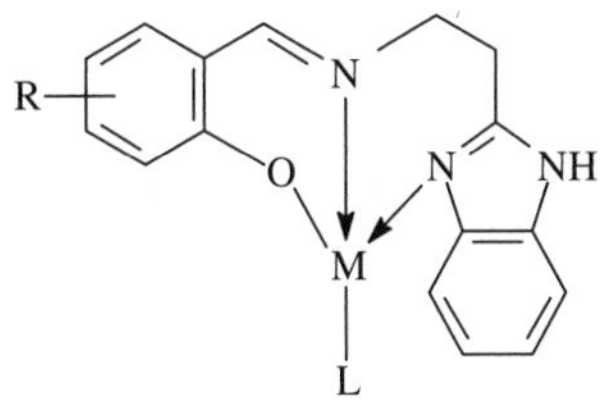

231

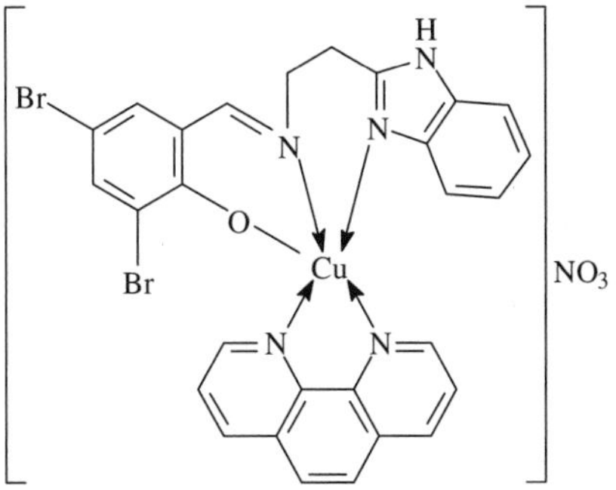

232

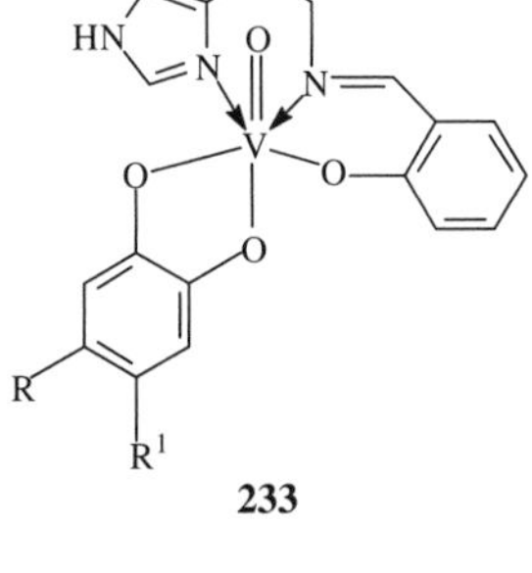

233

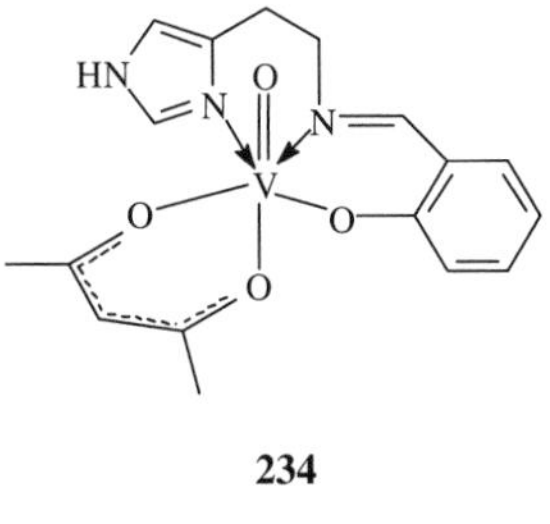

234

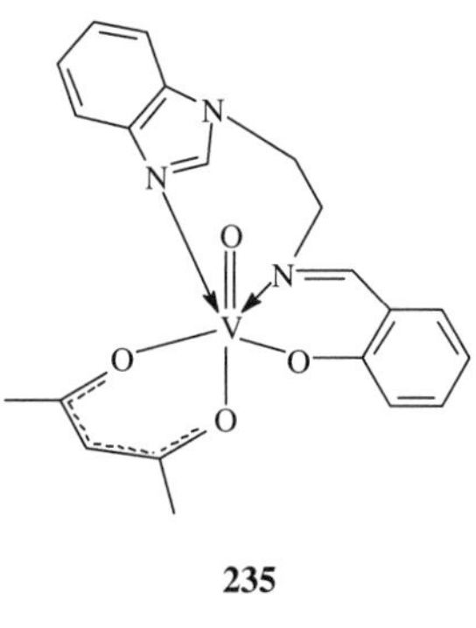

235

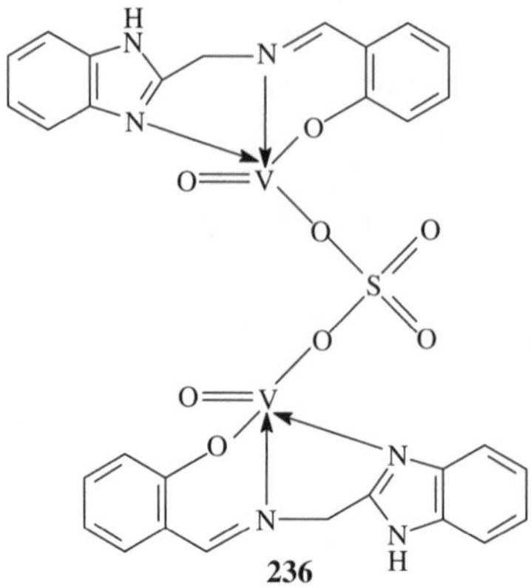

236

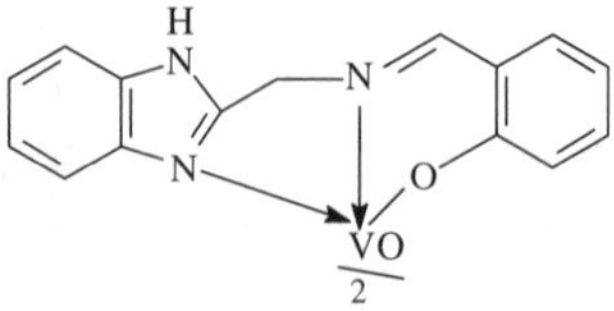

237

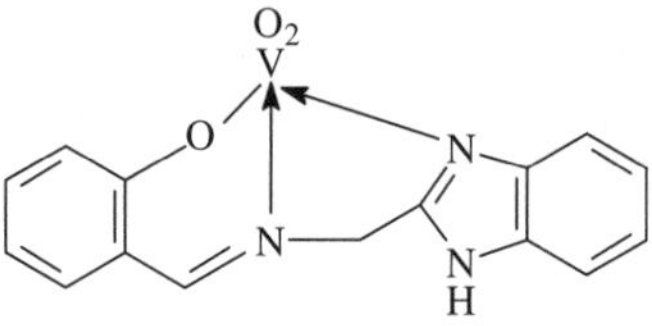

238

There exists a series of complexes with no coordination of the spatially remote imidazole **239** (05JCS(D)2312) and benzoisothiazole **240** (06ICA2321) moieties to a metal center. In tris-chelate **241** there are coordinated and non-coordinated benzimidazole N-substituents (05MI3). Oligonuclear complexes are represented by structures with two **242** (04ICA3407), **243** (06JCS(D)4260) and four **244** metal ions. Dinuclearity of the cobalt cationic complex of 1,3,4-thiadiazole **242** is related to the metal-containing anion. Dinuclear **243** is due to the bridging function of imidazolidine. One hydrogenated five-membered heterocycle along with the carbonate bridge ensures the formation of tetranuclear zinc **244** (02JCS(D)4746). Widespread intermolecular metal bonding where coordinatively unsaturated complex-forming agents and donor atoms participate is observed with imidazole moieties and leads to **245** (R = H, OEt) (93JCS(D)2157). Using the formate bridge, a manganese complex with the N-coordinated imidazole N-substituent can be prepared, **246** (95IC5252).

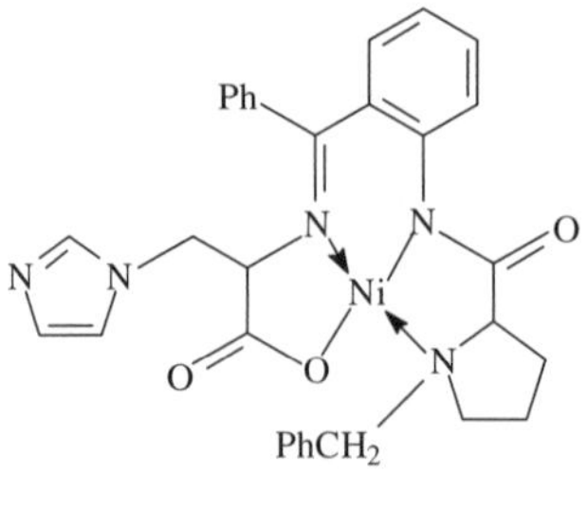

239

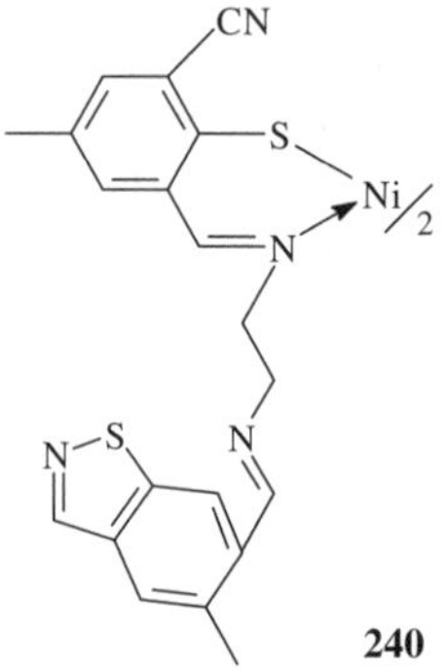

240

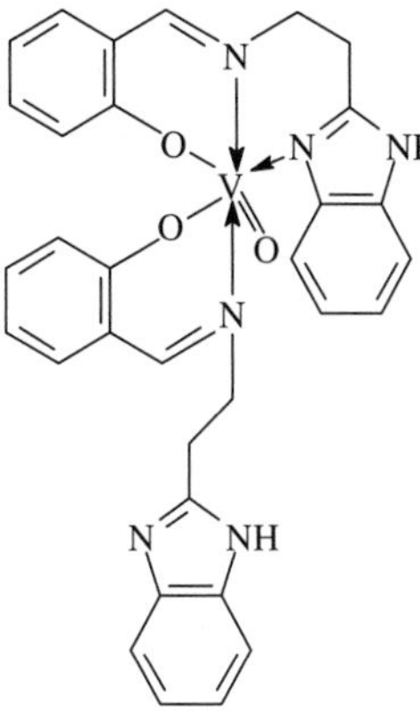

241

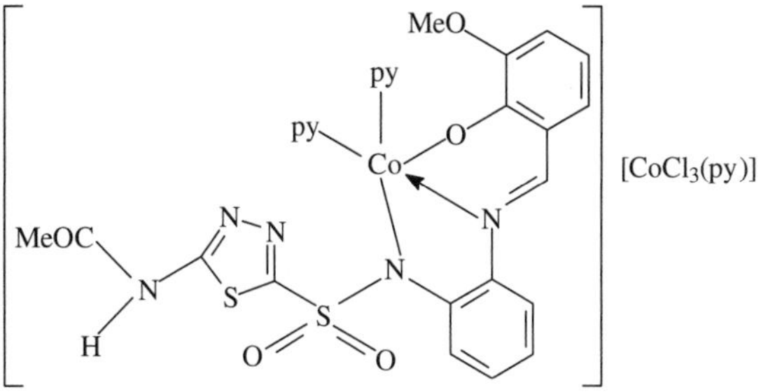

242

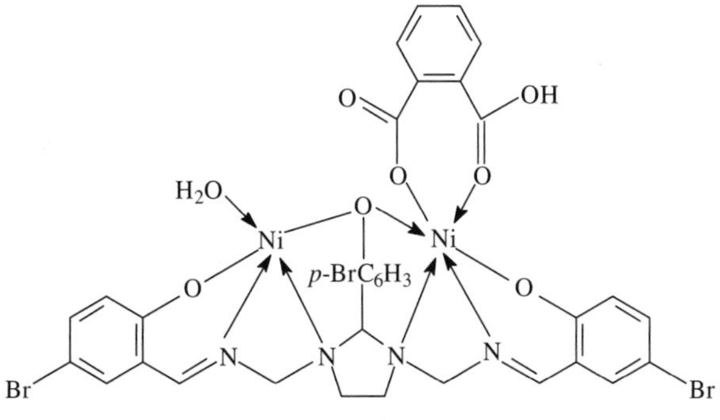

243

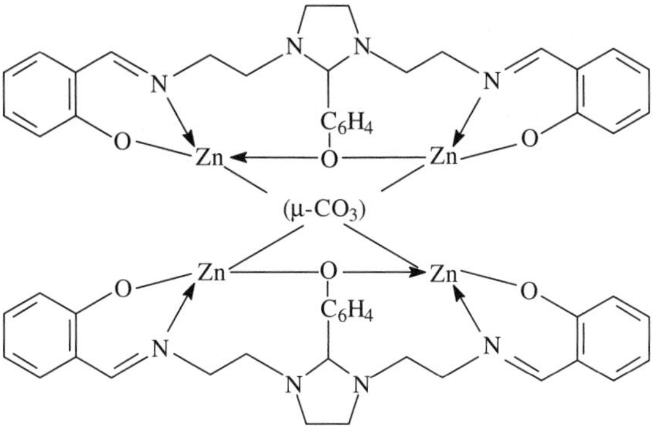

244

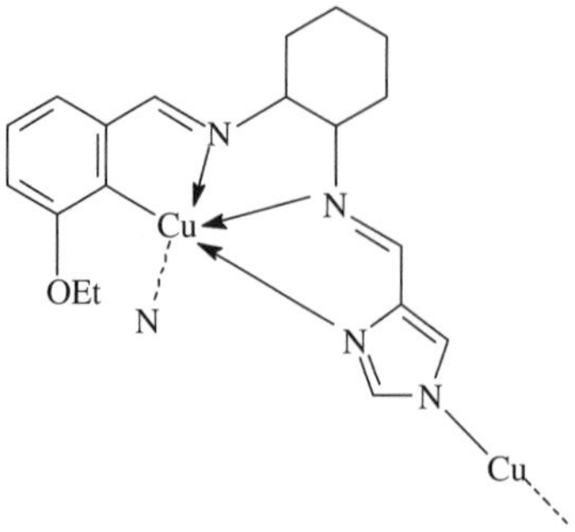

245

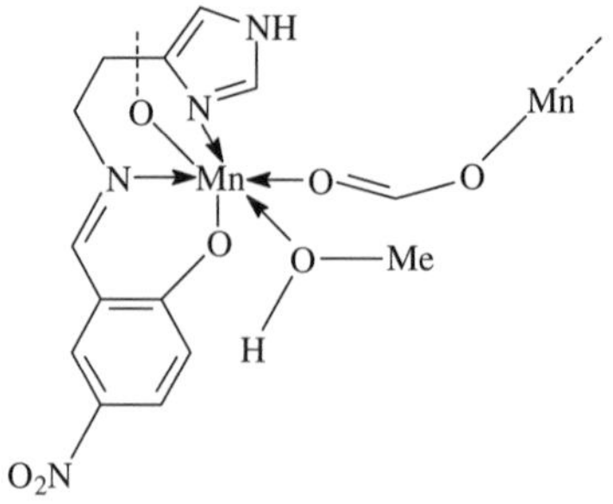

246

3.6.3 Azine N-substituents

Pyridine and quinoline rings may serve as N-substituents. Usually, an α-pyridyl fragment does not participate in coordination **247** (X = NTs, O, S), which is realized in cobalt **247** (M = Co, R = R^1 = H, X = NTs) (98POL1547), nickel (M = Ni, R = R^1 = H, X = S) (94KK824), copper (M = Cu, R = H, R^1 = 6-Me, X = O), (M = Cu, R = 5-OMe, R^1 = 3-Me, X = O) (89POL2543), and zinc (M = Zn, R = 2,3-*cyclo*-C$_4$H$_4$, R^1 = H) complexes (00JMS(523)61). The same number of metallacycles is contained in the metal complex of the Schiff base of 2-amino-3-benzyloxy-substituted pyridine **248** (04JOM936) with cobalt **249** (M = Co, R = Et) (04CJSC183) and zinc (M = Zn, R = H) (94KK824), complementary M–N coordination contacts arise. An interpretation of the lack of coordination of a pyridine N-substituent refers to the relatively low stability of the four-membered metallacycles (07MI1). Indeed, with more stable five- and six-membered chelates, metal complexes are formed with a coordinated pyridine. An extensive series of bis-chelates with a methylene bridge between azomethine and azine moieties is known, the structures of which include a coordinated pyridine nitrogen in the five-membered metallacycle.

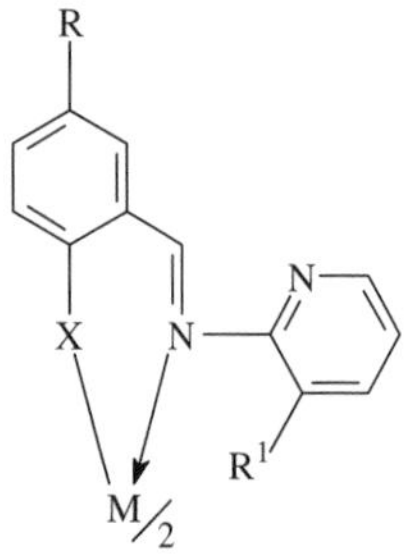

247

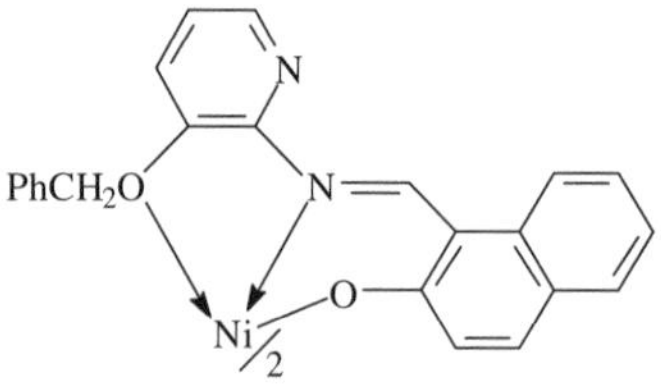

248

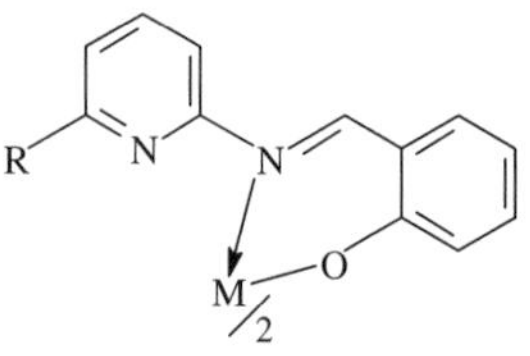

249

Illustrations include neutral dioxovanadium **250** (M = VO$_2$, R = H, no L or L^1) (04EJI1873), (M = VO$_2$, R = 6-Ph, no L or L^1) (04JCS(D)2314), iron (M = Fe, R = H, L = Cl$_2$, L^1 = MeOH) (05AX(E)2338), copper (M = Cu, R = H, L = NO$_3$, no L^1) (89AX(C)7), (M = Cu, R = 2,3-*cyclo*-C$_4$H$_4$, L = NO$_3$, no L^1) (04AX(E)1079), (M = Cu, R = 4-OMe, L = Cl, no L^1), (M = Cu, R = 3-OMe, L = Br, no L^1) (05ICA383), and zinc (M = Zn, R = 4-NO$_2$, L = NCS, L^1 = MeOH) (05AX(E)1571), (M = Zn, R = 3-COOH, L = NO$_3$, L^1 = H$_2$O) (01IC208), (M = Zn, R = 2,3-*cyclo*-C$_4$H$_4$, L = NO$_3$, no L^1) (05AX(E)247). Complex **251** exhibits high catalytic activity in alkene oxidation (08POL1556). Five-membered metallacycles are present in chromium complex with an exotic R substituent **252** (05JA11037), vanadyl chelates with carboxamide fragments and O-(95IC4213, 96IC357) or S-(96IC357) donor atoms **253** (X = O, S). Chelate rings of the same size are observed in the mixed-ligand **254** (02IC4502). Similar coordination sites with the five-membered metallacycles involving a pyridine nitrogen atom are present in the cationic complexes **255** (M = Cu, Zn; R = H) (04AX(E)884), (R = 4-Br) (05AX(E)370), and zinc (R = H) (04AX(E)1017), as well as in tetra-chelates **256** (M = Co, R = Br, A = NO$_3$) (05AX(E)335) and **256** (M = Fe, R = H, A = ClO$_4$) (07IC9558).

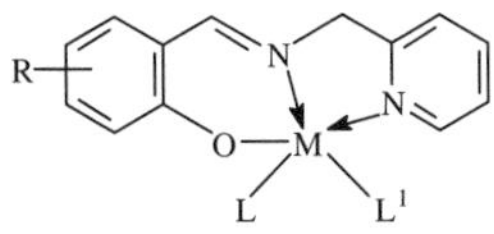

250 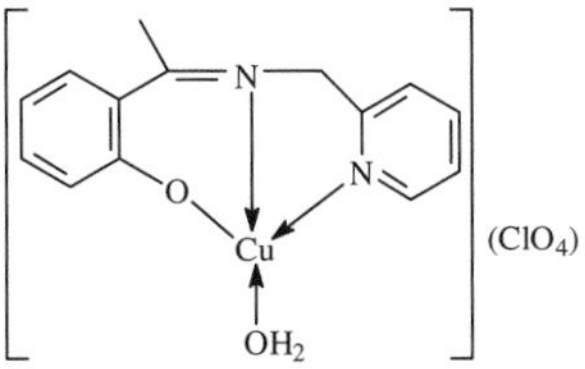   **251** 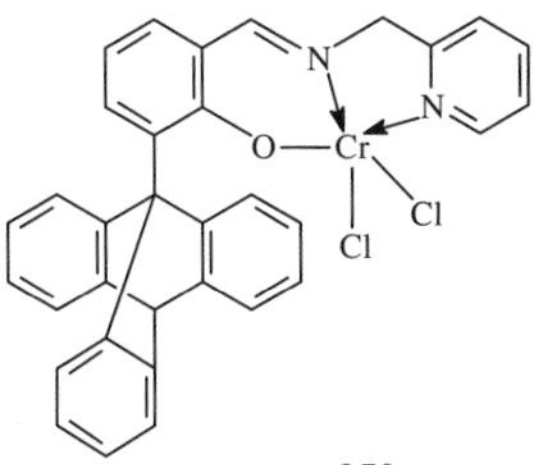   **252**

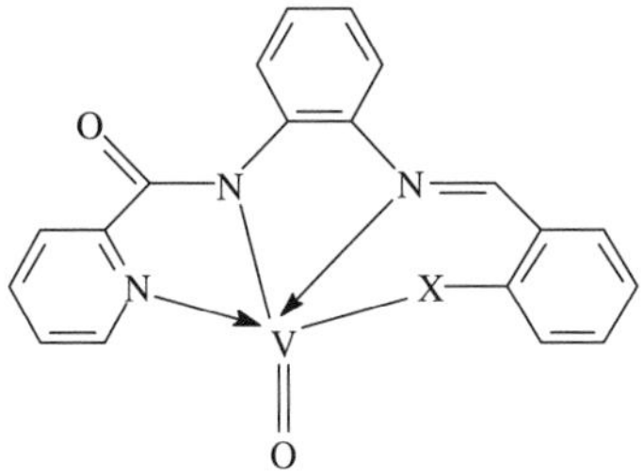

253 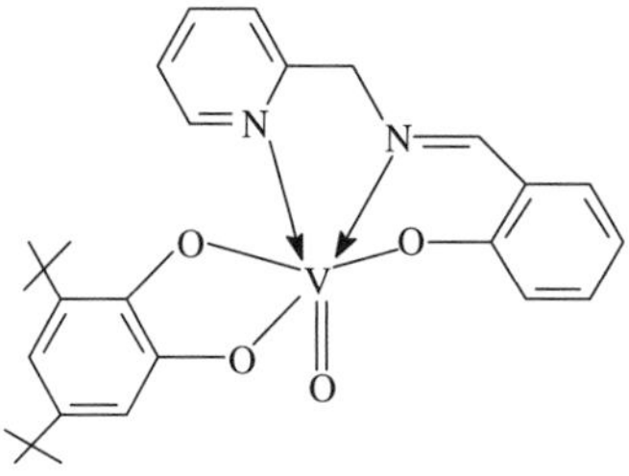   **254**

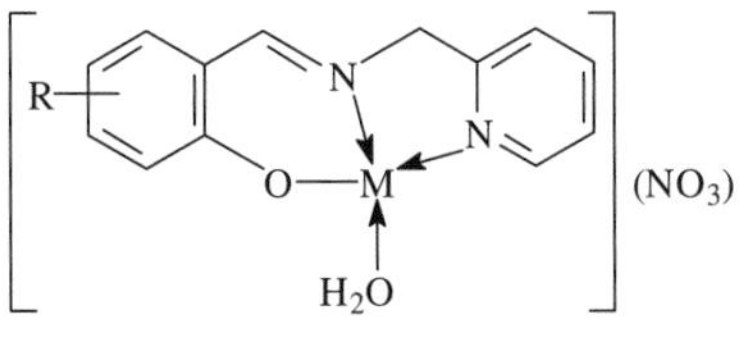

255 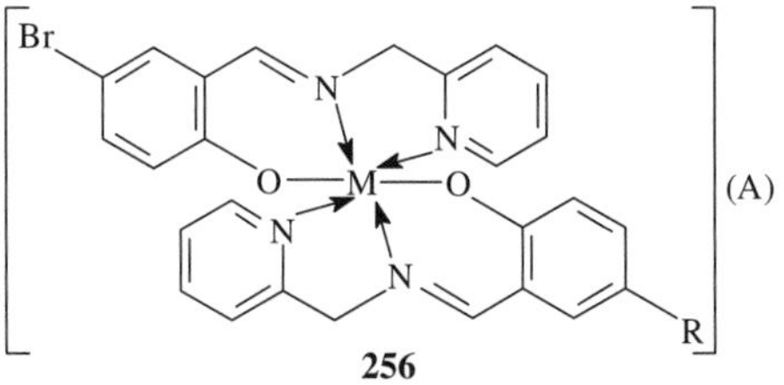   **256**

Bicyclic chelates containing a six-membered pyridine-coordinated ring are represented by neutral palladium **257** (92JCS(D)3083) and cationic copper **258** complexes (92BCJ1603). Similar coordination of a pyridine N-substituent can be found in neutral nickel tetra-chelate **259** (89ZAAC215) and cationic manganese **260** (M = Mn, A = ClO$_4$) (96BCJ1573) and iron **260** (M = Fe, A = BPh$_4$) (94ICA123). Pyridine-containing chelate moieties of different size are realized with manganese **261** (04MI2), cobalt **262** (M = Co), iron **262** (M = Fe) (78JCS(D)185), and zinc **263** (97JCS(D)161). Complexes **262** contain a seven-membered metallacycle with a coordinated pyridine. Pyridine azomethine complexes include chelates with nitrogen- (**264**) and sulfur-containing (**265**) electron-donor centers. Tris-chelates, including along with pyridine an amino-donor framework, are represented by neutral **264** (98IC1127). Nickel and copper **265** having a similar number of metallacycles can be prepared from azomethines of salicyl (M = Ni, X = O, R = N$_2$Ph, A = ClO$_4$) (03ACS207), (M = Cu, X = O, R = 4-Br, A = ClO$_4$) (04AX(E)1259), and thiosalicyl aldehydes (M = Ni, X = S, R = H, A = BF$_4$) (00ICA(310)183).

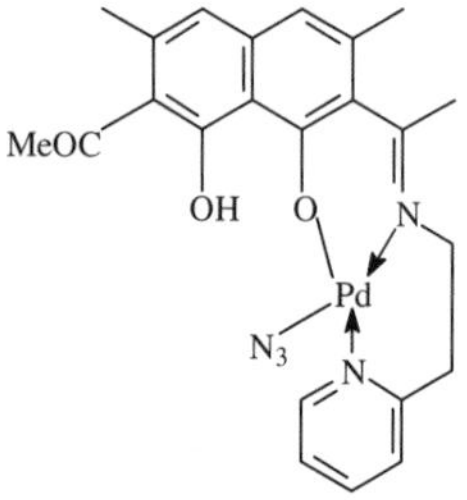

257 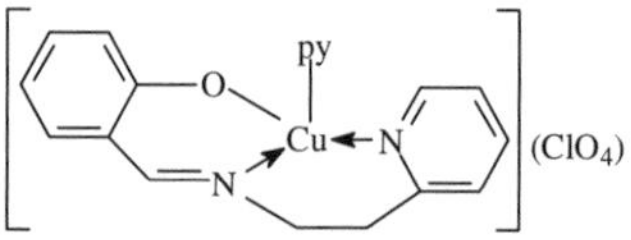   258 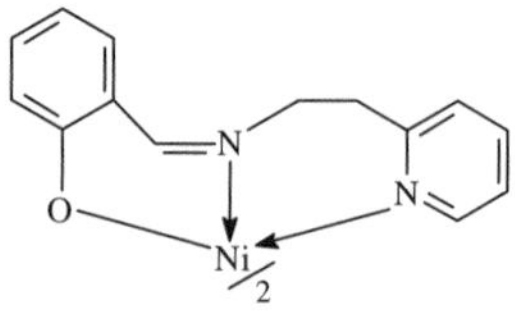   259

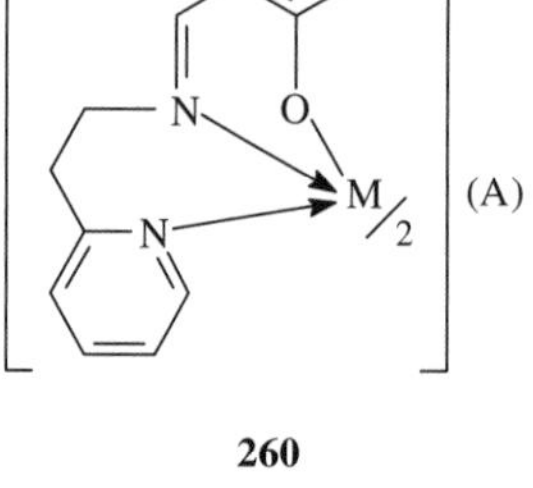

260 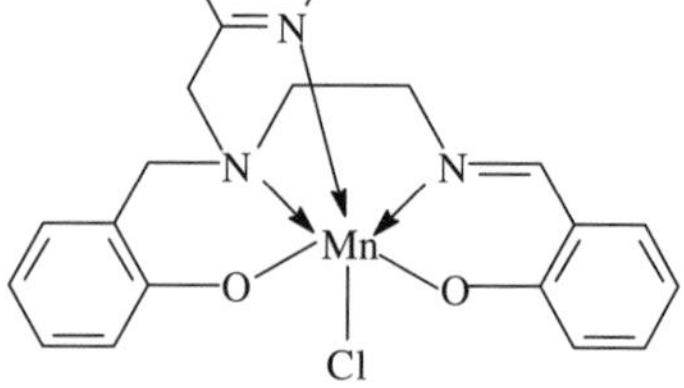   261

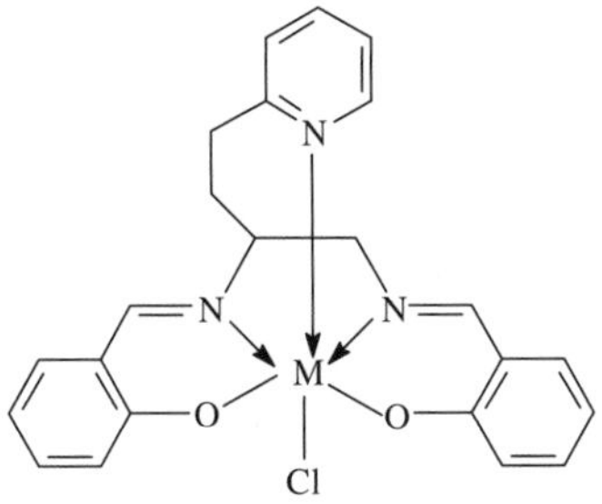

262

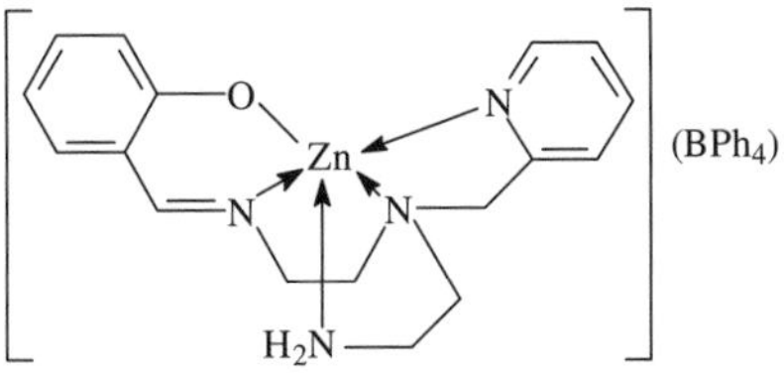

263

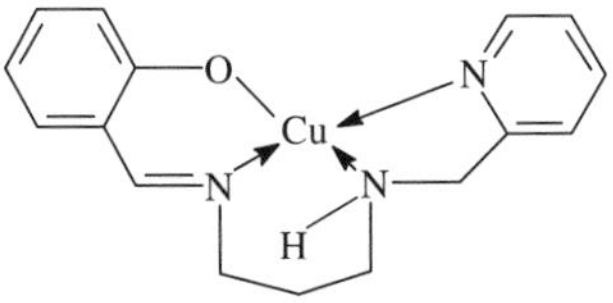

264

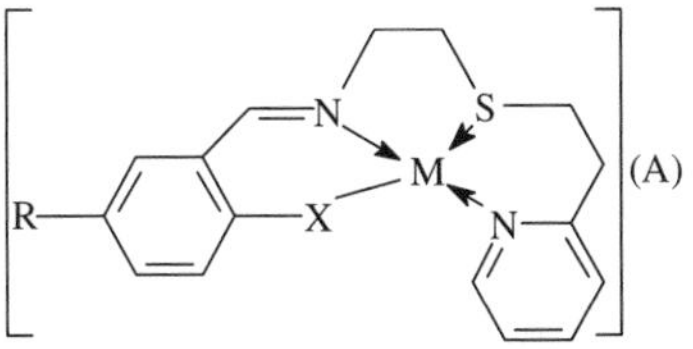

265

Pyridine azomethines with several coordinatively active heterofragments comprise neutral iron **266**, containing four six-membered metallacycles (91ICA19) and cationic manganese **267** (99ICA139), **268** (M = Mn, A = ClO$_4$) (06IC2373), and iron **268** (M = Fe, A = BPh$_4$) (02CC1460). The coordinatively active N-atom of the pyridine ring may be spatially unavailable for metal-binding, **269** (03IC8878), **270** (81ICA47), **271** (M = Ni) (98AX(C)725), **271** (M = Cu) (97AS519), and **272** (97IC6080). In manganese bis-chelate **273**, pyridine substituents are simultaneously coordinated in the six-membered metallacycle and in the non-coordinated fragment (93BCJ1675). The same coordination situation is observed in manganese **274** (90CL1181) and **275** (93BCJ1675). In **276**, both 2-ethylpyridine fragments are not coordinated (03IC8878). A complex of a pyridylazomethine ligand containing phosphorus of the diphenylphosphine moiety as the ligating atom has the less-common structure **277** (95CC331).

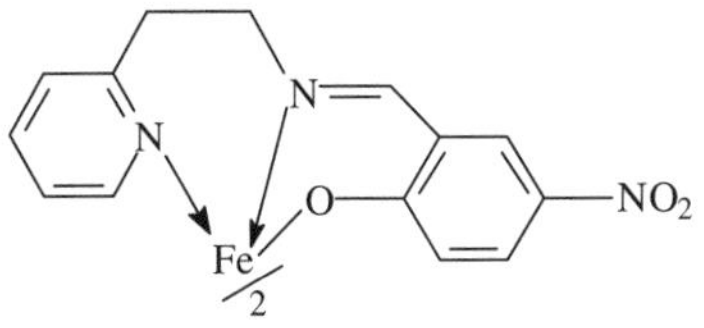

266

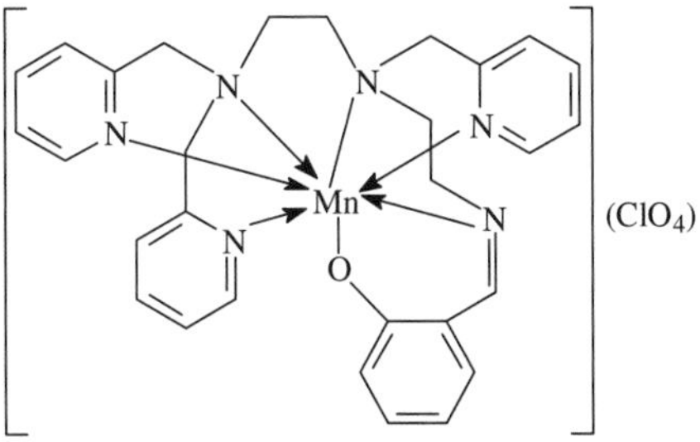

267

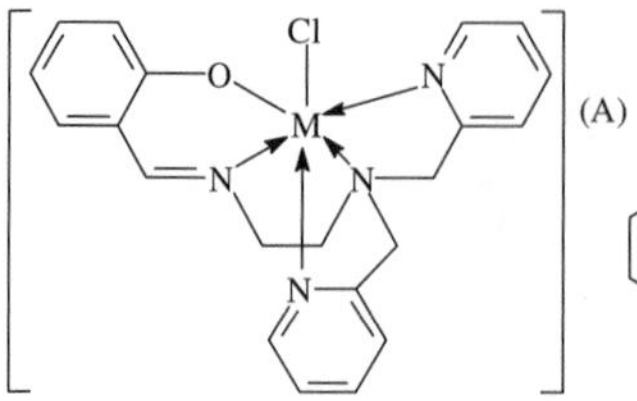

268

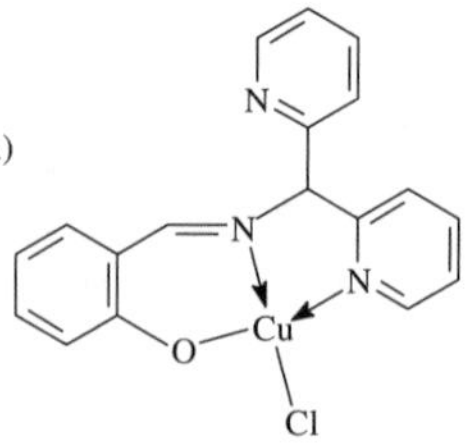

269

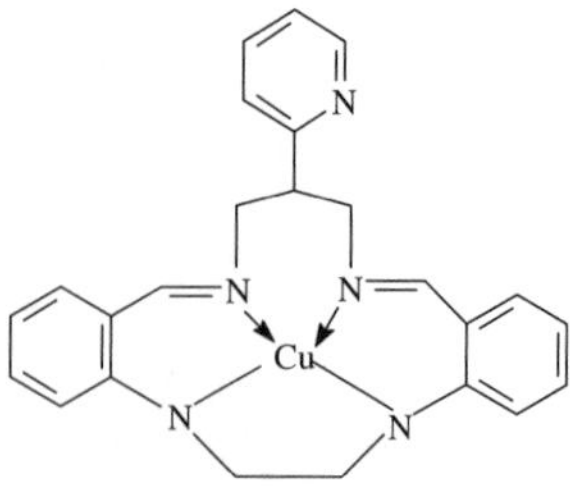

270

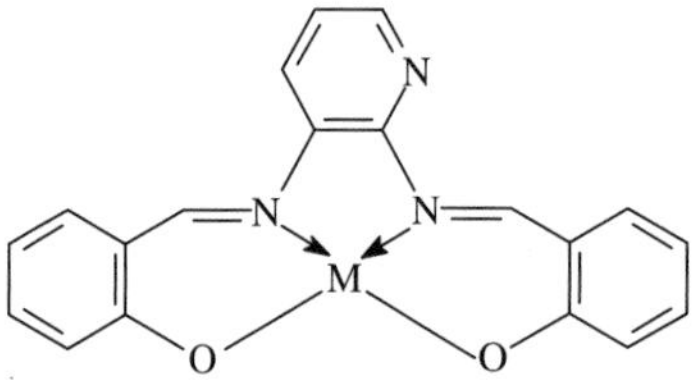

271

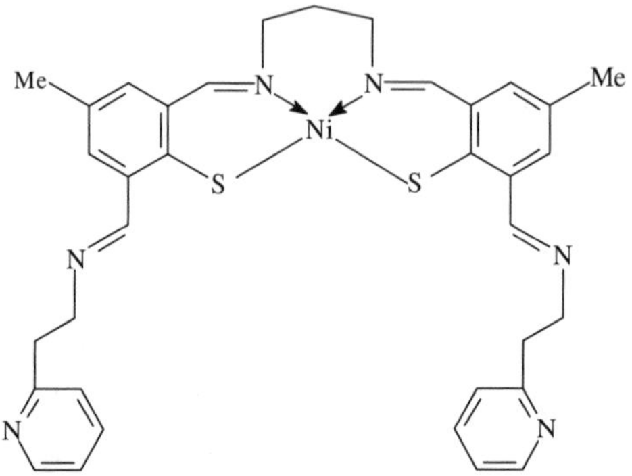

272

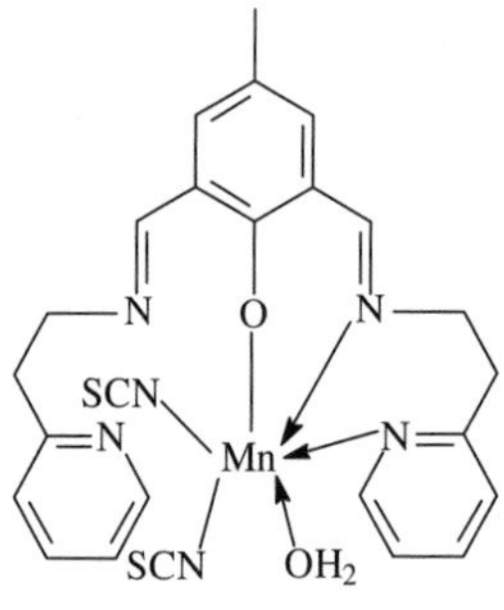

273

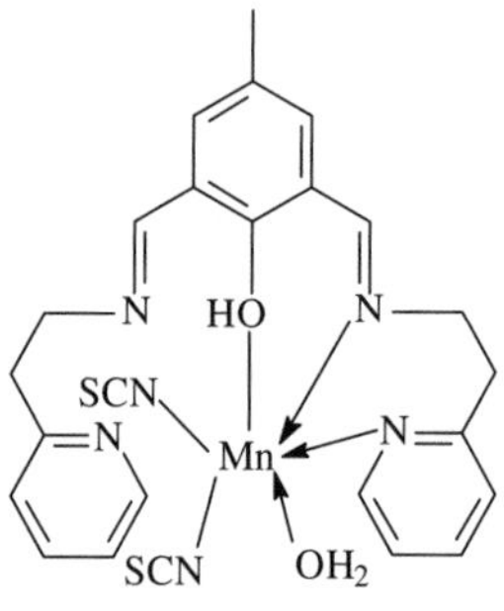

274

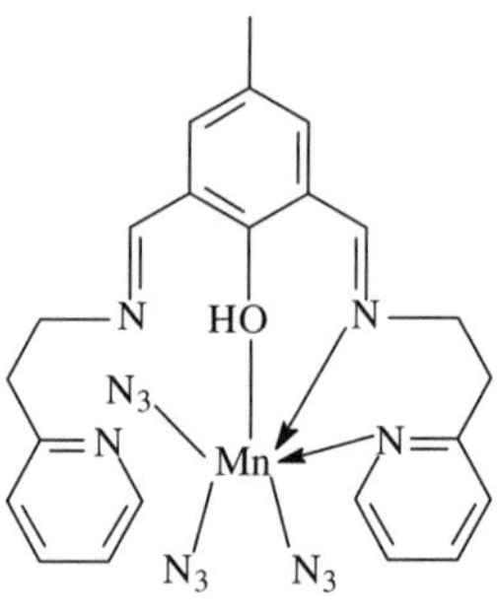

275

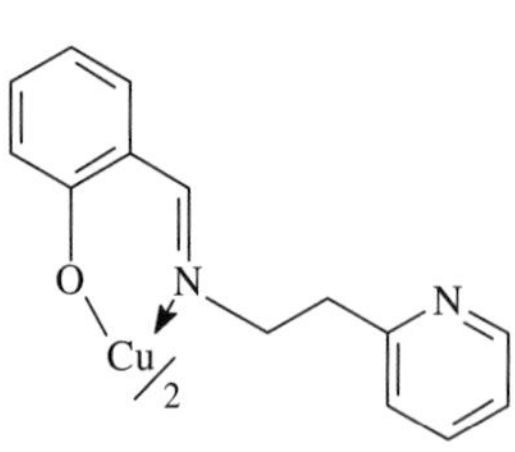

276

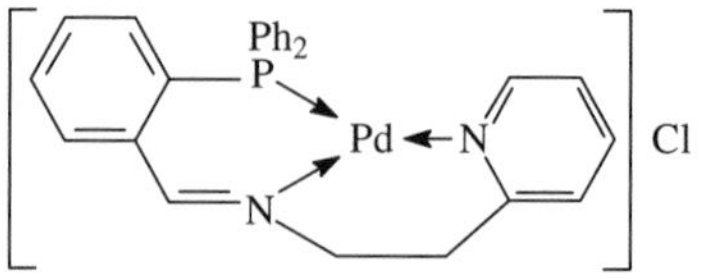

277

The disposition of the endocyclic N-atom of the quinoline and azomethine rings determines the coordination. In azomethine bis-chelates prepared from 3-aminoquinoline, the azine nitrogen is not coordinated in **278** (04AX(E)1552), but in the complexes of Schiff bases of 8-aminoquinoline such coordination is present. It is also observed in bis- **279** and tetra- **280**, **281** chelates of quinoline azomethine ligands. Neutral compounds **279** exist for dioxovanadium (M = V, R = H, *cyclo*-C_4H_4-5,6, X = O, L = L^1 = O, no L^2) (96JCS(D)93), iron (M = Fe, R = 5-NO_2, L = L^1 = Cl, L^2 = Me_2NCO) (02AX(E)334), copper (M = Cu, R = H, X = NTs, L = OCOMe, no L^1 and L^2) (99ZNK1278), and technetium (M = Tc, R = *cyclo*-C_4H_4-5,6, L = L^1 = Cl, L^2 = O) (90JCS(D)2225). Cationic complexes **280** of iron (M = Fe, A = NCSe) (01JA11644) and nickel (M = Ni, no A) (05JMS153) were reported. Complex **281** is regarded as a photoswitchable spin-crossover molecular conductor (06IC5739). In the tetra-chelate **282**, a four-membered metallacycle due to coordination of the quinoline fragment is realized along with five- and six-membered rings (05JA16776). Cationic chelate **283** contains N- and P-donor sites (06OM236).

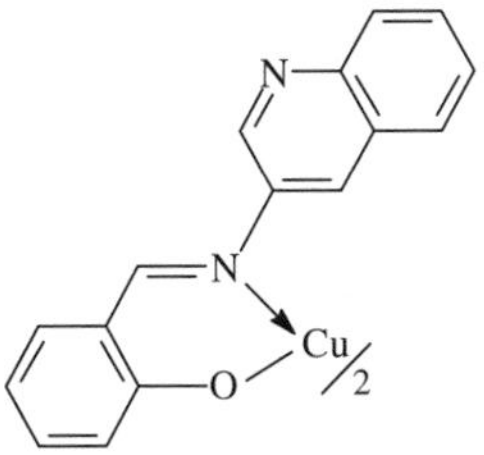

278 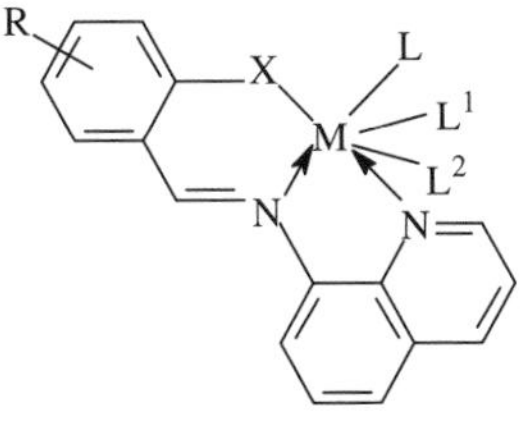   **279** 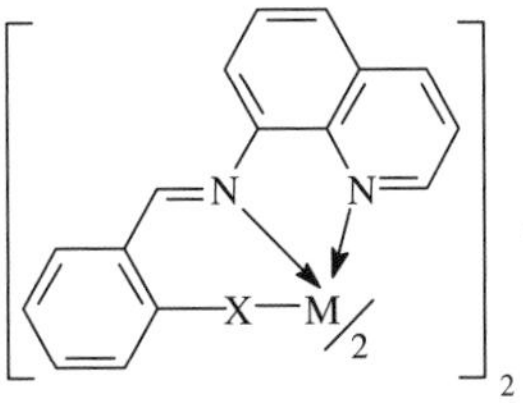   **280**

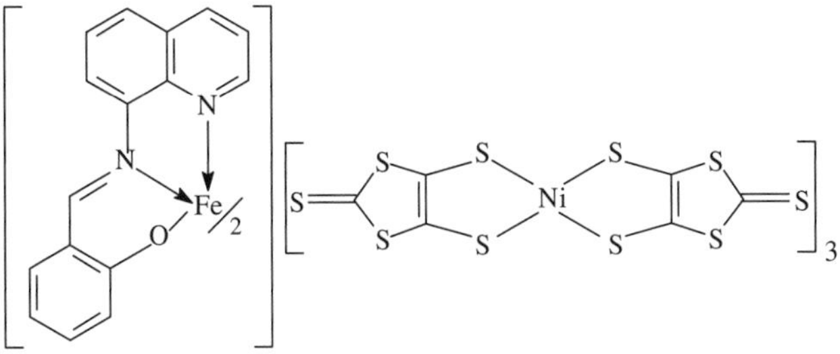

281

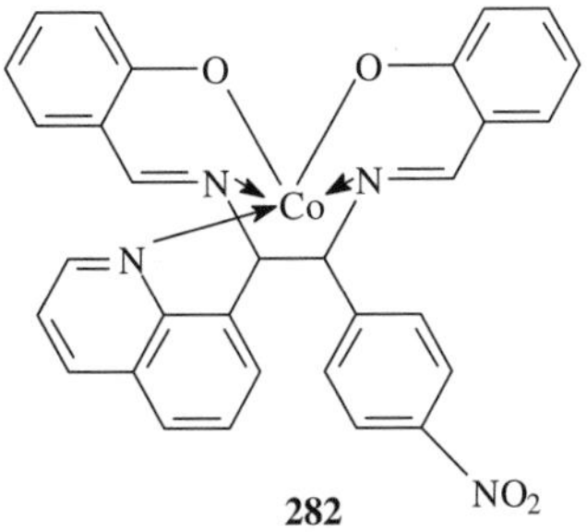

282 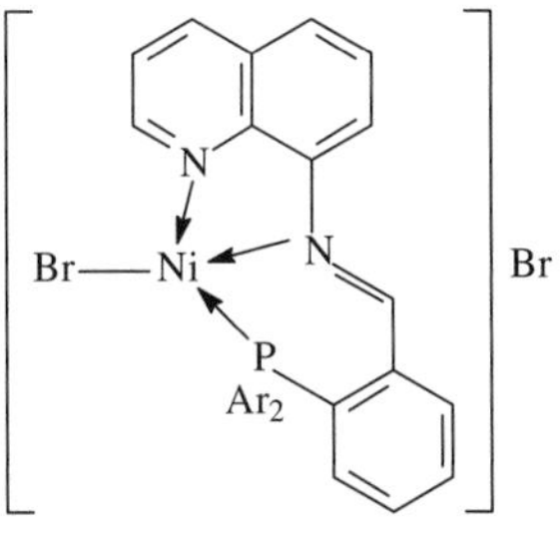   **283**

Azine-coordinated Schiff bases include oligonuclear structures. They are considered with the nature of intermetal fragments, which can involve N, O, S, and Hal (Cl, Br) donor centers and their combinations. Tetra-chelates with a nitrogen bridge may contain azide **284** (05AX(C)432) and N-coordinated isothiocyanate **285** (04ZAAC2754) intermetal groups. O-Containing bridges consist of neutral dinuclear complexes with interligand units such as atomic **286** and phenolic **287** (00JMS(T)289) oxygen atoms. Complex **286** contains non-coordinated quinoline groups. A similar structure is characteristic for a neutral penta-chelate with non-coordinated pyridine N-substituents, in spite of the favorable conditions (coordination number six) for intermetal bonding **288** (99JCS(D)539). The dicationic copper **289** (M = Cu, no L or L^1) (03ZK347), (M = Cu, L = H$_2$O, no L^1) (98JCU1546), and zinc (M = Zn, L = L^1 = MeOH) (02EJI1615) contain two coordinated pyridine moieties. Multi-chelate dinuclear complexes of pyridine azomethines contain a four-membered chelate O,O-intermetal group, as illustrated by neutral zinc **290** (01ZAAC857) and dicationic manganese **291** (06EJI4324).

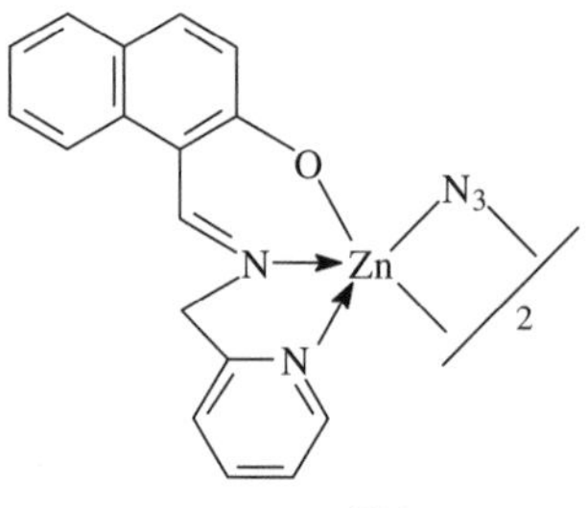

284

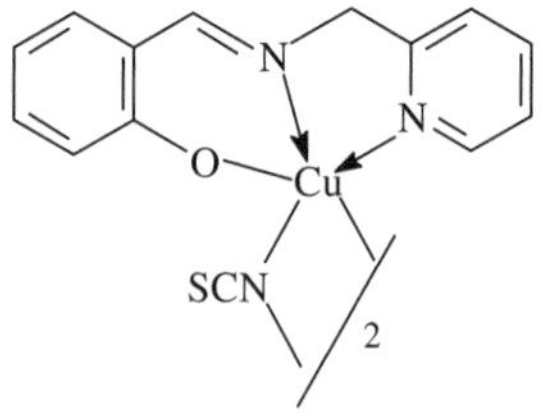

285

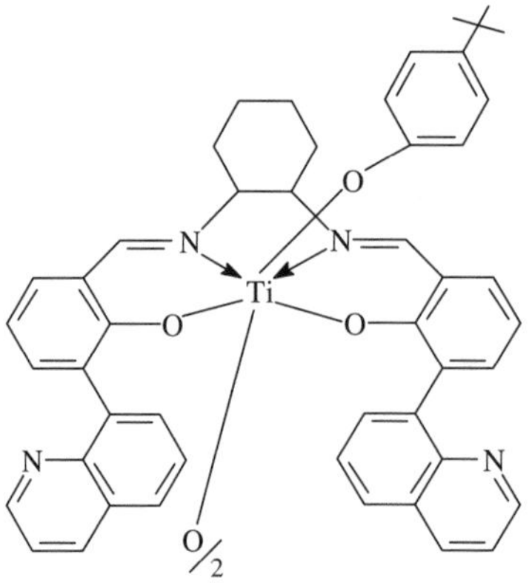

286

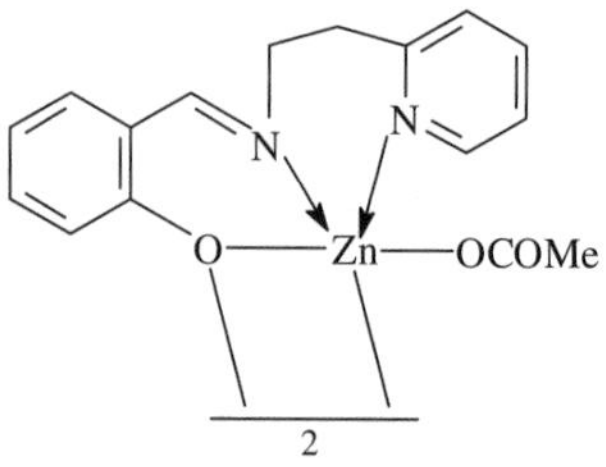

287

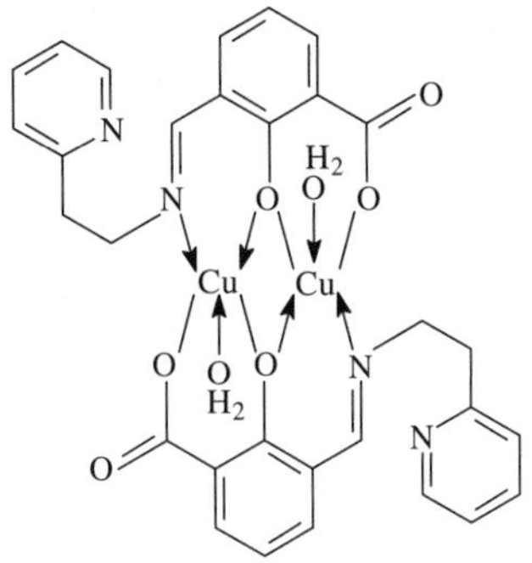

288

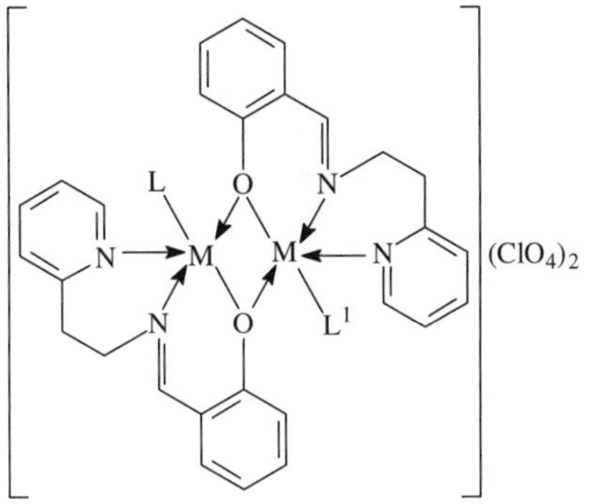

289

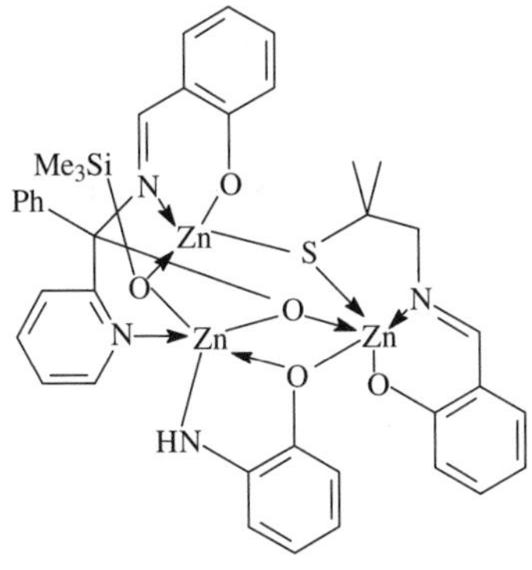

(ClO$_4$)$_2$

290

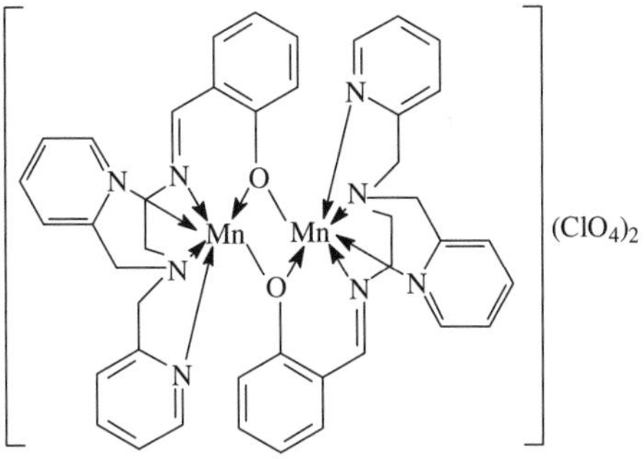

291

(ClO$_4$)$_2$

Cationic copper complexes can be prepared from aryl 2,6-diazomethines containing two pyridine moieties separated by CH$_2$-groups (Z) **292** (Z = CH$_2$) (87IC1375) and (Z = (CH$_2$)$_2$) (89JCS(D)1117). This series includes pyridine-bound chelates with two non-coordinated **293** (00ICA(305)83), one metal-bound **294** (02TMC469), and two coordinated **295** (00ICA(305)83) amino groups in the internitrogen ring-forming moieties. Tri- **296** (03ZK509) and tetranuclear complexes **297** (01IC208) contain oxygen intermetal bridges. In **296**, only two of four pyridine fragments take part in coordination, whereas in **297**, all four azine N-substituents are coordinated. Four coordinated pyridine N-substituents are present in tetranuclear complex **298** (02EJI1615). Carboxyl group plays an important role in the intermetal binding. A series of structurally characterized cationic copper di- and tetranuclear complexes includes tetra-chelates **299** (R = H, no L, A = BF$_4$) (98JCU1546), (R = 4-Me, L = MeOH, A = ClO$_4$) (02CRT1018, 04POL1115) and **300** (02IC4461). Heterotrinuclear poly-chelate **301** deserves attention (03IC8878). Intermetal chloride bridges are present in polymeric pyridine disalicylidene imines **302** (05AX(E)1055). In dimeric **303** having a distorted trigonal-bipyramidal structure, there are two bridging oxygen atoms (08POL709). The ytterbium dichloride **304** is monomeric and contains the coordinated DME ligand, making the central atom seven-coordinate. The organometallic monomeric species **305** has a distorted octahedral structure and complex **306** has an unusually low coordination number of five for the ytterbium center.

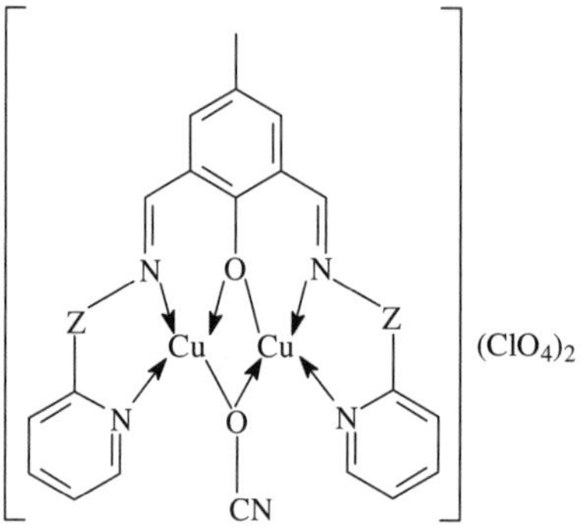

292

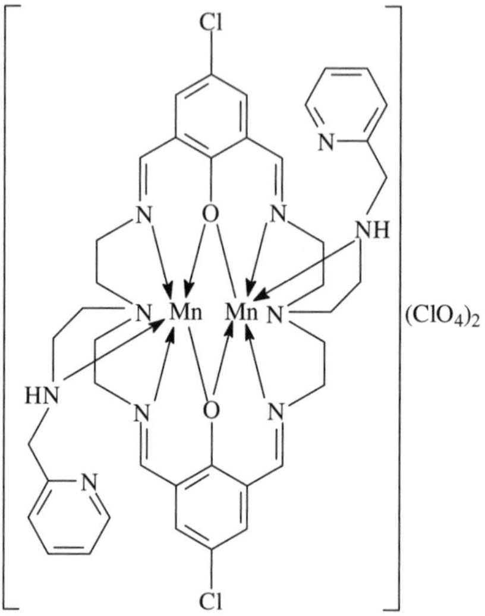

293

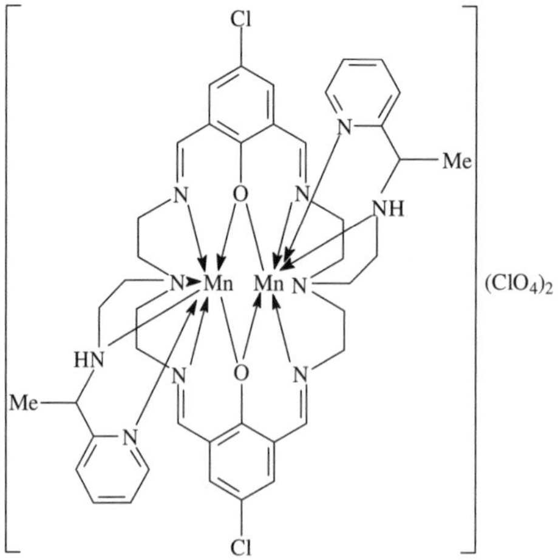

294

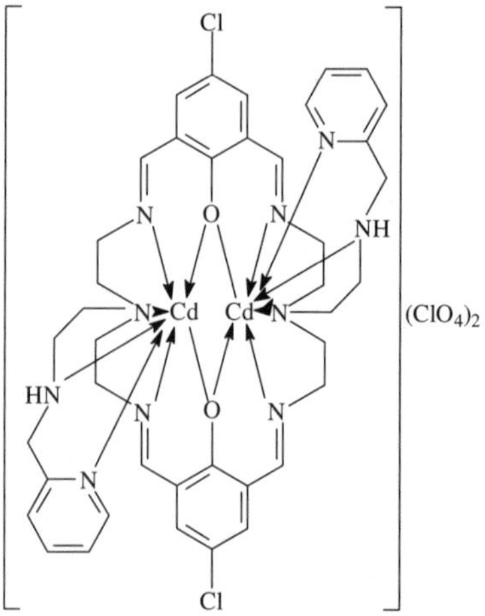

295

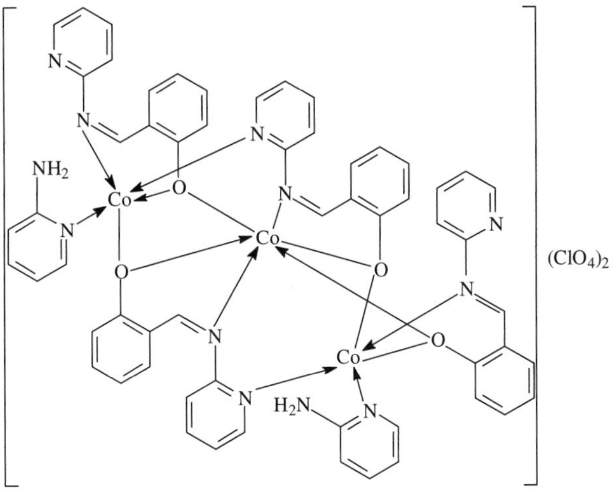

296

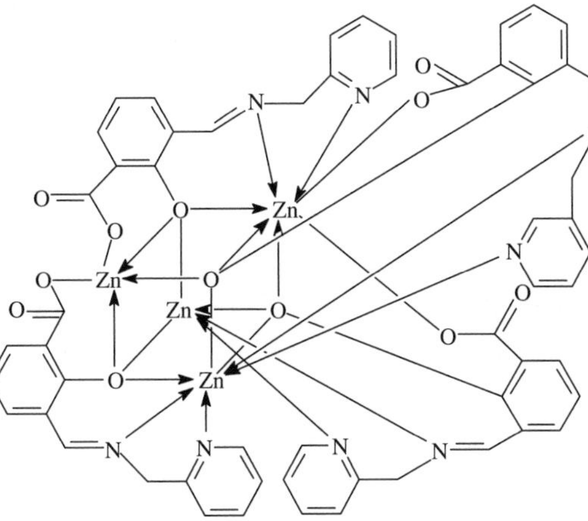

297

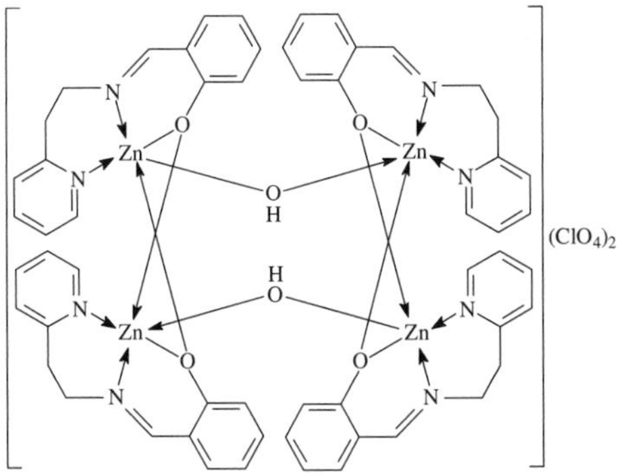

298

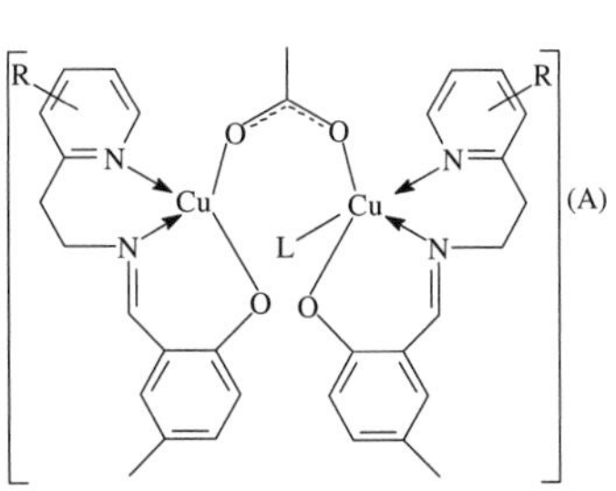

299

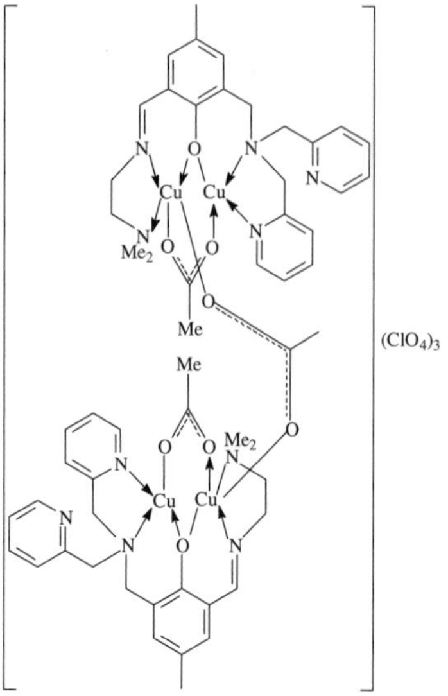

300

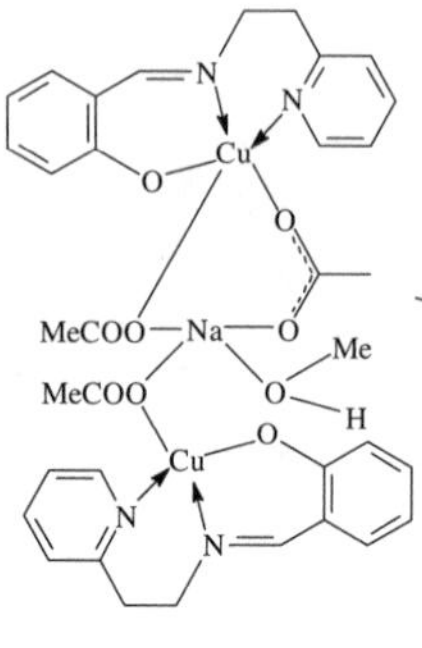

301

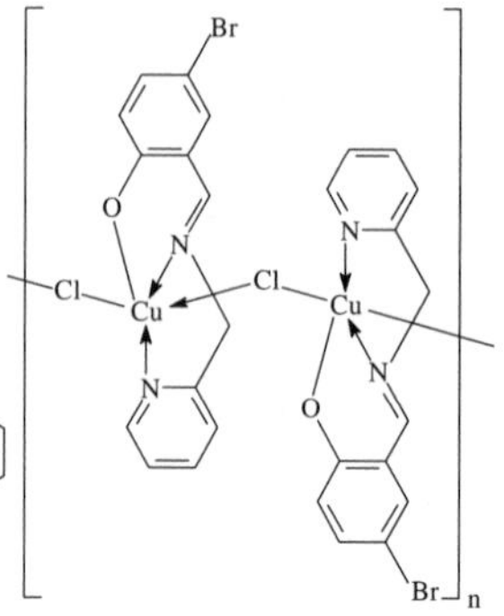

302

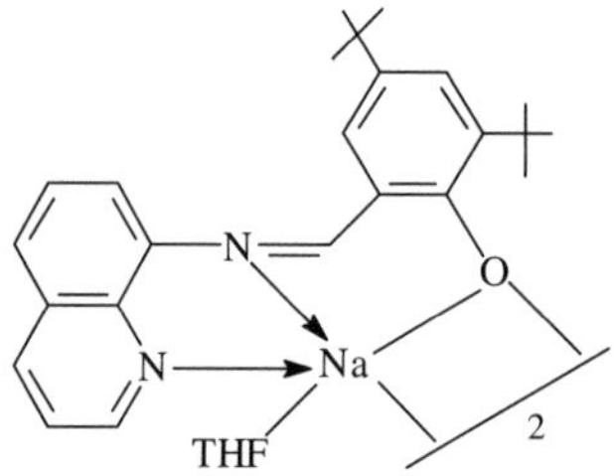

303

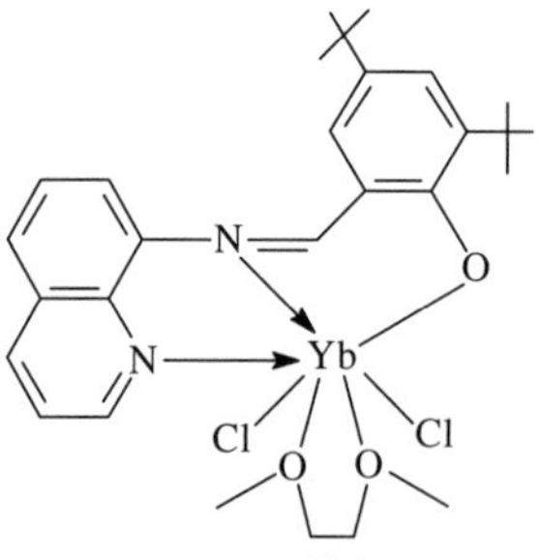

304

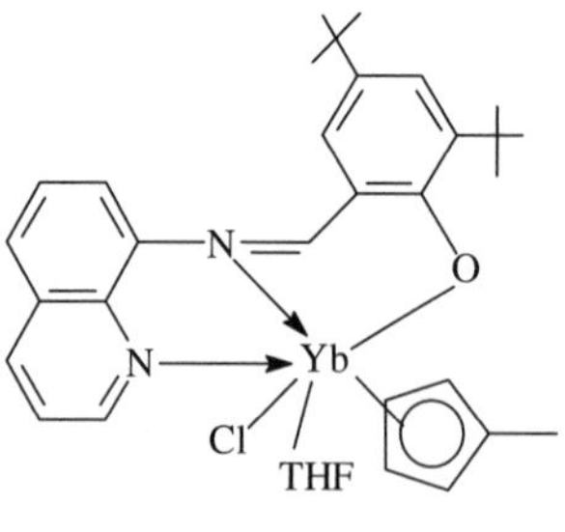

305

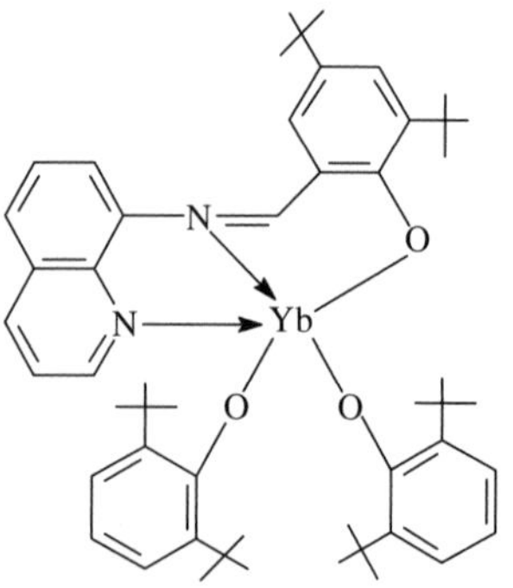

306

Apart from a carboxyl group, carbon–oxygen fragments with more oxygen donor atoms may participate in intermetal bonding. They include carbonate **307** (02EJI1615), oxalate **308** (03ICA(342)131), fumarate **309** (03POL1385), and acetylene dicarboxylate **310** (02ICA201). In **307** (L = MeOH), **309**, and **310**, there are hydroxyl and phenol intermetal bridges. There are few examples of dinuclear pyridine azomethine complexes containing sulfur **311** (90JCS(D)2637) and chloride **312** (99JCS(D)539) bridges. A series of oligonuclear pyridine azomethines with several intermetal bridges includes cationic copper structures containing bridging hydroxyl and phenol oxygen centers **313** (no L) (98IC2134), (L = H$_2$O) (02ACS2011), and **314** (90IC2816). In **314** with a Cu–Cu bond, a complementary intermetal bridge is formed by the oxygen atoms of the ClO$_4^-$-group.

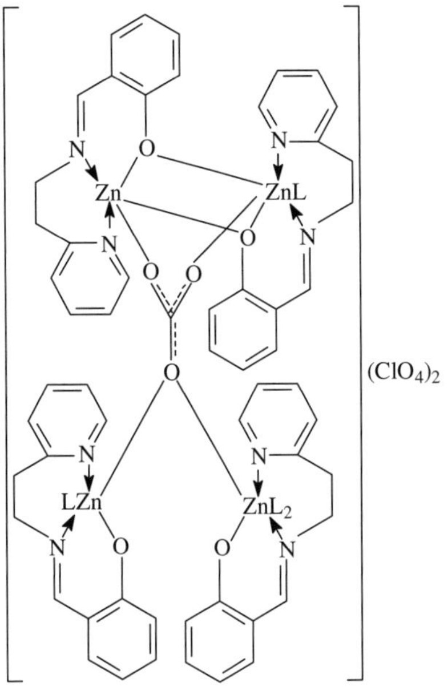

307

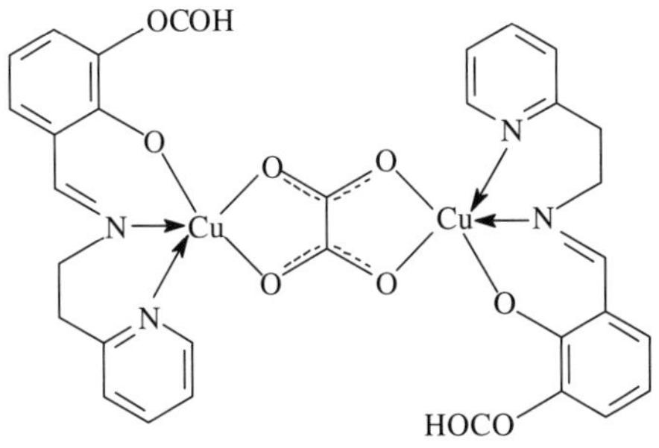

308

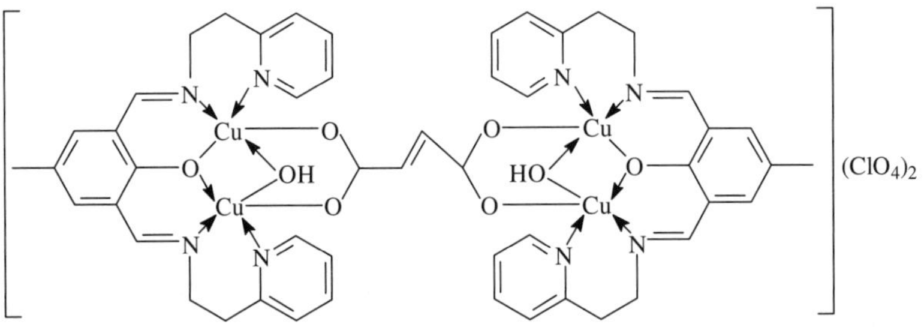

309

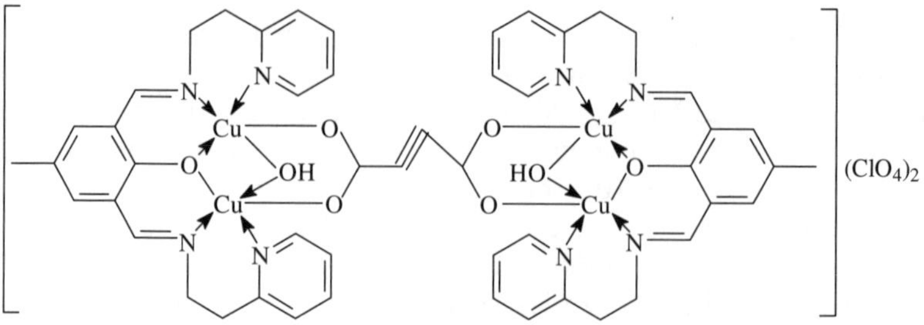

310

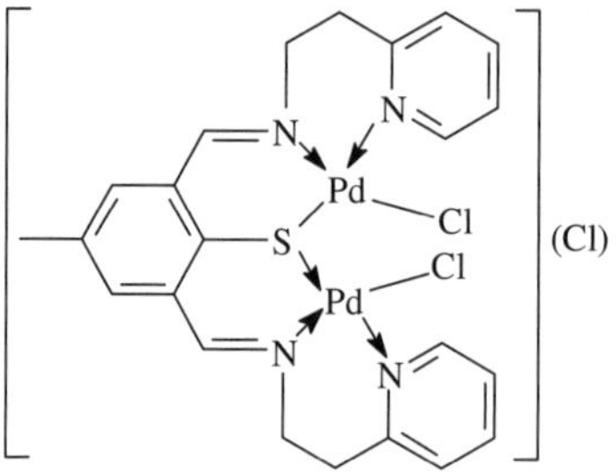

311

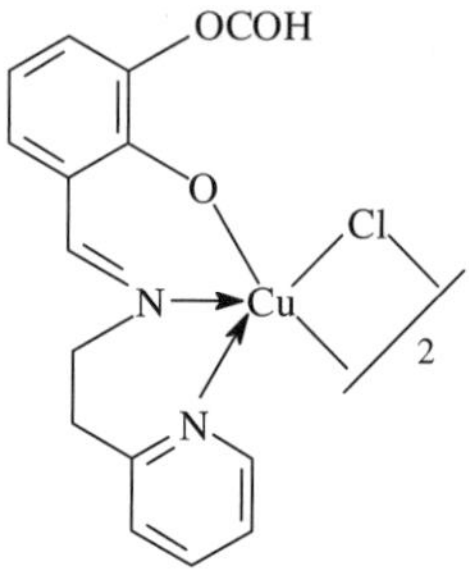

312

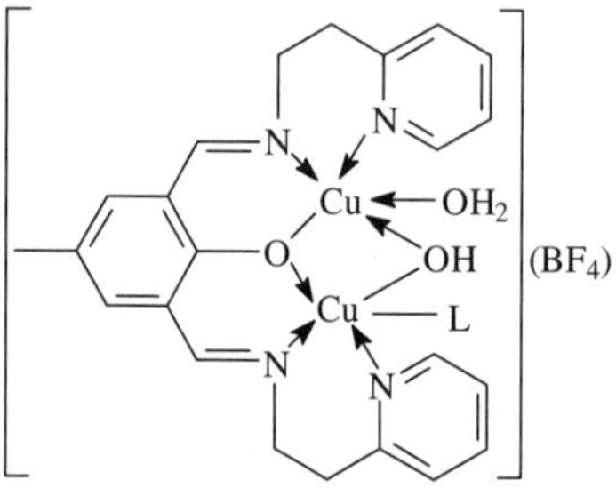

313

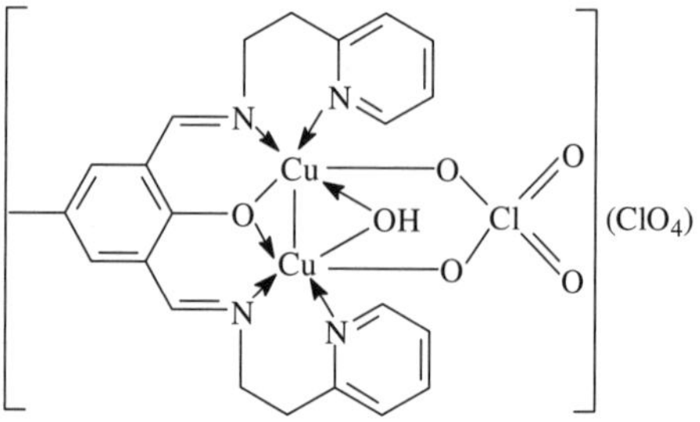

314

Tetra-chelate dinuclear zinc complex contains two acetate bridging groups **315** (05AX(E)466). Two acetate and one phenolate groups form neutral **316** due to intermetal O-coordination. This group may include dinuclear manganese chelates (no L, L^1 = NCS) (90CL1181, 93BCJ1675), (L = L^1 = N_3) (93BCJ1675). Neutral bis-acetate mono- **317** (L = MeOH) (02ICC1063) and di- **318** (02JCS(D)441) azine-coordinated nickel complexes can be prepared from pyridine-containing aldimines of 2,6-diformyl-4-methylphenol. A nickel cationic complex of a 2,6-bis-azomethine derivative contains two coordinated pyridine N-substituents **319** (M = Ni, R = H, R^1 = Me, L^1 = MeOH) (96T3521). Zinc chelate **319** (M = Zn, R = R^1 = Me, no L) has a similar structure (02IC6426). Tetranuclear manganese poly-chelate contains two coordinated pyridine

N-substituents and four acetate bridges in **320** (97IC6279). N,O-Bridging moieties are present in the dinuclear cationic complexes of azide and pyridine azomethines **321** ($Z = CH_2$) (87IC1375) and ($Z = (CH_2)_2$) (89JCS(D)1117). In the neutral **322**, pyrazole nitrogen atoms and a phenolate oxygen perform are bridging function, and the azine N-substituent is not coordinated (79JA6917). Complexes with N,S-bridges are relatively rare, **323** (05AX(E)466) and **324** (00POL1887). In the penta-chelate **324** a pyrazole fragment serves as an additional bridge.

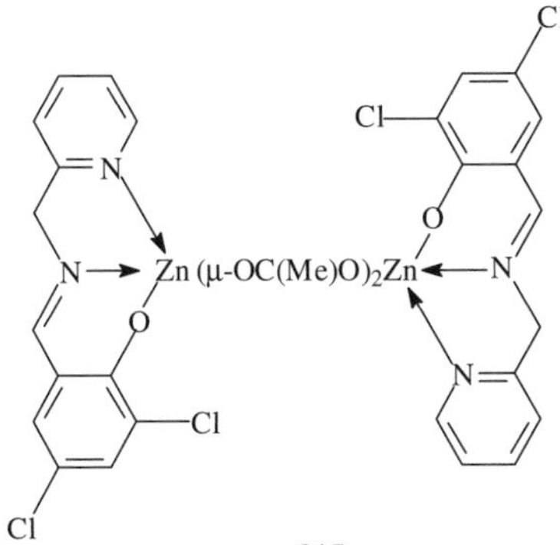

315

316

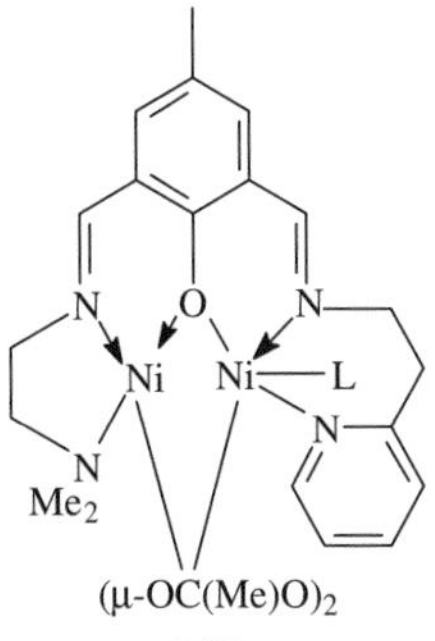

317

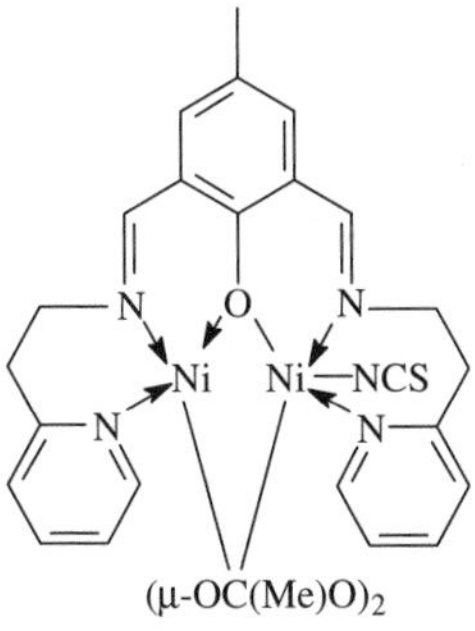

318

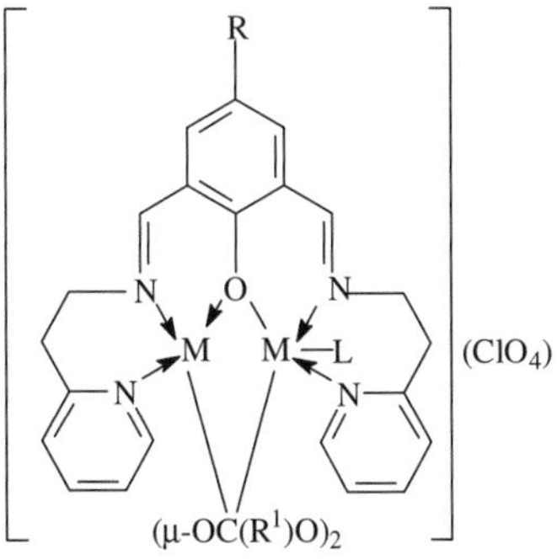

319

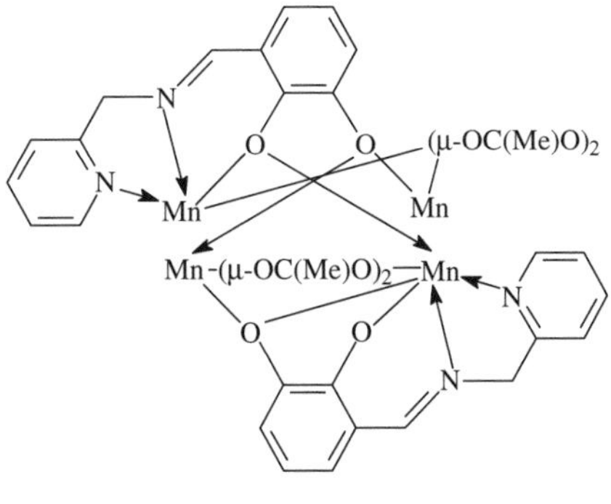

320

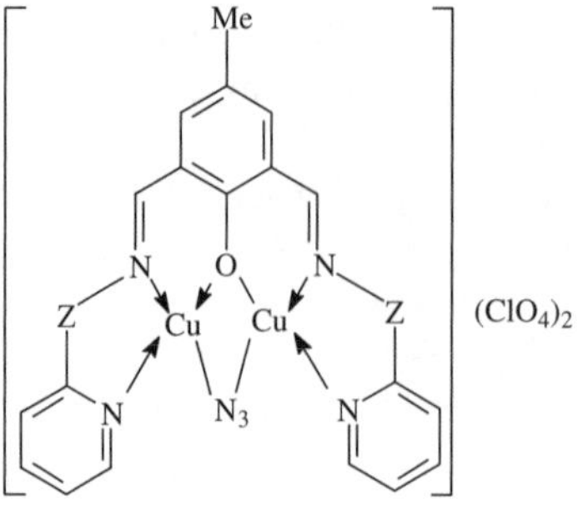

321 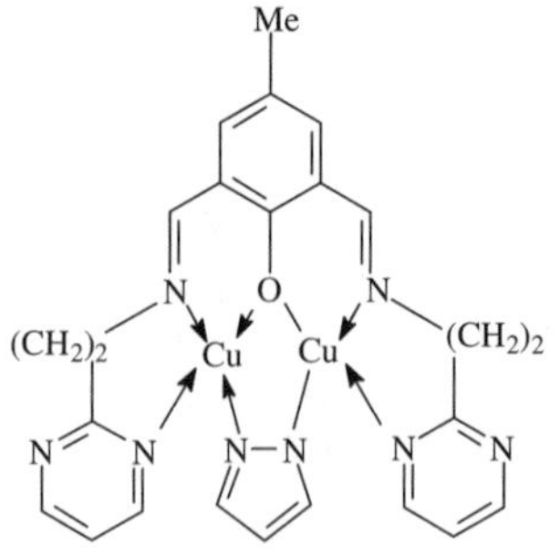   **322**

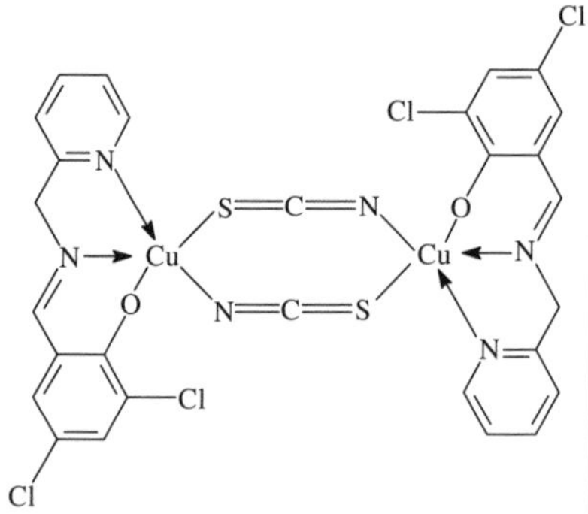

323 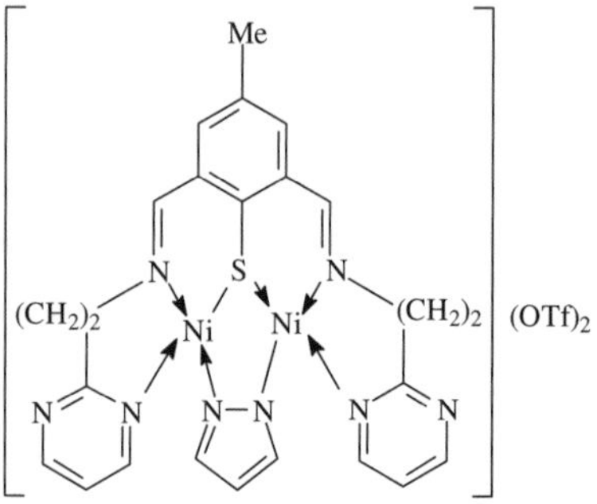   **324**

Halogen–oxygen bridges are present in the dinuclear complexes of copper(II) chloride **325** (L = Cl) (83AX(C)52) and bromide **325** (L = Br) (86IC2300). In the bromide complex, a Cu–Cu bond was postulated. The dicationic tetranuclear copper complex **326** with Cu_2Cl_6 anion contains two chloro-bridges (07POL115). Oligonuclear complexes may be formed by intermolecular association of the coordinatively unsaturated ligand donor sites and metal ions, **327** (M = Cu) (05AX(C)421), **327** (M = Cd) (05AX(C)532), and **328** (M = Zn) (05AX(C)456). Pyridine moieties may participate in intermetal bonding, **329** (00MCL231). The structures of the dicationic tetra-chelates **330** involve both an intermetal Cu-O benzene ring bridging the chelate fragments (A = BF_4, OTf) (06JCS(D)4914).

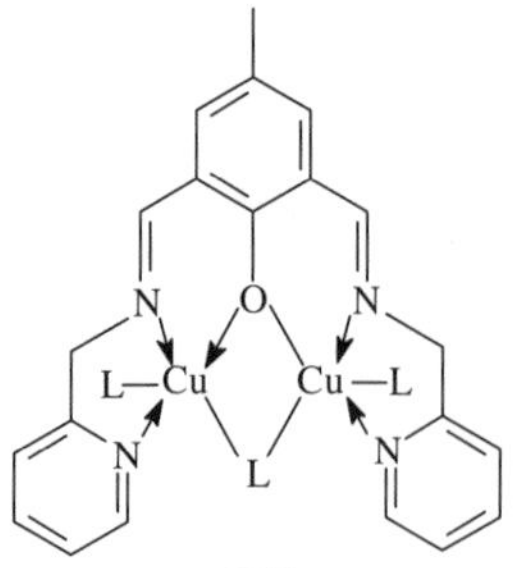

325 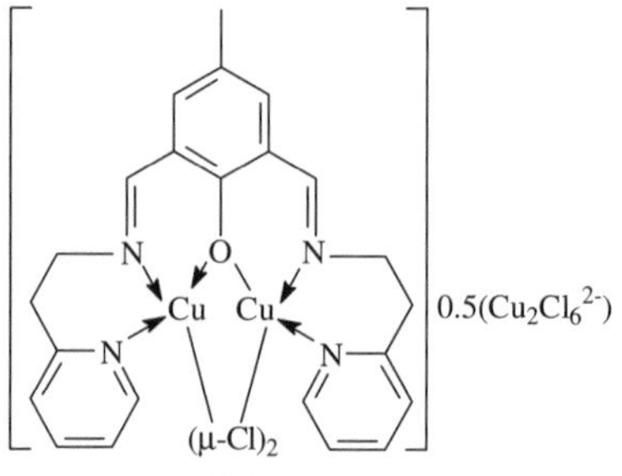   **326**

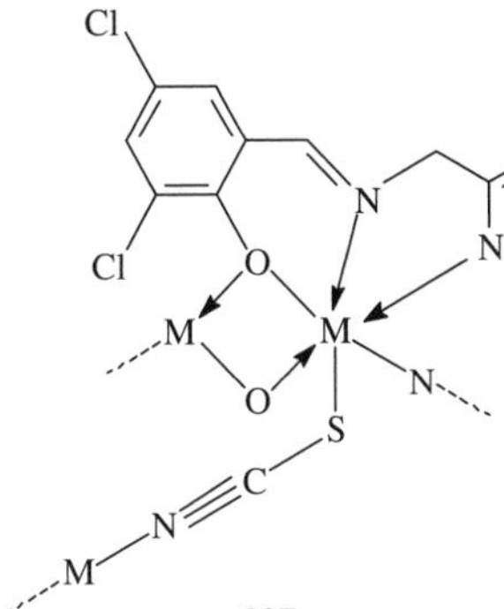

327 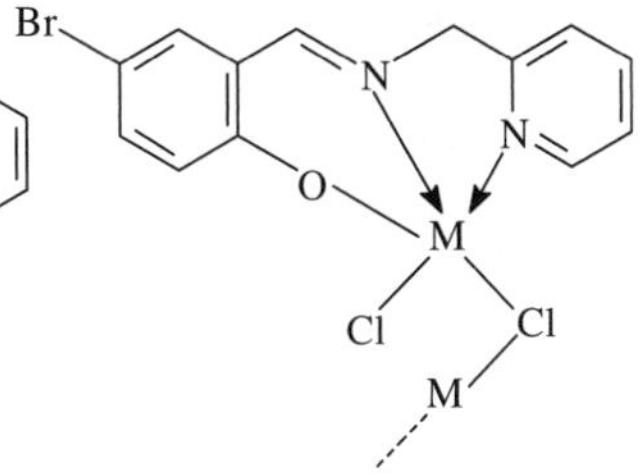   328

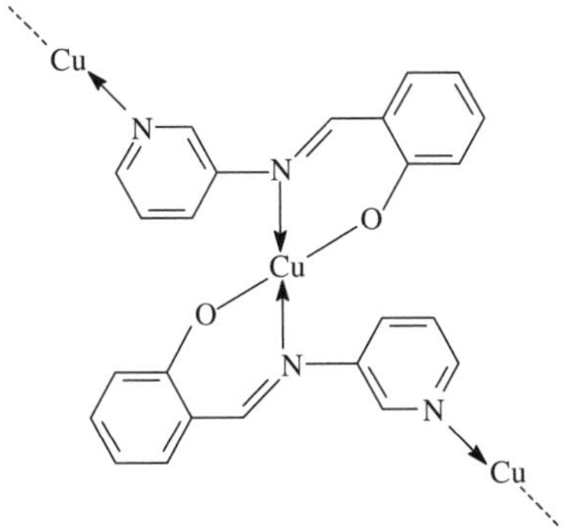

329 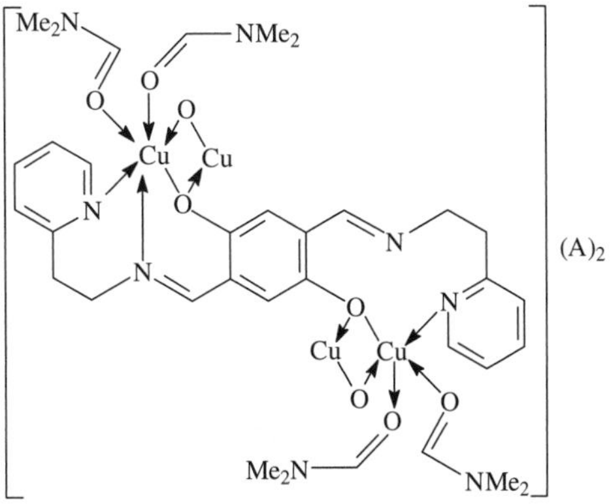   330

In the absence of intermolecular bridges, oligonuclear pyridine azomethine complexes may be formed by the bonding the ligand fragments **331** (06BCJ595), **332** (95ICA(235)273) or by the formation of metal–metal bonds **333** (00JCS(D)1649). In the chelates **331** and **332** the pyridine N-substituent participates in coordination, whereas in less-common **333** having an N,P-azomethine constituent, coordination of the azine fragment is absent. Dinuclear complexes with a quinoline N-substituent, in contrast with those containing pyridine, are represented by a limited number of structures **334–336** with a coordinated azine fragment. These include dinuclear penta-chelate copper complexes with acetate **334** (99ZNK1278) and chloride **335** (89IC2141) bridges, as well as the cationic heteronuclear tetra-chelate **336** (05CL1240). The latter contains Fe(III) and Ni(II) metal ions in the cationic and anionic parts.

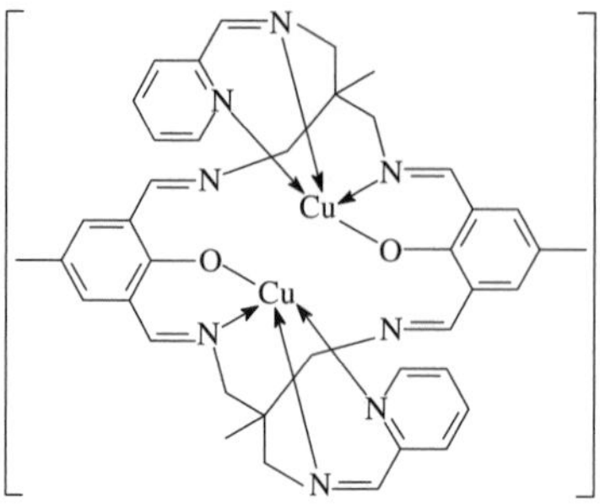

331

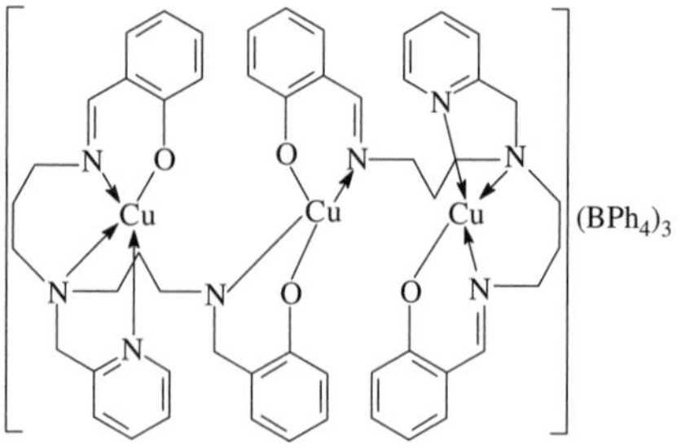

332

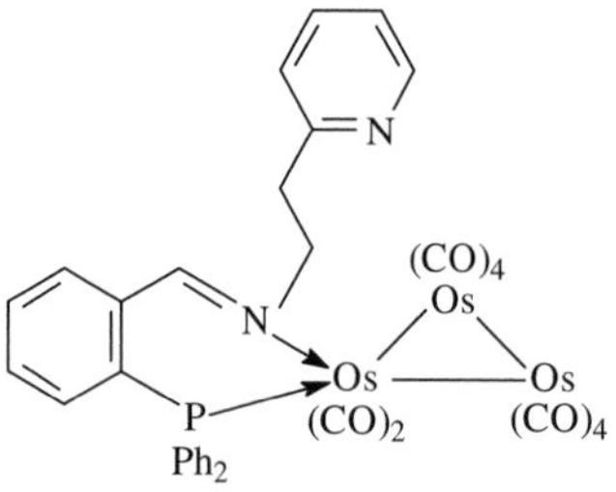

333

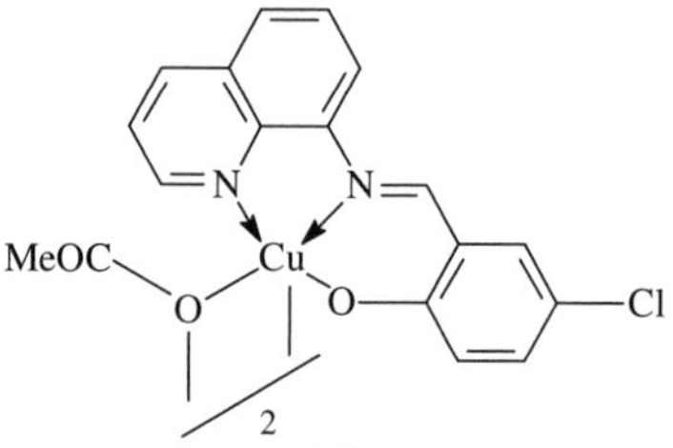

334

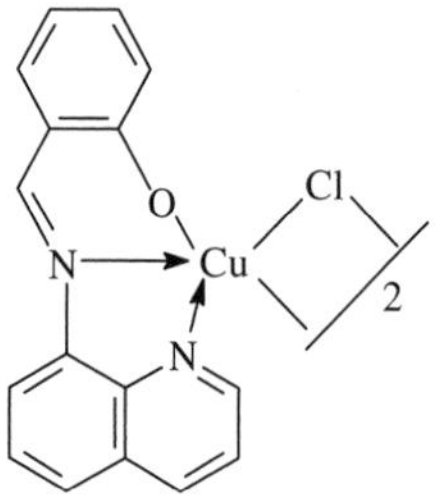

335

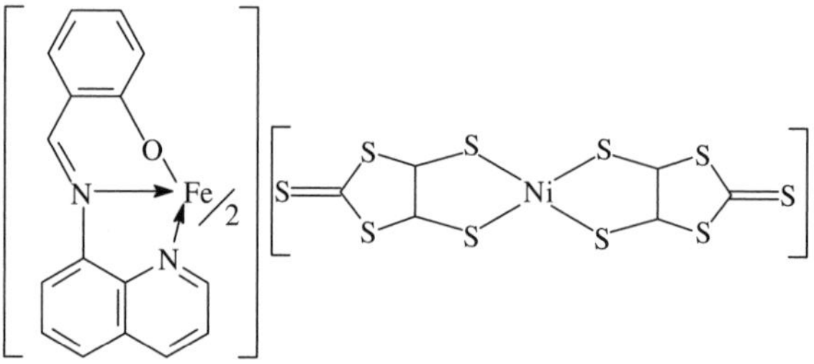

336

Coordination compounds of 2-(2-hydroxyphenyl)azines are repre-sented by the platinum chelates of 2,2′-bipyridine derivatives **337** ($R^1 = R^2 = H$, $X = Cl$; t-Bu, $R^1 = Me$, $R^2 = H$, $X = Cl$; $R^1 = t$-Bu, $R^2 = H$, $X = Cl$, Br, I, –C≡CPh; $R^1 = H$, t-Bu, $R^2 = F$, $X = Cl$) (05IC4442). 2,5-Bis(2-hydroxyphenyl)pyrazine tends to form dinuclear ruthenium(II) complexes **338** ($L = acac$, $n = 1+$; $L = $ 2-phenylazopyri-dine, $n = 0$) (07JCS(D)2411).

337   **338**

3.7 Coordination compounds of cyclic analogues of hetarylazomethines

This type of metal chelates contains six-membered coordination units. However, the azomethine $C=N$ bond is included into the endocyclic heteroligand. The representatives are mainly complex compounds of 2-amino- (hydroxy-, mercapto-) phenylazoles (76ZOB2706, 07CHC1359). Scarce examples are complexes of azines and 2-(1′,3′-dioxo-) and 2-(1′-oxo-3′-thio-) derivatives of indanone (07CHC1359). Material in this section is systematized by the character of the other (not the pyridine nitrogen) heteroatoms comprising the five-membered ring *e.g.* publication (05IC4270) in which the bis-chelates of 2-(2-hydroxyphenyl)benzazoles **339** (M = Be; X = NH, O, S) are described. Bis-chelates of 2-hydroxyphenyl derivatives of imidazole are illustrated by copper **340** (M = Cu, R = Ph, R^1 = 3′,5′-*t*-Bu$_2$) (03JCS(D)1975), (M = Cu, R = Ph, R^1 = 4′,6′-*t*-Bu$_2$) (01CC1824), and zinc (M = Zn, R = H, R^1 = H) (06AX(E)3042). A chromium(III) complex is similar (06AX(E)2886). Compounds **341** can be prepared from 2-hydroxyphenylbenzimidazole. They include cobalt (M = Co, R = R^1 = H, no L) (05AX(E)1986), (M = Co, R = Et, R^1 = 6-Me, no L) (99POL1527), manganese (M = Mn, R = H, R^1 = 5-Br, L = py) (02CJIC643), copper (M = Cu, R = R^1 = H, no L) (05AX(E)1953), (M = Cu, R = H, R^1 = Me, no L) (95ICA(237)181), (M = Cu, R = H, R^1 = 4,6-*t*-Bu$_2$, no L) (06JCS(D)258), and zinc (M = Zn, R = R^1 = H, no L) (06AX(E)2361), (M = Zn, R = H, R^1 = 4,6-*t*-Bu$_2$, no L) (06JCS(D)258). Complexes **342** (02JCS(D)3434) or **343** (99POL1527) contain one or three 2-hydroxyphenylbenzimidazole ligands, respectively. Mononuclear complexes containing a 4,5-imidazoline ring are represented by bis-chelates **344** of nickel (M = Ni) (06AX(E)3537) and zinc **344** (M = Zn) (07AX(E)344). Copper **345** (R = NO$_2$, Br) contains the nitroxyl moiety (03POL2499).

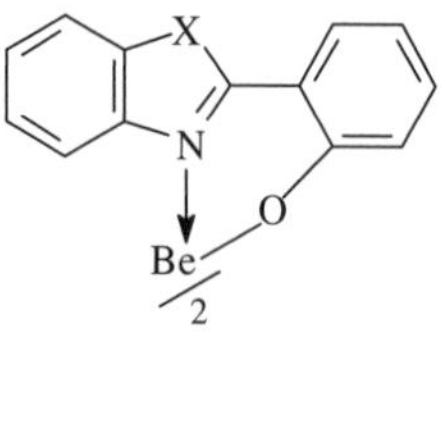

339 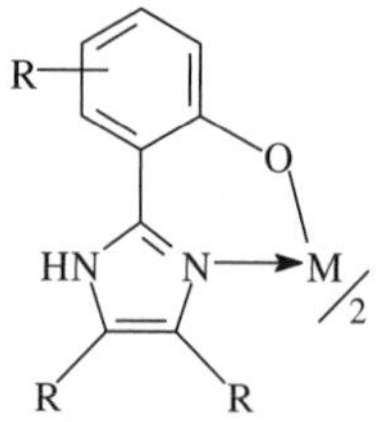   **340** 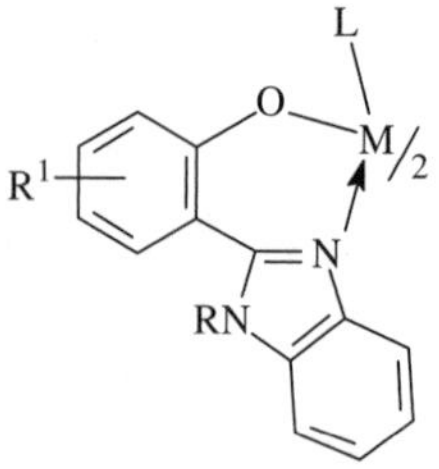   **341**

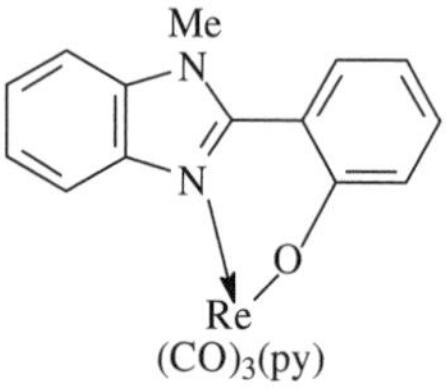

342 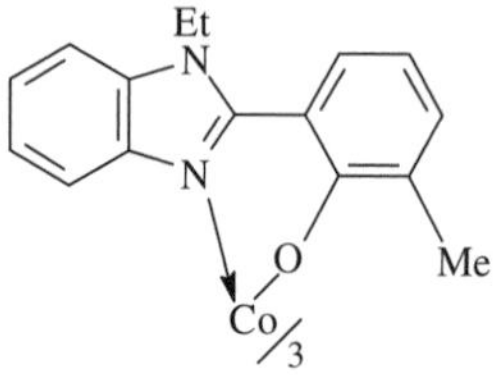   **343**

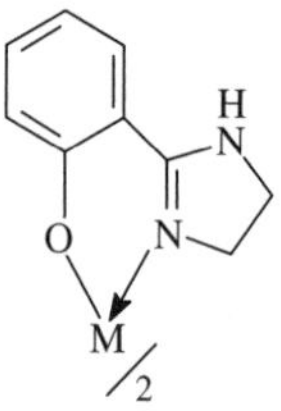

344 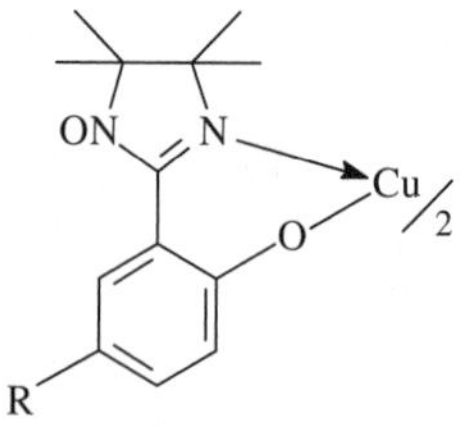   **345**

A dinuclear complex of derivatives of 2-aminophenylbenzimidazole is cationic **346** with the oxygen intermetal bridge (01IC4036). Oligonuclear complexes of 2-hydroxyphenylbenzimidazole ligands include homonuclear structures with oxygen donor sites, atomic **346** (01IC4036) and phenolate **347** (R = Et, n-Bu) (07JMS104), as well as the acetate **348** (02ICC211). In heterodinuclear copper-lanthanide complexes of the benzimidazole derivative **349** (M = Er, Yb, Nd) (06POL881), phenol hydroxyls participate both in intrachelate and intermetal bonding.

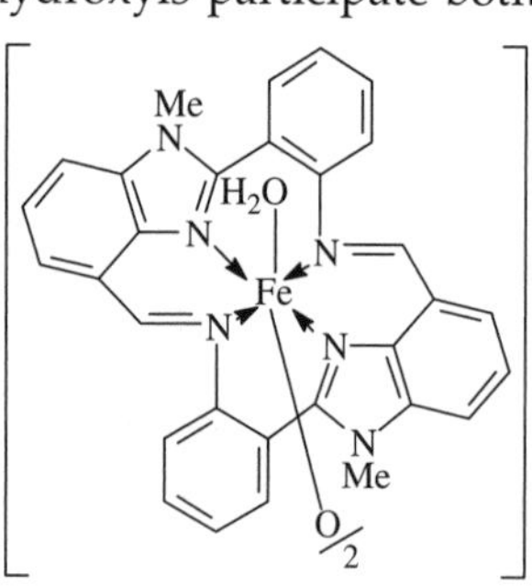

346 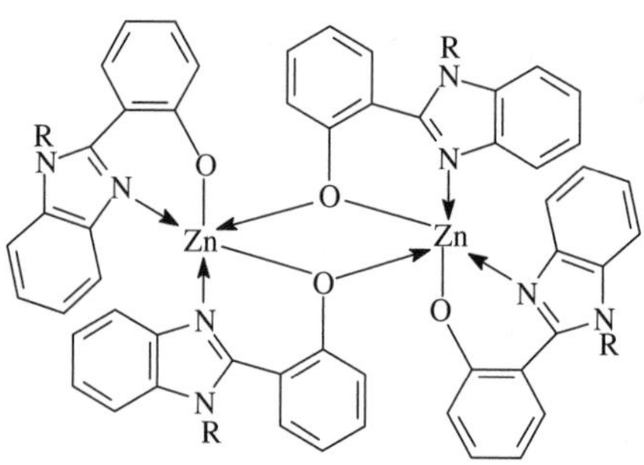   **347**

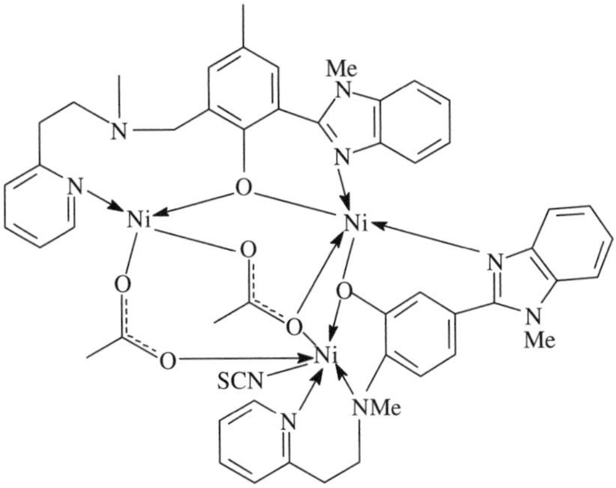

348

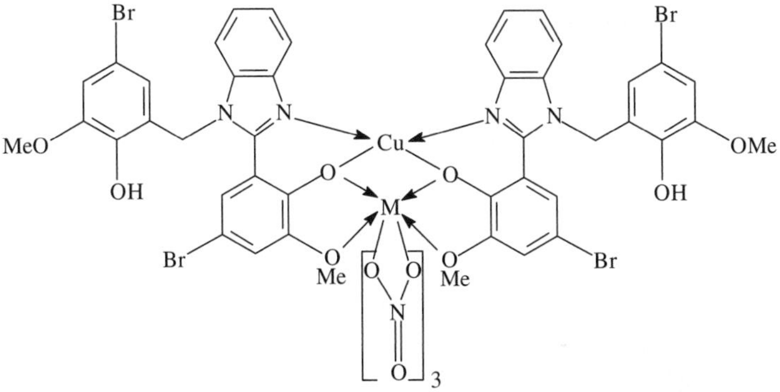

349

Coordination compounds of 2-hydroxyphenyloxazole are rarely studied, e.g., **350** (03CM4949). For a similar benzoxazole derivative, however, there are various metal complexes differing in the number of the chelate fragments. Mono-chelate coordination compounds include neutral platinum **351** (M = Pt, L = L^1 = Cl) (91BCJ149) and rhenium **352** (M = Re, L = CO, L^1 = py) complexes (02JCS(D)3434). Coordinated manganese **352** (M = Mn, L = py) (98ACS371) and palladium (M = Pd, no L) (77ACA(2)80) are bis-chelate structures. Tris-chelates are mixed-ligand metal complexes where 2-hydroxyphenylbenzoxazole is one of the fragments. Examples are neutral complexes with the N$_2$O$_3$ ligand environment **353** of manganese (M = Mn) (99CL525, 02POL1139) and oxorhenium (M = ReO) (01ICA(316)33), chelate **354**, containing N$_2$O$_2$S donor sites in the coordination unit (01ICA(316)33). 2-Hydroxyphenyl-benzoxazole ruthenium tris-chelate **355** contains 2,2'-bipyridyl chelating ligands (02IC5721). Cyclometalated heteroleptic Ir(III) complexes **356** also include 2-(2-hydroxyphenyl)-6-methylbenzoxazole, 2-(2-hydroxy-phenyl)naphthoxazole (08IC1476). 2-Amino-(hydroxy-) phenyloxazo-lines form a wide series of chelates exemplified by **357** (X = NTs, O, S, Se, Te). Bis-chelates with a 2-aminotosylphenyl fragment **357** and MN$_4$ ligand environment include complexes of cobalt (M = Co, R = R^1 = H, R^2 = Me, Et, i-Pr, no L) (01POL2329), copper (M = Cu, R = R^1 = R^2 = H; R = R^1 = H, R^2 = Me, no L) (01TMC709), and zinc (M = Zn, R = R^1 = R^2 = H, no L; R = R^1 = H, R^2 = Et, no L; R = R^2 = H, R^1 = Me, no L; R = R^1 = H, R^4 = i-Pr, no L) (02ZAAC1210). Nickel complex **358** contains chelate units similar in the composition of the donor sites (98CEJ818). A variety of coordination compounds of Group IV metals include bis-chelates **357** (X = O) of titanium (M = Ti, R = R^1 = H, R^2 = Me, L$_n$ = Cl$_2$) and zirconium (M = Zr, R = H, R^1 = Ph, R^2 = Me, L$_n$ = Cl$_2$) (95IC2921), (M = Zr, R = 2-t-Bu, R^1 = H, R^2 = Me, L$_n$ = Cl$_2$) (05JCS(D)3611), (M = Zr, R = R^1 = H, R^2 = Me, L$_n$ = (CH$_2$Ph)$_2$) (95OM4994), (M = Zr, R = 4,6-t-Bu$_2$, R^1 = H, R^2 = t-Bu, R^5 = H, L$_n$ = (CH$_2$Ph)$_2$) (06JOM2228). Hafnium chelates are exemplified by cationic **359** (95OM4994). Complex compounds of 2-hydroxyphenylox-azolines include bis-chelates **357** (X = O) of vanadyl (M = V, R = R^1 = H, R^2 = Et, i-Pr, L$_n$ = O) (97CB887), manganese (M = Mn, R = R^1 = R^2 = H, no L$_n$) (98ICA(279)217), (M = Mn, R = R^1 = H, R^2 = COOMe, L$_n$ = Br) (05IC9253), oxorhenium (M = Re, R = R^1 = R^2 = H, L$_n$ = O, Br) (96IC368), cobalt (M = Co, R = R^1 = H, R^2 = Me, no L$_n$) (01EJI669), nickel (M = Ni, R = R^1 = H, R^2 = i-Pr, no L$_n$) (91HCA717), (M = Ni, R = R^1 = H, R^2 = Me, L$_n$ = (MeCOOH)$_2$) (01EJI669), palladium (M = Pd, R = R^1 = R^2 = H, no L$_n$) (99IC4510), (M = Pd, R = R^1 = H, R^2 = Et, no L$_n$) (97JCS(D)3755), copper (M = Cu, R = R^1 = H, R^2 = i-Pr, no L$_n$) (91HCA717), (M = Cu, R = R^1 = H, R^2 = Me, no L$_n$) (01EJI669), (M = Cu, R = 2-t-Bu, R^1 = R^2 = Ph, no L$_n$) (01JOM204), zinc (R = R^1 = H, R^2 = i-Pr, no L$_n$) (91HCA717), (M = Zn, R = R^1 = H, R^2 = Me, no L$_n$) (01EJI669), and dioxomolybdenum (M = MoO$_2$, R = R^1 = H, R^2 = Et, no L$_n$) (01EJI1071). Cationic complexes are represented by the adduct-containing bis-chelates **360** of manganese (M = Mn, R = 3-Cl, L = L^1 = MeOH, A = ClO$_4$) (01ICA(320)117), (M = Mn, R = H, L = L^1 = 1-methylimidazole, A = ClO$_4$) (02EJI2897) and rhenium (M = Re, R = H, L = O, L^1 = H$_2$O, A = OTf) (01IC2185).

350   351   352

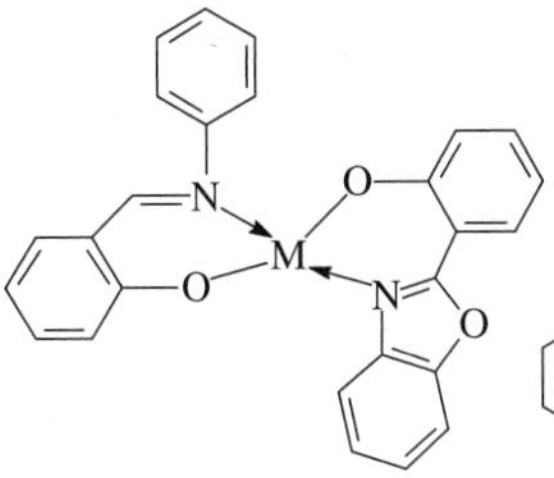

353 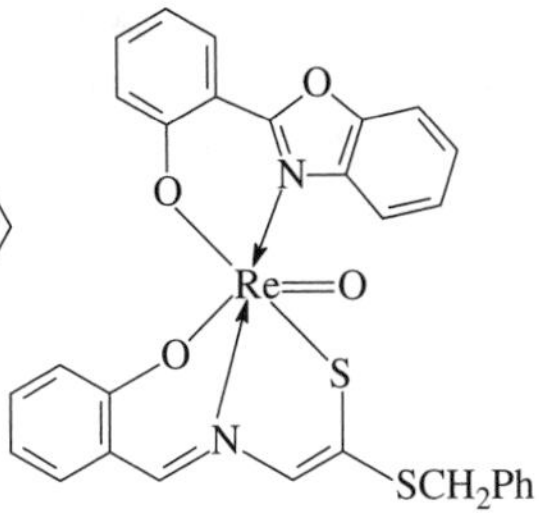   354 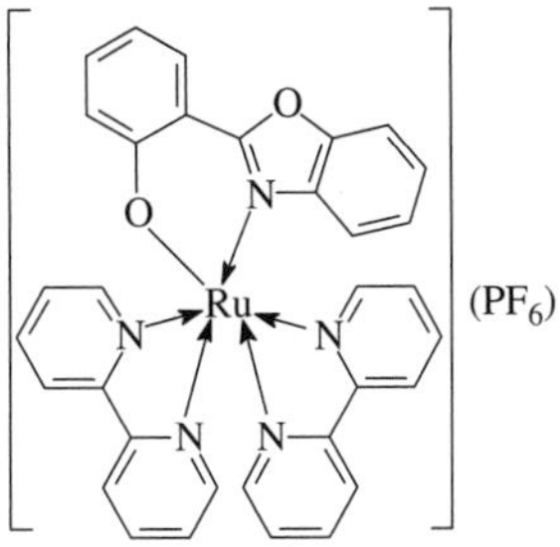   355

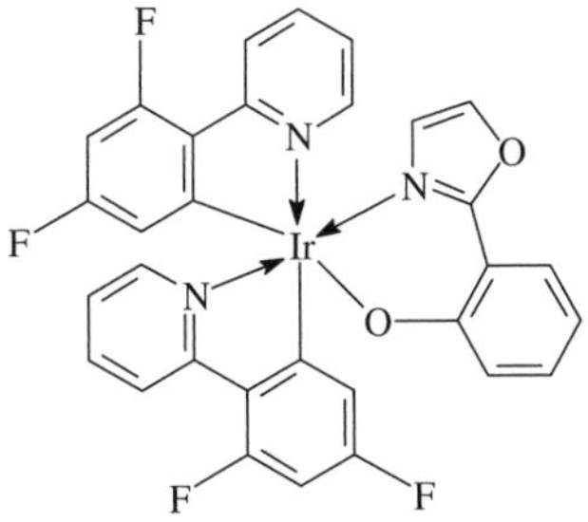

356 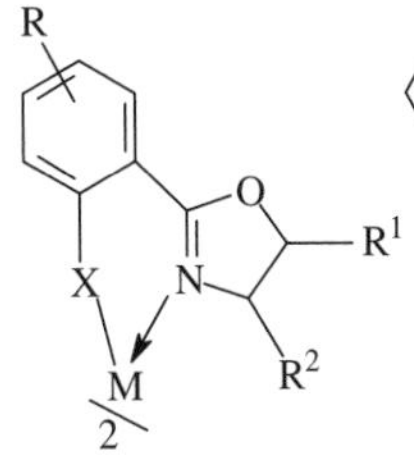   357 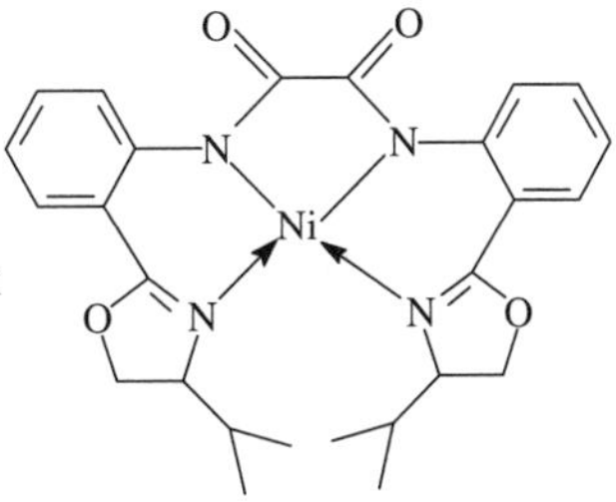   358

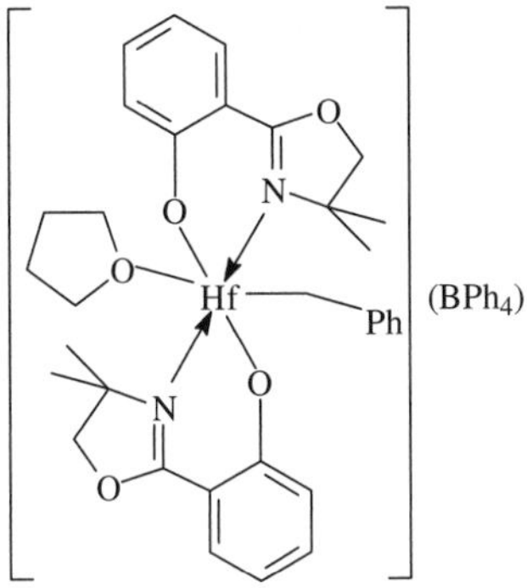

359 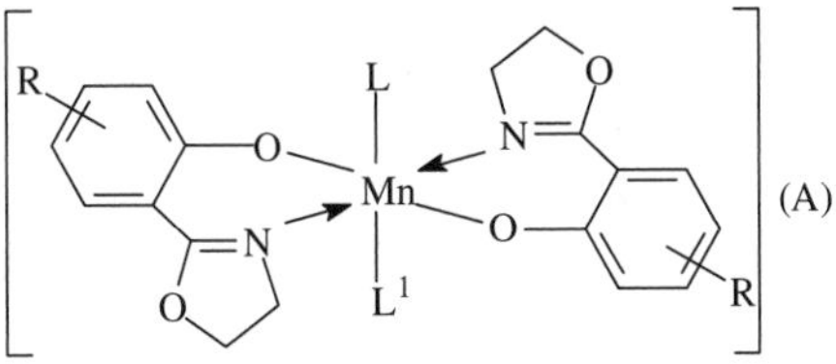   360

Coordination compounds of 2-hydroxyphenylbenzothiazole include monocyclic neutral chelate **361** (02JCS(D)3434) and anionic complex **362** (88IC4208). The same ligating moiety is included in the composition of mixed-ligand tris-chelates **363** of oxorhenium (M = Re, L = O) (00ICA(307)149, 03BKC504) and molybdenum (M = Mo, L = NO) (98ICA(269)350). Bis-chelate neutral complexes **364** of oxorhenium (M = Re, L = O, no L^1) (01ICA(316)33) and oxotechnetium (M = Tc, L = O, L^1 = Cl) (96IC368) are based on 2-hydroxyphenylthiazoline. Complex compounds of 2-hydroxyphenylazoles with three heteroatoms can be exemplified by the complexes of 1,2,4-triazole **365** (04EJI4177) and 1,3,4-oxadiazole **366** and **367** (01CCR79). Depending on the nature of the metal, tetra- (Fe), tri- (Al), or di- (Be) chelate structures are formed.

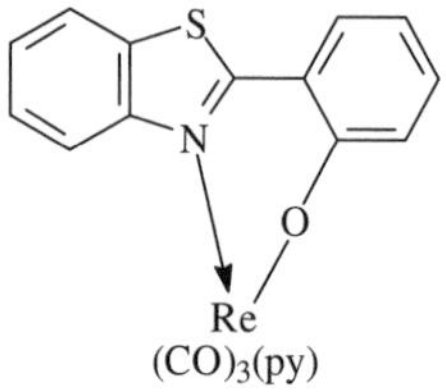

361

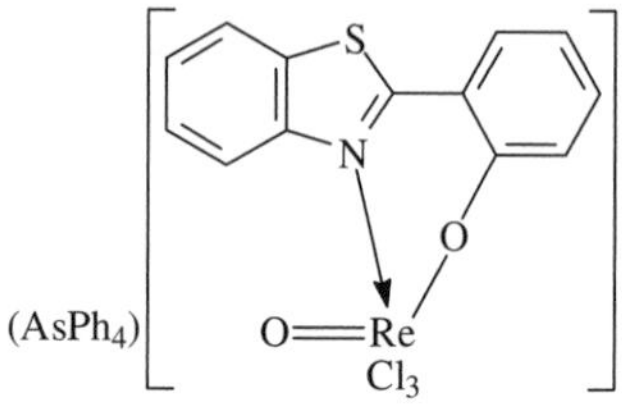

362

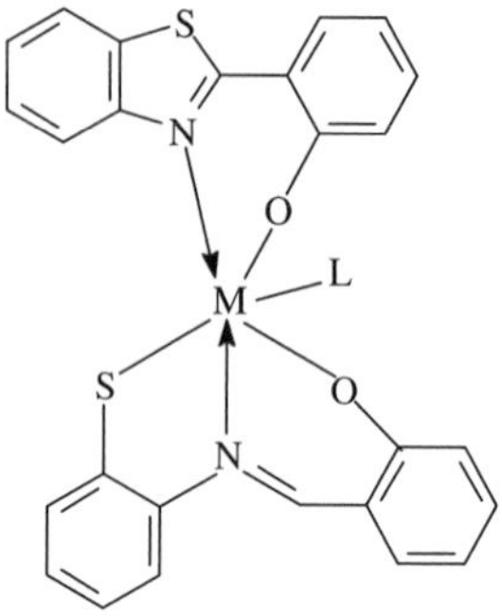

363

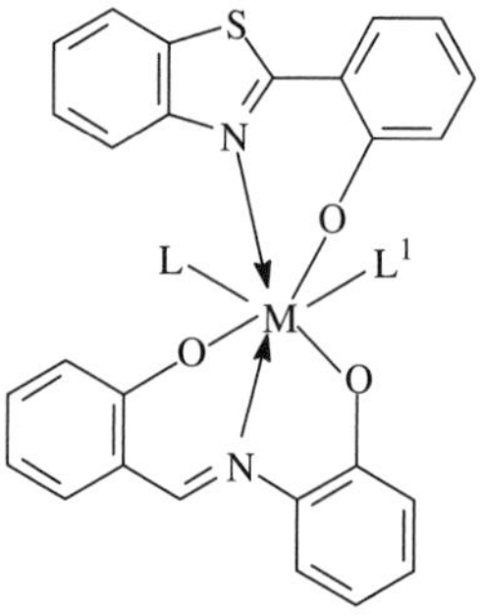

364

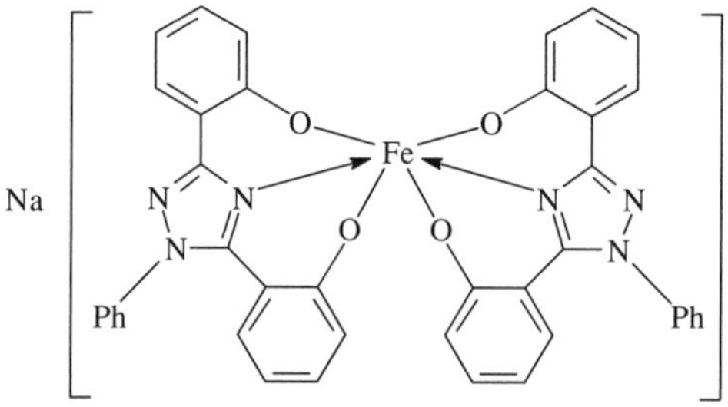

365

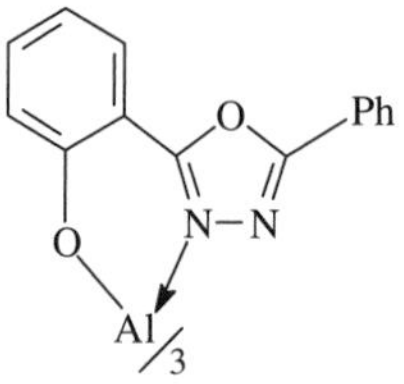

366

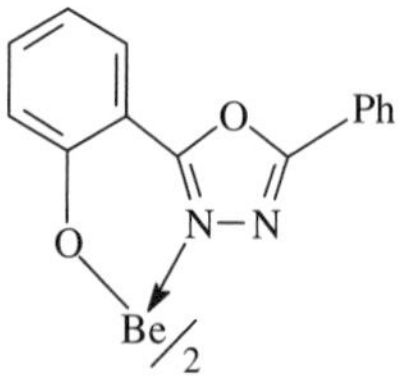

367

Several oligonuclear complexes with oxygen intermetal bridges have been structurally characterized in the case of 2-hydroxyphenyl-derivatives of five-membered nitrogen heterocycles (03JA14816). They include 2-hydroxyphenyl-derivatives of oxazole **368** (03CM4949), oxazoline **369** (06ZN(C)263), and 1,2,4-triazole **370** (04EJI4177). Trinuclear ruthenium carbonyl structure **371** is formed from an oxazoline derivative bearing a 2-aminophenyl substituent (06JCS(D)2450). This coordination compound has a hydride intermetal bridge and a triangular cluster with ruthenium–ruthenium bonds. Less-common coordination with a six-membered unit incorporating C- and N-donor sites exists in the heterodinuclear mono-chelate with oxazoline moiety **372** (R = H, Ph) (00OM5484).

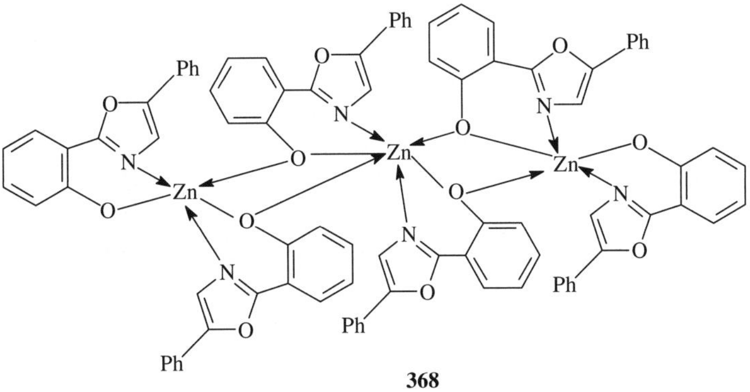

368

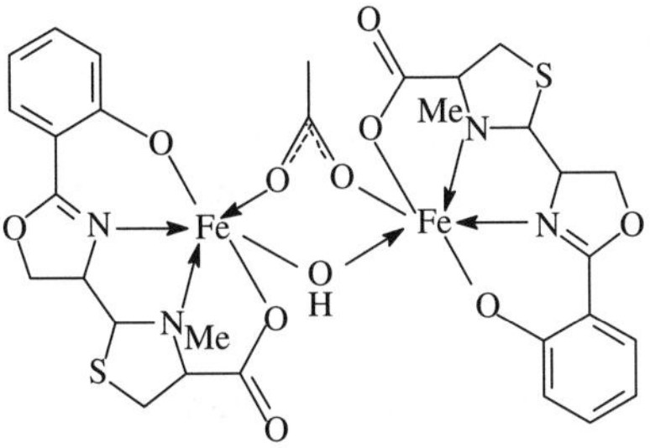

369

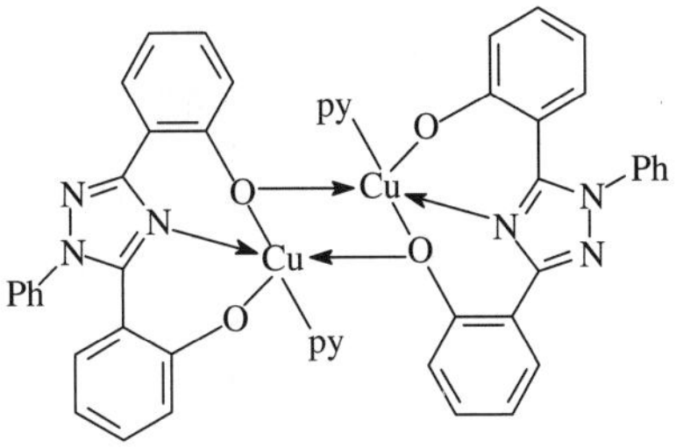

370

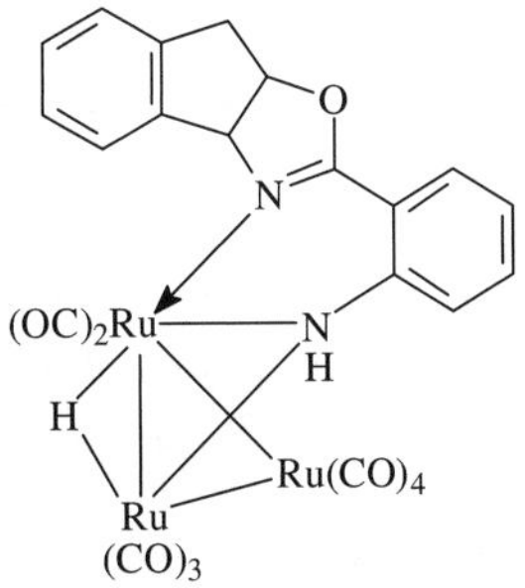

371

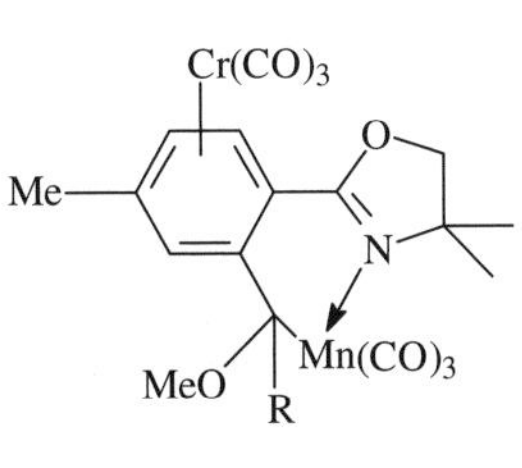

372

4. CONCLUSION

Other ligating compounds with azomethine $C=N$ bonds and their metal complexes involve hydrazine carbodithioates (08AX(E)328) heteryloximes (07ICA69, 07JCS(D)2658), hydrazones (07SA(A)852, 07JCC243, 08AX(C)137, 08POL1917), semi- and chalcogenosemicarbazones (07POL2603, 08IC1488, 08POL593). Their consideration is beyond the limits of this review, but some of the recent references are given.

The data compiled in this review recognize a wide range of reaction abilities of heteryl-containing ligating compounds that play an important role in the development of the fundamental aspects of modern chemistry of metal coordination compounds (96M11, 08M11, 03MI1), including the synthesis of new polyfunctional materials. A number of hetarylazomethine complexes display switchable ferro- or antiferromagnetic and photomagnetic properties (99JCS(D)539, 99RCR345, 05MC133, 06IC1277, 07IC4114, 07JCC1493, 07RJC176, 08CCR1871, 08ICA3519). Zinc coordination compounds prepared from 2-amino- (hydroxy-) phenyl-substituted benzazoles form luminescent materials usable in OLED (97TED1222, 01CCR79, 03JA14816, 06JKP1057). Erbium and terbium complexes of Schiff bases of indole-2-carbaldehyde exhibit effective fluorescence (05CP17). Complexes containing long-chain alkoxy groups possess Langmuir–Blodgett film-formation capability and NLO properties and show metal-ligand charge transfer optimal characteristics (07OM5423).

Hetarylazomethine chelates are used as catalysts (07RCR617, 08POL1556) and biochemical electrocatalysts (06MI1). Zeolite-encapsulated complexes of Schiff bases derived from salicylaldehyde and 2-amino-methylbenzimidazole of copper(II) and dioxovanadium(V) are active catalysts for the oxidation of phenol and styrene (06JMC(A)227). Similar metal complexes based on the Schiff bases of N,N-bis(3-hydroxyquinoxaline-2-carboxalidene)-o-phenylenediamine, 3-hydroxyquinoxaline-2-carboxalidene-o-aminophenol, 3-hydroxyquinoxaline-2-carboxalidene-2-aminobenzimidazole, salicylidene-2-aminobenzimidazole of ruthenium(III) catalyze oxidation of catechol (08JMC(A)92). Complexes of 2,6-bis(imino)-pyridine ligands are catalysts for epoxidation (99JMC(A)101) and cyclopropanation (00OM1833) of alkenes, reduction of ketones (90ICA(174)9), hydrogenation and hydrosilylation of alkenes (00JCS(D)2545, 04EJI1204, 04JA13794, 08OM1470), dimerization of α-alkenes (01OM5738, 03OM3178, 04CEJ1014, 05OM280), cycloisomerization of terminal dienes (06JA13340, 06JA13901), and oligomerization and polymerization of alkenes (ethylene and propene) (03OM2545, 04JOM1356, 04OM6087, 06CCR1391, 07CRV1745, 07OM726, 08CCR1420). Coordination compounds of 2,6-bis(imino)pyridines are especially widely used for the catalysis of oligomerization and polymerization of ethylene (98CC849, 98JA4049, 98JA7143, 99AGE428, 99JA8728, 00CRV1169, 04MI3).

The role of azomethine ligands in the formation of the supramolecular structures is illustrated by self-assembly of the aggregates of silver perchlorate with N,N'-bis(pyridin-2-ylmethylene)benzene-1,4-diamine having a one-dimensional zigzag polymeric structure [{Ag_2L_2}$(ClO_4)_2(AN)]_n$ (06IC295). Using 3,3'-dimethyl-N,N'-bis(pyridin-2-ylmethylene)biphenyl-4,4'-diamine one can prepare a molecular rectangle [{Ag_2L_2}$(ClO_4)_2$]. Both compounds are characterized by their unique luminescence properties. Another supramolecular example is the formation of nanostructured catalytic materials, successfully applied for cross-coupling reactions (08ICA1562).

The complex of 2,6-bis[1-(4-amino-1,2,3,6-tetrahydro-1,3-dimethyl-2,6-dioxopyrimidin-5-yl)imino]ethylpyridine with cadmium(II) of composition CdL_2 (05JBI924) reveals antitumor activity. A great number of hetarylazomethine complexes show remarkable antibacterial activity (05SA(A)1059).

LIST OF ABBREVIATIONS

Ac	Acyl
acac	Acetylacetonate
Alk	Alkyl
AN	Acetonitrile
Ar	Aryl
bipy	2,2'-bipyridine
Bu	Butyl
cod	Cyclooctadiene-1,5
COE	Cyclooctene
Cp	Cyclopentadienyl
Cp*	Pentamethylcyclopentadienyl
Cy	Cyclohexyl
DME	Dimethoxyethane
DMF	Dimethylformamide
DMSO	Dimethylsulfoxide
Et	Ethyl
Hal	Halide
Het	Heteryl
IHB	Intramolecular hydrogen bond
Me	Methyl
Mes	Mesityl
OLED	Organic light emitting diode
OTf	Triflate
Ph	Phenyl
Pr	Propyl

py	Pyridine
salen	N,N'-bis(salicylidene)ethylenediamine dianion
solv	Solvent
THF	Tetrahydrofuran
Tol	Tolyl
Ts	$SO_2C_6H_4Me-p$ (tosyl)

ACKNOWLEDGMENTS

The authors are grateful for financial support from the Ministry of Education and Science of the Russian Federation (grant RNP.2.1.1/2348), a grant from the President of the Russian Federation (NS-363.2008.3), RFBR (grants 08-03-00154, 08-03-00223) and an internal grant by the Southern Federal University.

REFERENCES

63AHC339   A. R. Katritzky and J. M. Lagowski, *Adv. Heterocycl. Chem.*, **1**, 339 (1963).
63BCJ1477   O. Tsuge, M. Nishinohara, and M. Tashiro, *Bull. Chem. Soc. Jpn.*, **36**, 1477 (1963).
63CRV489   R. W. Lauer, *Chem. Rev.*, **63**, 489 (1963).
63JCS4600   G. R. Bedford, A. R. Katritzky, and H. M. Wuest, *J. Chem. Soc.*, 4600 (1963).
66PIC66   R. H. Holm, G. M. Ewerett, and A. Chakravorty, *Prog. Inorg. Chem.*, **7**, 83 (1966).
67ICA365   L. F. Lindoy and S. E. Livingstone, *Inorg. Chim. Acta*, **1**, 365 (1967).
71AX(B)1644   R. Tewari and R. C. Srivastava, *Acta Crystallogr.*, **B27**, 1644 (1971).
71JCM139   A. Mangia, M. Nardelli, G. Pelizzi, and C. Pelizzi, *J. Cryst. Mol. Struct.*, **1**, 139 (1971).
71PIC214   R. H. Holm and M. J. O'Connor, *Prog. Inorg. Chem.*, **14**, 214 (1971).
71ZC81   H. Hennig, *Z. Chem.*, **11**, 81 (1971).
72CC683   L. E. Lindoy and D. H. Busch, *J. Chem. Soc. Chem. Commun.*, 683 (1972).
72IC1100   C. H. Wei, *Inorg. Chem.*, **11**, 1100 (1972).
72IC2315   C. H. Wei, *Inorg. Chem.*, **11**, 2315 (1972).
72ICA248   J. A. Bertrand and C. E. Kirkwood, *Inorg. Chim. Acta*, **6**, 248 (1972).
73IC2316   W. L. Goedken and G. C. Christoph, *Inorg. Chem.*, **12**, 2316 (1973).
73RCR89   A. D. Garnovskii, O. A. Osipov, L. I. Kuznetsova, and N. N. Bogdashev, *Russ. Chem. Rev.*, **42**, 89 (1973), [*Usp. Khim.*, **42**, 177 (1973)].
74CC727   N. W. Alcock, D. C. Liles, M. McPartlin, and P. A. Tasker, *J. Chem. Soc. Chem. Commun.*, 727 (1974).
74JPR971   V. S. Bogdanov, V. P. Litvinov, G. L. Goldfarb, N. N. Petuchova, and E. G. Ostapenko, *J. Prakt. Chem.*, **316**, 971 (1974).
74ZOR2210   L. N. Kurkovskaya, N. N. Shapetko, I. Y. Kvito, Y. N. Koshelev, and E. D. Samarzeva, *Zh. Org. Khim.*, **10**, 2210 (1974).

76AHC(S)1 J. Elguero, C. Marzin, A. R. Katritzky, and P. Linda, *Adv. Heterocycl. Chem., Suppl.,* **1**, 1 (1976).

76CSC447 N. A. Bailey and S. E. Hull, *Cryst. Struct. Commun.,* **5**, 447 (1976).

76ZOB675 N. N. Bogdashev, A. D. Garnovskii, O. A. Osipov, V. P. Grigoriev, and N. M. Gontmakher, *Zh. Obshch. Khim.,* **46**, 675 (1976).

76ZOB2706 A. D. Garnovskii, T. A. Yusman, B. M. Krasovitskii, O. A. Osipov, N. F. Levchenko, B. M. Bolotin, L. M. Afanasiadi, N. I. Chernova, and V. A. Alekseenko, *Zh. Obshch. Khim.,* **46**, 2706 (1976).

77ACA(2)80 S. H. Simonsen and C. E. Urdy, *ACA, Ser. 2,* **5**, 80 (1977).

78IC1288 P. G. Sim and E. Sinn, *Inorg. Chem.,* **17**, 1288 (1978).

78JCS(D)185 G. B. Jameson, F. C. March, W. T. Robinson, and S. S. Koon, *J. Chem. Soc. Dalton Trans.,* 185 (1978).

78KGS1677 I. Y. Kvitko, L. N. Kurkovskaya, R. B. Khozeeva, and A. V. Eltsov, *Khim. Geterotsikl. Soedin.,* 1677 (1978), [Chem. Heterocycl. Compd., 1364 (1978)].

79CC1033 M. G. B. Drew, J. O. Cabral, M. F. Cabral, F. S. Esho, and M. Nelson, *J. Chem. Soc. Chem. Commun.,* 1033 (1979).

79JA6917 R. R. Gagne, R. P. Kreh, and J. A. Dodge, *J. Am. Chem. Soc.,* **101**, 6917 (1979).

79JCS(D)1899 A. J. Deeming, I. P. Rothwell, M. B. Hursthouse, and K. M. A. Malik, *J. Chem. Soc. Dalton Trans.,* 1899 (1979).

79MI1 J. Tennant, in *"Comprehensive Organic Chemistry"* (D. Barton and W. D. Ollis, eds.), Vol. 2, Chapter 8, Pergamon Press, Oxford and New York (1979).

80IC891 H. Brunner and D. K. Rastogi, *Inorg. Chem.,* **19**, 891 (1980).

80UK1234 G. V. Panova, N. K. Vikulova, and V. M. Potapov, *Usp. Khim.,* **49**, 1234 (1980).

81CJC1205 F. A. Bottino, M. L. Longo, D. Sciotto, and M. Torre, *Can. J. Chem.,* **59**, 1205 (1981).

81ICA47 K. Henrick and I. A. Tasker, *Inorg. Chim. Acta,* **47**, 47 (1981).

81JA241 P. G. Sim and E. Sinn, *J. Am. Chem. Soc.,* **103**, 241 (1981).

81JCS(D)1524 R. W. Balk, G. Boxhoorn, T. L. Snoeck, G. C. Schoemaker, D. J. Stufkens, and A. Oskam, *J. Chem. Soc. Dalton Trans.,* 1504 (1981).

82ICA213 L. MacDonald, D. H. Brown, and W. E. Smith, *Inorg. Chim. Acta,* **62**, 213 (1982).

82POL617 J. D. Korp and I. Bernal, *Polyhedron,* **1**, 617 (1982).

83AX(C)52 R. J. Majeste, C. L. Klein, and E. D. Stevens, *Acta Crystallogr.,* **C39**, 52 (1983).

84CL573 C. Kabuto, T. Kikuchi, H. Yokoi, and M. Iwaizumi, *Chem. Lett.,* 573 (1984).

84JCS(D)2741 N. A. Bailey, A. Barrass, D. E. Fenton, M. S. L. Gonzalez, R. Moody, and C. O. R. de Barbarin, *J. Chem. Soc. Dalton Trans.,* 2741 (1984).

84JHC1603 P. H. Olsen, F. E. Nielsen, E. B. Pedersen, and J. Becher, *J. Heterocycl. Chem.,* **21**, 1603 (1984).

85CC1172 D. S. C. Black, D. C. Craig, N. Kumar, and L. C. H. Wong, *J. Chem. Soc. Chem. Commun.,* 1172 (1985).

85ICA7 H. Adams, N. A. Bailey, I. S. Baerd, D. E. Fenton, J. P. Costes, G. Cros, and J. P. Laurent, *Inorg. Chim. Acta,* **101**, 7 (1985).

85KGS921 L. M. Sitkina, A. D. Dubonosov, A. E. Lyubarskaya, V. A. Bren, and V. I. Minkin, *Khim. Geterotsikl. Soedin.,* **21**, 921 (1985).

86IC394	J. C. Davis, W. J. Kung, and B. A. Averill, *Inorg. Chem.*, **25**, 394 (1986).
86IC2300	C. J. O'Connor, D. Firmin, A. K. Pant, B. R. Babu, and E. D. Stevens, *Inorg. Chem.*, **25**, 2300 (1986).
86ICA(112)65	H. Brunner, G. Riepl, I. Bernal, and W. H. Ries, *Inorg. Chim. Acta*, **112**, 65 (1986).
86ICA(124)121	L. Casella, M. Gillotti, and R. Vigano, *Inorg. Chim. Acta*, **124**, 121 (1986).
87IC349	J. M. Fernandez, M. J. Rosales-Hoz, M. F. Rubio-Arroyo, R. Salcedo, R. A. Toscano, and A. Vela, *Inorg. Chem.*, **26**, 349 (1987).
87IC1375	T. Mallah, O. Kahn, J. Gouteron, S. Jeannin, Y. Jeannin, and C. J. O'Connor, *Inorg. Chem.*, **26**, 1375 (1987).
87JCS(D)513	L. Chan and A. J. Lees, *J. Chem. Soc. Dalton Trans.*, 513 (1987).
87KK869	Y. I. Ryabukhin, N. V. Shibaeva, A. S. Kuzharov, V. G. Korobkova, A. V. Khokhlov, and A. D. Garnovskii, *Koord. Khim.*, **13**, 869 (1987).
87MI1	G. Wilkinson, R. D. Gillard, and J. A. McCleverty (eds.), "*Comprehensive Coordination Chemistry*" pp. V.1–V.7, Elsevier-Pergamon Press, Oxford and New York (1987).
87MI2	M. Calligaris and L. Randaccio, in "*Comprehensive Coordination Chemistry*" (G. Wilkinson, R. D. Gillard, and J. A. McCleverty, eds.), Vol. 2, p. 715, Elsevier-Pergamon Press, Oxford and New York (1987).
88IC4208	A. Duatti, A. Marchi, R. Rossi, L. Magon, E. Deutsch, V. Bertolasi, and F. Bellucci, *Inorg. Chem.*, **27**, 4208 (1988).
88JCS(D)1943	N. Matsumoto, S. Yamashita, A. Ohyoshi, S. Kohata, and H. Okawa, *J. Chem. Soc. Dalton Trans.*, 1943 (1988).
89AX(C)7	J. M. Latour, S. S. Tandon, and D. C. Povey, *Acta Crystallogr.*, **C45**, 7 (1989).
89IC2141	B. Chiari, O. Piovesana, T. Tarantelli, and P. F. Zanazzi, *Inorg. Chem.*, **28**, 2141 (1989).
89JCC273	P. Gill, P. M. Zarza, P. Nunez, A. Medina, M. C. Diaz, M. G. Martin, J. M. Arrieta, M. Vlassi, G. Germain, M. Vermeire, and L. Dupont, *J. Coord. Chem.*, **20**, 273 (1989).
89JCS(D)1117	O. Kahn, T. Mallah, J. Gouteron, S. Jeannin, and Y. Jeannin, *J. Chem. Soc. Dalton Trans.*, 1117 (1989).
89JCS(D)1979	M. Gullotti, L. Casella, A. Pintar, E. Suardi, P. Zanello, and S. Mangani, *J. Chem. Soc. Dalton Trans.*, 1989 (1979).
89JOM303	A. I. Uraev, A. L. Nivorozhkin, A. S. Frenkel, A. S. Antsyshkina, M. A. Porai-Koshits, L. E. Konstantinovsky, G. K. I. Magomedov, and A. D. Garnovskii, *J. Organomet. Chem.*, **368**, 303 (1989).
89POL491	A. S. Rothin, H. J. Banbery, F. J. Berry, T. A. Hamor, C. J. Jones, and J. A. McCleverty, *Polyhedron*, **8**, 491 (1989).
89POL2543	A. Castineiras, J. A. Castro, M. L. Duran, J. A. Garcia-Vazquez, A. Macias, J. Romero, and A. Sousa, *Polyhedron*, **8**, 2543 (1989).
89ZAAC215	M. L. Duran, J. A. Garcia-Vazquez, A. Macias, J. Romero, A. Sousa, and E. B. Rivero, *Z. Anorg. Allg. Chem.*, **573**, 215 (1989).
90AX(C)1414	M. L. Rodriguez, C. Ruiz-Perez, F. V. Rodriguez-Romero, M. S. Palacios, and P. Martin-Zarza, *Acta Crystallogr.*, **C46**, 1414 (1990).
90AX(C)1434	M. Parvez and W. J. Birdsall, *Acta Crystallogr.*, **C46**, 1434 (1990).

90CL1181   M. Mikuriya, T. Fujii, S. Kamisawa, Y. Kawasaki, T. Tokii, and
                H. Oshio, *Chem. Lett.*, 1181 (1990).
90IC2816   O. J. Gelling, A. Meetsma, and B. L. Feringa, *Inorg. Chem.*, **29**, 2816
                (1990).
90ICA(174)9   S. De Martin, G. Zassinovich, and G. Mestroni, *Inorg. Chim. Acta*,
                174, 9 (1990).
90ICA(176)261   P. Gill, P. Martin-Zarza, M. S. Palacios, M. L. Rodriguez, C. Ruiz-
                Perez, and F. V. Rodriguez-Romero, *Inorg. Chim. Acta*, **176**, 261
                (1990).
90JCS(D)2225   F. Tisato, F. Refosco, A. Moresco, G. Bandoli, U. Mazzi, and
                M. Nicolini, *J. Chem. Soc. Dalton Trans.*, 2225 (1990).
90JCS(D)2637   B. F. Hoskins, C. J. McKenzie, R. Robson, and L. Zhenrong,
                J. Chem. Soc. Dalton Trans., 2637 (1990).
90KK395   Y. T. Struchkov, T. V. Timofeeva, and A. S. Batsanov, *Koord. Khim.*,
                16, 395 (1990).
90POL615   H. J. Banbery, F. S. McQuillan, T. A. Hamor, C. J. Jones, and J. A.
                McCleverty, *Polyhedron*, **9**, 615 (1990).
90ZOR2389   V. P. Rybalkin, L. M. Sitkina, Z. V. Bren, V. A. Bren, and V. I.
                Minkin, *Zh. Org. Khim.*, **26**, 2389 (1990).
91BCJ149   A. Furuhashi, I. Ono, A. Ouchi, and A. Yamasaki, *Bull. Chem. Soc.
                Jpn.*, **64**, 149 (1991).
91HCA717   C. Bolm, K. Weickhardt, M. Zehnder, and D. Glasmacher,
                Helv. Chim. Acta, **74**, 717 (1991).
91IC42   L. J. Larson, A. Oskam, and J. J. Zink, *Inorg. Chem.*, **30**, 42
                (1991).
91IC4754   R. M. Dominey, B. Hauser, J. Hubbard, and J. Dunham, *Inorg.
                Chem.*, **30**, 4754 (1991).
91ICA19   A. Rakotonandrasana, D. Boinnard, J. M. Savariault, J. P.
                Tuchagues, V. Petrouleas, C. Cartier, and M. Verdaguer, *Inorg.
                Chim. Acta*, **180**, 19 (1991).
91MC78   A. L. Nivorozhkin, L. E. Nivorozhkin, L. E. Konstantinovskii, and
                V. I. Minkin, *Mendeleev Commun.*, **1**, 78 (1991).
92BCJ1603   S. Ohkubo, K. Inoue, H. Tamaki, M. Ohba, N. Matsumoto,
                H. Okawa, and S. Kida, *Bull. Chem. Soc. Jpn.*, **65**, 1603 (1992).
92CCR67   J. Costamagna, J. Vargas, R. Latore, A. Alvarado, and G. Mena,
                Coord. Chem. Rev., **119**, 67 (1992).
92IC1981   C. R. Comman, J. Kampf, and V. L. Pecoraro, *Inorg. Chem.*, **31**,
                1981 (1992).
92IC2035   C. R. Comman, J. Kampf, M. S. Lah, and V. L. Pecoraro, *Inorg.
                Chem.*, **31**, 2035 (1992).
92ICA213   W. J. Birdsall, B. A. Weber, and M. Parvez, *Inorg. Chim. Acta*, **196**,
                213 (1992).
92IZV917   I. I. Chuev, O. S. Filipenko, V. G. Ryzhikov, S. M. Aldoshin, and
                L. O. Atovmyan, *Izv. Ross. Akad. Nauk, Ser. Khim.*, 917 (1992).
92JA9925   C. R. Comman, G. J. Colpas, J. D. Hoeschele, J. Kampf, and V. L.
                Pecoraro, *J. Am. Chem. Soc.*, **114**, 9925 (1992).
92JCS(D)3083   B. F. Hoskins, C. J. McKenzie, and R. Robson, *J. Chem. Soc. Dalton
                Trans.*, 3083 (1992).
92MC30   A. E. Mistryukov, I. S. Vasilchenko, V. S. Sergienko, A. L.
                Nivorozhkin, S. G. Kochin, M. A. Porai-Koshits, L. E. Nivor-
                ozhkin, and A. D. Garnovskii, *Mendeleev Commun.*, **2**, 30 (1992).

92POL235   J. A. Castro, J. Romero, J. A. Garcia-Vazquez, M. L. Duran, A. Sousa, E. E. Castellano, and J. Zukerman-Schpector, *Polyhedron*, **11**, 235 (1992).

92SR321   V. P. Litvinov, A. F. Vaisburg, and V. Y. Mortikov, *Sulf. Rep.*, **11**, 321 (1992).

92ZK61   C. Ruiz-Perez, F. V. Rodriguez-Romero, P. Martin-Zarza, and P. Gill, *Z. Kristallogr.*, **198**, 61 (1992).

93BCJ1675   M. Mikuriya, T. Fujii, T. Tokii, and A. Kawamori, *Bull. Chem. Soc. Jpn.*, **66**, 1675 (1993).

93CCR1   A. D. Garnovskii, A. L. Nivorozhkin, and V. I. Minkin, *Coord. Chem. Rev.*, **126**, 1 (1993).

93CCR237   A. P. Sadimenko, A. D. Garnovskii, and N. Retta, *Coord. Chem. Rev.*, **126**, 237 (1993).

93DOK54   A. S. Antsyshkina, M. A. Porai-Koshits, I. S. Vasilchenko, A. L. Nivorozhkin, and A. D. Garnovskii, *Dokl. Ross. Akad. Nauk*, **330**, 54 (1993).

93IC2670   A. J. Conti, R. K. Chadha, K. M. Sena, A. L. Rheingold, and D. N. Hendrickson, *Inorg. Chem.*, **32**, 2670 (1993).

93IC5528   B. de Klerk-Engels, H. W. Fruhauf, K. Vrieze, H. Koojman, and A. L. Spek, *Inorg. Chem.*, **32**, 5528 (1993).

93JCC125   J. A. Castro, J. Romero, J. A. Garcia-Vazquez, A. Sousa, E. E. Castellano, and J. Zukerman-Schpector, *J. Coord. Chem.*, **28**, 125 (1993).

93JCS(D)2157   N. Matsumoto, T. Nozaki, H. Ushio, K. Motada, M. Ohba, G. Mago, and H. Okawa, *J. Chem. Soc. Dalton Trans.*, 2157 (1993).

93JCS(D)1927   V. De Felice, A. De Renzi, F. Giordano, and D. Tesauro, *J. Chem. Soc. Dalton Trans.*, 1927 (1993).

93JCSR473   E. Kwiatkowski, M. Kwiatkowski, A. Olechnowicz, and G. Bandoli, *J. Crystallogr. Spectrosc. Res.*, **23**, 473 (1993).

93OM3677   Z. Zhou, C. Jablonski, and J. Bridson, *Organometallics*, **12**, 3677 (1993).

94CRV279   F. C. J. M. van Veggel, W. Verbootom, and D. N. Reinhoudt, *Chem. Rev.*, **94**, 279 (1994).

94ICA123   M. Lubben, A. Meetsma, F. van Bolhuis, B. L. Feringa, and R. Hage, *Inorg. Chim. Acta*, **215**, 123 (1994).

94JCR2469   J. Castro, J. Romero, J. A. Garcia-Vazquez, A. Sousa, and A. Castineiras, *J. Chem. Crystallogr.*, **24**, 2469 (1994).

94JCS(D)1145   B. Crociani, S. Antonaroli, F. De Bianca, L. Canovese, F. Visentin, and P. Uguagliati, *J. Chem. Soc. Dalton Trans.*, 1145 (1994).

94KK824   I. S. Vasilchenko, A. S. Antsyshkina, D. A. Garnovskii, G. G. Sadikov, M. A. Porai-Koshits, S. G. Sigeikin, and A. D. Garnovskii, *Koord. Khim.*, **20**, 824 (1994).

94POL2055   W. J. Birdsall, D. P. Long, S. P. E. Smith, M. E. Kastner, K. Tang, and C. Kirk, *Polyhedron*, **13**, 2055 (1994).

94RJCC74   D. A. Garnovskii, A. D. Garnovskii, A. P. Sadimenko, and S. G. Sigeikin, *Russ. J. Coord. Chem.*, **20**, 74 (1994).

94ZOB657   I. Y. Kvitko, L. V. Alam, M. N. Bobrovnikov, G. B. Zvetikyan, and S. L. Panasyuk, *Zh. Obshch. Khim.*, **64**, 657 (1994), [*Russ. J. Gen. Chem.*, **64**, 597 (1994)].

95AX(C)1522   S. Brooker and B. M. Carter, *Acta Crystallogr.*, **C51**, 1522 (1995).

95CC331	P. Wehman, R. E. Rulke, V. E. Kaasjager, P. C. J. Kamer, H. Kooijman, A. L. Spek, C. J. Elsevier, K. Vrieze, and P. W. N. M. van Leeuwen, *Chem. Commun.*, 331 (1995).
95IC2108	T. Nozaki, N. Matsumoto, H. Okawa, H. Miyasaka, and G. Mago, *Inorg. Chem.*, **34**, 2108 (1995).
95IC2921	P. G. Cozzi, C. Floriani, A. Chiesi-Villa, and C. Rizzoli, *Inorg. Chem.*, **34**, 2921 (1995).
95IC4213	C. R. Cornman, E. P. Zovinka, Y. D. Boyajian, K. M. Geiser-Bush, P. D. Boyle, and P. Singh, *Inorg. Chem.*, **34**, 4213 (1995).
95IC4467	H. Frydendahl, H. Toftlund, J. Becher, J. C. Dutton, K. S. Murray, L. F. Taylor, O. P. Anderson, and E. R. T. Tiekink, *Inorg. Chem.*, **34**, 4467 (1995).
95IC5252	M. J. Baldwin, J. W. Kampf, M. L. Kirk, and V. L. Pecoraro, *Inorg. Chem.*, **34**, 5252 (1995).
95ICA(235)273	N. A. Bailey, D. E. Fenton, Q. Y. He, N. Terry, W. Haase, and R. Werner, *Inorg. Chim. Acta*, **235**, 273 (1995).
95ICA(237)181	J. D. Crane, R. Hughes, and E. Sinn, *Inorg. Chim. Acta*, **237**, 181 (1995).
95OM4994	P. G. Cozzi, E. Gallo, C. Floriani, A. Chiesi-Villa, and C. Rizzoli, *Organometallics*, **14**, 4994 (1995).
95UK705	S. I. Yakimovich and K. N. Selenin, *Usp. Khim.*, **65**, 705 (1995).
96AHC1	R. E. Valters, F. Fulop, and D. Korbonits, *Adv. Heterocycl. Chem.*, **66**, 1 (1996).
96BCJ1573	M. Mikuriya, R. Nukada, W. Tokami, Y. Hashimoto, and T. Fujii, *Bull. Chem. Soc. Jpn.*, **69**, 1573 (1996).
96CB1195	C. Bruckner, E. K. Mar, S. J. Rettig, and D. Dolphin, *Chem. Ber.*, **129**, 1195 (1996).
96IC357	A. D. Keramidas, A. B. Papaioannou, A. Vlahos, T. A. Kabanos, G. Bonas, A. Makriyannis, C. P. Rapropoulou, and A. Terzis, *Inorg. Chem.*, **35**, 357 (1996).
96IC368	E. Shuter, H. R. Hoveyda, V. Karunaratne, S. J. Rettig, and C. Orvig, *Inorg. Chem.*, **35**, 368 (1996).
96JCS(D)93	G. Asgedom, A. Sreedhara, J. Kivikoski, E. Kolehmainen, and C. P. Rao, *J. Chem. Soc. Dalton Trans.*, 93 (1996).
96JOM109	R. E. Rulke, J. G. P. Delis, A. M. Croot, C. J. Elsevier, P. W. N. M. van Leeuwen, K. Vrieze, K. Goubitz, and H. Schenk, *J. Organomet. Chem.*, **508**, 109 (1996).
96MCJ19	A. D. Garnovskii, D. A. Garnovskii, A. S. Burlov, and I. S. Vasilchenko, *Mendeleev Chem. J.*, **40**(4–5), 19 (1996).
96MCJ209	V. A. Bren, *Mendeleev Chem. J.*, **40**(4–5), 209 (1996).
96M11	A. R. Katritzky, C. W. Rees, and E. E. V. Scriven (eds.), "*Comprehensive Heterocyclic Chemistry II*", 2nd ed., Pergamon, Oxford (1996).
96OM2108	C. J. Levy, J. J. Vittal, and R. J. Puddephatt, *Organometallics*, **15**, 2108 (1996).
96POL1723	E. C. Alyea and V. K. Jain, *Polyhedron*, **15**, 1723 (1996).
96POL4407	A. Garcia-Raso, J. J. Foil, F. Badenas, and M. Quiros, *Polyhedron*, **15**, 4407 (1996).
96T3521	M. T. Rispens, O. J. Gelling, A. H. M. de Vries, A. Meetsma, F. van Bolhuis, and B. L. Feringa, *Tetrahedron*, **52**, 3521 (1996).

96ZN(B)757   A. Bach, L. Beyer, T. Gelbrich, K. H. Hallmeier, C. Hennig, M. Mobius, R. Richter, R. Szargan, V. Fernandez, and J. Losada, Z. Naturforsch., **B51**, 757 (1996).

96ZOB1546   D. A. Garnovskii, A. S. Burlov, A. D. Garnovskii, I. S. Vasilchenko, and A. Sousa, Zh. Obshch. Khim., **66**, 1546 (1996).

97AS519   O. Atakol, H. Nazir, M. N. Tahir, and D. Ulku, Anal. Sci., **13**, 519 (1997).

97BCJ2461   N. Matsumoto, M. Mimura, Y. Sunatsuki, S. Eguchi, Y. Mizuguchi, H. Miyasaka, and T. Nakashima, Bull. Chem. Soc. Jpn., **70**, 2461 (1997).

97CB887   C. Bolm, T. K. K. Luong, and R. Harms, Chem. Ber., **130**, 887 (1997).

97CC1711   A. L. Nivorozhkin, A. I. Uraev, G. I. Bondarenko, A. S. Antsyshkina, V. P. Kurbatov, A. D. Garnovskii, C. I. Turta, and N. D. Brashoveanu, Chem. Commun., 1711 (1997).

97CCR191   S. Madal, G. Das, R. Singh, R. Shukla, and P. K. Bhardwaj, Coord. Chem. Rev., **160**, 191 (1997).

97DOK212   A. I. Uraev, A. L. Nivorozhkin, A. S. Antsyshkina, O. Y. Korshunov, G. I. Bondarenko, I. S. Vasilchenko, V. P. Kurbatov, and A. D. Garnovskii, Dokl. Ross. Akad. Nauk., **356**, 212 (1997).

97IC6080   A. Christensen, H. S. Jensen, V. McKee, C. J. McKenzie, and M. Munch, Inorg. Chem., **36**, 6080 (1997).

97IC6279   S. Theil, R. Yerande, R. Chikate, F. Dahan, A. Bousseksou, S. Padhye, and J. P. Tuchagues, Inorg. Chem., **36**, 6279 (1997).

97JCS(D)111   O. P. Anderson, A. La Cour, M. Findeisen, L. Hennig, O. Simonsen, L. F. Taylor, and H. Toftlund, J. Chem. Soc. Dalton Trans., 111 (1997).

97JCS(D)121   A. La Cour, M. Findeisen, A. Hazell, R. Hazell, and G. Zdobinsky, J. Chem. Soc. Dalton Trans., 121 (1997).

97JCS(D)161   C. O. R. de Barbarin, N. A. Bailey, D. E. Fenton, and Q. Y. He, J. Chem. Soc. Dalton Trans., 161 (1997).

97JCS(D)2045   A. La Cour, M. Findeisen, K. Hansen, R. Hazell, L. Hennig, C. E. Olsen, L. Pedersen, and O. Simonsen, J. Chem. Soc. Dalton Trans., 2045 (1997).

97JCS(D)2155   M. Barz, M. U. Rauch, and W. R. Thiel, J. Chem. Soc. Dalton Trans., 2155 (1997).

97JCS(D)3755   M. Gomez-Simon, S. Jansat, G. Muller, D. Panyella, M. Font-Bardia, and X. Solans, J. Chem. Soc. Dalton Trans., 3755 (1997).

97JCS(D)4539   B. Adhikhari, O. P. Anderson, A. La Cour, R. Hazell, S. M. Miller, C. E. Olsen, L. Pedersen, and O. Simonsen, J. Chem. Soc. Dalton Trans., 4539 (1997).

97JOM255   H. Brunner, P. Faustmann, A. Dietl, and B. Nuber, J. Organomet. Chem., **542**, 255 (1997).

97T11936   M. S. Silva, N. Jagerovic, and J. Elguero, Tetrahedron, **53**, 11936 (1997).

97TED1222   Y. Kijima, N. Asai, N. Kishi, and S. Tamura, IEEE Trans. Electron Dev., **44**, 1222 (1997).

98AHC1   A. D. Garnovskii and A. P. Sadimenko, Adv. Heterocycl. Chem., **72**, 1 (1998).

98ACS371   J. H. Zhou, S. L. Li, H. H. Jiang, D. X. Liu, and Z. H. Yang, Acta Chim. Sinica, **56**, 371 (1998).

98AX(C)725   D. Ulku, M. N. Tahir, H. Nazir, and H. Yilmaz, Acta Crystallogr., **C54**, 725 (1998).

98CC849	G. J. P. Britovsek, V. C. Gibson, B. S. Kimberley, P. J. Maddox, S. J. McTavish, G. A. Solan, A. J. P. White, and D. J. Williams, *Chem. Commun.*, 849 (1998).
98CCR31	A. D. Garnovskii, A. P. Sadimenko, M. I. Sadimenko, and D. A. Garnovskii, *Coord. Chem. Rev.*, **173**, 31 (1998).
98CEJ818	N. End, L. Macko, M. Zehnder, and A. Pfaltz, *Chem. Eur. J.*, **4**, 818 (1998).
98IC1127	H. Luo, P. E. Fanwick, and M. A. Green, *Inorg. Chem.*, **37**, 1127 (1998).
98IC2134	S. Ryan, H. Adams, D. E. Fenton, M. Becker, and S. Schindler, *Inorg. Chem.*, **37**, 2134 (1998).
98IC2663	G. Mugesh, H. B. Singh, R. P. Patel, and R. J. Butcher, *Inorg. Chem.*, **37**, 2663 (1998).
98IC5968	M. Shivakumar, K. Pramanik, P. Ghosh, and A. Chakravorty, *Inorg. Chem.*, **37**, 5968 (1998).
98ICA(269)350	W. Bansse, J. Fliegner, S. Sawusch, U. Schilde, and E. Uhlemann, *Inorg. Chim. Acta*, **269**, 350 (1998).
98ICA(279)217	M. Hoogenraad, K. Ramkisoensing, H. Kooijman, A. L. Spek, E. Bouwman, J. G. Haasnoot, and J. Reedijk, *Inorg. Chim. Acta*, **279**, 217 (1998).
98JA4049	B. L. Small, M. Brookhart, and M. A. Bennett, *J. Am. Chem. Soc.*, **120**, 4049 (1998).
98JA7143	B. L. Small and M. J. Brookhart, *J. Am. Chem. Soc.*, **120**, 7143 (1998).
98JCU1546	Y. G. Yin, C. K. Cheung, and W. T. Wong, *J. Chin. Univ.*, **19**, 1546 (1998).
98KK164	V. I. Minkin, M. S. Korobov, L. E. Nivorozhkin, O. E. Kompan, G. S. Borodkin, and R. Y. Olekhnovich, *Koord. Khim.*, **24**, 164 (1998).
98M455	S. I. Mostafa and S. A. Abd El-Maksoud, *Monatsh. Chem.*, **129**, 455 (1998).
98MI1	C. C. Leznoff and A. B. P. Lever (eds.), "*Phthalocyanines: Properties and Applications*" Vols 1–4, Wiley, New York (1998).
98MI2	N. B. Keown, Phthalocyanine Materials, Cambridge University Press, Cambridge (1998).
98POL1547	J. A. Garcia-Vazquez, J. Romero, M. L. Duran, A. Sousa, A. D. Garnovskii, A. S. Burlov, and D. A. Garnovskii, *Polyhedron*, **17**, 1547 (1998).
98T(A)3763	H. Brunner, B. Nuber, and T. Tracht, *Tetrahedron: Asymm.*, **9**, 3763 (1998).
99AGE428	G. J. P. Britovsek, V. C. Gibson, and D. F. Wass, *Angew. Chem. Int. Ed. Engl.*, **38**, 428 (1999).
99CL525	H. Asada, M. Ozeki, M. Fujiwara, and T. Matsushita, *Chem. Lett.*, 525 (1999).
99EJI1229	G. Mugesh, H. B. Singh, and R. J. Butcher, *Eur. J. Inorg. Chem.*, 1229 (1999).
99EJI1393	S. Knoblauch, R. Benedix, M. Ecke, T. Gelbrich, J. Sieler, F. Somoza, and H. Hennig, *Eur. J. Inorg. Chem.*, 1393 (1999).
99IC4510	K. J. Miller, J. H. Baag, and M. M. Abu-Omar, *Inorg. Chem.*, **38**, 4510 (1999).
99IC5633	N. N. Gerasimchuk, A. Gerges, T. Clifford, A. Danby, and K. Bowman-James, *Inorg. Chem.*, **38**, 5633 (1999).

99ICA139	O. Homer, J. J. Jirerd, C. Philouze, and L. Tchertanov, *Inorg. Chim. Acta*, **290**, 139 (1999).
99JA6956	A. J. Stemmler and C. J. Burrows, *J. Am. Chem. Soc.*, **121**, 6956 (1999).
99JA8728	G. J. P. Britovsek, M. Bruce, V. C. Gibson, B. S. Kimberley, P. J. Maddox, S. Mastroianni, S. J. McTavish, C. Redshaw, G. A. Solan, S. Stromberg, A. J. P. White, and D. J. Williams, *J. Am. Chem. Soc.*, **121**, 8728 (1999).
99JCC219	A. D. Garnovskii, L. M. Blanco, B. I. Kharisov, D. A. Garnovskii, and A. S. Burlov, *J. Coord. Chem.*, **48**, 219 (1999).
99JCS(D)539	F. Tuna, L. Patron, Y. Journeaux, M. Andruh, W. Plass, and J. C. Trombe, *J. Chem. Soc. Dalton Trans.*, 539 (1999).
99JCS(D)2031	H. Adams, M. R. J. Elsegood, D. E. Fenton, S. L. Heath, and S. J. Ryan, *J. Chem. Soc. Dalton Trans.*, 2031 (1999).
99JMC(A)101	B. Cetinkaya, E. Cetinkaya, M. Brookhart, and P. S. White, *J. Mol. Catal.*, **A142**, 101 (1999).
99JOM(572)271	J. Nijhoff, M. J. Bakker, F. Hartl, and D. J. Stufkens, *J. Organomet. Chem.*, **572**, 271 (1999).
99JOM(573)121	J. Nijhoff, F. Hartl, J. W. M. van Outersterp, D. J. Stufkens, M. J. Calhorda, and L. F. Veiros, *J. Organomet. Chem.*, **573**, 121 (1999).
99JOM(577)243	G. Mugesh, H. B. Singh, and R. J. Butcher, *J. Organomet. Chem.*, **577**, 243 (1999).
99MI1	A. D. Garnovskii and B. I. Kharisov, Direct Synthesis of Coordination and Organometallic Compounds, p. 244, Elsevier, Amsterdam (1999).
99MI2	N. V. Gerbeleu, V. B. Arion, and J. Burgess, Template Synthesis of Macrocyclic Compounds, p. 565, Wiley-VCH, Weinheim (1999).
99MI3	K. M. Kadish, K. M. Smith, and R. Guilard (eds.), "*The Porphyrin Handbook*" Academic Press, Boston (1999).
99OM4373	C. R. Barr, M. C. Jennings, R. J. Puddephatt, and K. W. Muir, *Organometallics*, **18**, 4373 (1999).
99OM4380	J. Nijhoff, F. Hartl, D. J. Stufkens, and J. Fraanje, *Organometallics*, **18**, 4380 (1999).
99POL1527	J. D. Crane, E. Sinn, and B. Tann, *Polyhedron*, **18**, 1527 (1999).
99POL2651	M. A. Mendez-Rojas, F. Cordova-Lozano, G. Gojon-Zorila, E. Gonzalez-Verigara, and M. Quiroz, *Polyhedron*, **18**, 2651 (1999).
99RCR345	V. I. Ovcharenko and R. S. Sagdeev, *Russ. Chem. Rev.*, **68**, 345 (1999).
99T12045	V. A. Soloshonok, C. Cai, V. J. Hruby, and L. Van Meervelt, *Tetrahedron*, **55**, 12045 (1999).
99ZNK1278	I. S. Vasilchenko, A. S. Antsyshkina, A. S. Burlov, G. G. Sadikov, A. I. Uraev, A. L. Nivirozhkin, D. A. Garnovskii, V. S. Sergienko, V. P. Kurbatov, A. Y. Korshunov, and A. D. Garnovskii, *Zh. Neorg. Khim.*, **44**, 1278 (1999).
99ZOB1533	D. A. Garnovskii, I. S. Vasilchenko, A. Y. Eliseeva, A. Sousa, A. P. Sadimenko, and A. D. Garnovskii, *Zh. Obshch. Khim.*, **69**, 1533 (1999).
00AA52	R. D. Rieje, *Aldrichchim. Acta*, **33**, 52 (2000).
00AHC1	J. Elguero, A. R. Katrizky, and O. V. Denisko, *Adv. Heterocycl. Chem.*, **76**, 1 (2000).

00AHC85	W. Friedrichsen, T. Taulsen, J. Elguero, and A. R. Katrizky, *Adv. Heterocycl. Chem.*, **76**, 85 (2000).
00AHC157	V. I. Minkin, A. D. Garnovskii, J. Elguero, A. R. Katritzky, and O. V. Denisko, *Adv. Heterocycl. Chem.*, **76**, 157 (2000).
00ARK382	R. A. Jones, G. Quintanilla-Lopez, S. A. N. Taheri, M. M. Hania, O. Ozturk, and H. Zuilhof, *Arkivoc*, 382 (2000).
00AS553	T. Hokelek, N. Akduran, M. Yildiz, and Z. Kilic, *Anal. Sci.*, **16**, 553 (2000).
00CL1114	Y. Matsuo, K. Mashima, and K. Tani, *Chem. Lett.*, 1114 (2000).
00CL1270	Y. Yoshida, S. Matsui, Y. Takagi, M. Mitani, N. Nitabaru, T. Nakano, H. Tanaka, and T. Fujita, *Chem. Lett.*, 1270 (2000).
00CRV1169	S. D. Ittel, L. K. Johnson, and M. Brookhart, *Chem. Rev.*, **100**, 1169 (2000).
00EJI1717	M. Bigioni, P. Ganis, A. Panunzi, F. Ruffo, C. Salvatore, and A. Vito, *Eur. J. Inorg. Chem.*, 1717 (2000).
00IC5878	H. Keypor, S. Saledzadeh, R. G. Pritchard, and R. V. Parish, *Inorg. Chem.*, **39**, 5878 (2000).
00ICA(298)256	J. P. Costes, F. Dahan, A. Dupuis, and J. P. Lorent, *Inorg. Chim. Acta*, **298**, 256 (2000).
00ICA(305)83	M. Kiang, S. Gou, S. Chantrapromma, S. S. S. Raj, H. K. Fun, Q. Zeng, Z. Yu, and X. You, *Inorg. Chim. Acta*, **305**, 83 (2000).
00ICA(310)183	V. E. Kaasjager, L. Puglisi, E. Bowman, W. Driessen, and J. Reedijk, *Inorg. Chim. Acta*, **310**, 183 (2000).
00ICA(307)149	X. Chen, F. J. Femia, J. W. Babich, and J. Zubieta, *Inorg. Chim. Acta*, **307**, 149 (2000).
00IZV1891	A. I. Uraev, A. L. Nivorozhkin, G. I. Bondarenko, K. A. Lyssenko, O. Y. Korshunov, V. G. Vlasenko, A. T. Shuvaev, V. P. Kurbatov, M. Y. Antipin, and A. D. Garnovskii, *Izv. Ross. Akad. Nauk, Ser. Khim.*, 1891 (2000).
00JCS(D)459	D. M. Dawson, D. A. Walker, M. Thomion-Pett, and M. Bochmann, *J. Chem. Soc. Dalton Trans.*, 459 (2000).
00JCS(D)1649	W. Y. Yeh, C. C. Yang, S. M. Peng, and G. H. Lee, *J. Chem. Soc. Dalton Trans.*, 1649 (2000).
00JCS(D)2545	C. Boriello, M. L. Ferrara, I. Orabona, A. Panunzi, and F. Ruffo, *J. Chem. Soc. Dalton Trans.*, 2545 (2000).
00JMS(523)61	T. Hokelek, Z. Kilik, M. Isiklan, and M. Toy, *J. Mol. Struct.*, **523**, 61 (2000).
00JMS(524)241	H. Nazir, M. Yilmaz, M. N. Tahir, and D. Ulku, *J. Mol. Struct.*, **524**, 241 (2000).
00JMS(T)289	M. Brauer, M. Kunert, E. Drijus, M. Kussmann, M. Doning, H. Goris, and E. Anders, *J. Mol. Struct.: THEOCHEM*, **505**, 289 (2000).
00JOC2322	S. H. Kim and R. D. Rieke, *J. Org. Chem.*, **65**, 2322 (2000).
00JOM86	C. Stem, F. Franceschi, E. Solari, C. Floriana, N. Re, and R. Scopelliti, *J. Organomet. Chem.*, **593–594**, 86 (2000).
00K850	A. S. Antsyshkina, G. G. Sadikov, A. I. Uraev, O. Y. Korshunov, A. L. Nivorozkin, and A. D. Garnovskii, *Kristallografiya*, **45**, 850 (2000).
00MCL231	S. Noro, M. Kondo, S. Kitagawa, T. Ishii, H. Matsuzaka, and M. Yamashita, *Mol. Cryst. Liq. Sci. Technol.*, **A42**, 231 (2000).
00OM1833	C. Bianchini and H. M. Lee, *Organometallics*, **19**, 1833 (2000).
00OM2482	C. F. Baar, L. P. Cabray, M. C. Jennings, and R. J. Puddephatt, *Organometallics*, **19**, 2482 (2000).

00OM4310	M. V. Bakker, F. Hartl, D. J. Stufkens, O. S. Jina, X. Z. Sun, and M. W. George, *Organometallics*, **19**, 4310 (2000).
00OM5484	J. P. Djukic, A. Maiss-Francois, M. Pfeffer, K. H. Dotz, A. de Cian, and J. Fisher, *Organometallics*, **19**, 5484 (2000).
00POL1519	H. Brunner, A. Kollnberg, T. Burgemeister, and M. Zabel, *Polyhedron*, **19**, 1519 (2000).
00POL1887	S. Brooker, P. D. Croucher, T. C. Davidson, and P. D. Smith, *Polyhedron*, **19**, 1887 (2000).
00STC361	N. Galic, D. Matkovac-Calogovic, and Z. Cimerman, *Struct. Chem.*, **11**, 361 (2000).
00ZN(B)1037	C. Dietz, F. W. Heinemann, and A. Grihmann, *Z. Naturforsch.*, **B55**, 1037 (2000).
00ZSK1095	Y. M. Chumakov, B. A. Antosyak, M. D. Mazus, V. I. Tsapkov, and N. M. Samus, *Zh. Strukt. Khim.*, **41**, 1095 (2000).
01AX(C)117	T. Hokelek, M. Isiklan, and Z. Kilic, *Acta Crystallogr.*, **C57**, 117 (2001).
01CAR133	A. K. Sah, C. P. Rao, P. K. Searenkelo, E. Kolehmainen, and K. Rissanen, *Carbohydr. Res.*, **335**, 33 (2001).
01CC1824	L. Benisvy, A. J. Blake, D. Collison, E. S. Davies, C. D. Gamer, E. J. L. Mcines, J. McMaster, G. Whittaker, and C. Wilson, *Chem. Commun.*, 1824 (2001).
01CCR79	S. Wang, *Coord. Chem. Rev.*, **215**, 79 (2001).
01CEJ1468	F. Franceschi, G. Guillemot, E. Solari, C. Floriani, N. Re, H. Birkedal, and P. Pattison, *Chem. Eur. J.*, **7**, 1468 (2001).
01CEJ2922	K. Heinze, *Chem. Eur. J.*, **7**, 2922 (2001).
01CL566	T. Matsugi, S. Matsui, S. Kojoh, Y. Tavagi, Y. Inoui, T. Fujita, and K. Nashiva, *Chem. Lett.*, 566 (2001).
01CRV37	D. A. Atwood and M. J. Harvey, *Chem. Rev.*, **101**, 37 (2001).
01EJI669	G. Mugesh, H. B. Singh, and R. J. Butcher, *Eur. J. Inorg. Chem.*, 669 (2001).
01EJI1071	M. Gomez, S. Jansat, G. Müller, G. Noguera, H. Teruel, V. Moliner, E. Cerrada, and M. B. Hursthouse, *Eur. J. Inorg. Chem.*, 1071 (2001).
01ICA(315)172	B. Crociani, C. Antonaroli, L. Canovese, F. Visentin, and P. Uguagliati, *Inorg. Chim. Acta*, **315**, 172 (2001).
01ICA(316)33	X. Chen, F. J. Femia, J. W. Babich, and J. Zubieta, *Inorg. Chim. Acta*, **316**, 33 (2001).
01ICA(317)45	L. Valencia, H. Adams, R. Bastida, D. E. Fenton, and A. Macías, *Inorg. Chim. Acta*, **317**, 45 (2001).
01ICA(320)117	M. Hoogenraad, K. Ramkisoensing, W. L. Driessen, H. Kooijman, A. L. Spek, E. Bouwman, J. G. Haasnoot, and J. Reedijk, *Inorg. Chim. Acta*, **320**, 117 (2001).
01EJI223	M. J. Calhorda, E. Hunstock, L. F. Veiros, and F. Hartl, *Eur. J. Inorg. Chem.*, 223 (2001).
01IC208	A. Erxleben, *Inorg. Chem.*, **40**, 208 (2001).
01IC2185	J. Arias, C. R. Newlands, and M. M. Abu-Omar, *Inorg. Chem.*, **40**, 2185 (2001).
01IC4036	P. Payra, S. C. Hung, W. H. Kwok, D. Johnson, J. Galluci, and M. K. Chan, *Inorg. Chem.*, **40**, 4036 (2001).
01IZV2334	V. Y. Chemyadev, Y. A. Ustynyuk, O. V. Yazov, E. A. Kataev, M. D. Reshetova, A. A. Sidorov, G. G. Aleksandrov, V. N. Ikorskii,

V. M. Novotortsev, S. E. Nefedov, I. L. Eremenko, and I. I. Moiseev, *Izv. Ross. Acad. Nauk, Ser. Khim.*, 2334 (2001).

01JA751   C. H. Jun, H. Lee, and S. G. Lim, *J. Am. Chem. Soc.*, **123**, 751 (2001).

01JA11644   S. Hayami, Z. Z. Gu, H. Yoshiki, A. Fujishima, and O. Sato, *J. Am. Chem. Soc.*, **123**, 11644 (2001).

01JOM204   Y. Peng, X. Feng, K. Yu, Z. Li, Y. Jiang, and C. H. Yeung, *J. Organomet. Chem.*, **619**, 204 (2001).

01OM408   C. R. Baar, L. P. Carbay, M. C. Jennings, R. J. Puddephatt, and J. J. Vittal, *Organometallics*, **20**, 408 (2001).

01OM3510   Y. Matsuo, K. Mashima, and K. Tani, *Organometallics*, **20**, 3510 (2001).

01OM3635   A. Kundu and B. P. Buffin, *Organometallics*, **20**, 3635 (2001).

01OM4345   S. Nuckel and P. Burger, *Organometallics*, **20**, 4345 (2001).

01OM4793   Y. Yoshida, S. Matsui, Y. Takagi, M. Mitani, T. Nakano, H. Tanaka, N. Kashiwa, and T. Fujita, *Organometallics*, **20**, 4793 (2001).

01OM5738   B. L. Small and A. J. Marucci, *Organometallics*, **20**, 5738 (2001).

01POL2329   J. Castro, S. Cabaleiro, P. Perez-Lourido, J. Romero, J. A. Garcia-Vazquez, and A. Sousa, *Polyhedron*, **20**, 2329 (2001).

01POL2877   A. Garcia-Raso, J. J. Fiol, F. Badenas, E. Lago, and E. Molins, *Polyhedron*, **20**, 2877 (2001).

01TMC709   S. Cabaleiro, P. Perez-Lourido, J. Castro, J. Romero, J. Garcia-Vazquez, and A. Sousa, *Transit. Met. Chem.*, **26**, 709 (2001).

01ZAAC857   C. Sudbrake and H. Vahrenkamp, *Z. Anorg. Allg. Chem.*, **627**, 857 (2001).

01ZK251   B. A. Uzoukwu, K. Gloe, H. Duddeck, and O. Rademacher, *Z. Kristallogr.*, **216**, 251 (2001).

02ACS2011   Y. G. Yin, D. Li, and T. Wu, *Acta Chim. Sinica*, **60**, 2011 (2002).

02AX(E)334   J. L. Xie, W. J. Tong, Y. Zou, X. M. Ren, Y. Z. Li, and Q. J. Meng, *Acta Crystallogr.*, **E58**, m334 (2002).

02AX(E)1365   J. L. Wang, S. M. Zhang, and F. M. Miao, *Acta Crystallogr.*, **E58**, o1365 (2002).

02CAR79   A. K. Sah, C. P. Rao, P. K. Searenkelo, E. Kolehmainen, and K. Rissanen, *Carbohydr. Res.*, **337**, 79 (2002).

02CAR1477   G. Rajsekhar, U. B. Gangadharmath, C. P. Rao, P. Guionneau, P. K. Searenkelo, and K. Rissanen, *Carbohydr. Res.*, **337**, 1477 (2002).

02CC1460   G. Roges, A. Marvilliers, P. Sarr, S. Parsons, S. J. Teat, L. Ricard, and T. Mallah, *Chem. Commun.*, 1460 (2002).

02CJIC643   J. Li, F. X. Zhang, and Q. Z. Shi, *Chin. J. Inorg. Chem.*, **18**, 643 (2002).

02CRT1018   L. Govindasamy, V. Rajakannan, D. Velmurugan, S. Shanmuga, S. Raj, T. M. Rajendran, R. Venkatesan, R. Srinivasan, and H. K. Fun, *Cryst. Res. Technol.*, **37**, 1018 (2002).

02CRV3031   L. Bouget-Merle, M. F. Rappert, and J. R. Severn, *Chem. Rev.*, **102**, 3031 (2002).

02EJI1060   H. Hao, S. Bhandari, Y. Ding, H. W. Roesky, J. Magull, H. G. Schmidt, M. Nollemeyer, and C. Cui, *Eur. J. Inorg. Chem.*, 1060 (2002).

02EJI1615   M. Doring, M. Ciesielski, O. Walter, and H. Gorls, *Eur. J. Inorg. Chem.*, 1615 (2002).

02EJI2179   A. Bacchi, M. Corcelli, C. Pelizzi, G. Pelizzi, P. Pelagatti, and S. Ugolotti, *Eur. J. Inorg. Chem.*, 2179 (2002).

02EJI2897	M. Hoogenraad, H. Kooijman, A. L. Spek, E. Bouwman, J. G. Haasnoot, and J. Reedijk, *Eur. J. Inorg. Chem.*, 2897 (2002).
02IC4461	K. Abe, K. Matsufuji, M. Ohba, and H. Okawa, *Inorg. Chem.*, **41**, 4461 (2002).
02IC4502	B. Baruah, S. Das, and A. Chakravorty, *Inorg. Chem.*, **41**, 4502 (2002).
02IC5721	T. E. Keyes, D. Leane, R. J. Forster, C. G. Coates, J. J. McGarvey, M. N. Nieuwenhuyzen, E. Figgemeier, and J. S. Vos, *Inorg. Chem.*, **41**, 5721 (2002).
02IC6426	D. H. Ye, X. Y. Li, I. D. Williams, and X. M. Chen, *Inorg. Chem.*, **41**, 6426 (2002).
02ICA201	D. Visinescu, G. I. Pascu, M. Andruh, J. Magull, and H. W. Roesky, *Inorg. Chim. Acta*, **340**, 201 (2002).
02ICC211	H. Adams, S. Clunas, D. E. Fenton, T. J. Gregson, P. E. McHugh, and S. E. Spey, *Inorg. Chem. Commun.*, **5**, 211 (2002).
02ICC1063	H. Adams, S. Clunas, and D. E. Fenton, *Inorg. Chem. Commun.*, **5**, 1063 (2002).
02JA12268	H. Sugiyama, G. Aharonian, S. Gambarotta, G. P. A. Yap, and P. H. M. Budzelaar, *J. Am. Chem. Soc.*, **124**, 12268 (2002).
02JCS(D)212	F. De Bianca, G. Bandoli, A. Dolmella, S. Antonaroli, and B. Crociani, *J. Chem. Soc. Dalton Trans.*, 212 (2002).
02JCS(D)441	H. Adams, S. Clunas, D. E. Fenton, and S. E. Spey, *J. Chem. Soc. Dalton Trans.*, 441 (2002).
02JCS(D)2379	K. Heinze and V. Jacob, *J. Chem. Soc. Dalton Trans.*, 2379 (2002).
02JCS(D)3434	R. Czerwieniec, A. Kapturkiewicz, R. Anulewiec-Ostrowska, and J. Nowacki, *J. Chem. Soc. Dalton Trans.*, 3434 (2002).
02JCS(D)3696	L. Canovese, F. Visentin, G. Chessa, C. Santo, P. Uguagliati, L. Maini, and M. Polito, *J. Chem. Soc. Dalton Trans.*, 3696 (2002).
02JCS(D)4746	M. Fondo, A. M. Garcia-Delbe, M. R. Bermejo, J. Sanmartin, and A. L. Llamas-Saiz, *J. Chem. Soc. Dalton Trans.*, 4746 (2002).
02JOM43	M. Chandra, A. N. Sahay, S. M. Mobin, and D. S. Pandey, *J. Organomet. Chem.*, **658**, 43 (2002).
02JOM62	L. Wang, W. H. Sun, L. Han, H. Yang, Y. Hu, and X. Jin, *J. Organomet. Chem.*, **658**, 62 (2002).
02KGS1428	P. G. Morozov, S. V. Kurbatov, and L. P. Olekhnovich, *Khim. Geterotsikl. Soedin.*, 1611 (2002).
02OM1462	G. Tian, P. D. Boyle, and B. M. Novak, *Organometallics*, **21**, 1462 (2002).
02OM4891	C. M. Standfest-Hauser, K. Mereiter, R. Schmid, and K. Kirchner, *Organometallics*, **21**, 4891 (2002).
02POL1139	H. Asada, M. Ozeki, M. Fujiwara, and T. Matsushita, *Polyhedron*, **21**, 1139 (2002).
02RCR943	A. D. Garnovskii and I. S. Vasilchenko, *Russ. Chem. Rev.*, **71**, 943 (2002), [*Usp. Khim.*, **71**, 1064 (2002)].
02TMC469	Y. Luo, J. Zhang, L. Lu, M. Qian, X. Wang, and X. Yang, *Transit. Met. Chem.*, **27**, 469 (2002).
02ZAAC1210	J. Castro, S. Cabaleiro, P. Perez-Lourido, J. Romero, J. A. Garcia-Vazquez, and A. Sousa, *Z. Anorg. Allg. Chem.*, **628**, 1210 (2002).
03ACS207	N. Daneshvar, L. A. Saghatforoush, A. A. Khandar, and A. A. Entezami, *Acta Chim. Slov.*, **50**, 207 (2003).
03AGE1632	S. Nuckel and P. Burger, *Angew. Chem. Int. Ed. Engl.*, **42**, 1632 (2003).

03AGE3271	Z. Wu, Q. Chen, S. Xiong, B. Xin, Z. Zhao, L. Jiang, and J. S. Ma, *Angew. Chem. Int. Ed. Engl.*, **42**, 3271 (2003).
03AX(E)430	J. L. Wang, Y. Yang, X. Zhang, and F. M. Miao, *Acta Crystallogr.*, **E59**, o430 (2003).
03BKC504	J. Y. Kim, Y. J. Ji, H. J. Ha, and H. K. Chae, *Bull. Korean Chem. Soc.*, **24**, 504 (2003).
03CL756	H. Tsurugi, T. Yamagata, K. Tani, and K. Mashima, *Chem. Lett.*, 756 (2003).
03CM4949	T. S. Kim, T. Okubo, and T. Mitani, *Chem. Mater.*, **15**, 4949 (2003).
03CRV283	V. C. Gibson and S. K. Spizmesser, *Chem. Rev.*, **103**, 283 (2003).
03CR3071	C. Gennari and U. Piarulli, *Chem. Rev.*, **103**, 3071 (2003).
03EJI1883	C. M. Standfest-Hausner, K. Mereiter, R. Schmid, and K. Kirchner, *Eur. J. Inorg. Chem.*, 1883 (2003).
03EJI3193	R. Pedrido, M. R. Bermejo, A. M. Garcia-Deibe, A. M. Gonzalez-Noya, M. Maneiro, and M. Vazquez, *Eur. J. Inorg. Chem.*, 3193 (2003).
03IC8878	M. E. Bluhm, M. Ciesielski, H. Gorls, O. Walter, and M. Doring, *Inorg. Chem.*, **42**, 8878 (2003).
03IC1663	B. J. Kraft, N. L. Coalter, M. Nath, A. E. Clark, A. R. Siedle, J. C. Huffman, and J. M. Zaleski, *Inorg. Chem.*, **42**, 1663 (2003).
03ICA(342)131	F. Tuna, G. I. Pascu, J. P. Sutter, M. Andruh, S. Golhen, J. Guillevic, and H. Pritzkow, *Inorg. Chim. Acta*, **342**, 131 (2003).
03ICA(342)229	A. Bacchi, M. Carcelli, L. Gabba, S. Ianelli, P. Pelagatti, G. Pelizzi, and D. Rogolino, *Inorg. Chim. Acta*, **342**, 229 (2003).
03ICA(355)286	H. Keypor, H. Khanmohammadi, K. P. Wainwright, and M. R. Taylor, *Inorg. Chim. Acta*, **355**, 286 (2003).
03ICC1140	H. Sasabe, S. Nakanishi, and T. Takata, *Inorg. Chem. Commun.*, **6**, 1140 (2003).
03JA12674	C. N. Iverson, C. A. G. Carter, R. T. Baker, J. D. Scollard, J. A. Labinger, and J. E. Bercaw, *J. Am. Chem. Soc.*, **125**, 12674 (2003).
03JA14816	G. Yu, S. Yin, L. Liu, Z. Shu, and D. Zhu, *J. Am. Chem. Soc.*, **125**, 14816 (2003).
03JCC975	C. M. Wansapura, C. Juyoung, J. L. Simpson, D. Szymanski, G. R. Eaton, S. S. Eaton, and S. Fox, *J. Coord. Chem.*, **56**, 975 (2003).
03JCS(D)221	V. C. Gibson, S. J. McTavish, C. Redshaw, G. A. Solan, A. J. P. White, and D. J. Williams, *Dalton Trans.*, 221 (2003).
03JCS(D)1975	L. Benisvy, A. J. Blake, D. Collison, E. S. Davies, C. D. Gamer, E. J. L. McInnes, J. McMaster, G. Whittaker, and C. Wilson, *Dalton Trans.*, 1975 (2003).
03JCS(D)4387	S. D. Reid, A. J. Blake, W. Kockenberger, C. Wilson, and J. B. Love, *Dalton Trans.*, 4387 (2003).
03JCS(D)4431	R. M. Bellabarba, P. T. Gomez, and S. I. Pascu, *Dalton Trans.*, 4431 (2003).
03JOM(667)185	Y. S. Li, Y. R. Li, and X. F. Li, *J. Organomet. Chem.*, **667**, 185 (2003).
03JOM(673)110	G. J. P. Britovsek, G. Y. Y. Woo, and N. Assavathorn, *J. Organomet. Chem.*, **673**, 110 (2003).
03MI1	J. A. McCleverty and T. J. Meyer (eds.), *"Comprehensive Coordination Chemistry II"* pp. V.1–V.10, Elsevier-Pergamon Press, Oxford and New York (2003).
03MI2	A. D. Garnovskii and B. I. Kharisov, Synthetic Coordination and Organometallic Chemistry, p. 513, Marcel Dekker, New York and Basel (2003).

03MI3	R. Hernandez-Molina and A. Mederos, in *"Comprehensive Coordination Chemistry II"* (J. A. McCleverty and T. J. Meyer, eds.), Vol. 2, p. 411, Elsevier-Pergamon Press, Oxford and New York (2003).
03MI4	K. M. Smith, in *"Comprehensive Coordination Chemistry II"* (J. A. McCleverty and T. J. Meyer, eds.), Vol. 1, p. 493, Elsevier-Pergamon Press, Oxford and New York (2003).
03MI5	N. B. McKeown, in *"Comprehensive Coordination Chemistry II"* (J. A. McCleverty and T. J. Meyer, eds.), Vol. 1, p. 507, Elsevier-Pergamon Press, Oxford and New York (2003).
03M16	H. Hennig, S. Knoblauch, B. Roland, M. Ecke, J. Sieler, S. Jelonek, and F. Somoza, http://www.ccdc.cam.uk (2003).
03MRC61	E. Kwiatkowski and G. Romanowski, *Magn. Reson. Chem.*, **41**, 61 (2003).
03OM434	C. D. Berube, S. Cambarotta, G. P. A. Yap, and G. Cozzi, *Organometallics*, **22**, 434 (2003).
03OM850	E. F. DiMauro, A. Mamai, and M. C. Kozlowski, *Organometallics*, **22**, 850 (2003).
03OM1047	L. Gonsalvi, J. A. Gaunt, H. Adams, A. Castro, G. J. Sunley, and A. Haynes, *Organometallics*, **22**, 1047 (2003).
03OM2545	G. Bianchini, G. Mantovani, A. Meli, and F. Migliacci, *Organometallics*, **22**, 2545 (2003).
03OM3178	B. L. Small, *Organometallics*, **22**, 3178 (2003).
03POL1385	D. Visinescu, A. M. Madalan, V. Kravtsov, Y. A. Simonov, M. Schmidtmann, A. Muller, and M. Andruh, *Polyhedron*, **22**, 1385 (2003).
03POL2499	E. V. Tretyakov, I. V. Eltsov, S. V. Fokin, Y. G. Shvedenkov, G. V. Romanenko, and V. I. Ovcharenko, *Polyhedron*, **22**, 2499 (2003).
03POL2639	H. Brunner, A. Kollnberger, and M. Zabel, *Polyhedron*, **22**, 2639 (2003).
03T10037	L. Y. Yang, Q. Q. Chen, G. Q. Yang, and J. S. Ma, *Tetrahedron*, **59**, 10037 (2003).
03ZK347	M. Bums, B. Moubaraki, K. S. Murray, and E. R. T. Tiekink, *Z. Kristallogr.*, **218**, 347 (2003).
03ZK492	S. G. Oztas, E. Sahin, N. Ancin, S. Ide, and M. Tuz, *Z. Kristallogr.*, **218**, 492 (2003).
03ZK509	H. L. Zhu, S. C. Shao, J. L. Ma, X. Y. Qiu, and L. Sun, *Z. Kristallogr.*, **218**, 509 (2003).
04ACS2329	J. L. Wang, A. X. Li, Y. J. Jia, and S. M. Zhang, *Acta Chim. Sinica*, **62**, 2329 (2004).
04AX(E)884	Z. L. You, B. Chen, H. L. Zhu, and W. S. Liu, *Acta Crystallogr.*, **E60**, m884 (2004).
04AX(E)1017	Z. X. Li and X. L. Zhang, *Acta Crystallogr.*, **E60**, m1017 (2004).
04AX(E)1079	Z. L. You and H. L. Zhu, *Acta Crystallogr.*, **E60**, m1079 (2004).
04AX(E)1259	L. A. Saghatforous, M. H. Sadr, W. Lewis, J. Wikaira, W. T. Robinson, and S. W. Ng, *Acta Crystallogr.*, **E60**, m1259 (2004).
04AX(E)1552	T. Akitsu and Y. Einaga, *Acta Crystallogr.*, **E60**, m1552 (2004).
04AX(E)1855	X. H. Qiu and H. Y. Wu, *Acta Crystallogr.*, **E60**, m1855 (2004).
04CCR1717	P. A. Vigato and S. Tamburini, *Coord. Chem. Rev.*, **248**, 1717 (2004).
04CEJ1014	B. L. Small and R. Schmidt, *Chem. Eur. J.*, **10**, 1014 (2004).
04CJAC49	A. X. Li, Y. Yang, and J. L. Wang, *Chin. J. Appl. Chem.*, **21**, 49 (2004).

04CJSC183	H. J. Yang, H. G. Wang, and W. H. Sun, *Chin. J. Struct. Chem.*, **23**, 183 (2004).
04CJIC439	S. M. Zhang, P. F. Li, M. Yu, and J. L. Wang, *Chin. J. Inorg. Chem.*, **20**, 439 (2004).
04CSR410	P. G. Cozzi, *Chem. Soc. Rev.*, **33**, 410 (2004).
04DOK(395)626	L. X. Minacheva, I. S. Ivanova, E. N. Pyatova, A. V. Dorokhov, A. V. Bicherov, A. S. Burlov, A. D. Garnovskii, and V. S. Sergienko, *Dokl. Ross. Akad. Nauk*, **395**, 626 (2004).
04DOK(398)62	L. X. Minacheva, I. S. Ivanova, A. V. Dorokhov, A. V. Bicherov, A. S. Burlov, A. D. Garnovskii, V. S. Sergienko, and A. Y. Tsivadze, *Dokl. Ross. Akad. Nauk*, **398**, 62 (2004).
04EJI1204	Q. Knijnenburg, A. D. Horton, H. van der Heijden, T. M. Kooistra, D. G. H. Hettershield, J. M. M. Smits, B. de Bruin, P. H. M. Budzelaar, and A. W. Gal, *Eur. J. Inorg. Chem.*, 1204 (2004).
04EJI1478	L. Yang, Q. Chen, Y. Li, S. Xiong, G. Li, and J. S. Ma, *Eur. J. Inorg. Chem.*, 1478 (2004).
04EJI1873	B. Baruah, S. P. Rath, and A. Chakravorty, *Eur. J. Inorg. Chem.*, 1873 (2004).
04EJI2053	K. Heinze, V. Jacob, and C. Feige, *Eur. J. Inorg. Chem.*, 2053 (2004).
04EJI3424	B. Siggelkow, M. B. Meder, C. H. Galka, and L. H. Gade, *Eur. J. Inorg. Chem.*, 3424 (2004).
04EJI3498	K. Heinze and J. D. B. Toro, *Eur. J. Inorg. Chem.*, 3498 (2004).
04EJI4177	S. Steinhauser, U. Heinz, M. Bartholoma, T. Weyhermüller, H. Nick, and K. Hegetschweiler, *Eur. J. Inorg. Chem.*, 4177 (2004).
04IC622	R. J. Ball, T. R. Shtoyko, J. A. K. Bauer, W. J. Oldham, and W. B. Connick, *Inorg. Chem.*, **43**, 622 (2004).
04IC1220	J. M. Veauthier, W. S. Cho, V. M. Linch, and J. L. Sessier, *Inorg. Chem.*, **43**, 1220 (2004).
04ICA1161	S. Di Bella, I. Fragala, A. Guerri, P. Dapporto, and K. Nakatani, *Inorg. Chim. Acta*, **357**, 1161 (2004).
04ICA1283	H. Keypor, H. Khanmohammadi, K. P. Wainwright, and M. R. Taylor, *Inorg. Chim. Acta*, **357**, 1283 (2004).
04ICA3407	J. D. Crane, L. C. Emeleus, D. Harrison, and P. A. Nilsson, *Inorg. Chim. Acta*, **357**, 3407 (2004).
04ICA3697	L. D. Pachon, A. Golobic, B. Kozlevcar, P. Gamez, H. Kooijman, A. L. Spek, and J. Reedijk, *Inorg. Chim. Acta*, **357**, 3697 (2004).
04ICC249	Z. Wu, G. Yang, Q. Chen, J. Liu, S. Yang, and J. S. Ma, *Inorg. Chem. Commun.*, **7**, 249 (2004).
04ICC1269	B. I. Kharisov, C. E. C. Coronado, K. P. C. Cerda, U. Ortiz-Mendez, J. A. J. Guzman, and L. A. R. Patlan, *Inorg. Chem. Commun.*, **7**, 1269 (2004).
04JA13794	S. C. Bart, E. Lobkovsky, and P. J. Chirik, *J. Am. Chem. Soc.*, **126**, 13794 (2004).
04JCS(D)1731	K. H. Seng, C. C. Huang, Y. H. Liu, Y. H. Hu, P. T. Chou, and Y. C. Lin, *Dalton Trans.*, 1731 (2004).
04JCS(D)2314	P. P. Plitt, H. Pritzkov, and R. Kramer, *Dalton Trans.*, 2314 (2004).
04JOM936	F. Chang, D. Zhang, G. Xu, H. Yang, J. Li, H. Song, and W. H. Sun, *J. Organomet. Chem.*, **689**, 936 (2004).
04JOM1249	S. S. Keisham, Y. A. Mozharivskyj, P. J. Carroll, and M. R. Kollipara, *J. Organomet. Chem.*, **689**, 1249 (2004).
04JOM1356	C. Bianchini, G. Giambastiani, G. Mantovani, A. Meli, and D. Mimeau, *J. Organomet. Chem.*, **689**, 1356 (2004).

04JOM3612	S. K. Singh, M. Chandra, D. S. Pandey, M. C. Puerta, and P. Valerga, *J. Organomet. Chem.*, **689**, 3612 (2004).
04JUCR71	J. L. Simpson, D. J. Lombardo, and S. Fox, *J. Undergrad. Chem. Res.*, **3**, 71 (2004).
04MI1	S. Parsons, A. Marvilliers, R. Winpenny, and P. Wood, http://www.ccdc.cam.uk (2004).
04MI2	S. Parsons, A. Smith, P. Tasker, R. Gould, and P. Wood, http://www.ccdc.cam.uk (2004).
04MI3	V. C. Gibson and E. L. Marshall, in *"Comprehensive Coordination Chemistry II"*, Vol. 9, p. 1 (2004).
04OM2797	H. Tsurugi, Y. Matsuo, T. Yamagata, and K. Mashima, *Organometallics*, **23**, 2797 (2004).
04OM5375	K. P. Brilyakov, N. V. Semikolenova, V. A. Zakharov, and E. P. Talsi, *Organometallics*, **23**, 5375 (2004).
04OM5503	K. P. Tellman, M. J. Humphries, H. S. Rzepa, and V. C. Gibson, *Organometallics*, **23**, 5503 (2004).
04OM6087	C. Bianchini, G. Giambastiani, I. G. Rios, A. Meli, E. Passaglia, and T. Gragnoli, *Organometallics*, **23**, 6087 (2004).
04POL1115	R. Srinivasan, R. Venkatesan, B. Verghese, I. Sougandi, and P. S. Rao, *Polyhedron*, **23**, 1115 (2004).
04POL1619	H. Liang, J. Liu, X. Li, and Y. Li, *Polyhedron*, **23**, 1619 (2004).
04POL1909	A. D. Garnovskii, B. I. Kharisov, E. L. Anpilova, A. V. Bicherov, O. Y. Korshunov, M. A. Mendez-Rojas, L. M. Blanko, G. S. Borodkin, I. E. Uflyand, and U. Ortiz-Mendez, *Polyhedron*, **23**, 1909 (2004).
04RCR5	A. Y. Tsivadze, *Russ. Chem. Rev.*, **73**, 5 (2004).
04RKZ30	V. I. Minkin, T. N. Gribanova, A. D. Dubonosov, V. A. Bren, R. M. Minyaev, E. N. Shepelenko, and A. V. Zukanov, *Ross. Khim. Zh.*, **48**(1), 30 (2004).
04RKZ38	A. I. Uraev, S. E. Nefedov, A. V. Dorokhov, R. N. Borisenko, I. S. Vasilchenko, A. D. Garnovskii, and A. Y. Tsivadze, *Ross. Khim. Zh.*, **48**(1), 38 (2004).
04RKZ103	L. P. Olekhnovich, E. P. Ivakhnenko, S. N. Lyubchenko, V. I. Simakov, G. S. Borodkin, A. V. Lesin, I. N. Shcherbakov, and S. V. Kurbatov, *Ross. Khim. Zh.*, **48**(1), 103 (2004).
04ZAAC91	H. Brunner, F. Henning, and M. Zabel, *Z. Anorg. Allg. Chem.*, **630**, 91 (2004).
04ZAAC2754	Z. L. You and H. L. Zhu, *Z. Anorg. Allg. Chem.*, **630**, 2754 (2004).
04ZNK1696	A. N. Chekhlov, *Zh. Neorg. Khim.*, **49**, 1696 (2004).
05AHC126	A. P. Sadimenko, *Adv. Heterocycl. Chem.*, **89**, 126 (2005).
05AX(C)318	F. R. Perez, J. Belmar, C. Jimenez, Y. Moreno, P. Hermosilla, and R. Baggio, *Acta Crystallogr.*, **C61**, 318 (2005).
05AX(C)421	Z. L. You and H. L. Zhu, *Acta Crystallogr.*, **C61**, 421 (2005).
05AX(C)432	Z. L. You, *Acta Crystallogr.*, **C61**, 432 (2005).
05AX(C)456	Z. L. You, *Acta Crystallogr.*, **C61**, 456 (2005).
05AX(C)532	Z. L. You, *Acta Crystallogr.*, **C61**, 532 (2005).
05AX(E)247	De. S. Yang, *Acta Crystallogr.*, **E61**, m247 (2005).
05AX(E)335	Y. X. Sun, *Acta Crystallogr.*, **E61**, m335 (2005).
05AX(E)370	Y. X. Sun, G. Z. Gao, H. X. Pei, and R. Zhang, *Acta Crystallogr.*, **E61**, m370 (2005).
05AX(E)466	Z. L. You, *Acta Crystallogr.*, **E61**, m466 (2005).
05AX(E)1055	Y. X. Sun, Y. Z. Gao, H. L. Zhang, D. S. Kong, and Y. Yu, *Acta Crystallogr.*, **E61**, m1055 (2005).

05AX(E)1571	Z. L. You, *Acta Crystallogr.*, **E61**, m1571 (2005).
05AX(E)1953	Y. Xi, J. Li, and F. Zhang, *Acta Crystallogr.*, **E61**, m1953 (2005).
05AX(E)1986	H. Y. Bu, Y. J. Liu, Q. F. Liu, and J. F. Jia, *Acta Crystallogr.*, **E61**, m1986 (2005).
05AX(E)2302	X. H. Qiu and X. I. Tong, *Acta Crystallogr.*, **E61**, m2302 (2005).
05AX(E)2338	J. H. Zhang, Y. Xi, C. Wang, J. Li, F. X. Zhang, and S. W. Ng, *Acta Crystallogr.*, **E61**, m2338 (2005).
05CC3150	D. A. Pennington, S. J. Coles, M. B. Hursthouse, M. Bochmann, and S. J. Lancaster, *Chem. Commun.*, 3150 (2005).
05CC4935	J. Lewinski, M. Dranka, I. Kraszewska, W. Sliwinski, and I. Justyniak, *Chem. Commun.*, 4935 (2005).
05CC5217	J. M. Benito, E. de Jesus, F. J. de la Mata, J. C. Flores, and R. Gomez, *Chem. Commun.*, 5217 (2005).
05CCR1857	B. J. Baker and C. Jones, *Coord. Chem. Rev.*, **249**, 1857 (2005).
05CCR2156	W. Radecka-Paryzek, V. Patroniak, and J. Lisowski, *Coord. Chem. Rev.*, **249**, 2156 (2005).
05CJSC909	J. Wang, H. Xu, H. Zhou, and P. Wei, *Chin. J. Struct. Chem.*, **24**, 909 (2005).
05CL1240	K. Takahashi, H. Cui, H. Kobayashi, Y. Einaga, and O. Sato, *Chem. Lett.*, **34**, 1240 (2005).
05CP17	T. L. Yang, W. W. Qin, Z. F. Xiao, and W. S. Liu, *Chem. Pap.*, **59**, 17 (2005).
05EJI4626	A. W. Kleij, D. M. Tooke, A. L. Spek, and J. N. H. Reek, *Eur. J. Inorg. Chem.*, 4626 (2005).
05IC1187	I. Vidyaratne, S. Gambarotta, I. Korobkov, and P. H. M. Budzelaar, *Inorg. Chem.*, **44**, 1187 (2005).
05IC4270	Y. P. Tong, S. L. Zheng, and X. M. Chen, *Inorg. Chem.*, **44**, 4270 (2005).
05IC4442	C. C. Kwok, H. M. Y. Ngai, S. C. Chan, I. H. T. Sham, C. M. Che, and N. Zhu, *Inorg. Chem.*, **44**, 4442 (2005).
05IC6476	B. L. Dietrich, J. Egbert, A. M. Morris, M. Wicholas, O. P. Anderson, and S. M. Miller, *Inorg. Chem.*, **44**, 6476 (2005).
05IC8916	N. Brefuel, C. Lepetit, S. Shova, F. Dahan, and J. P. Tuchagues, *Inorg. Chem.*, **44**, 8916 (2005).
05IC9253	M. D. Godbole, A. C. G. Hotze, R. Hage, A. M. Mills, H. Kooijman, A. L. Spek, and E. Bouwman, *Inorg. Chem.*, **44**, 9253 (2005).
05ICA383	R. Kannappan, S. Tanase, L. Mutikainen, U. Turpeinen, and J. Reedijk, *Inorg. Chim. Acta*, **358**, 383 (2005).
05JA11037	D. J. Jones, V. C. Gibson, S. M. Green, P. J. Maddox, A. J. P. White, and D. J. Williams, *J. Am. Chem. Soc.*, **127**, 11037 (2005).
05JA13019	J. Scott, S. Gambarotta, I. Korobkov, and P. H. M. Budzelaar, *J. Am. Chem. Soc.*, **127**, 13019 (2005).
05JA16776	H. J. Kim, W. Kim, A. J. Lough, B. M. Kim, and J. Chin, *J. Am. Chem. Soc.*, **127**, 16776 (2005).
05JBI924	N. A. Illan-Cabeza, R. A. Vilaplana, Y. Alvarez, K. K. S. Akdi, F. Hueso-Urena, M. Quiros, F. Gonzalez-Vílchez, and M. N. Moreno-Carretero, *J. Biol. Inorg. Chem.*, **10**, 924 (2005).
05JCS(D)2312	J. C. Pessoa, I. Correia, A. Galvao, A. Cameiro, V. Felix, and E. Fluza, *Dalton Trans.*, 2312 (2005).
05JCS(D)3611	R. K. J. Bott, M. Hammond, P. N. Horton, S. J. Lancaster, M. Bochmann, and P. Scott, *Dalton Trans.*, 3611 (2005).

05JMS153	H. Xu, Z. P. Ni, X. M. Ren, and Q. J. Meng, *J. Mol. Struct.*, **752**, 153 (2005).
05MAC2559	J. Y. Liu, Y. S. Li, J. Y. Liu, and Z. S. Li, *Macromolecules*, **38**, 2559 (2005).
05MC133	A. I. Uraev, I. S. Vasilchenko, V. N. Ikorskii, T. E. Shestakova, A. S. Burlov, K. A. Lyssenko, V. G. Vlasenko, T. A. Kuzmenko, L. N. Divaeva, I. V. Pirog, G. S. Borodkin, I. E. Uflyand, M. Y. Antipin, V. I. Ovcharenko, A. D. Garnovskii, and V. I. Minkin, *Mendeleev Commun.*, **15**, 133 (2005).
05MI1	P. Taker, A. Smith, and D. Messenger, http://www.ccdc.cam.uk (2005).
05MI2	P. Taker, A. Smith, S. Parsons, and D. Messenger, http://www.ccdc.cam.uk (2005).
05MI3	M. R. Maurya, D. Rehder, and M. Ebel, http://www.ccdc.cam.uk (2005).
05MI4	P. Taker, A. Smith, S. Parsons, and D. Messenger, http://www.ccdc.cam.uk (2005).
05NJC283	F. R. Perez, J. Belmar, Y. Moreno, R. Baggio, and O. Pena, *New J. Chem.*, **29**, 283 (2005).
05OM280	K. P. Tellmann, V. C. Gibson, A. J. P. White, and D. J. Williams, *Organometallics*, **24**, 280 (2005).
05OM2039	M. J. Humphries, K. P. Tellman, V. C. Gibson, A. J. P. White, and D. J. Williams, *Organometallics*, **24**, 2039 (2005).
05OM3375	T. Yasumoto, T. Yamagata, and K. Mashima, *Organometallics*, **24**, 3375 (2005).
05OM3664	P. M. Castro, P. Lahtinen, K. Axenov, J. Viidanoja, T. Kotiaho, M. Leskela, and T. Repo, *Organometallics*, **24**, 3664 (2005).
05OM4878	J. Campora, A. M. Naz, P. Palma, E. Alvarez, and M. L. Reyes, *Organometallics*, **24**, 4878 (2005).
05OM5015	K. A. Pittard, T. R. Cundari, T. B. Gunnoe, C. S. Day, and J. L. Petersen, *Organometallics*, **24**, 5015 (2005).
05POL1710	P. Govindaswamy, Y. A. Mozharivskyj, and M. R. Kollipara, *Polyhedron*, **34**, 1710 (2005).
05PR(B)214408	P. Guionneau, M. Marchivie, Y. Garcia, J. A. K. Howard, and D. Chasseau, *Phys. Rev.*, **B72**, 214408 (2005).
05RCR193	A. D. Garnovskii and I. S. Vasilchenko, *Russ. Chem. Rev.*, **74**, 193 (2005), [*Usp. Khim.*, **74**, 211 (2005)].
05SA(A)1059	G. G. Mohamed and Z. H. A. El-Wahab, *Spectrochim. Acta*, **A61**, 1059 (2005).
05SR755	B. I. Kharisov, L. A. Gazza-Rodriguez, H. M. L. Gutierrez, U. Ortiz-Mendez, R. Garcia-Caballero, and A. Y. Tsivadze, *Synth. React. Inorg. Met. Org. Chem.*, **35**, 755 (2005).
05ZSK388	A. N. Chekhlov, *Zh. Strukt. Khim.*, **46**, 388 (2005).
06AHC1	B. Stanovik, M. Tisler, A. R. Katritzky, and O. V. Denisko, *Adv. Heterocycl. Chem.*, **91**, 1 (2006).
06AX(E)977	X. H. Qiu, J. Zhao, and X. L. Tong, *Acta Crystallogr.*, **E62**, m977 (2006).
06AX(E)2361	Y. H. Zhao, Z. M. Su, Y. Wang, X. R. Hao, and K. Z. Shao, *Acta Crystallogr.*, **E62**, m2361 (2006).
06AX(E)2668	J. W. Ran, D. J. Gong, and Y. H. Li, *Acta Crystallogr.*, **E62**, m2668 (2006).
06AX(E)2886	H. S. He, *Acta Crystallogr.*, **E62**, m2886 (2006).

06AX(E)3042	H. S. He, *Acta Crystallogr.*, **E62**, m3042 (2006).
06AX(E)3537	H. S. He, *Acta Crystallogr.*, **E62**, m3537 (2006).
06BCJ595	Y. Suetsugu, Y. Mitsuka, Y. Miyasato, and M. Ohba, *Bull. Chem. Soc. Jpn.*, **79**, 595 (2006).
06CJSC1343	X. H. Qiu, J. Zhao, and X. L. Tong, *Chin. J. Struct. Chem.*, **25**, 1343 (2006).
06CCR1391	C. Bianchini, G. Giambastiani, I. G. Rios, G. Mantovani, A. Meli, and A. M. Segarra, *Coord. Chem. Rev.*, **250**, 1391 (2006).
06EJI4324	L. Sabater, C. Hureau, G. Blain, R. Guillot, P. Thuery, E. Riviere, and A. Aukauloo, *Eur. J. Inorg. Chem.*, 4324 (2006).
06EPJ928	F. Bao, X. Liu, B. Kang, and Q. Wu, *Eur. Polym. J.*, **42**, 928 (2006).
06IC295	H. C. Wu, P. Thanasekaran, C. H. Tsai, J. Y. Wu, S. M. Huang, Y. S. Wen, and K. L. Lu, *Inorg. Chem.*, **45**, 295 (2006).
06IC636	S. D. Reid, A. J. Blake, C. Wilson, and J. B. Love, *Inorg. Chem.*, **45**, 636 (2006).
06IC1277	G. Margraf, T. Kretz, F. F. de Biani, F. Laschi, S. Losi, P. Zanello, J. W. Bats, B. Wolf, K. Removic-Langer, M. Lang, A. Prokofiev, W. Assmus, H. W. Lerner, and M. Wagner, *Inorg. Chem.*, **45**, 1277 (2006).
06IC1445	N. C. Habermehl, P. M. Angus, N. L. Kilah, L. Noren, A. D. Rae, A. C. Willis, and S. B. Wild, *Inorg. Chem.*, **45**, 1445 (2006).
06IC2373	L. Sabater, C. Hureau, R. Guillot, and A. Aukauloo, *Inorg. Chem.*, **45**, 2373 (2006).
06IC5739	K. Takahashi, H. B. Cui, Y. Okano, H. Kobayashi, Y. Einaga, and O. Sato, *Inorg. Chem.*, **45**, 5739 (2006).
06IC5924	M. R. Maurya, A. Kumar, M. Ebel, and D. Rehder, *Inorg. Chem.*, **45**, 5924 (2006).
06IC7994	S. K. Padhi and V. Manivannan, *Inorg. Chem.*, **45**, 7994 (2006).
06IC9890	Y. D. M. Champouret, J. Fawcett, W. J. Nodes, K. Singh, and G. A. Solan, *Inorg. Chem.*, **45**, 9890 (2006).
06IC10293	H. M. Foucault, D. L. Bryce, and D. E. Fogg, *Inorg. Chem.*, **45**, 10293 (2006).
06ICA1103	R. Balamurugan, M. Palaniandavar, H. Stoeckli-Evans, and M. Neuburger, *Inorg. Chim. Acta*, **359**, 1103 (2006).
06ICA2321	C. M. Fierro, B. P. Murphy, P. D. Smith, S. J. Coles, and M. B. Hursthouse, *Inorg. Chim. Acta*, **359**, 2321 (2006).
06JA9610	P. L. Arnold, D. Patel, A. J. Blake, C. Wilson, and J. B. Love, *J. Am. Chem. Soc.*, **128**, 9610 (2006).
06JA13340	M. W. Bouwkamp, A. C. Bowman, E. Lobkovsky, and P. J. Chirik, *J. Am. Chem. Soc.*, **128**, 13340 (2006).
06JA13901	S. C. Bart, K. Chlopek, E. Bill, M. W. Bouwkamp, E. Lobkovsky, F. Neese, K. Weighardt, and P. J. Chirik, *J. Am. Chem. Soc.*, **128**, 13901 (2006).
06JCCR61	W. Qin, F. Cao, Z. Li, H. Zhou, H. Wan, P. Wei, Y. Shi, and P. Quyang, *J. Chem. Cryst.*, **36**, 61 (2006).
06JCS(D)258	L. Benisvy, E. Bill, A. J. Blake, D. Collison, E. S. Davies, C. D. Garner, G. M. McArdle, E. J. L. Mcinnes, J. McMaster, S. H. K. Ross, and C. Wilson, *Dalton Trans.*, 258 (2006).
06JCS(D)2450	J. A. Cabeza, I. da Silva, I. del Rio, R. A. Gossage, D. Miguel, and M. Suarez, *Dalton Trans.*, 2450 (2006).
06JCS(D)4260	M. Fondo, A. M. Garcia-Deibe, N. Ocampo, J. Sanmartin, M. R. Bermejo, and A. L. Llamas-Saiz, *Dalton Trans.*, 4260 (2006).

06JCS(D)4914   J. W. Bats, S. Losi, B. Wolf, H. W. Lerner, M. Lang, P. Zanello, and M. Wagner, *Dalton Trans.*, 4914 (2006).

06JCS(D)5362   C. E. Anderson, A. S. Batsanov, P. W. Dyer, J. Fawcett, and J. A. K. Howard, *Dalton Trans.*, 5362 (2006).

06JCS(D)5442   Q. Knijnenburg, S. Gambarotta, and P. H. M. Budzelaar, *Dalton Trans.*, 5442 (2006).

06JKP1057   Y. K. Jang, D. E. Kim, O. K. Kwon, and B. J. Lee, *J. Kor. Phys. Soc.*, **49**, 1057 (2006).

06JMC(A)227   M. R. Maurya, A. K. Chandrakar, and S. Chand, *J. Mol. Catal. A*, **263**, 227 (2006).

06JOM1321   E. Labisbal, L. Rodriguez, A. Sousa-Pedrares, M. Alonso, A. Visozo, J. Romero, J. A. Garcia-Vazquez, and A. Sousa, *J. Organomet. Chem.*, **691**, 1321 (2006).

06JOM2228   I. Westmoreland, I. J. Munslow, A. J. Clarke, G. Clarkson, R. J. Deeth, and P. Scott, *J. Organomet. Chem.*, **691**, 2228 (2006).

06JOM3434   L. A. Garcia-Escudero, D. Miguel, and J. A. Turiel, *J. Organomet. Chem.*, **691**, 3434 (2006).

06JOM4196   W. H. Sun, S. Zhang, S. Jie, W. Zhang, Y. Song, and H. Ma, *J. Organomet. Chem.*, **691**, 4196 (2006).

06JOM4573   G. Giancaleoni, G. Bellachioma, G. Cardaci, G. Ricci, R. Ruzzcioni, D. Zucaccia, and A. Machioni, *J. Organomet. Chem.*, **691**, 165 (2006).

06MI1   A. Pailleret and F. Bedioui, in "*N₄-Macrocyclic Metal Complexes*" (J. H. Zagal, F. Bedioui, and J. P. Dodelet, eds.), p. 363, Springer, New York (2006).

06OM236   J. Hou, W. H. Sun, S. Zhang, H. Ma, Y. Deng, and X. Lu, *Organometallics*, **25**, 236 (2006).

06OM666   W. H. Sun, S. Jie, S. Zhang, W. Zhang, Y. Song, and H. Ma, *Organometallics*, **25**, 666 (2006).

06OM3045   J. M. Benito, E. de Jesus, F. J. de la Mata, J. C. Flores, and R. Gomez, *Organometallics*, **25**, 3045 (2006).

06OM3876   J. M. Benito, E. de Jesus, F. J. de la Mata, J. C. Flores, R. Gomez, and P. Gomez-Sal, *Organometallics*, **25**, 3876 (2006).

06OM5210   H. Tsurugi and K. Mashima, *Organometallics*, **25**, 5210 (2006).

06OM5800   J. H. H. Ho, D. S. C. Black, B. A. Messerle, J. K. Clegg, and P. Turner, *Organometallics*, **25**, 5800 (2006).

06POL881   X. Yang, R. A. Jones, R. J. Lai, A. Waheed, M. M. Oye, and A. L. Holmes, *Polyhedron*, **25**, 881 (2006).

06RJCC879   A. S. Burlov, E. N. Shepelenko, I. S. Vasilchenko, A. S. Antsyshkina, G. G. Sadikov, P. V. Matuev, S. A. Nikolaevskii, G. S. Borodkin, V. S. Sergienko, V. A. Bren, and A. D. Garnovskii, *Russ. J. Coord. Chem.*, **32**, 879 (2006).

06SA(A)188   G. G. Mohamed, *Spectrochim. Acta*, **A64**, 188 (2006).

06ZN(C)263   K. Schlegel, J. Lex, K. Taraz, and H. Budzikiewicz, *Z. Naturforsch.*, **C61**, 263 (2006).

07AGE584   G. Givaja, M. Volpe, M. A. Edwards, A. J. Blake, C. Wilson, M. Schroder, and J. B. Love, *Angew. Chem. Int. Ed. Engl.*, **46**, 584 (2007).

07AHC(93)185   A. P. Sadimenko, *Adv. Heterocycl. Chem.*, **93**, 185 (2007).

07AHC(94)107   A. P. Sadimenko, *Adv. Heterocycl. Chem.*, **94**, 107 (2007).

07AX(E)330   A. La Cour and A. Hazeil, *Acta Crystallogr.*, **E62**, m330 (2007).

07AX(E)344   H. S. He, *Acta Crystallogr.*, **E63**, m344 (2007).

07CCR1311	P. A. Vigato, S. Tamburini, and L. Bertolo, *Coord. Chem. Rev.*, **251**, 1311 (2007).
07CHC1359	A. D. Garnovskii and E. V. Sennikova, *Chem. Heterocycl. Compd.*, 1359 (2007), [*Khim. Geterotsikl. Soedin.*, 1603 (2007)].
07CRV1745	V. C. Gibson, C. Redshaw, and G. A. Solan, *Chem. Rev.*, **107**, 1745 (2007).
07EJI3816	M. Zhang, S. Zhang, P. Hao, S. Jie, W. H. Sun, P. Li, and X. Lu, *Eur. J. Inorg. Chem.*, 3816 (2007).
07IC3548	Y. Wang, H. Fu, F. Shen, X. Sheng, A. Peng, Z. Gu, H. Ma, J. S. Ma, and J. Yao, *Inorg. Chem.*, **46**, 3548 (2007).
07IC4114	J. S. Costa, C. Balde, C. Carbonera, D. Denux, A. Wattiaux, C. Desplanches, J. P. Ader, P. Gutlich, and J. F. Letard, *Inorg. Chem.*, **46**, 4114 (2007).
07IC5465	E. A. Katayev, K. Severin, R. Scopelliti, and Y. A. Ustynyuk, *Inorg. Chem.*, **46**, 5465 (2007).
07IC5829	A. W. Kleij, M. Kuil, D. M. Tooke, A. L. Spek, and J. N. H. Reek, *Inorg. Chem.*, **46**, 5829 (2007).
07IC6880	S. A. Carabineiro, L. C. Silva, P. T. Gomes, L. C. J. Pereira, L. F. Veiros, S. I. Pascu, M. T. Duarte, S. Namorado, and R. T. Henriques, *Inorg. Chem.*, **46**, 6880 (2007).
07IC7040	I. Vidyaratne, J. Scott, S. Gambarotta, and P. H. M. Budzelaar, *Inorg. Chem.*, **46**, 7040 (2007).
07IC7055	S. C. Bart, E. Lobkovsky, E. Bill, K. Wieghardt, and P. J. Chirik, *Inorg. Chem.*, **46**, 7055 (2007).
07IC8170	H. Nakamura, Y. Sunatsuki, M. Kojima, and N. Matsumoto, *Inorg. Chem.*, **46**, 8170 (2007).
07IC9558	M. S. Shongwe, B. A. Al-Rashdi, H. Adams, M. J. Morris, M. Mikuriya, and G. R. Hearne, *Inorg. Chem.*, **46**, 9558 (2007).
07ICA69	K. V. Pringouri, C. P. Raptopoulou, A. Escuer, and T. C. Stamatatos, *Inorg. Chim. Acta*, **360**, 69 (2007).
07ICA273	M. Volpe, S. D. Reid, A. J. Blake, C. Wilson, and J. B. Love, *Inorg. Chim. Acta*, **360**, 273 (2007).
07ICA557	F. Cisnetti, G. Pelosi, and C. Policar, *Inorg. Chim. Acta*, **360**, 557 (2007).
07ICA2298	H. Keypor, H. Goudarziafshar, A. K. Brisdon, and R. G. Pritchard, *Inorg. Chim. Acta*, **360**, 2298 (2007).
07ICC925	B. Pedras, H. M. Santos, L. Fernandes, B. Covelo, A. Tamayo, E. Bertolo, J. L. Capelo, T. Aviles, and C. Lodeiro, *Inorg. Chem. Commun.*, **10**, 925 (2007).
07JCC243	K. Shashidhar, K. Shivakumar, P. V. Reddy, and M. B. Halli, *J. Coord. Chem.*, **60**, 243 (2007).
07JCC1435	B. I. Kharisov, A. D. Garnovskii, O. V. Kharisova, U. O. Mendez, and A. Y. Tsivadze, *J. Coord. Chem.*, **60**, 1435 (2007).
07JCC1493	A. D. Garnovskii, V. N. Ikorskii, A. I. Uraev, I. S. Vasilchenko, A. S. Burlov, D. A. Garnovskii, K. A. Lyssenko, V. G. Vlasenko, T. E. Shestakova, Y. V. Koshchienko, T. A. Kuzmenko, L. N. Divaeva, M. P. Bubnov, V. P. Rybalkin, O. Y. Korshunov, I. V. Pirog, G. S. Borodkin, V. A. Bren, I. E. Uflyand, M. Y. Antipin, and V. I. Minkin, *J. Coord. Chem.*, **60**, 1493 (2007).
07JCS(D)1101	M. Broring and C. Kleeberg, *Dalton Trans.*, 1101 (2007).
07JCS(D)2411	S. Maji, B. Sarkar, S. M. Mobin, J. Fiedler, W. Kaim, and G. K. Lahiri, *Dalton Trans.*, 2411 (2007).

07JCS(D)2658	T. Afrati, C. N. Zaleski, C. Dendrinou-Samara, J. W. Kampf, V. L. Pencoraro, and D. P. Kessissglou, *Dalton Trans.*, 2658 (2007).
07JCS(D)4565	Y. D. M. Champouret, W. J. Nodes, J. A. Scrimshire, K. Singh, G. A. Solan, and I. Young, *Dalton Trans.*, 4565 (2007).
07JCS(D)4644	J. Tang, P. Gamez, and J. Reedijk, *Dalton Trans.*, 4644 (2007).
07JCS(D)4788	D. Prema, A. V. Wiznycia, B. M. T. Scott, J. Hillborn, J. Desper, and C. J. Levy, *Dalton Trans.*, 4788 (2007).
07JMC(A)85	S. Jie, S. Zhang, W. H. Sun, X. Kuang, T. Liu, and J. Guo, *J. Mol. Catal.*, **A269**, 85 (2007).
07JMS104	Y. P. Tong, S. L. Zheng, and X. M. Chen, *J. Mol. Struct.*, **826**, 104 (2007).
07JOM3532	S. Adewuyi, G. Li, S. Zhang, W. Wang, P. Hao, W. H. Sun, N. Tang, and J. Li, *J. Organomet. Chem.*, **692**, 3532 (2007).
07JOM4506	W. H. Sun, P. Hao, J. Li, S. Zhang, W. Wang, J. Li, M. Asma, and N. Tang, *J. Organomet. Chem.*, **692**, 4506 (2007).
07M175	A. A. Soliman and W. Linert, *Monatsh. Chem.*, **138**, 175 (2007).
07MC164	K. A. Lyssenko, A. O. Borissova, A. S. Burlov, I. S. Vasilchenko, A. D. Garnovskii, and V. I. Minkin, *Mendeleev Commun.*, **17**, 164 (2007).
07MI1	V. V. Skopenko, A. Y. Tsivadze, L. I. Savranskii, and A. D. Garnovskii, Koordinatsionnaya khimiya, p. 487, Akademkniga, Moskva (2007).
07OM726	C. Bianchini, D. Gatteschi, G. Giambastiani, I. G. Rios, A. Ienco, F. Laschi, C. Mealli, A. Meli, L. Sorace, A. Toti, and F. Vizza, *Organometallics*, **26**, 726 (2007).
07OM1104	J. Campora, C. M. Perez, A. Rodriguez-Delgado, A. M. Naz, P. Palma, and E. Alvarez, *Organometallics*, **26**, 1104 (2007).
07OM2439	P. Hao, S. Zhang, W. H. Sun, Q. Shi, S. Adewuyi, X. Lu, and P. Li, *Organometallics*, **26**, 2439 (2007).
07OM2720	W. H. Sun, P. Hao, G. Li, S. Zhang, Q. Shi, W. Zuo, and X. Tang, *Organometallics*, **26**, 2720 (2007).
07OM3201	I. Vdyarante, J. Scott, S. Gambarotta, and R. Duchateau, *Organometallics*, **26**, 3201 (2007).
07OM4639	P. Barbaro, C. Bianchini, G. Giambastiani, I. G. Rios, A. Meli, W. Oberhauser, A. M. Segarra, L. Sorace, and A. Toti, *Organometallics*, **26**, 4639 (2007).
07OM4781	W. H. Sun, K. Wang, K. Wedeking, D. Zhang, S. Zhang, J. Cai, and Y. Li, *Organometallics*, **26**, 4781 (2007).
07OM5406	K. Heinze and S. Reinhardt, *Organometallics*, **26**, 5406 (2007).
07OM5423	J. Zhang, B. W. K. Chu, N. Zhu, and V. W. W. Yam, *Organometallics*, **26**, 5423 (2007).
07POL115	J. Manzur, A. M. Garcia, A. Vega, and A. Ibanez, *Polyhedron*, **26**, 115 (2007).
07POL2433	P. Datta and C. Sinha, *Polyhedron*, **26**, 2433 (2007).
07POL2603	S. Roy, T. N. Mandal, A. K. Barik, S. Pal, S. Gupta, A. Hazra, R. J. Butcher, A. D. Hunter, M. Zeller, and S. K. Kar, *Polyhedron*, **26**, 2603 (2007).
07RCR617	S. S. Ivanchev, *Russ. Chem. Rev.*, **76**, 617 (2007).
07RJC176	I. S. Vasilchenko, T. E. Shestakova, V. N. Ikorskii, T. A. Kuzmenko, V. G. Vlasenko, L. N. Divaeva, A. S. Burlov, A. I. Uraev, I. V. Pirog, G. S. Borodkin, P. B. Chepurnoi, I. G.

	Borodkina, O. A. Beletskii, O. A. Karpov, I. E. Uflyand, and A. D. Garnovskii, *Russ. J. Coord. Chem.*, **33**, 176 (2007).
07SA(A)852	A. A. Soliman, *Spectrochim. Acta*, **A67**, 852 (2007).
08AHC221	A. P. Sadimenko, *Adv. Heterocycl. Chem.*, **95**, 221 (2008).
08AX(C)137	M. V. Plutenko, Y. S. Moroz, T. Yu. Sliva, M. Haukka, and I. O. Fritsky, *Acta Crystallogr.*, **C64**, 137 (2008).
08AX(E)328	X. X. Zhou, Z. Y. Zhou, J. Q. Chen, X. M. Lin, and Y. P. Cai, *Acta Crystallogr.*, **E64**, m328 (2008).
08CCR1420	K. C. Gupta and A. K. Sutar, *Coord. Chem. Rev.*, **252**, 1420 (2008).
08CCR1871	P. A. Vigato and S. Tamburini, *Coord. Chem. Rev.*, **252**, 1871 (2008).
08EJI1475	N. Meyer, M. Kuzdrowska, and P. W. Roesky, *Eur. J. Inorg. Chem.*, 1475 (2008).
08IC896	J. Scott, I. Vidyaratne, I. Korobkov, S. Gambarotta, and P. H. M. Budzelaar, *Inorg. Chem.*, **47**, 896 (2008).
08IC1476	Y. You, J. Seo, S. H. Kim, K. S. Kim, T. K. Ahn, D. Kim, and S. Y. Park, *Inorg. Chem.*, **47**, 1476 (2008).
08IC1488	T. S. Lobana, G. Bawa, and R. J. Butcher, *Inorg. Chem.*, **47**, 1488 (2008).
08ICA1415	H. Keypor, H. Goudarziafshar, A. K. Brisdon, R. G. Pritchard, and M. Rezaeivala, *Inorg. Chim. Acta*, **361**, 1415 (2008).
08ICA1562	D. Kovala-Demertzi, N. Kourkoumelis, K. Derlat, J. Michalak, F. J. Andreadaki, and J. D. Kostas, *Inorg. Chim. Acta*, **361**, 1562 (2008).
08ICA1843	L. Wang and J. Sun, *Inorg. Chim. Acta*, **361**, 1843 (2008).
08ICA2236	T. K. Panda, D. Petrovich, T. Bannenberg, C. G. Hrib, P. G. Jones, and M. Tamm, *Inorg. Chim. Acta*, **361**, 2236 (2008).
08ICA3519	F. Le Gac, P. Guionneau, J. F. Letard, and P. Rosa, *Inorg. Chim. Acta*, **361**, 3519 (2008).
08ICC349	B. Y. Li, Y. M. Yao, Y. R. Wang, Y. Zhang, and Q. Shen, *Inorg. Chem. Commun.*, 349 (2008).
08JCS(D)667	U. McDonnell, J. M. C. A. Kerchoffs, R. P. M. Castineiras, M. R. Hicks, A. C. G. Hotze, M. J. Hannon, and A. Rodger, *Dalton Trans.*, 667 (2008).
08JCS(D)887	D. Petrovic, L. M. R. Hill, P. G. Jones, W. B. Tolman, and M. Tamm, *Dalton Trans.*, 887 (2008).
08JMC(A)92	P. S. Chittilappilly, N. Sridevi, and K. K. M. Yusuff, *J. Mol. Catal.*, **A286**, 92 (2008).
08JOM278	J. M. Benito, E. de Jesus, F. J. de la Mata, J. C. Flores, and R. Gomez, *J. Organomet. Chem.*, **693**, 278 (2008).
08JOM483	M. Zhang, P. Hao, W. Zuo, S. Jie, and W. H. Sun, *J. Organomet. Chem.*, **693**, 483 (2008).
08JOM619	R. S. Herrick, C. J. Ziegler, M. Precopio, K. Crandall, J. Shaw, and R. M. Jarret, *J. Organomet. Chem.*, **693**, 619 (2008).
08JOM750	Y. Chen, W. Zuo, P. Hao, S. Zhang, K. Gao, and W. H. Sun, *J. Organomet. Chem.*, **693**, 750 (2008).
08JOM1829	Y. Chen, P. Hao, W. Zuo, K. Gao, and W. H. Sun, *J. Organomet. Chem.*, **693**, 1829 (2008).
08M11	A. R. Katritzky, C. A. Ramsden, E. F. V. Scriven, and R. J. K. Taylor (eds.), *"Comprehensive Heterocyclic Chemistry III"*, V. 1–V. 13, 3rd ed., Pergamon, Oxford (2008).
08OM109	I. Fernandez, R. J. Trovitch, E. Lobkovsky, and P. Chirik, *Organometallics*, **27**, 109 (2008).

08OM1470	R. J. Trovitch, E. Lobkovsky, E. Bill, and P. J. Chirik, *Organometallics*, **27**, 1470 (2008).
08POL709	B. Y. Li, Y. M. Yao, Y. R. Wang, Y. Zhang, and Q. Shen, *Polyhedron*, **27**, 709 (2008).
08POL593	S. Roy, T. N. Mandal, A. K. Barik, S. Gupta, R. J. Butcher, M. Nethaji, and S. K. Kar, *Polyhedron*, **27**, 593 (2008).
08POL1556	C. Adhikary, R. Bera, B. Dutta, S. Jana, G. Bocelli, A. Cantoni, S. Chaudhuri, and S. Koner, *Polyhedron*, **27**, 1556 (2008).
08POL1917	V. Mahalingam, N. Chitrapriya, F. R. Fronczek, and K. Natarajan, *Polyhedron*, **27**, 1917 (2008).
08ZOB1002	A. S. Burlov, A. I. Uraev, V. N. Ikorskii, S. A. Nikolaevskii, V. V. Koshchienko, I. S. Vasilchenko, D. A. Garnovskii, V. G. Vlasenko, Y. V. Zubavichus, L. N. Divaeva, G. S. Borodkin, and A. D. Garnovskii, *Zh. Obshch. Khim.*, **78**, 1002 (2008).